Applications of Advanced Technology to Ash-Related Problems in Boilers

Applications of Advanced Technology to Ash-Related Problems in Boilers

Edited by

L. Baxter

Sandia National Laboratories
Livermore, California

and

R. DeSollar

Central Illinois Public Service
Springfield, Illinois

Plenum Press • New York and London

CHEM

Library of Congress Cataloging-in-Publication Data

Applications of advanced technology to ash-related problems in boilers
    / edited by L. Baxter and R. DeSollar.
        p.   cm.
    "Proceedings of a conference on applications of advanced
technology to ash-related problems in boilers, held July 16-21,
1995, in Waterville Valley, New Hampshire"--T.p. verso.
    Includes bibliographical references and index.
    ISBN 0-306-45376-2
    1. Boilers--Combustion--Congresses. 2. Ash disposal--Congresses.
I. Baxter, L. II. DeSollar, R. (Richard)
TJ263.5.A67 1996
621.1'83--dc20   96-35756
                                                         CIP

Proceedings of an Engineering Foundation conference on Applications of Advanced Technology to
Ash-Related Problems in Boilers, held July 16 – 21, 1995, in Waterville Valley, New Hampshire

ISBN 0-306-45376-2

© 1996 Plenum Press, New York
A Division of Plenum Publishing Corporation
233 Spring Street, New York, N. Y. 10013

Printed in the United States of America

PREFACE

This book addresses the behavior of inorganic material in combustion systems. The past decade has seen unprecedented improvements in understanding the rates and mechanisms of inorganic transformations and in developing analytical tools to predict them. These tools range from improved fuel analysis procedures to predictive computer codes. While this progress has been met with great enthusiasm within the research community, the practices of the industrial community remain largely unchanged. The papers in this book were selected from those presented at an Engineering Foundation Conference of the same title. All have been peer reviewed. The intent of the conference was to illustrate the application of advanced technology to ash-related problems in boilers and, by so doing, engage the research and industrial communities in more productive dialog. Those attending the conference generally felt that we were successful on these counts. We also engaged the industrial community to a greater extent than ever before in the conference discussion and presentation. We hope these proceedings will facilitate a continued and improved interaction between industrial and research communities.

Behavior of inorganic material has long been recognized as one of the major considerations affecting the design and operation of boilers that burn ash-producing fuels. The practical problems associated with the behavior are sometimes catastrophic and spectacular, ranging from major slag falls that damage the bottom of furnaces to complete plugging of convection passes. More commonly, the problems are subtle and are associated with less dramatic deposits coating most of the furnace surface. These more subtle issues result in loss of availability, over-designed systems, derating, or excessive maintenance. The ASME Research Committee on Corrosion and Deposits from Combustion Gases recently completed a position paper that estimates the annual cost of ash-related problems in US boilers amounts to many hundreds of millions of dollars a year, most probably around $1–3 billion. An accurate understanding of ash deposition issues would also expand the range of coals appropriate for a boiler, leading to similar or possibly greater savings in fuel costs. While coal properties vary significantly around the world, these statistics from the US are probably representative of most countries. Clearly, ash-related problems in boilers represent an economic issue of national and international scope and importance.

Aside from economic issues, inorganic material affects boiler operation and design in major but sometimes poorly anticipated ways. This has led operators at some power plants to resort to shooting shotguns across on-line boilers in an attempt to control ash deposition. Boiler size, tube spacing, and materials are the principal variables affecting cost and all are determined by ash behavior more than any other single variable. Even with all the attention paid to ash issues during boiler design and fuel selection, ash management remains one of the most significant issues during boiler operation.

In the past, boiler operators could rely on experience to solve many ash-related problems. It was common that a utility would establish fuel contracts with a single coal suppliers for 20 years or more.

Fuel properties varied only over the relatively narrow range experienced by a mine and many mining companies were capable of blending coals from several mines to deliver an extremely consistent product. Under such conditions, operators can learn ash management strategies by experience, even if very little engineering or science is applied to the problem. In the past six years and for the foreseeable future, the constancy of fuel properties and long-term fuel contracts has disappeared. In the 1990s, new pollution regulations precipitated a move away from design fuels and into untested fuels. The introduction of low-sulfur fuels, fuel blending and switching, spot market fuel purchasing, low-NOx burners, fuel and air staging, reburning, and other pollution control measures has introduced an unprecedented level of variation in the day-to-day operation of boilers. Most of these fuel property and boiler design changes increase the ash management problems. Furthermore, the changes happen too frequently for operators to develop expertise in ash management by experience alone.

The future promises to be even more challenging. Utility operations in the past put a premium on reliability. The government regulatory boards and utilities cooperated to establish practices that best served the public, and in many cases both bodies saw reliability as a more important consideration than economics. Changes in operations often threaten reliability and therefore were not always embraced at utility power stations or demanded by regulators. Many governments now believe that electrical power rates could be lowered if the utility monopoly structure were replaced by a deregulated, free market structure for power generation. The United Kingdom has demonstrated such price reductions and the State of California is leading the US into similar practices. A great many other state and national governments are at various stages of deregulating their utilities. The economic benefits of deregulation cannot be achieved without changes in operation and such changes will assuredly include boiler operation and fuel selection, as is amply demonstrated by the United Kingdom experience. Furthermore, the environmental concerns of the future, such as global warming and trace metal emissions, cannot be addressed without involving utility power stations. These issues are primarily addressable through major changes in boiler design and operation. Combined-cycle gasification and coal-biomass cofiring represent two of the most promising means of addressing global warming issues at power stations, and both have major implications for ash management.

These economic, political, and environmental forces promise to make the next several decades an era of rapid change in the electrical power generation industry around the world. Most of the anticipated changes to boiler operation bring ash management concerns with them. There may never have been a greater need for the research community to work effectively with the industrial community to address these issues. This book is intended as one step toward improving this interaction.

Many institutions and individuals have contributed to this successful conference. We would like particularly to acknowledge the Department of Energy's Pittsburgh Energy Technology Center, Brigham Young University, and Sandia National Laboratories for their generous financial and logistical support. We would also like to acknowledge Linda Sager at Sandia and Bertha O'Keefe at Plenum Publishing for their patient and persistent help in publishing this volume.

Larry Baxter
Richard DeSollar

CONTENTS

ASH FORMATION, DEPOSITION, CORROSION, AND EROSION IN CONVENTIONAL BOILERS

Steven A. Benson, Edward N. Steadman,
Christopher J. Zygarlicke, and Thomas A. Erickson

Energy & Environmental Research Center
University of North Dakota
PO Box 9018
Grand Forks, ND 58202-9018

ABSTRACT

The inorganic components (ash-forming species) associated with coals significantly affect boiler design, efficiency of operation, and life span of boiler parts. During combustion in conventional pulverized fuel boilers, the inorganic components are transformed into inorganic gases, liquids, and solids. This partitioning depends upon the association of the inorganic components in the coal and upon combustion conditions. The inorganic components are associated as mineral grains and as organically associated elements, and these associations of inorganic components in the fuel directly influence their fate upon combustion. Combustion conditions, such as temperature and atmosphere, influence the volatility and the interaction of inorganic components during combustion and gas cooling, which influences the state, size, and composition distributions of the particulate and condensed ash species. The intermediate species are transported with the bulk gas flow through the combustion system, during which time the gases and entrained ash are cooled. Deposition, corrosion, and erosion occur when the ash intermediate species are transported to the heat-transfer surface, react with the surface, accumulate, sinter, and develop strength. Research over the past decade has significantly advanced understanding of ash formation, deposition, corrosion, and erosion mechanisms. Many of the advances in understanding and predicting ash-related issues can be attributed to the advanced analytical methods used to determine the inorganic composition of fuels and resulting ash materials. These new analytical techniques have been the key to elucidation of the mechanisms of ash formation and deposition. This information has been used to develop algorithms and computer models to predict the effects of ash on combustion system performance.

INTRODUCTION

The effects of ash on the performance of conventional boilers depends upon the inorganic composition of the fuel and upon operating conditions. Ash in conventional power systems is known to be a major problem, resulting in the loss of millions of dollars annually because of decreased efficiency, unscheduled outages, equipment failures, and cleaning.

Applications of Advanced Technology to Ash-Related Problems in Boilers
Edited by L. Baxter and R. DeSollar, Plenum Press, New York, 1996

The many ways in which the detrimental effects of ash are manifest in a boiler system include fireside ash deposition, corrosion and erosion of boiler parts, and production of fine particulates that are difficult to collect. The literature on ash-related issues is immense. Overviews and compilations of work by many investigators can be found by referring to the work of Couch [1994], Williamson and Wigley [1994], Benson and others [1993b], Benson [1992], Bryers and Vorres [1990], and Raask [1988; 1985]. There are many more sources too numerous to cite in this paper.

Significant advances have been made over the past 10 years in understanding and improving methods to predict ash behavior in conventional combustion systems. Increased understanding comes from being better able to determine the inorganic components in the coal, ash, and slag materials. This understanding has been made possible through the development of advanced methods of coal and coal ash analysis that have been specifically tailored to characterize fuel and fuel-derived ash components [Skorupska and Carpenter, 1993; Jones and others, 1992]. These advanced methods are based primarily on computer-controlled scanning electron microscopy (CCSEM), which allows for determination of the chemical and physical characteristics of coal minerals, fly ash, and deposits. These methods provide a mechanistic basis for understanding ash formation and deposition and provide information that can be used to assess corrosion and erosion of boiler parts. Ash formation and deposition mechanisms have been converted to algorithms and included in computer models to predict ash formation and deposition [Benson and others, 1993a].

This paper provides an overview of analytical techniques to characterize coals and coal-ash-related materials, mechanisms of ash formation and deposition, and their relationship to corrosion and erosion of boiler parts. Application of computer codes to predict coal ash deposition will also be discussed.

COAL INORGANIC COMPOSITION

The inorganic components in coals have been referred to as "mineral matter," "minerals," "inherent/extraneous ash," and other names by the many individuals who work with coal. Descriptions of the inorganic constituents in coal depend upon the perspective of the individuals describing them. The term "inorganic components" best describes all ash-forming constituents, including both organically associated inorganic species and mineral grains. Strictly speaking, coal does not contain ash. The inorganic part of the coal is transformed into ash during combustion.

The inorganic components in coal occur as discrete minerals, organically associated cations, and cations dissolved in pore water. The fraction of inorganic components that are organically associated varies with coal rank. Lower-rank subbituminous and lignitic coals have high levels of oxygen. Approximately 25% of the oxygen is in the form of carboxylic acid groups. These groups act as bonding sites for cations such as sodium, magnesium, calcium, potassium, strontium, and barium. Other minor and trace elements may also be associated in the coal in this form. In addition, some elements may be in the form of chelate coordination complexes with pairs of adjacent organic oxygen functional groups.

Mineral grains are usually the most abundant inorganic component in coals. The major mineral groups found in coals include silicates and oxides, carbonates, sulfides, sulfates, and phosphates. In order to predict the behavior of inorganic constituents during combustion, detailed information must be obtained on the abundance, size, and association of mineral grains in the coal. In addition, the association of the mineral grain with the coal matrix must be determined and classified. A mineral associated with the organic part of a coal particle is said to be "included". A mineral that is not associated with organic material is referred to as "excluded." Figure 1 illustrates the associations of inorganic components in coal.

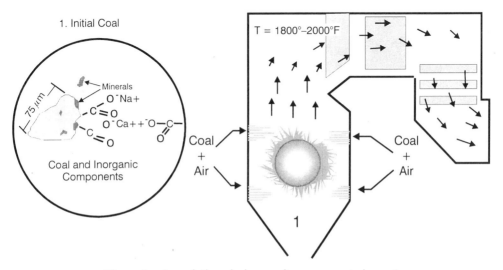

Figure 1. Associations in inorganic components in coal.

In determining the size, abundance, and association of mineral grains in both high- and low-rank coals, CCSEM and automated image analysis (AIA) are the preferred techniques to analyze polished cross sections of coal epoxy plugs [Steadman and others, 1991]. The CCSEM technique is used to determine the size, shape, quantity, and semiquantitative composition of mineral grains in coals [Jones and Benson, 1987; Zygarlicke and others, 1990]. Table 1 illustrates the range of mineral abundance and type for several coals from the United States. Quantification of the type and abundance of organically associated inorganic elements in lower-ranked subbituminous and lignitic coals is currently performed by chemical fractionation [Benson and Holm, 1985]. Chemical fractionation is used to selectively extract elements from the coal based on solubility, which reflects their association in the coal. Briefly, the technique involves extracting the

TABLE 1

Distribution of Minerals in Selected U.S. Coals as Determined by CCSEM Analysis, wt%, mineral basis

Mineral	U.S. Lignite				U.S. Western Subbituminous				U.S. Eastern Bituminous			
	Max, wt%	Min, wt%	Avg, wt%	Standard Deviation, wt%	Max, wt%	Min, wt%	Avg, wt%	Standard Deviation, wt%	Max, wt%	Min, wt%	Avg, wt%	Standard Deviation, wt%
Quartz	40.4	12.4	22.7	10.6	60.9	17.3	33.3	14.6	18.1	13.2	15.2	1.7
Iron oxide	1.8	0.4	1.1	0.7	1.7	0.0	0.5	0.5	2.6	0.0	0.9	0.9
Periclase	0.0	0.0	0.0	0.0	0.0	0.0	0.0	0.0	0.0	0.0	0.0	0.0
Rutile	1.2	0.1	0.7	0.4	4.9	0.1	1.1	1.4	0.6	0.0	0.2	0.2
Alumina	0.5	0.0	0.1	0.2	0.2	0.0	0.0	0.1	0.3	0.0	0.1	0.1
Calcite	12.7	0.1	3.4	5.4	11.2	0.0	1.7	3.3	9.5	0.0	3.1	3.7
Dolomite	0.1	0.0	0.0	0.0	0.2	0.0	0.0	0.1	2.9	0.0	0.5	1.1
Ankerite	0.0	0.0	0.0	0.0	0.0	0.0	0.0	0.0	0.0	0.0	0.0	0.0
Kaolinite	45.9	3.7	18.9	16.6	51.1	10.9	28.6	13.3	22.7	11.2	14.9	3.7
Montmorillonite	5.9	0.1	2.0	2.3	21.6	0.2	5.4	6.2	8.0	1.7	4.5	1.9
K Al-silicate	3.9	0.1	2.0	1.7	4.1	0.5	2.5	1.2	27.4	9.2	14.0	6.3
Fe Al-silicate	0.4	0.1	0.2	0.1	1.0	0.0	0.2	0.3	12.0	0.1	2.6	4.3
Ca Al-silicate	6.0	0.2	2.0	2.3	2.9	0.2	1.4	0.9	3.6	0.0	0.7	1.3
Na Al-silicate	0.2	0.0	0.1	0.1	0.1	0.0	0.0	0.0	0.5	0.0	0.1	0.2
Aluminosilicate	4.1	0.2	1.7	1.5	10.4	0.2	2.3	2.9	3.5	0.9	1.8	0.9
Mixed Al-silicate	0.3	0.0	0.1	0.1	1.3	0.1	0.5	0.3	2.9	0.0	0.8	1.0
Fe Silicate	0.0	0.0	0.0	0.0	0.1	0.0	0.0	0.0	0.4	0.0	0.1	0.1
Ca Silicate	0.3	0.0	0.1	0.1	0.3	0.0	0.1	0.1	0.6	0.0	0.1	0.2
Ca Aluminate	0.0	0.0	0.0	0.0	0.0	0.0	0.0	0.0	0.0	0.0	0.0	0.0
Pyrite	58.9	1.8	27.0	21.0	12.3	2.6	7.2	3.1	35.9	0.0	23.2	13.0
Pyrrhotite	3.0	0.1	0.8	1.3	0.5	0.0	0.1	0.1	0.5	0.0	0.2	0.2
Oxidized Pyrrho.	2.7	0.1	0.8	1.1	0.1	0.0	0.0	0.0	2.0	0.0	0.6	0.7
Gypsum	3.5	0.0	1.2	1.4	0.1	0.0	0.0	0.0	5.8	0.0	1.7	1.9
Barite	7.0	0.7	4.3	2.3	2.5	0.0	1.2	0.9	0.0	0.0	0.0	0.0
Apatite	0.0	0.0	0.0	0.0	0.8	0.0	0.2	0.3	0.1	0.0	0.0	0.0
Ca-Al-P	5.6	0.4	2.1	2.1	12.6	0.1	5.5	4.3	0.0	0.0	0.0	0.0
KCl	0.0	0.0	0.0	0.0	0.1	0.0	0.0	0.0	0.0	0.0	0.0	0.0
Gypsum–Barite	0.1	0.0	0.1	0.0	0.4	0.0	0.1	0.1	0.0	0.0	0.0	0.0
Gypsum–Al-silic.	1.0	0.1	0.4	0.3	1.1	0.0	0.4	0.4	0.5	0.0	0.2	0.2
Si-Rich	3.0	1.3	2.0	0.6	5.7	0.1	2.0	1.5	3.9	1.8	2.7	0.7
Ca-Rich	1.6	0.0	0.5	0.7	0.3	0.0	0.0	0.1	0.5	0.0	0.3	0.2
Ca-Si-Rich	0.0	0.0	0.0	0.0	0.1	0.0	0.0	0.0	0.2	0.0	0.1	0.1
Unknown	7.1	4.4	5.7	1.0	13.6	2.2	5.6	3.3	22.1	3.8	11.7	6.4

coal with water to remove water-soluble elements such as Na in sodium sulfate or those elements that were most likely associated with the groundwater in the coal. This is followed by extraction with 1M ammonium acetate to remove elements such as Na, Ca, and Mg that may be bound as salts of organic acids. The residue of the ammonium acetate extraction is then extracted with 1M HCl to remove acid-soluble species such as Fe and Ca that may be in the form of hydroxides, oxides, carbonates, and organically coordinated species. The components remaining in the residue after all three extractions are assumed to be associated with the insoluble mineral species such as clays, quartz, and pyrite. Table 2 summarizes the results from the analysis of a lower-rank coal containing high levels of organically associated cations.

ASH FORMATION

Inorganic coal components undergo complex chemical and physical transformations during combustion to produce ash in the form of vapors, liquids, and solids sometimes referred to as "intermediates." Partitioning of the inorganic components during combustion to form ash intermediates depends upon the association and chemical characteristics of the inorganic components, the physical characteristics of the coal particles, the physical characteristics of the coal minerals, and combustion conditions. Coal composition and combustion conditions directly influence the size and composition of intermediate ash species; the size and composition of the intermediate ash species directly influence slagging, fouling, corrosion, and erosion problems in combustion systems.

The physical transformation of inorganic components depends upon the inorganic composition of the coal and upon combustion conditions. The inorganic components illustrated in Fig. 1 can consist of organically associated cations, mineral grains that are included in coal particles, and excluded mineral grains. A wide range of combinations of associations in coal is possible: mineral–mineral, mineral–coal, mineral–cation–coal, and mineral–mineral–cation–coal. These associations are unique to each coal sample. The physical transformations involved in the formation of ash intermediates illustrated in Fig. 2 include coalescence of individual mineral grains within a char particle, shedding of the ash particles from the surface of the chars, incomplete coalescence owing to disintegration of the char, and convective transport of ash from the surface of burning char. Transformations of the included mineral grains depend upon the combustion characteristics of the char particle. During combustion, most of the nonvolatile inorganic species remain with the char. A very small amount of ash is found associated with the volatiles [Sarofim and others, 1977]. Therefore, the mode of char combustion influences the mechanism of fly ash formation. The size distribution of the resulting fly ash is multimodal [Sarofim and others, 1977; Benson and others, 1993b; Kauppinen, 1992].

The interaction of inorganic components during combustion is likely a combination of the following two extreme behaviors: one case is that each mineral grain forms one ash particle; the second is that one ash particle is formed per coal particle [Field and others, 1967]. The differences exhibited for different coals are likely related to

TABLE 2

Chemical Fractionation Results for a Selected Subbituminous Coal Composite

Element	Initial, $\mu g/g$	Removed H_2O, %	Removed NH_4OAc, %	Removed HCl, %	Remaining, %
Silicon	11,211	7	16	0	77
Aluminum	8312	6	0	43	51
Iron	2795	15	0	42	43
Titanium	841	4	20	12	64
Phosphorous	765	16	33	51	0
Calcium	12,427	7	70	22	2
Magnesium	3257	13	72	10	5
Sodium	416	100	0	0	0
Potassium	175	8	0	0	92

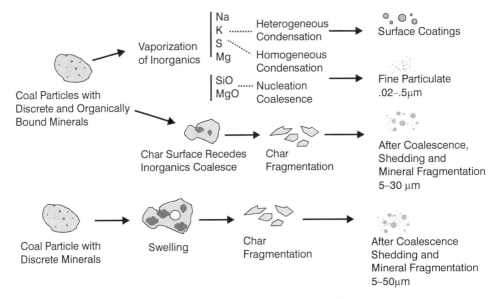

Figure 2. Mechanisms of ash formation.

characteristics. For example, the formation of one ash particle per mineral grain may be exhibited for coal particles that swell and become hollow and porous during combustion. The degree of swelling is likely dependent upon coal composition or maceral distribution and combustion conditions. A highly vesicular char particle is likely to burn by several expanding combustion fronts, causing the char particle to disintegrate, producing smaller char particles. Ash particles have sufficiently high surface tension on the char particles and do not wet the surface. Depending upon the distribution of minerals and other inorganic components, little or no coalescence may occur. Thus, the ash particles can be similar in size to the original minerals in the coal, producing one ash particle per mineral grain. In the case of the formation of one ash particle per coal particle, coalescence of mineral grains plays a significant role, particularly for nonswelling coal particles that burn as shrinking spheres. The shrinking-sphere model would produce one fly ash particle per coal particle. Limiting cases can be defined for fly ash particle-size evolution. These cases include a fine limit and a coarse limit (Loehden and others, 1989). The fine limit is the case where each mineral grain forms a fly ash particle. The coarse limit is where each coal particle forms a fly ash particle. In most cases, the actual fly ash size falls between these limits. In cases where mineral fragmentation occurs, the fine limit on size is exceeded. In addition, shedding or convective transport of small ash particles originating from organic associations or submicron mineral grains can also contribute to fine-particle formation.

Vaporization and condensation of inorganic elements contribute to the formation of fine particulate when the vapors condense homogeneously. In addition, these vapors can condense on surfaces of entrained ash particles and ash deposits, producing low-melting-point phases. The volatility of an inorganic species depends upon coal particle temperature, relative volatility of the species, and the form of the element in the coal. During the combustion of a coal particle, the atmosphere surrounding the particle contains reducing and oxidizing zones that can influence the volatility of an element. Near the surface of the particle, reducing conditions in the gas phase are expected because of the burning of the carbon material. The reducing condition produces more volatile reduced species than would occur under oxidizing conditions.

In lower-rank lignitic and subbituminous coals, organically associated inorganic elements such as sodium, calcium, magnesium, and potassium have the potential to vaporize during combustion. Evidence for the vaporization and condensation of sodium and potassium is abundant. The reactions of calcium and magnesium are less clear. During combustion of the coal and char particle, a boundary layer at the surface of the particle is highly reducing. This zone may be sufficiently low in oxygen to allow for the vaporization of calcium and magnesium. If the calcium and magnesium were to vaporize, it is likely that once they reached an environment that contained some oxygen ($\approx 3\%$ O_2) they would rapidly oxidize, producing submicron particulates.

ASH DEPOSITION

The transport of intermediate ash species (i.e., inorganic vapors, liquids, and solids) is a function of the state and size of the ash species and system conditions such as gas flow patterns, gas velocity, and temperature. Several processes are involved in the transport of ash particles. These processes have been described by Raask [1985] and Rosner and others [1986]. The primary transport mechanisms are illustrated in Fig. 3. The small particles (<1 μm) and vapor-phase species are transported by vapor-phase and small-particle diffusion. These species are characteristically rich in flame-volatilized species that condense upon gas cooling in the bulk gas or in the gas boundary layer next to the tube.

The mechanism of ash particle transport in the intermediate size range of particles as illustrated in Fig. 3 is that of thermophoresis. Thermophoresis is a transport force that is produced as a result of a temperature gradient in the direction from hot to cold. This transport mechanism is important for particles of < 10 μm. Electrophoresis is another transport mechanism that may be important with respect to the formation of deposits. The initial deposit layers are more pronounced when coals are fired that produce an abundance of intermediate- and small-sized particles.

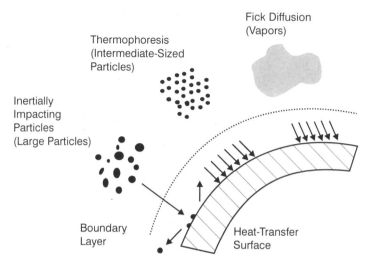

Figure 3. Primary ash transport mechanisms to heat-transfer surfaces in utility boilers [Benson and others, 1993b].

Coals that produce an abundance of fine particles characteristically contain low levels of mineral grains and high levels of organically associated inorganic components. The particles in the initial white-colored layer formed on all sides of the tube are less than 5 μm in diameter. The massive deposit on the upstream side of the tube is a result of particles greater than 5–10 μm being transported to the surface by inertial impaction. Inertial impaction accounts for the bulk of the deposit growth. Particles that inertially impact have sufficient momentum to leave gas stream lines and impact the tube. For small particles, the drag effect will be great enough to change the direction of the particles, allowing them to flow past the tube. The chances of a particle impacting a surface depend upon inertial momentum, drag force on the particle, and position in the flow stream. Gas velocity has a significant effect on the size of the ash particles that will impact the surface. For example, in a gas turbine with a gas velocity on the order of 100 m/s, particles with diameters greater than 1 μm will impact. In typical utility boilers, the gas velocity is 10–25 m/s, and particles with diameters of 5–10 μm and larger will impact.

The characteristics of a deposit depend upon the chemical and physical characteristics of the intermediate ash species, the geometry of the system (gas flow patterns), gas temperature, gas composition, and gas velocity. Deposits that form in the radiant section, called "slag deposits," are illustrated in Fig. 4. Deposits that form in the convective pass on steam tubes are called "fouling deposits" (Fig. 5). Slag deposits are usually associated with a high level of liquid-phase components and are exposed to radiation from the flame. Slag deposits are usually dominated by silicate liquid phases, but may also contain moderate-to-high levels of reduced iron phases. In addition, the initiating layers of slag deposits may consist of very fine particulate that can produce a reflective ash layer. Fouling deposits form in the convective passes of utility boilers and, in most cases, do not contain the high levels of liquid phases that are usually associated with slagging-type deposits. The fouling deposits contain low levels of liquid phases, usually consisting of a combination of silicates and sulfates that bind the particles together. The formation of these deposits on heat-transfer surfaces can significantly reduce heat transfer. Heat transfer through a deposit is related to the temperature, thermal history, and physical and chemical properties of the deposited material. Heat transfer through a deposit of a given thickness is affected by thermal conductivity, emissivity, and absorptivity.

EROSION

Erosion of boiler fireside heat-transfer surfaces is dependent upon coal composition and system operating conditions. Fireside surfaces are exposed to the transport of particulate and gaseous species carried with the flue gas through the system. Metal wastage occurs by two primary mechanisms: erosion and corrosion. In the case of erosion, coal mineral properties have been related to sliding abrasion wear and particle impaction wear [Raask, 1988]. The particle properties that influence erosion include size, density, shape, hardness, and cohesive strength (fragmentation upon impact). The system characteristics and operating conditions that affect erosion in power plants include 1) gas flow patterns and velocity, which influence particle trajectories, impaction efficiencies, and impaction angle; 2) gas and metal temperatures; 3) calorific value; and 4) load conditions.

Relationships have been developed by Raask [1988] to predict erosion. The abrasion index (AI) has been found to be related to the quartz content (coal basis) of the coal, q, pyrite content (coal basis) of the coal, p, and the ash content, a.

$$AI = q + 0.5p + 0.2a \qquad \text{[Eq. 1]}$$

The extent of damage to fuel-handling and combustion equipment is dependent upon the mass flow rate of coal through the system as well as the characteristics of the coal. The wear propensity, Wp, is written in the form

$$Wp = AI \times M \qquad \text{[Eq. 2]}$$

where M is the coal mass feed rate to the system. Erosion rates, We, [Creelman and others, 1994] can be determined by

$$We = IaCV^{3.5} \qquad \text{[Eq. 3]}$$

where Ia is the erosion index, C is the concentration of ash particles in the gas stream at the point at which erosion takes place, and V is the velocity. The erosion index, Ia [Raask, 1985] is defined as

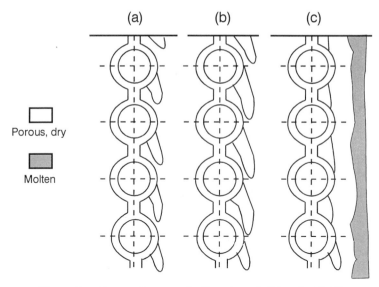

Figure 5.   Convective pass fouling deposit [Couch, 1994].

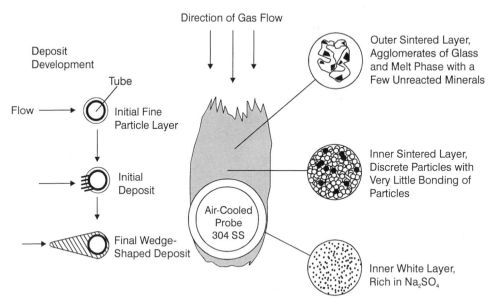

Figure 4.   Waterwall slag deposit a) initial stages, b) intermediate, and c) large deposit
with molten phases [Couch, 1994]. (This figure is published here with Electric
Power Research Institute permission.)

$$Ia = (x1 + 0.5)Ig + (\times 1q1 + 0.5 \times 2q2)Iq \qquad \text{[Eq. 4]}$$

where Ia, Ig, and Iq are the erosion indices for ash, spherical glass fly ash, and quartz particles, respectively; x1 and x2 are the weight fractions of ash in the >45 μm and 5- to 45-μm ash size fractions; and q1 and q2 are the quartz contents in those fractions.

PREDICTING ASH BEHAVIOR

A review of methods currently used by the EERC to predict ash behavior in utility boilers was conducted by Benson and others [1993a]. The following is a brief overview of the key components of that publication.

Assessing Ash Behavior

Three scanning electron microscope/electron microprobe analysis (SEM/EMPA) techniques—CCSEM, scanning electron microscopy point count (SEMPC), and AIA—are presently used in ash behavior research in combustion and gasification systems at the Energy & Environmental Research Center (EERC) [Jones, and others, 1992]. These techniques permit the study of transformations of inorganic components from the initial stages of coal conversion through the transformations that occur during ash deposition and slag formation. Their specific applications include 1) determination of the size, composition, and association of minerals in coals; 2) determination of the size and composition of intermediate ash components; 3) determination of the degree of interaction (sintering) in ash deposits; and 4) identification and quantification of the components of ash deposits and slags, including liquid-phase composition, reactivity, and crystallinity. Using SEM/EMPA for coals throughout all stages of utilization achieves a continuity of data.

Advanced Indices

The EERC has derived indices that rank low-rank coals according to their fouling propensity in utility boilers [Weisbecker and others, 1992]. The advanced indices are based on a thorough characterization of minerals and organically associated inorganic components from advanced methods of analysis, CCSEM and chemical fractionation, combined with their relationship to ash behavior in full-scale systems. Full-scale boiler performance was based on the general degree of fouling and the ability of the unit to maintain load. An index was derived that correlates certain applicable parameters of the inorganic composition and the general performance or experience at the full-scale utility boiler. The usefulness of these indices has been demonstrated by their repeated use by several utilities in the midwest United States that are screening Powder River Basin coals for use in coal-fired boilers. A low-temperature index has been developed that pertains more to fouling that would occur in lower-temperature regions of a boiler, such as in the primary superheater or economizer. A high-temperature index has been formulated that pertains to fouling in higher-temperature zones, such as in the secondary superheater or reheater. This work was recently expanded to include advanced indices for grindability, slagging, high-temperature fouling (HTF), sootblowing, erosion, low-temperature fouling (LTF), slag tapping, and opacity. This has been incorporated into a tool called PCQUEST (Predictive Coal Quality Effects Screening Tool).

Ash Particle-Size and Composition Distribution

A computer model entitled ATRAN (Ash Transformations) was developed to predict the particle-size and composition distribution (PSCD) of ash produced during the combustion of coal [Zygarlicke and others, 1992a, 1992b; Benson and others, 1992]. This technique uses advanced analytical characterization data, boiler parameters, and a detailed knowledge of the chemical and physical transformations of inorganic components during combustion to predict the PSCD of the resulting ash. The PSCD of the ash directly impacts deposit growth, deposit strength development, and ash collectability.

The PSCD of the ash is predicted from the abundance and distribution of the inorganic components in the coal. Bulk ash analysis, CCSEM, and ultimate analysis are used to characterize the coal. The resultant balance provides the compositions of the minerals with their associations to the coal, organically associated constituents, and submicron particles. The discrete minerals are represented by a database (size and composition) of individual particles as determined from the CCSEM analysis, while the organically associated and submicron constituents are represented by a bulk composition analysis only.

The ATRAN code allows for the fragmentation and coalescence of minerals and the vaporization and condensation of volatile organical and mineral-associated inorganic components. The three resultant data sets—locked fly ash, liberated fly ash, and submicron fly ash particles—are characterized and combined on a mass basis, producing the resultant PSCD of the ash.

Ash Deposition

LEADER (Low-Temperature Engineering Algorithm of Deposition Risk) is a computer code designed to predict the low-temperature fouling potential of a coal based on analysis of the inorganic components of the coal, boiler design, and firing rate [Hurley and others, 1991]. The code requires specific inputs for the amount and associations of the inorganic components in the coal. The CCSEM and chemical fractionation data, along with the standard ash composition and proximate–ultimate data, are used as inputs. The code first determines the ash size and composition distributions using the ATRAN code. These distributions are then used to determine the deposition rate of the upstream inner layer and downstream deposits; strength development in these deposits; and the loss of heat-exchange efficiency because of the deposits. The deposition rates are a function of the particle-size distribution of the ash and of the gas velocity and temperature. Strength development rate is a function of the composition of the deposited ash. Heat-exchange efficiency loss is a function of deposit depth and thermal conductivity. The deposit growth and strength development rates are also expressed as a shedding index (SI) equal to the ratio of the deposition rate and strength development rate.

CONCLUSIONS

Over the past 10 years, significant progress has been made in developing a more precise and quantitative understanding of the chemical and physical transformations of the inorganic components in coal during combustion. The progress has largely been due to significant efforts conducted by many investigators worldwide on bench-, pilot-, and full-scale systems. These efforts have focused not only on coal quality effects, but also on the effects of system design and operating conditions on ash formation and deposition. The results of this work have led to the elucidation of many of the ash formation and deposition mechanisms. The elucidation of these mechanisms has been made as a result of better analyses of the inorganic composition of coal, intermediate materials, and deposits. The acquired knowledgebase has led to the formulation of advanced indices and phenomenological models used to predict ash behavior. All of these more advanced methods or tools for predicting ash behavior in power system must continue to be tested and validated in full-scale power systems.

REFERENCES

Abbott, M.F., Moza, A.K., and Austin, L.G. (1981). "Studies on Slag Deposit Formation in Pulverized Coal Combustors: 2. Results, on the Wetting and Adhesion of Synthetic Ash Drops on Different Steel Substrates." *Fuel*, *60*, 1065–1072.

Benson, S.A. (Ed.) (1992). *Inorganic Transformations and Ash Deposition During Combustion*. New York: American Society of Mechanical Engineers for the Engineering Foundation.

Benson, S.A., and Austin, L.G. (1990). "Study of Slag Deposit Initiation Using a Laboratory-Scale Furnace." In R. W. Bryers, and K.S. Vorres (Eds.), *Mineral Matter and Ash Deposition*. New York: Engineering Foundation, pp. 261–278.

Benson, S.A., and Holm, P.L. (1985). "Comparison of Inorganic Constituents in Three Low-Rank Coals," *Ind. Eng. Chem. Prod. Res. Dev.*, *24*, 145.

Benson, S.A., Hurley, J.P., Zygarlicke, C.J., Steadman, E.N., and Erickson, T.A. (1993a). "Predicting Ash Behavior in Utility Boilers." *Energy & Fuels*, *7* (6), 746–754.

Benson, S.A., Hurley, J.P., Zygarlicke, C.J., Steadman, E.N., and Erickson, T.A. (1992). "Predicting Ash Behavior in Utility Boilers: Assessment of Current Status." In *Proceedings of the 3rd International Conference on the Effects of Coal Quality on Power Plants*. LaJolla, CA, August 25–27.

Benson, S.A., Jones, M.L., and Harb, J.N. (1993b). "Ash Formation and Deposition." In D. L. Smoot (Ed.), *Fundamentals of Coal Combustion for Clean and Efficient Use*. New York: Elsevier, Chapter 4, pp. 299–373.

Bryers, R.W., and Vorres, K.S. (Eds.) (1990). *Proceedings of the Engineering Foundation Conference on Mineral Matter and Ash Deposition from Coal*. Santa Barbara, CA: Unit Engineering Trustees Inc.

Couch, G. (1994). "Understanding Slagging and Fouling During pf Combustion." In J. Williamson, and F. Wigley (Eds.), *The Impact of Ash Deposition on Coal Fired Plants: Proceedings of the Engineering Foundation Conference*. London: Taylor & Francis.

Creelman (1994). In J. Williamson, and F. Wigley (Eds.), *The Impact of Ash Deposition on Coal Fired Plants: Proceedings of the Engineering Foundation Conference*, London: Taylor & Francis.

Field, M.H., Gill, D.W., Morgan, B.B., and Hawksley, P.G.W. (1967). *Combustion of Pulverized Coal*, Leatherhead: BCURA, pp. 1045–1065.

Hurley, J.P., Erickson, T.A., Benson, S.A., and Brobjorg, J.N. (1991). "Ash Deposition at Low Temperatures in Boilers Firing Western U.S. Coals." Presented at the International Joint Power Generation Conference, San Diego, CA, 8 p.

Jones, M.L., and Benson, S.A. (1987). "An Overview of Fouling/Slagging with Western Coals." Presented at the EPRI-Sponsored Conference on Effects of Coal Quality on Power Plants, Atlanta, GA, October 13–15, 22 p.

Jones, M.L., Kalmanovitch, D.P., Steadman, E.N., Zygarlicke, C.J., and Benson, S.A. (1992). "Application of SEM Techniques to the Characterization of Coal and Coal Ash Products." In H.L.C. Meuzelar (Ed.), *Advances in Coal Spectroscopy*. New York: Plenum Press.

Kauppinen, E.I. (1992). *Experimental Studies on Aerosol Size Spectroscopy with Multijet Low-Pressure Inertial Impactors*. Espoo, Finland: Julkaisija-Utgivare, pp. 3–40.

Loehden, D.O., Walsh, P.M., Sayre, A.N., Beér, J.M., and Sarofim, A.F. (1989). "Generation and Deposition of Fly Ash in the Combustion of Pulverized Coal." *J. of Inst. Energy*, 119–127.

Raask, E. (1988). *Erosion Wear in Coal Utility Boilers*. Washington: Hemisphere Publishing Corporation.

Raask, E. (1985). *Mineral Impurities in Coal Combustion*. Washington: Hemisphere Publishing Corporation.

Rosner, D.E. (1986). *Transport Processes in Chemically Reacting Flow Systems*. Butterworth Publishers.

Sarofim, A.F., Howard, J.B., and Padia, A.S. (1977). "The Physical Transformation of the Mineral Matter in Pulverized Coal Under Simulated Combustion Conditions." *Combust. Sci. Technol.*, *16*, 187–204.

Skorupska, N.M., and Carpenter, A.M. (1993). "Computer-Controlled Scanning Electron Microscopy of Minerals in Coal," IEA Coal Research report, December.

Steadman, E.N., Benson, S.A., and Zygarlicke, C.J. (1991). "Digital Scanning Electron Microscopy Techniques for the Characterization of Coal Minerals and Coal Ash." Presented at the Low-Rank Fuels Symposium, Billings, MT, May 20–23, 16 p.

Weisbecker, T., Zygarlicke, C.J., and Jones, M.L. (1992). "Correlation of Inorganics in Powder River Basin Coals in Full-Scale Combustion." In S.A. Benson (Ed.), *Inorganic Transformations of Ash Deposition During Combustion*. New York: ASME for the Engineering Foundation, pp. 699–711.

Williamson, J., and Wigley, F. (Eds.) (1994). *The Impact of Ash Deposition on Coal Fired Plants: Proceedings of the Engineering Foundation Conference.* London: Taylor & Francis.

Zygarlicke, C.J., Jones, M.L., Steadman, E.N., and Benson, S.A. (1990). "Characterization of Mineral Matter in ACERC Coals." Prepared for the Advanced Combustion Engineering Research Center Brigham Young University, Provo, UT.

Zygarlicke, C.J., McCollor, D.P., Benson, S.A., and Holm, P.L. (1992a). "Ash Particle-Size and Composition Evolution During Combustion of Synthetic Coal and Inorganic Mixtures." In *Proceedings of the 24th International Symposium on Combustion.* Pittsburgh, PA: The Combustion Institute, pp. 1171–1177.

Zygarlicke, C.J., Ramanathan, M., and Erickson, T.A. (1992b). "Fly Ash Particle-Size Distribution and Composition: Experimental and Phenomenological Approach." In S.A. Benson (Ed.), *Inorganic Transformations and Ash Deposition During Combustion.* New York: ASME, for the Engineering Foundation, pp. 469–711.

RESEARCH NEEDS OF THE POWER INDUSTRY

Richard W. DeSollar

Senior Fuels Engineer
Central Illinois Public Service Company
Springfield, Illinois 62739

ABSTRACT

The Clean Air Act of 1990 has changed the way utilities burn coal.
Utilities are now studying scrubbers, low NOx burners, fuel switching
and fuel blending just to name a few changes. The Clean Air Act
Amendments have also changed the way utilities do business.
Utilities are now looking to buy and sell emission credits, which may
be tied to fuel purchases and fuel contracts. The purchasing or
selling of credits may also be tied to interchange of electricity and
they are being traded on the open market to other utilities and
industries.

Most utilities have boilers that were designed to burn a specific
fuel and, in most cases, the fuel was a high sulfur, high Btu,
bituminous coal. With fuel switching many boilers are now being
required to burn a fuel that is drastically different than that for
which the boiler was designed. This is leading to a whole range of
new problems. Fuel engineers now are more concerned with the
slagging, fouling, corrosion and erosion that can take place in the
boiler, and not only how the fuel burns.

Utilities now look not only at the Btu of the fuel but are concerned
with the ash chemistry, grindability, and the ultimate analysis,
especially nitrogen and oxygen that is inherent in the coal. Many
utilities are not geared for and do not have the people and expertise
necessary for all of the studies and evaluation that must be done.
Some areas that need to be addressed by research are slagging and
fouling indices for western coals and blends of eastern and western
coals. Corrosion indices are needed which pertain not only to the
high temperature superheater and reheat areas of the boiler, but also
to the backpasses, the economizer, air heater, and especially the
precipitator. The effects of chlorine in a boiler and hazardous air
pollutants need to be addressed.

Fuel switching has also caused precipitator problems. Most
precipitators are designed to handle a low resistivity, high sulfur
coal ash. Most low sulfur coal produces a high resistivity ash which
is very difficult to precipitate. Work needs to be done on
identifying coals that are difficult to collect as well as those

Applications of Advanced Technology to Ash-Related Problems in Boilers
Edited by L. Baxter and R. DeSollar, Plenum Press, New York, 1996

17

coals that will not accept flue gas conditioning, and why the ash won't condition.

Conferences of this type are very important for the open exchange of information between researchers and industry. Many times the research that is done is very interesting. It produces a lot of answers academically but is of very little value to industry because it can not be related to a large industrial boiler consuming 3,000-10,000 tons of coal per day. The open exchange of information between what industry needs and what research can provide is essential to provide everyone involved with direction. Industry needs answers today, academics need to pursue long term approaches but they also need to address the short term, what industry needs today to answer immediate questions.

INTRODUCTION

The Clean Air Act of 1990 has changed the way utilities burn coal. Utilities are now studying scrubbers, low NOx burners, fuel switching and fuel blending just to name a few of the changes.

Clean Air Act Amendments have also changed the way utilities do business. Utilities are now looking to buy and sell emission credits as well as power. These emission credits may be tied to fuel purchases and fuel contracts. The purchasing or selling of credits may also be tied to interchange sales of electricity and credits are also being traded on the open market between utilities and other industries.

In the past, utilities had boilers that were designed to burn a specific coal and in most cases that coal was a high sulfur, high volatile bituminous coal. There were few boiler problems and the utilities knew how to handle the problems because of years of experience.

With the Clean Air Act Amendments came the need for fuel switching and conversion of boilers to fuels that are drastically different than that for which they were designed. Utility fuel engineers are now concerned with slagging, fouling, corrosion, and erosion that takes place in the boiler and not just how well the fuel burns. The utility fuel engineer today is not only looking at the heating value of the coal but he is concerned with ash chemistry, grindability, and the ultimate analysis, especially the nitrogen and oxygen that are inherent in the coal. Many utilities are not geared to do this, they do not have the people and the expertise necessary to do the studies and evaluation that are required.

The Clean Air Act Amendments have also changed the research that needs to be done in order to help the coal consuming industry. Some of the areas that the research community needs to address in the near future are described below.

Chlorine In Coal

It has been known for many years that chlorine can be a major contributor to corrosion in a coal combustor. In 1991, the Electric Power Research Institute held a conference on Chlorine In Coal in Chicago, Illinois. A great many of the papers presented at that conference outlined and identified the problems with chlorine and its effect on combustors. Much of the work presented at this conference had been done in the United Kingdom.

The British coals tend to be high in chlorine when compared to most U.S. coals. However, most of the research presented at this EPRI conference had been directed at the high temperature corrosion problems in the superheater and reheater or in the combustion area of fluidized bed combustors. Much of that research has been directed at the types of materials to use in the combustors. The need for fuel switching to lower sulfur coals has increased the use of high chlorine U.S. coals, some of which tend to contain low sulfur.

Chlorine information on United States coals is limited.
Linda J. Bragg, Robert B. Finkelman, and Susan J. Tewalt with the U. S. Geological Survey did a study of over 5,000 coal samples and published the results "Distribution of Chlorine in United States coal". This paper was presented as a manuscript at the EPRI Chlorine in Coal Conference.[1] Most of the information on chlorine in U. S. coals is in the U. S. Geological Survey's National Coal Resources Data System.

High temperature corrosion is a problem but the effects of low temperature chlorine corrosion are also of concern. The economizer, air heater, and the duct work from the air heater to the electrostatic precipitator (ESP), as well as the precipitator and stack linings all have the potential for low temperature corrosion. These are all areas that at times reach the acid due point for both sulfuric acid and hydrochloric acid.

With the changes that are taking place in the utility industry, more and more units are being cycled on a daily basis. Cycling includes daily load cycling and the extreme of cycling the unit off line and back on line. In the case where a unit is cycled off, the entire backpass area of the unit will go through the acid due point twice in the cycle. Research needs to be done on corrosion rates that are taking place in units that operate in this manner. Chlorine levels in the coal used in these units need to be included in any studies.

Also, additional research is needed on the form chlorine takes in the coal. In reviewing the presentation at the Chlorine In Coal Conference, it appears that the way chlorine is bound in the coal matrix, can have a critical impact on its behavior and impact in a combustor. (2, 3, 4)

Coal Switching And Blending Models And Indices

Fuel switching and blending is becoming more important in the U.S. coal consuming industry. As a result models are needed that will predict boiler performance on a new or blended fuel. In order for the new models to give accurate predictions, new slagging, fouling, corrosion, and erosion indices are needed.

Indices have always been important to the fuel engineer, helping to predict fuel behavior in the combustor. Most of the conventional indices were developed for eastern bituminous coal and have worked well for many years. (5) As the U.S. coal consuming industry switches to the low sulfur sub-bituminous coal from the Powder River Basin of Wyoming and the low sulfur bituminous coal of eastern Kentucky and West Virginia, new indices are needed.

Combustion models are also needed that will address the performance of these fuels in a boiler designed for high sulfur, high Btu eastern bituminous coal. The indices, and especially the models, must be able to predict boiler behavior on the new fuel alone as well as on blends of the old and new fuels in any and all proportions.

Another shortcoming of many of the old indices is that they use the percentage of ash or mineral matter in the coal to predict behavior. A boiler is a material handling device and it does not see a percent of anything, it sees pounds of material. The new indices and or models should make their predictions based on pounds of material per million Btu. (6, 7)

The models and indices should be user friendly and not require major detailed boiler inputs and they should use conventional coal and ash analyses as much as possible. If detailed computer control scanning electron microscopy (CCSEM) or some other completely new analysis is needed this will tend to limit the effective use of the new models and indices. The new models or indices should predict slagging, fouling, corrosion, erosion, deposit strength, deposit growth rates, and ease of deposit removal. Lower furnace (water walls) deposits need to be considered also. In developing these new indices and models it is essential that the cyclone boiler not be forgotten. There are a lot of cyclones still in service and these boilers do behave differently than pulverized coal boilers.

Currently, several institutions are undertaking work in developing some of the models and indices that are needed. They are:

Electric Power Research Institute (EPRI)

EPRI has developed the Coal Quality Impact Model (CQIM). This model provides a state of the art tool for evaluating the effects of coal quality on power plant performance and a means for quantifying those effects in a form that can be used by both power plant personnel and fuel purchasers.

CQIM uses engineering calculations to obtain energy and material balances around each major equipment system in a power plant. For a given coal, CQIM calculates the performance of each piece of equipment and determines a cost for changes in equipment performance based on any changes in capacity, auxiliary power requirements, availability or maintenance expenses.

CQIM relies on a large number of specific inputs to build the model for a power plant. There are approximately 1,000 entries required for a complete model. These inputs include specific information about the plant and its equipment. CQIM also requires detailed coal quality information, including full proximate analysis, ultimate analysis, ash minerals and grindability. Detailed fuel delivery and plant maintenance costs are also required.

The output from CQIM can provide a relative ranking of coal purchase options, as well as plant or unit bus bar costs. CQIM does use some of the traditional indices that were developed for bituminous coals. Currently, the developer of CQIM is working with EPRI and others to update and enhance the model's capability. There is also work being done to make the model more user friendly.

Physical Sciences, Inc. (PSI)

PSI has developed Slagging Advisor and Compliance Advisor for use by the utility industry. Slagging Advisor software helps utilities evaluate fuels by predicting the tendency of ash deposits to form at a specific boiler location. The model also calculates the tendency for deposit growth and development of sintering strength of the ash. Slagging Advisor has been developed from many years of research at PSI Technologies in conjunction with, and with support from, the Department of Energy and several utilities.

The software includes an ash formation model, a furnace aerodynamic and thermoprofile model, and an ash viscosity model. The inputs required for Slagging Advisor include boiler design and operating parameters, standard coal analyses and a detailed coal mineral characterization using CCSEM. By using fly ash composition, ash viscosity, and furnace temperature profiles, the model determines which ash particles may be sticky at particular furnace locations. The aerodynamic model and the ash size distribution predict which particles will actually reach a water wall surface. Based on tracking the paths of particles and knowing their stickiness, an estimate of relative deposit growth and deposit composition is made. The composition of the deposit can then be used as a rough indicator of deposit strength, which will determine its ease of removal by sootblowers.

Slagging Advisor is used to help select coals for test burns. It can be used in the overall planning of test burns of new coals and help in determining coal blend ratios and optimization of sootblowing.

Compliance Advisor, an extension of Slagging Advisor, is a new model that has just been developed by PSI to help solve coal switching and blending problems. The software in this model takes into account SO_2/SO_3 formation and interactions between fly ash and tube metal surfaces. The model also includes an ESP model and makes predictions of ESP performance based on fly ash size distribution, fly ash resistivity and the effects SO_3 has on fly ash resistivity.

Currently, work is being done by PSI and EPRI to try to include Slagging Advisor and Compliance Advisor in CQIM to enhance the capabilities of both models.

Sandia National Laboratories

Sandia has been working on the Ash Deposit Local Viscosity Index of Refraction and Composition (ADLVIC) model that will make quantitative predictions of the elemental composition of ash deposits, quantitative estimates of deposition rates and the mechanisms by which particles deposit. It also makes a quantitative indication of deposit strength, morphology, removability, and emissivity.

The inputs required for ADLVIC include a description of the boiler, in terms of both dimensions and operating conditions, a detailed description of the coal, including mineral matter, as well as chemical species and information about selected mineral components. This model differs from the traditional indices and models for fouling, slagging, and the other deposition models that have been mentioned. ADLVIC makes explicit predictions of the effects of boiler operating conditions on deposit behavior, predictions of variations of deposit properties with respect to location within a boiler and which are dependent on the total amount of inorganic material flowing through the boiler.

There are four deposition process that are treated by ADLVIC. They include inertial impaction, thermophoresis, condensation, and heterogeneous reaction. All heat transfer surfaces in the boiler are involved in these processes. The four processes of ash deposition mentioned are assumed to have additive influences on deposit composition.

The outputs from this model attempt to give a rate of deposition, morphology and strength of the deposit, removability of the deposit and emissivity of the deposit. Currently, Sandia is working to develop this model for commercial use and hopefully it will be made available to the utility industry.

Energy & Environmental Research Center, Grand Forks

The Energy & Environmental Research Center is currently working on models that will predict the impact of ash deposition in a pulverized coal fired boiler. The models under development are being incorporated into a Coal Quality Expert software package that is being developed for a DOE clean coal technology program. There are plans to include these models in an updated CQIM model.

In the programs being developed, slagging and fouling are addressed by two separate sub-models that interact with each other and with a boiler performance model to account for effects of changes in coal input properties and boiler operation. The input for these models will include a detailed coal analysis obtained using CCSEM coal size distribution, detailed boiler design data, and detailed boiler operating conditions. The output from the model will be an assessment of boiler performance associated with the coal characterized in the model.

The sub-model for slagging will outline deposit growth and it's effect on the heat transfer surfaces as well as the performance of the unit. The fouling sub-model will use the calculation fly ash particle size distribution as well as boiler operating conditions to calculate deposit growth in the backpass area of the boiler. The slagging and fouling sub-models will both predict heat loss associated with the deposit for a given coal under the specified operating conditions. In addition, both sub-models address deposit strength and ease of removal.

Currently, pilot scale tests are being conducted and information is being obtained from commercial units in an effort to verify the outputs from both of these models.

The EERC has also been working on advanced indices for low rank coals and has derived indices that will predict fouling tendencies of low rank coals in utility boilers. The approach taken on this project was to gain a complete inorganic characterization of discreet minerals and organically associated inorganic components in selected coals and to compare these with the performance of the coals, if possible, in a utility scale boiler. Full scale boiler performance was based on the general degree of fouling and the ability of the unit to maintain load. These indices take into account the sodium content, specifically organically bound sodium, which is easily vaporized. Also, the indices use the calcium content, again, the organically bound portion, which reacts quickly, and total mineral matter, along with the total organically bound inorganic matter. The indices also take into consideration the small quartz grains, Juxtaposition of minerals, calcite content, and clay content. All these components and their relationships are utilized to give a prediction of the high temperature slagging and low temperature fouling characteristics of the coal. So far, confirmation of the indices has been performed on several Powder River Basin coals and some of the North Dakota lignites. The indices that have been developed so far tend to work well and do give a prediction of behavior of the coal ash in the investigated boilers.

EERC is also working on an ash transformation model that will help predict the particle size and composition distribution of ash produced during the combustion process. This will help in giving predictions of the ash composition and ash characteristics present at different locations within the boiler. The advances made over the past several years by EERC in predicting ash behavior have been made possible by a more detailed and better analysis of coal and ash minerals. The advanced techniques such as CCSEM and quantitative determination of the chemical and physical characteristics of inorganic compounds in the coal and ash have facilitated the development of these models and indices. Future efforts will be focused on a more comprehensive model and indices that will include waterwall slagging and convection pass fouling. (7, 8, 9)

The work that these institutions are doing is work that needs to continue and be expedited.

Boiler Cleanup

Additional research needs to be conducted in methods of boiler cleaning. With fuel switching and fuel blending boilers are not always designed with the proper tube spacing or the appropriate numbers of cleaning devices, (steam lances, water lances, etc.) to keep the heat transfer surfaces clean.

Research needs to be conducted in improving the operation and placement of the cleaning devices.

Generally, the Fuels Engineer is reluctant to use coal additives. In the past, some of the "wonder" additives have been less than successful. However, an area of potential research is for an additive that would affect the slagging and fouling characteristics of the ash. Also, would the additive help in removing an ash deposit once it is formed?

Ash Chemistry

Ash chemistry impacts the quality of fly ash and bottom ash produced in a boiler. There are cases where two coals that are almost identical in standard ASTM Coal and Ash Analyses produce ashes that are markedly different. In one case the ash may be very hard and usable, in the other the ash may be very soft and unusable, and not easily disposed of. Also, the pH of the two ashes can be significantly different.

A review of the standard ASTM Coal and Ash Analyses needs to be done and perhaps these analyses need to be expanded or modified. Additional fuel identification methods need to be investigated.

Generally, a metallic oxide in the ash is reported as a percent of the ash. Perhaps the non-metallic makeup of the coals' inherent ash needs to be identified. Research into the behavior of the chlorides, carbonates, bicarbonates, sulfates, etc. and their relationship to the properties of the ash that is produced needs to be undertaken. This may answer some of the questions as to why ashes of very similar appearing coals behave differently in a boiler. The results of this research could have a major impact on coal indices, and coal blending models.

23

Hazardous Air Pollutants

The Clean Air Act has also raised questions about trace components of coal and coal ash. The U.S. Environmental Protection Agency has identified 189 potentially hazardous air pollutants. Twelve of the 189 are elements or compounds that are found in coal. Research needs to be done in helping coal consumers identify these materials and develop methods to effectively remove them from the coal, or from the flue gas.

The elements of concern are antimony, arsenic, beryllium, cadmium, chlorine, chromium, cobalt, lead, manganese, mercury, nickel and selenium. Almost all coals contain trace amounts of some or all of these elements or their compounds.

Currently, the belief in the industry is that the U.S. EPA will set stringent limits on arsenic, cadmium, lead and mercury. Research needs to be done to determine ways to effectively remove these from the flue gas stream, and a way to handle the materials after they are collected.

Wet scrubbers do remove some of the identified elements from flue gas streams. Ways need to be developed to handle the concentrations of the hazardous air pollutants that collect in the scrubber discharge. However, some arsenic and mercury may remain behind in the boiler and concentrate in deposits. Research is needed to address the mechanisms that cause these materials to concentrate in the deposits, and effective methods are needed for the removal and neutralization of these materials.

Currently, mercury presents some additional problems. The analytical procedure to determine mercury in coal is difficult, time consuming, and of questionable accuracy. A quick and relatively easy method to measure the mercury in coal needs to be developed, as well as ways to identify mercury in the fly ash.

Other Environmental Areas

Also, in the environmental area, biomass combustion and the burning of alternate fuels is becoming more prevalent in industry. Work needs to be conducted in addressing environmental discharge from some of these alternate technologies, for example; if tires are being burned, there may be large qualities of tin, mercury, zinc and nickel in the flue gas streams.

The burning of municipal waste has presented many problems. Some of these areas have been discussed such as chlorine, mercury, and nickel. Municipal Waste can also produce other hazardous air pollutants - mainly organic compounds. At times, this waste material can be attractively priced, but the problems affect the price advantage. Effective cleaning of the gas stream must be available before alternate fuels become the rule.

Again, addressing an environmental concern, fuel switching has caused a lot of electrostatic precipitator problems in the industry. Most older ESPs were designed to handle a low resistivity, high sulfur fuel and in many cases were only installed to protect the induced draft fans by cutting down the erosion of the rotors. However, currently the precipitators are being upgraded and used to reduce particulate and bring the stack discharge into compliance. Fuel switching to low sulfur coals has caused problems in older ESPs. Industry has been working on the technology of upgrading these older units, and trying to modify the ash to lower the resistivity so that

ESPs would function properly. But research needs to be done in electrostatic precipitator design, and ways to lower coal ash resistivity. There are some low sulfur coals that are currently being used that do not condition with existing technology. Ways to identify these coals need to be developed.

CONCLUSION

Conferences of this type are important to industry and the research community. The open exchange of information between the two areas is a benefit to both. It is imperative that the industrial community give direction to research. Pure research is important and is needed, but all research needs a direction; it is industry that must give that direction.

Many times "nice to know" research is essential, but the thing researchers should keep in mind is that industry is looking for an immediate answer to today's problem. Another item to keep in mind is industry is looking for a cheap answer. Any research that is done must keep the primary objective in mind, which is to solve a problem in a cost effective manner.

Many times research will find an apparent solution to a problem that works in the laboratory and/or in a bench scale. Most of the time it is difficult to find a company that is willing to commit a commercial unit to advance research. In my opinion, industry is wrong if they take this approach. The solution to a lot of problem areas that have been outlined can only be achieved through the cooperation of industry with the research community. It is my hope in the future that industry will be more receptive to trying some of the new applications that develope.

REFERENCES

1. J. Strenger and D. Banerjee (1991). "Chlorine In Coal Proceeding Of An International Conference": EPRI & Center for Research On Sulfur In Coal, Elsevier.

2. Chou, C. L. (1991). "Distribution and Forms of Chlorine in Illinois Basin Coals". Chlorine in Coal: Elsevier.

3. Cox, J. A. (1991). "Chemical, Extraction – Based, and Ion Chromatographic Methods for the Determination of Chlorine in Coal". Chlorine in Coal: Elsevier.

4. Huggins, Frank E. and Huffman, Gerald P. (1991). "An Xafs Investigation of the form – of – occurrence of chlorine in U.S. Coal". Chlorine in Coal: Elsevier.

5. Attig, R. C. and Duzy, A. F. (1969). "Coal Ash Deposition Studies and Application to Boiler Design; Proceedings of the American Power Conference Vol. 31.

6. Hensel, R. P. (1975). "Coal Combustion" First Engineering Foundation Conference on Coal Preparation for Coal Conversion.

7. Hurley, J. P., Erickson, T. A., Benson, S. A. and Biobjorg, J. N. (1991). "Ash Deposition at Low Temperatures in Boiler Firing Western U.S. Coals", International Joint Power Conference.

8. Hurley, J. P., Erickson, T. A., Benson, S. A. and Biobjorg, J. N. (etal) (1993) Modify of Fouling and Slagging Coal – Fired Utility Boiler Proposal for Fuel Processing Technology.

9. Hurley, J. P., Erickson, T. A., Benson, S. A. and Biobjorg, J. N. (etal) (1993) Predicting Ash Behavior in Utility Boilers, Energy & Fuels. 7, 746-754

EVOLUTIONARY CHANGES IN FURNACE COMBUSTION CONDITIONS WHICH AFFECT ASH DEPOSITION IN MODERN BOILERS

David Fitzgerald

Foster Wheeler Energy Corp.
Perryville Corp. Park
Clinton, N.J. 08809-4000, USA

Abstract

Extensive literature and research results are availible which are directed at analyzing or predicting the tendency of coal ash to deposit in radiant furnaces and onto convective surfaces. However, these results are generally based upon conventional pulverized coal (PC) furnace designs ,combustion systems, and fuels utilized over the last 40 years. Such conventional systems are being supplanted on some new units by combustion systems designed to respond to recent changes in the market place.

Some recent changes in the market which ultimately affect the furnace combustion processes can be cited. Environmental laws have placed greater importance on reducing emissions of NO_x, CO, SO_2 and air toxics. Similarly, there is some political support to reduce global emissions of CO_2 and N_2O, and one response has been the introductory commercial usage of PC-based combined cycles. And finally, there is an increase in the practice of world sourcing for coal supplies for a single power station, leading to the need to design a single furnace to reliably combust as many as 20 disparate fuels. This trend may accelerate in Pacific Rim nations due to the reduction in imported fuel tariffs associated with recent open trade agreements.

This paper outlines some changes in operation and design of new pulverized coal fired radiant utility furnaces which affect coal ash deposition, including the following:
>Deep staging, substochiometric combustion
>Rotating dynamic classsifiers
>Individual coal burner flame analyzers
>World sourcing of coal supplies
>Fully fired (hot windbox) combined cycles, with and without pyrolysis
>Pulsating combustion
>Preheating of coal-air mixture upstream of burner nozzle
>Miscellaneous effects (waterwall tube orientation, indirect bin fired system, water deslaggers)

1. INTRODUCTION

There are several areas of change and innovation which affect the sale, design, and operation of new, large PC fired utility power stations. Some of these areas which are relevant to ash deposition in a radiant furnace include the following :

Analytical---LMR (laser microreactor),CCSEM (computer controlled scanning electron microscope), XRD (X-ray diffraction), EMPA (electron microprobe analyzer), in situ laser anemometry

Applications of Advanced Technology to Ash-Related Problems in Boilers
Edited by L. Baxter and R. DeSollar, Plenum Press, New York, 1996

27

Predictive---CQIM (coal quality impact model), macerals analysis, deposition models, 3-D numerical simulation of radiant furnaces (finite difference, finite element/Galerkin solution of combustion, heat transfer, and momentum), empirical statistical regression analyses of relevant coal properties.

Monitoring---acoustic or infrared pyrometers, infrared television monitoring of slagging conditions, signal analysis(light frequency) and acoustic analysis of flame signatures

Regulatory--- Local emission characteristics of station (NO_x, SO_2, Hg, CO, particulate), global emissions of station (CO_2, N_2O, normalized on a per KWe basis), slag and ash disposal requirements, limits on imported fuel tariffs

Design and Process---combined cycles, coal gasification or pyrolysis, staged combustion, low Nox burners, pulsating combustion, preheating coal /air mixture upstream of burner nozzle, reburn technologies, dynamic classifiers, mechanistic models of pulverizer performance, load cycling operation, slagging combustors

Market Place--- The primary market for new PC units has shifted to the Pacific Rim and Asia, with repowering opportunities as a possible domestic market, free trade opportunities associated with more open international markets , improvements in real time communications reducing the need to base engineering or support operations within developed nations, increased exposure to trends in foreign technology

One can postulate that the governing changes are in the regulatory and market place areas. The analytical and predictive areas represent the tools which the engineer has availible to design the furnace to burn the coal within the new process envelope, and their main value is usually seen in their ability to reduce the risk of guaranteeing the performance of the new process. Therefore, it is important to recognize the direction of changes in the design of the PC furnace which is occurring in response to new changes in the market place, and to use the full menu of analytical and predictive tools to reduce the risk associated with the use of new combustion technology.

In reviewing the following discussion of recent combustion process changes ,one may wish to keep in mind the following generally accepted design assumptions of the past which are gradually being supplanted:

*A single design coal for each station
*base loaded operation
*20% excess air, oxidizing atmosphere in the lower furnace
*static mill classifiers, with a 70% thru 200 mesh and 98.5% thru 50 mesh
*21% windbox O_2 content
*+/- 15% burner to burner stochiometry unbalance
*steady state local combustion conditions
*avoidance of reducing environment and high burner zone heat input rates
*vertical waterwall tube orientation

2. NO_x Related Changes

Developed areas represented by Japan, Western Europe, and North America began modifying their methods of combusting pulverized coal so as to reduce the in-furnace NO_x formation during the late 1970's. Most domestic radiant furnaces were designed prior to 1978, however, and many of the recent international innovations developed to reduce NO_x have not been widely implemented in the USA. Thus, much of the slagging behavior reported in the US literature does not neccesarily represent the same problems anticipated in a new unit designed to generate low NO_x levels using the latest technology.

If we limit discussion to furnace combustion changes, we can summarize these changes as follows. Since the tendency of coal ash to cause unacceptable slagging behavior is related to all of the process effects during the char's combustion and the ash's furnace transit history, the following changes should be viewed with an eye toward their effect on these processes.

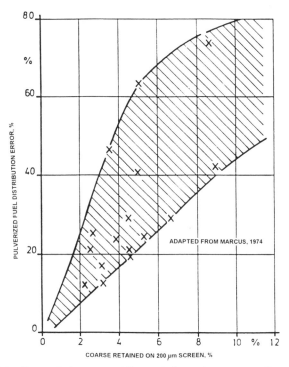

FIGURE 1. Burner-to-burner coal flow unbalance increases as the amount of coarse particles in the classifier outlet coal-air mixture increases. Shown is the results for a 4 burner per pulverizer classifier design. Adapted from Marcus, 1974.

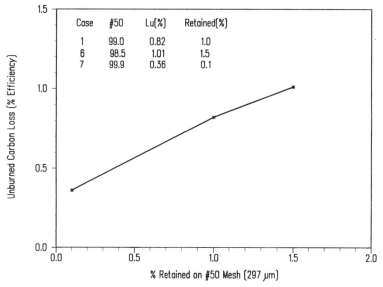

FIGURE 2. The unburned carbon loss is most strongly correlated to the amount of coarse particles vented to the burners, for a given fuel, load, and combustion system utilized. This has been confirmed by field testing, and advanced numerical modeling of the combustion process.Adapted from Fiveland, 1992

MBF COAL PULVERIZER WITH ROTARY CLASSIFIER

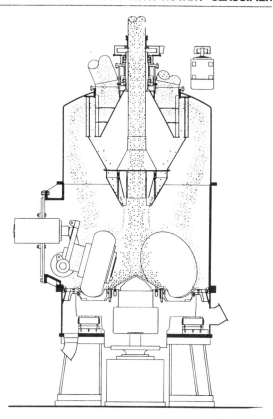

FIGURE 3. Rotating (dynamic) classifiers are finding increasing applications on PC units which utilize staged combustion for NOx reduction. Their use results in a decrease in coarse transport to the burners, which directly improves unburned carbon loss, burner coal flow unbalance, and related slagging. Additional benefits include reduced specific mill power consumption, increased mill capacity, and improved mill dynamic response .

FW ROTARY CLASSIFIER MECHANICAL DESIGN

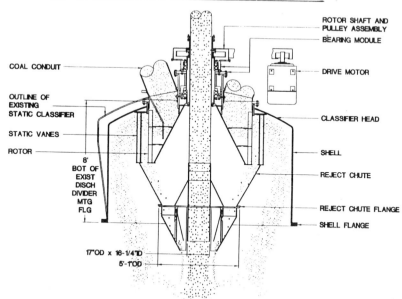

ROTOR SHAFT AND PULLEY ASSEMBLY
BEARING MODULE
COAL CONDUIT
DRIVE MOTOR
OUTLINE OF EXISTING STATIC CLASSIFIER
CLASSIFIER HEAD
STATIC VANES
ROTOR
SHELL
8' BOT OF EXIST DISCH DIVIDER MTG FLG
REJECT CHUTE
REJECT CHUTE FLANGE
SHELL FLANGE
17"OD x 16-1/4"ID
5'-T"OD

FW PLAN OF STATIC AND ROTOR VANES

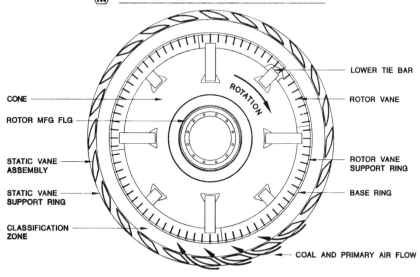

LOWER TIE BAR
CONE
ROTATION
ROTOR VANE
ROTOR MFG FLG
STATIC VANE ASSEMBLY
ROTOR VANE SUPPORT RING
STATIC VANE SUPPORT RING
BASE RING
CLASSIFICATION ZONE
COAL AND PRIMARY AIR FLOW

31

The first direction chosen to reduce NO_x was toward increasing the size of the furnace, or reducing the nominal parameters of heat input per plan area and heat input per burner zone effective area. This change resulted in reducing the equilibrium flame temperature, and was directed at reducing thermal NO_x. The reduction in flame temperature results in a corresponding reduction in net wall radiation heat flux in the burner zone, with a beneficial effect on reducing the equilibrium surface temperature of wall ash deposits, and increasing the liklihood of avoiding a wet slag coating. The resulting friable ash coverage is usually capable of being removed with conventional steam or air wall blowers.

The second direction chosen was a partial staging of the combustion process, using uncontrolled overfire air. Initially, oxidizing conditions (greater than 0% excess air) in the burner zone were retained. This method of operation increases the flame equilibrium temperature, and large variations in NO_x and CO can occur due to burner to burner coal or air flow unbalances. When the degree of staging was increased to provide substoichiometric conditions in the burner zone, there was a reduction in NOx, due to reducing the flame temperature and the formation of NOx scavenging radicals (HCN), but the unburned carbon loss often became unacceptable.

The unburned carbon loss is primarily related to the coarse particles which are not refluxed to the mill's grinding table by the standard, low efficiency static classifier. The coarse particles, generally represented by the fraction retained on a #50 mesh screen (297 μm), led to two related causes of lower furnace slagging. First, the larger particles can increase the burner to burner coal flow unbalance, due to the increased importance of two phase flow behavior with large particles. Refer to figure 1 for a typical variation in burner to burner coal flow unbalance (Marcus, 1974) as a function of particle fraction larger than 200 μm(approx #70 mesh). There is no currently reliable method of directly measuring the individual coal burner coal flows, and such unbalances lead to unbalances in burner stoichiometry, with a resulting large variation in NOx and CO.

Secondly, the large particles take much longer to burn out, because their surface area to volume ratio is substantially lower than for fines. If a char particle's pores fuse closed due to prolonged periods at high char surface temperature, then the particle's effective reactivity drops, and a large char/ash particle then becomes availible for impaction onto the waterwall surface. This is likely to occur at the sidewall, and at the target wall, depending upon the burner jet streamlines. Once on the wall, the unburned char can cause high local reducing conditions, and promote slagging and corrosion. Refer to figure 2 for a typical correlation between #50 mesh retention (100% passage thru #50) and unburned carbon loss, for a fixed 75% thru #200 mesh fines charactersitic, coal type, furnace, and combustion process conditions (Fiveland, 1992). Similar results were reported by Singh (1994), who observed that the majority of the unburned carbon loss is accounted for in collected ash particles greater than 100 μm diameter.

One solution to this problem, which has been employed in Europe and Japan, is to install a dynamic, rotating classifier atop the pulverizer. These classifiers have a much improved separation efficiency for coarse particles than do standard static classifiers, and have acceptable pressure drops. By eliminating the transport of coarse particles to the burners, the unburned carbon loss during deep staged (85% of stochiometric air in the burner zone) combustion is reduced to acceptable levels (0.5%), and lower furnace slagging conditions can be improved. Refer to figure 3 for a schematic of a typical dynamic classifier.

Fairly reliable statistical correlations of unburned carbon vs. raw coal ultimate and proximate analyses are availible, without resorting to the use of advanced analytical techniques or char burnout models. Refer to figure 4 (Miyamae, 1987) for a correlation of char oxidation rate at 600^0 C vs. combustion index (combination of inherent moisture and fixed carbon) and unburned carbon in ash vs. combustion index. These correlations are only valid for identical furnaces, burners, and coal size distributions, and are useful for determining the unburned carbon loss in units which must fire a wide range of coals.

The problem of burner to burner stoichiometry unbalances is not completely solved by reducing the coarse transport. If one accepts that a finite air and coal flow unbalance will occur, then modern monitoring and signals analysis techniques can be used to estimate the individual

$Ic = a (IM)^x (FC)^y + b (IM)^z$ Ic = Combustion Index   a,b =constant multipliers

x,y,z = constant exponents

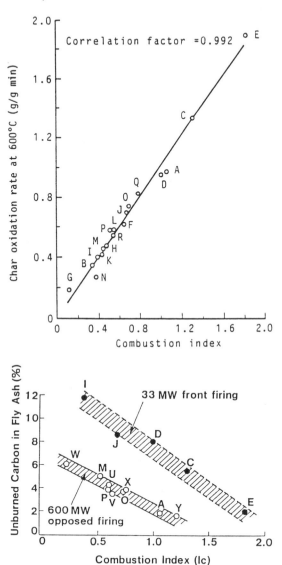

FIGURE 4. The variation in unburned carbon loss can be predicted with good accuracy based upon the fuels inherent moisture (IM) and fixed carbon (FC) content, without resorting to first principles or numerical modeling of the combustion process. Such correlations are valid for a particular furnace, combustion system, and coal fines size distribution, and are helpful in evaluating a single unit's combustion behavior for a range of coals.Adapted from Miyamae,1987.

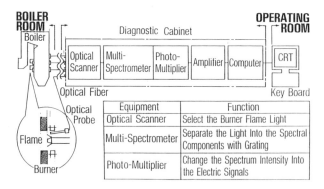

FIGURE 5. System integration of advancements in computers, signals analysis techniques, and monitoring elements has made possible the commercial use of individual coal flame analyzers. Fiber optic monitoring probes installed at each burner are multiplexed at a light spectrum analyzer. Noise is filtered, and appropriate realtime signal analysis hardware is used to provide input to the auto-correlative program, which provides a realtime estimate of each burner's stochiometry. Adapted from Ito, 1990.

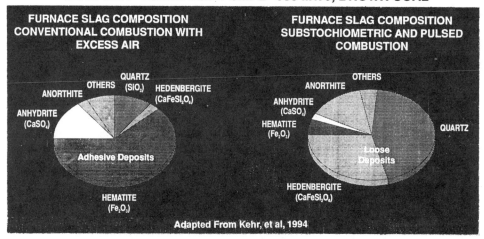

FIGURE 6. VEAG units similar to plant Schwarze Pumpe had their conventional combustion systems modified for Low NOx combustion, and the results included a fundamental change to the lower furnace slag behavior. The adhesive hematite deposits associated with conventional , steady oxidizing conditions were replaced by friable deposits of quartz (Cristobalite) and Hedenbergite when the lower furnace combustion conditions were modified to reducing conditions , with pulsating combustion and an increased burner zone heat liberation rate. Adapted from Kehr, 1994.

burner stoichiometries, in real time. With this estimate as a source of feedback, the individual burner air flows can be adjusted to provide more equalized burner to burner stoichiometries, which permits a lower overall excess air for improved boiler efficiency, while limiting NOx and CO emissions. Figure 5 represents one such individual coal flame (Hashimoto, 1992)monitoring system, using the following components:
-fiber optic flame sensor
-fiber optic multiplexor
-visible light and IR frequency analyzer
-signal analysis processor, using auto correlation techniques
-output display, auxiliaries, etc.

This improvement in equalized burner stoichiometries has been reported to also reduce burner eyebrow formation.

Low NO_x burners which use internal staging are also finding widespread use. The variation in burner jet behavior can affect wall impingement characteristics and char burnout rates, which can effect furnace slagging. In most cases, the improvement in burner to burner air flow distribution and independent control of flame swirl characteristics associated with low NOx burners has led to reduced furnace slagging. The point to be made here is that this change also results in a change to the combustion history of the char and ash particles.

One of the most remarkable changes to the lower furnace slagging conditions due to staged combustion was reported recently associated with VEAG's units similar to plant Schwarze Pumpe (Kehr, 1994). These 800 MWg, brown coal-fired tower units, use furnace exit gases to dry and transport the coal from the beater mills to the burners. In their original design, which used oxidizing conditions in the burner zone and conventional burner zone heat release rates, the furnace slagging was excessive, and managed only by the regular use of water deslaggers.

After redesign and modification for low Nox combustion, the new furnace performance parameters included deep staging (low burner zone stoichiometry, reducing conditions), reburning of fines in the upper furnace, an increased burner zone heat release rate, and pulsating combustion. Most of these changes would be expected to worsen the lower furnace slagging, yet the actual slagging performance greatly improved. Refer to figure 6 for a summary of the changes to the lower furnace's slag constituents due to the fundamentally changed combustion conditions. It was reported that such a modified furnace has operated for 20 months without requiring furnace waterwall cleaning, which is in stark contrast to the previous need for regular water deslagging. The adhesive deposits of Hematite associated with conventional combustion were replaced by friable deposits of Hedenbergite and quartz.

3. ADVANCED AND COMBINED CYCLES

The impetus to develop advanced cycles is driven both by standard economic considerations (eg, improving the return on investment by reducing the fuel costs) and by the need to reduce the costs to the environment. These costs are either quantified by externalities, or by a carbon or pollution tax, or by progressively tightening the emission requirements by regulation. The relevant pollutants may be the local pollutants Hg, SO_2, NOx, As, and CO or the global pollutants CO_2 and N_2O.

On an as-received basis, possibly the most effective way to improve plant economics is to reduce the unburned carbon loss, both to reduce fuel costs and to reduce ash disposal cost. Modifying the combustion process and/or the flue gas treatment system is an effective method to reduce most categories of emissions. Regarding CO_2, increasing plant net cycle efficiency or switching to a fuel mix with less carbon content are currently the only effective strategies, until CO_2 removal technologies become viable.

Recent advances in metalurgical science, for example, tungsten-modified ferritic alloys (HCM12a, HCM2s, NF616, or ASME P112, T23, and P92, respectively), permit the use of advanced steam cycles, such as the ultra-supercritical cycle (USC). Such a change in technology does not impact the fireside combustion process, except for the secondary effect of higher waterwall tube metal temperatures, and yet can reduce the specific emissions of all pollutants by about 7.8% (kg pollutant/kWe hr) soley by virtue of the improvement in net plant efficiency. Refer to figure 7 for the effect of increasing the steam cycle conditions vs. net plant heat rate improvement (Bennert, 1982) .

In Germany, Holland, and Japan, a substantial reduction in specific emissions of CO_2 has been demonstrated by converting fossil-fired units to operate as fully fired, hot windbox combined cycles (FFCC). Refer to figure 8 for a cycle schematic (Joyce, 1993 and Termuehlen, 1992). Approximately 20% of the plant's fuel heat input is provided by natural gas, fired in a combustion turbine, which exhausts its vitiated air (16.5%O_2 content, 425-590^0 C) into the PC furnace's windbox, to be used as secondary air. Because this results in a much lower equilibrium coal flame temperature in the burner zone (due to the 27% increase in kg flue gas/kg coal), less thermal NOx, and a reduced waterwall radiant heat flux result, which reduces the tendency for wet slag to form in the burner zone. The FFCC has about a 6% better net plant cycle efficiency (LHV basis) then a simple Rankine steam cycle, and the addition of low carbon content natural gas to the fuel mix further reduces the specific emission of (kg CO_2/KWe hr) . This conversion of PC units has potential for use in the USA due to our secure source of natural gas, but the economics would need to be augmented by a carbon tax before it is viable. It is reported that the FFCC is not suitable for units which must utilize a wet scrubber, however, due to the stack gas reheating penalty associated with the higher flue gas flowrate.

Regarding the use of an all-coal-based combined cycle, FWEC is currently studying the feasibility of repowering an original PC-fired unit in Delaware, using the combination of an air-blown pyrolyzer, feeding a cleaned fuel gas to a combustion turbine, and firing a mixture of char and coal in the original PC furnace in a fully fired, hot windbox mode. Obviously, the combustion and furnace transit history of the resulting char particles will be substantially different than the original combustion process, with consequent changes to the furnace slagging behavior. The original plant's 33.5% net cycle efficiency is expected to be improved to 39%, reducing the specific emission of CO_2 by 16%. Refer to figure 8 for a cycle configuration of an all-coal FFCC.

4.   MISCELLANEOUS CHANGES

There are several miscellaneous changes to modern boiler furnaces which were not common on units built prior to 1978, and they deserve some mention, as they can affect lower furnace slagging.

Inclined waterwall tubes--- After 1978, most once thru units (OTU) built overseas utilized spiral wound, variable pressure waterwalls. These membrane walls are oriented at about a 20 degree inclination above horizontal, as opposed to conventional vertical waterwall tubes. While sidestepping issues of support and thermal stress, it can be admitted that the inclined tube orientation accumulates slag more readily than a vertical tube. Due to the support arrangement of these panels, extra attention must be applied to the design of the waterwall supports and the operation of water deslaggers, if such methods must be used to manage slag. Due to the higher cost to repair these walls, combustion conditions which encourage fireside corrosion should be avoided.

Indirect fired, bin system--- In China, most domestically fabricated and designed boilers use an indirect fired system, where the pulverizers exhaust to external cyclone classifiers and to storage hoppers (bins). The coal is transported from the hopper to the burners using eductors and transport air. In low volatile coal applications, very hot transport air is used, which preheats the coal, reduces unburned carbon loss, and improves flame stability. One slagging problem associated with the indirect firing method is that the cyclone hopper slope behaves as a particle classifier, with coarse particles migrating toward the hopper surface, and fines migrating to the hopper centerline. When the coal is split into individual streams at the eductors, some burner feed lines receive a high fraction of coarse particles, and high slagging is associated with those burners.

Water deslaggers--- The practical research conducted by the vendors of deslaggers has resulted in reliable designs, and useful design guidelines for the boiler manufacturer. In retrofit applications where excessive slagging cannot be reduced by modifications to the combustion system (mills, burners, air supply), these deslaggers can be of great assistance. However, modifications to the mills and burners should be a primary method to control slagging, since the effect of coarse particles , burner-to-burner stochiometry unbalance, and burner clearances to the sidewalls, target wall, and to the hopper knuckle are well known today.

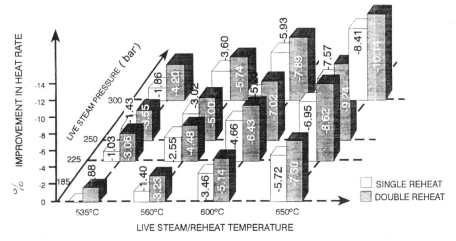

FIGURE 7. The recent metallurgical improvements of tungsten modified ferritic alloys has enabled the commercial implementation of advanced steam cycle conditions, such as the ultrasupercritical (USC) cycle to 300 bar/600°C/620°C/620°C. Net plant heat rate improvements of about 7.8% result in a uniform reduction in all pollutants, on a normalized basis of (kg poll./kwe hr). Adapted from Bennert. 1982.

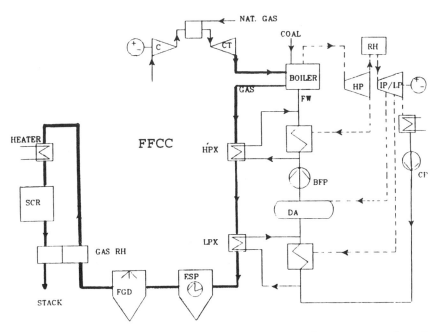

FIGURE 8. The use of a fully fired (hot windbox) combined cycle(FFCC) can be configured to improve the plant's LHV net station cycle efficiency by about 6%. Additional reduction in specific emissions of CO_2 can be realized if low carbon content natural gas is used in the fuel mix. The resulting lower equilibrium flame temperature in the burner zone of the PC radiant furnace results in lower NOx, reduced peak waterwall radiant heat absorption rate, and a reduced slag surface temperature. Adapted from Joyce. 1993.

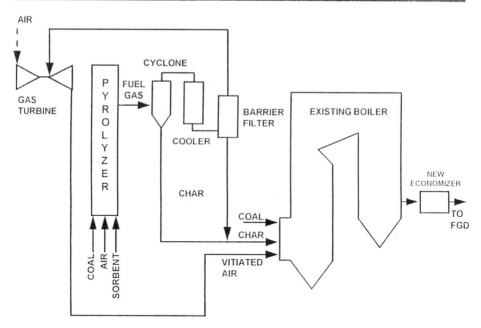

FIGURE 9. The fully fired combined cycle can also be configured as an all-coal based cycle. An air blown pyrolizer generates a low BTU fuel gas, which is cooled, cleaned , and burned in a combustion turbine (CT). The CT exhausts to the PC furnace windbox, while the PC burners combust a mixture of char (from the pyrolizer) and raw coal.

5.   CONCLUSION

Many of the slagging problems encountered in domestic units are associated with the combustion system designs availible 30 years ago. International changes in the method of combusting pulverized coal have provided new design options for countering slagging problems, but have also introduced new unknowns. The advanced analytical and predictive tools used to explain the problems of conventional units can be brought to bear on the new combustion systems being proposed, to help reduce the risk when using new technologies.

6.   REFERENCES

1)Fiveland, W., et al, " An Efficient Method for Predicting Unburned Carbon in Boilers", Comb.Sci.&Tech.,V81 no4-6, 1992, p147-167

2)Marcus,H., "Grosse Dampferzeuger mit Steinkohlen-Staubfeuerung und Trockenentaschung", Energie und Technik, Heft 7/8,1974, p179-186

3)Singh, B.,"Pulverized Fuel Size Distribution Influence in a Power Generating Utility Boiler", proc. 1994 ACTC Workshop on the Impact of Coal Quality on Power Plant Performance,Brisbane, Australia, 15May94.

4)Miyamae,S., et al, "Nox and unburned Carbon Simulation Technologies on Pulverized Coal firing Boiler", IHI Eng. Rev.,V20no2 Apr87,p56-60

5)Hashimoto,H., et al, "Developement of a Flame Diagnostic System for a coal- firing Boiler", ASME paper 92-JPGC-EC-2, presented at the IJPGC conference, Atlanta, 18Oct92.

6)Kehr, M. et al, "800MW Brown coal fired Steam Generator Schwarze Pumpe",17May94 PowerGen conference, Cologne, Germany, Penwell Pub, Houston, Tx.

7)Bennert,J.,"Auslegungskriterion fur grosse Steinkohleblocke",VDI-Berichte, Nr454,1982,p95-101

8)Joyce, J. S. "How Gas Turbines can Improve the Efficiency and Environmental Compatability of new and old Steam Power Plants", Proc. ASME-JSME Int Conf on Pow Eng(ICOPE-93),Sept 1993(Tokyo)p469-474.

9)Termuehlen, H.,et al, "Repowering Existing Power Stations with Heavy Duty Gas Turbines-An Economical Approach", Proc. PowerGen, 5th int'l Conf, Orlando, 1992,v11p235-244.

THE UK COLLABORATIVE RESEARCH PROGRAMME ON SLAGGING PULVERISED COAL-FIRED BOILERS: SUMMARY OF FINDINGS

W.H.Gibb

PowerGen
Power Technology
Ratcliffe-on-Soar
Nottingham NG11 0EE
England

ABSTRACT

A major UK collaborative research programme to address all aspects of slagging in bituminous coal-fired boilers has recently been completed. The 3½-year programme comprised an integrated package of full-scale plant trials; rig and laboratory-scale tests; coal, flyash, and deposit characterisation by Computer Controlled scanning Electron Microscopy (CCSEM); and the development of computer models. The major findings are presented in this paper.

It is concluded that the key to the behaviour of ash deposits lies in understanding their consolidation by a viscous flow sintering mechanism. For UK coals, iron and calcium (derived from pyrite and calcite respectively) are the main fluxing elements that affect ash viscosity and it is, therefore, the degree of assimilation of these elements into the aluminosilicate glass that ultimately determines the slagging characteristics of the coal.

CCSEM can be used to quantify the mineral matter distribution in coals and a novel slagging index based on CCSEM analysis that takes account of the effectiveness of the fluxing elements present, in particular the pyrite-derived iron, is proposed. A methodology for studying the ash deposition characteristics of coals using a laboratory-scale Entrained Flow Reactor (EFR) has also been established. CCSEM analysis of deposits formed in the EFR shows the potential for coalescence and provides the basis for an alternative slagging index.

Applications of Advanced Technology to Ash-Related Problems in Boilers
Edited by L. Baxter and R. DeSollar, Plenum Press, New York, 1996

41

BACKGROUND

Slagging has long been recognised as one of the most serious potential problems in operating power station boilers. Most boilers are equipped with an array of sootblowers that are designed to remove deposits by routine on-load cleaning. The uncontrollable build-up of slag deposits can lead to substantial financial penalties. Initially this may simply be due to a reduction in boiler efficiency, but severe slagging can ultimately cause unscheduled outages for off-load cleaning. Apart from the cleaning costs, large replacement generation costs are incurred.

When the UK Collaborative Slagging Research Project was first conceived in 1987 much research effort had already been devoted to the subject worldwide. However, slagging is an extremely complex phenomenon influenced by a combination of factors - boiler design, boiler operation, and the nature of the coal fired, in particular the composition of the ash minerals in the coal. Consequently the level of understanding at that time was still insufficient to prevent the major losses that were being incurred. Furthermore, in 1986 a programme of retrofitting low-NOx burners to all the 500 and 660MW coal-fired boilers had just been initiated in the UK. The principle of low-NOx burners is to create sub-stoichiometric conditions in the near flame zone and, because of the known adverse effect of reducing conditions on slagging, there were major fears that the conversion to low-NOx burners would exacerbate slagging problems. It was against this background that this 3½-year project was started in May 1991.

PROJECT OUTLINE AND OBJECTIVES

The project was conceived as a comprehensive study of all aspects of furnace slagging and its structure and overall objectives have been described in a paper presented at the last Engineering Foundation Conference in this series (Gibb et al, 1993). A diagrammatic representation of the complete project showing the links between the programme elements is given in Fig 1.

There were two key features that it was felt would increase the chances of success of this particular project. First, the fundamental element in the programme should be tests at full-scale in a power station boiler; these would be backed up by additional testing at smaller scale, but under more controlled conditions, using samples of the same coal. This gives a fully integrated set of comparative experimental results.

Second, to tackle such a complex problem, it was considered essential that the resources and expertise of the generators, plant manufacturers, coal supplier, and Universities with experience in this area should be pooled in a fully collaborative project. In both these respects it is considered the project has proved successful.

The principal objective of the project was to develop an improved method for predicting the slagging behaviour of individual coals. Historically a major part of the research effort has been geared towards developing suitable slagging indices based on either chemical composition or some physical parameter, eg ash fusion temperature or ash viscosity. At the time the project was conceived, it was increasingly being recognised that slagging indices based on bulk ash composition had limitations and that any improvement would require a knowledge of the composition and distribution of the minerals within the coal.

Accordingly, another key aspect of the project was the provision for the purchase of a Computer Controlled Scanning Electron Microscope (CCSEM) to be housed within the Department of Materials at Imperial College. This equipment was considered essential for the detailed characterisation of the coal, flyash, and deposit samples generated by the various tests within the programme. Provision was also made to establish a unique, laboratory-scale facility for the assessment of the deposition characteristics of coal ashes. This rig or Entrained Flow Reactor (EFR) was also the responsibility of the Department of Materials at Imperial College.

The complexity of the slagging process means that many factors beyond the inherent properties of the coal must be taken into account in determining the risk of slagging in an individual boiler. Subsidiary objectives of the project were, therefore, to gain further information on the influences of boiler design, boiler operation, and on-line cleaning equipment, ie sootblowers. A particular objective was to assess the effect of low-NOx burners on boiler slagging and the plant trials at Ratcliffe PS were based on Unit 2 which had just been converted to low-NOx burners in 1991.

FINDINGS UNDER INDIVIDUAL ACTIVITIES

Some of the detailed findings under the individual programme elements were published in the proceedings of the previous conference in Solihull (Williamson and Wigley, 1994) and further details of the CCSEM studies (Wigley and Williamson, 1995), Entrained Flow Reactor studies (Hutchings et al, 1995), and the computer modelling activity (Lee et al, 1995) are presented at this conference. Accordingly, only selected highlights are presented in this summary report.

Plant Trials

The plant trials at Ratcliffe PS were the focal point for the whole project and were intended to provide actual slagging data against which all other experiments and predictions on the same coals would be calibrated. The compositions of the three main coals are shown in Table 1. These were selected on the following basis:

Bentinck	High tonnage coal supply to Ratcliffe with low slagging propensity.
Daw Mill	Above average calcium content coal that had caused slagging problems at Ratcliffe prior to the low NOx burner retrofit.
Silverdale	Typical high iron content coal of assumed high slagging propensity.

The most significant conclusion from these trials was that the conversion to low-NOx burners has been beneficial by increasing the tolerance of the boiler to slagging coals. No measurable effect on boiler operation due to slagging was detected during any of the trials on Unit 2. During each trial the level of excess oxygen was varied, but no effect of this parameter on slagging was detected. This apparent insensitivity to coal type and excess oxygen is discussed later.

Although there was no significant operational effect, marked differences in the nature of the deposits formed by the three coals were observed. Daw Mill gave highly fluid, but relatively thin deposits. In contrast, Silverdale yielded more massive, but more friable deposits. The least extensive and most friable deposits were found when firing Bentinck coal. These differences can be related to the minerals in the individual coals and their effect on the microstructure of the consolidated deposits.

A second planned activity at full-scale was to carry out "slag mapping" in an effort to understand where in an operational boiler deposits tend to form. Given the restricted access, dust loading and extreme light intensity in the furnace, this is virtually impossible to achieve on-load. Accordingly, it was decided to attempt this (as the opportunity arose) by inspection of boiler furnaces immediately after coming off load, but with a few oil burners still in service to provide illumination.

This approach proved fairly successful and it is concluded that, as far as heat transfer is concerned, the sensitive area for ash deposition is the rear wall from top burner level up to the nose. The sidewalls and front wall above the burners are generally clean, particularly with low-NOx burners. Massive deposits can form on the burner quarls and may cause problems regarding burner performance and ash handling, but will have little effect on overall heat transfer.

Single Burner Rig Tests

Each of the 3 main test coals was fired on Babcock Energy's 60 MW$_{th}$ single burner rig at Renfrew, Scotland. The full-scale burner used was identical to the low-NOx burners installed on Unit 2 at Ratcliffe PS. The objective of this element of the programme was to study mineral transformations, particularly for pyrite (FeS_2), that occur during the early stages of pf combustion. Sampling ports and probes were available for the measurement of gas composition and temperature profiles along the flame and the collection of water-quenched particulate samples. To check the findings against real plant, similar measurements were made by probing a bottom row wing burner at Ratcliffe PS during the trials.

concerned, the nature of slag deposits depends on the occurrence and distribution of the aluminosilicate, pyrite, and calcite minerals in the coal.

This determines the viscosity of the deposited ash particles and, hence, the ultimate degree of consolidation of the deposit. This in turn will determine the nature of that deposit, its effect on boiler heat transfer, and ease of removal by sootblowing.

A new slagging index is proposed based on CCSEM analysis that takes account of mineral occurrences in the coal and, thereby, the degree of assimilation of the fluxing oxides, CaO and Fe_2O_3. The proposed index is based on CCSEM analysis of the coal to provide data on the

TABLE 1. Coal Ash Compositions and Slagging Indices.

	Main Test Coals (ex Ratcliffe PS)		Additional Test Coals (EFR Tests at Imperial college)								
	BENTINCK	DAW MILL	SILVERDALE	ASFORDBY	BUTTERWELL	DALQUHANDY	HARWORTH	NADINS	OXCROFT	POINT OF AYR	PITTSBURGH No 8
Ash Content, % dry	17.4	15.8	15.1	7.5	9.8	8.9	16.1	13.9	10.8	12.9	6.9
SiO_2	54.5	55.1	45.5	51.8	41.9	52.2	53.2	39.1	37.8	48.4	47.6
Al_2O_3	26.3	21.7	25.1	25.4	24.3	32.2	26.8	21.6	21.9	24.3	24.0
Fe_2O_3	10.1	11.1	22.5	8.8	20.9	9.6	10.9	21.8	31.7	13.6	16.9
CaO	2.0	5.2	2.5	6.6	7.1	1.8	1.1	9.4	2.4	6.7	6.2
MgO	1.3	2.0	1.0	1.7	3.3	0.8	1.3	4.5	1.5	3.0	1.1
K_2O	3.0	2.6	2.0	2.1	1.0	1.0	3.8	1.7	3.0	2.2	1.7
Na_2O	1.4	1.2	0.6	1.5	0.1	0.1	1.3	0.4	0.2	0.3	0.8
TiO_2	1.2	1.2	1.0	1.3	1.0	1.3	1.0	0.8	0.6	0.8	1.0
B/A	0.22	0.28	0.40	0.26	0.48	0.16	0.23	0.61	0.64	0.35	0.37
Rs	0.44	0.46	1.35	0.21	1.00	0.17	0.46	1.79	2.40	0.55	0.73
New Indices											
CCSEM	9.6	17.3	15.2	21.2	16.9	8.8	7.0	22.0	15.1	14.5	11.7
EFR	41	68	38	73	72	61	21	89	55	48	74

45

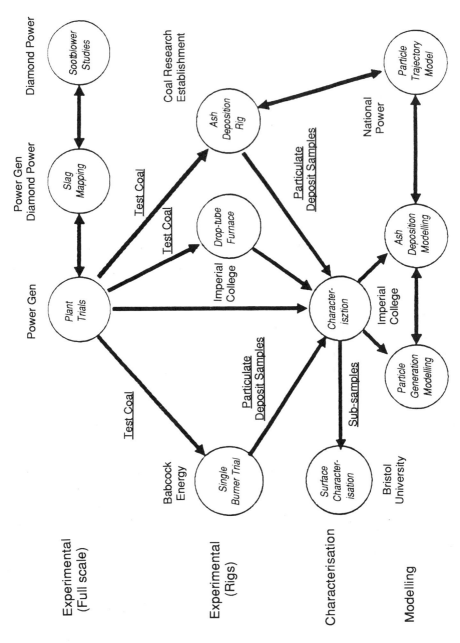

FIGURE 1. Interaction between Programme Elements.

46

composition and size distribution of the individual mineral matter particles present. These may be pure minerals (eg quartz, clay, pyrite) or mixtures.

It has been shown that all the calcium in the coal is assimilated into the aluminosilicate glass and it, therefore, has an "effectiveness" of 1.0 as a fluxing agent. For iron, it has been shown that the degree of assimilation and "effectiveness" is potentially variable and will depend on size and distribution of the pyrite within the coal. Any pyrite that occurs with the clay as a mixed mineral particle has more chance of coalescing to form an iron-rich glass than a pure pyrite particle on its own. For calculating the index it is assumed that pure pyrite particles have an "effectiveness" of 0.5 and any clay/pyrite mix an "effectiveness" between 0.5 and 1.0 in line with the amount of clay present in the mix. The CCSEM-based index is then defined as follows:

$$\text{Index} = \sum_{\text{All mineral occurences}} \left[\text{mass fraction} \times \left(CaO + Fe_2O_3 \times \left(1 - 0.5 \times \frac{Fe_2O_3}{Fe_2O_3 + Al_2O_3 + SiO_2} \right) \right) \right]$$

The results obtained for the 11 coals that have been analysed are shown in Table 1, together with their full ash compositions, base/acid ratios, and the derived slagging indices, Rs. This ash composition-based index is the most widely used and is defined as the product of base/acid ratio and sulphur content (dry basis), where the basic oxides are Fe_2O_3, CaO, MgO, K_2O, and Na_2O, and the acidic oxides SiO_2, Al_2O_3, and TiO_2. Base/acid ratio alone is used as a guide to slagging propensity and, as the key basic oxides are Fe_2O_3 and CaO, is similar in concept to the proposed CCSEM-based index. Indeed, a plot of base/acid ratio against the new index shows a reasonable trend for many of the coals examined - see Fig 7. The outliers are due to differences in Ca:Fe ratio and the proportion of the pyrite present in the coal as mixed pyrite/clay occurrences. For example, the CCSEM-based index predicts that the high calcium content and high Ca:Fe ratio coals Asfordby and Daw Mill have higher slagging propensities than indicated by base/acid ratio. Conversely the CCSEM-based index for Oxcroft is lower than indicated by Base/Acid ratio due to the low Ca:Fe ratio and relatively small proportion of pyrite present as mixed pyrite/clay occurrences.

The CCSEM-based indices for the 3 main test coals place the coals in order of increasing severity as follows:

Bentinck (9.6) < Silverdale (15.2) < Daw Mill (17.3)

This prediction agrees well with the nature of the slagging observed during the trials at Ratcliffe PS where Daw Mill gave the most fluid deposits and Bentinck the least. In contrast, based on either base/acid ratio or Rs (see Table 1) Silverdale would be considered much worse than Daw Mill. The slagging index Rs emphasises the importance of pyrite as this will influence both the base and sulphur terms, ie it is effectively proportional to $(Fe)^2$. It has long been recognised that this index underestimates the effect of calcium as exemplified for example by the virtually identical values for Bentinck and Daw Mill. In the absence of CCSEM data, based on this study, Base/Acid ratio is probably a more reliable ash composition-based index than Rs. However, based on both theory and experience, the proposed CCSEM-based index will provide a more reliable prediction.

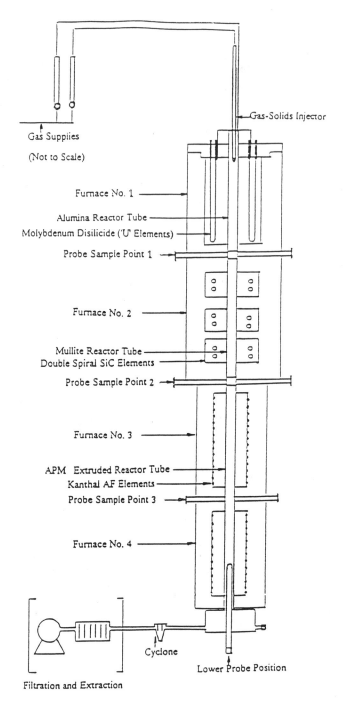

Gas Supplies
(Not to Scale)

Gas-Solids Injector

Furnace No. 1

Alumina Reactor Tube

Molybdenum Disilicide ('U' Elements)

Probe Sample Point 1

Furnace No. 2

Mullite Reactor Tube
Double Spiral SiC Elements

Probe Sample Point 2

Furnace No. 3

APM Extruded Reactor Tube
Kanthal AF Elements

Probe Sample Point 3

Furnace No. 4

Cyclone

Lower Probe Position

Filtration and Extraction

FIGURE 2. Schematic of the EFR.

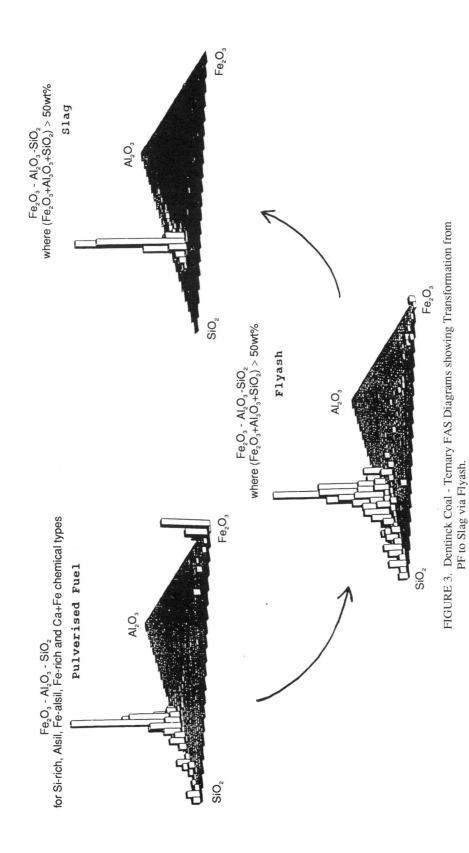

FIGURE 3. Dentinck Coal - Ternary FAS Diagrams showing Transformation from PF to Slag via Flyash.

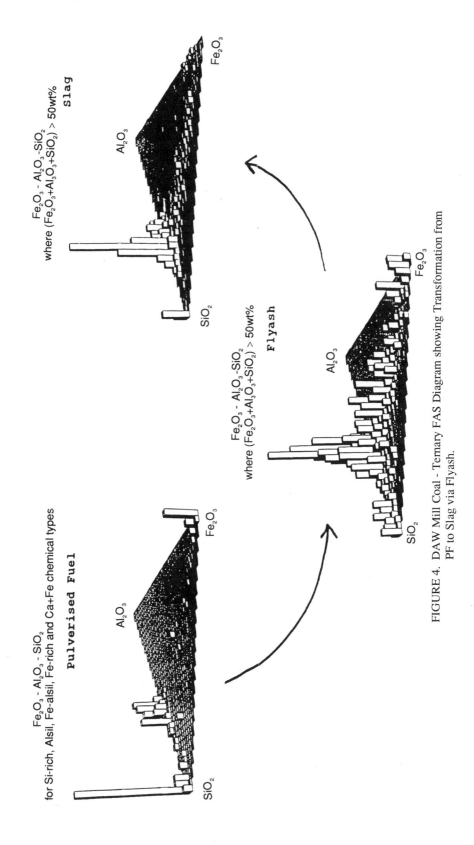

FIGURE 4. DAW Mill Coal - Ternary FAS Diagram showing Transformation from PF to Slag via Flyash.

Under this project, just 11 coals have been studied, 10 UK coals and a single US coal from the Pittsburgh No 8 seam. However, the principles on which the CCSEM index is based should be applicable to any bituminous coal. Pyrite is the main mineral contributing to slagging in such coals and the "effectiveness" factor built into the index should be generally applicable. Pending studies of a wider range of coals, there is more uncertainty concerning the effectiveness of fluxes occurring as carbonates. Based on the coals studied here calcium derived from calcite appears to be totally effective, ie is totally assimilated into the aluminoaticate glass. It must be assumed, therefore, that where iron is mainly present as the carbonate mineral siderite, it is also likley to be totally effective. Further work is planned to test this assumption.

By Laboratory-Scale Testing (the EFR)

The Entrained Flow Reactor was designed to enable accurate predictions of slagging behaviour to be made from laboratory-scale testing with a few kilograms of coal. Though small in scale, it has been shown that deposits with a similar morphology to those found in real plant can be collected. The predictions based on EFR testing should, therefore, be more reliable than those based on coal analysis, however comprehensive. At the chosen test conditions little variation in deposit type was seen, but a novel method of ranking coals based on CCSEM analysis of the microstructure of the deposits looks promising.

The deposit sample is mounted in resin and a polished cross-section prepared. An area of deposit is then selected for point analysis on a 16x16 square grid and the results plotted as FAS and CAS ternary diagrams. The purpose of this analysis is to quantify the proportion of the deposit having a composition range that is likely to promote sintering by viscous flow and, thereby, deposit consolidation. Based on liquidus temperatures, the composition ranges for the two ternary systems that were chosen are given below:

FAS System $Fe_2O_3 > 10\%$ and $< 50\%$

CAS System $CaO > 5\%$ and $< 40\%$

The proposed EFR-derived index is then taken as the sum of the numbers of point analyses falling within these composition ranges and the results obtained for deposits collected on an uncooled ceramic probe at 1200°C (Port 2) are included in Table 1. The use of an uncooled ceramic probe, rather than a cooled metal probe, was preferred as this accelerates the critical deposit growth and consolidation process.

The relationship between this index and the proposed CCSEM-based index is shown in Fig 8 and for most of the coals tested agreement is good. The outliers above the line are Dalquhandy and Pittsburgh No 8, ie the EFR indicates a worse slagging behaviour than the CCSEM analysis. In the case of Dalquhandy this is certainly surprising as this coal would appear to be particularly inert. The other clear outlier (below the line) is Silverdale with an EFR index of just 38 compared to 41 for Bentinck and 68 for Daw Mill.

The apparently unrealistic predictions for Dalquhandy and Silverdale may be related to test conditions. The type of deposit obtained depends on gas temperature at the sampling port as this will directly affect the degree of assimilation of the iron phases into the aluminosilicate melt. It is possible, therefore, that the ranking order obtained from the EFR index will vary with gas temperature. For example, at a higher temperature, for Silverdale, the degree of assimilation may be higher relative to the other coals. Similarly, for Dalquhandy it may be lower. It follows that the relative full-scale behaviour of different coals may vary depending on which area of the furnace is being considered; to obtain a full characterisation it will be necessary to undertake EFR tests over a range of gas temperatures. This further illustrates the complexity of the slagging process and the inherent difficulties in attempting to apply any single index.

By Rig-Scale Testing

For coal assessment, the next step up in scale is to go to rig testing. This obviously requires special facilities that are considerably more expensive to operate. However, a rig provides a real combustion environment and, therefore, much more realistic test conditions. The $50kW_{th}$ Ash Deposition Rig at CRE probably represents the smallest scale at which this type of testing can be carried out and the results obtained within this Project have shown that this facility can be used to reproduce the slagging behaviour observed at full scale.

The most realistic prediction of slagging behaviour was obtained from the nature of the deposits found on the refractory walls of the combustion chamber, rather than the air-cooled metal deposition panel and probe. It is concluded that, for relatively short term tests, the best approach is to use an uncooled ceramic probe for slagging assessment. This greatly accelerates phases 1 and 2 of slagging (initiation and deposition - see Section 4.1) and enables the critical consolidation phase to be realistically simulated. This comment applies to short term testing at any scale from plant tests to laboratory-scale tests such as those performed on the EFR at Imperial College.

Pyrite decomposes and oxidises to form iron oxides during combustion, forming pyrrhotite, FeS, and Fe-O-S species as intermediates. The results obtained indicate that this process is extremely rapid for the smaller (<20mm) pyrite particles and is complete within the flame. It follows that only the relatively coarse pyrite particles can escape the flame in the form of potentially sticky Fe-O-S intermediates and can, therefore, participate in ash deposit initiation.

The detailed examination of pf, in-flame and back end flyash samples by CCSEM (see Section 3.4) has provided clear evidence that a significant degree of coalescence of the mineral occurrences within the pf occurs during combustion. The major consequences of mineral coalescence are a decrease in the abundance of the Fe-rich, Ca-rich, and Si-rich mineral types in favour of the aluminosilicate species, and increases in the Fe_2O_3 and CaO contents of the AlSi and Si-rich species and of the CaO content of the Fe-AlSi species. All of these expected changes were observed in the in-flame samples and there is evidence of increasing coalescence with a decrease in the carbon content of the samples. This suggests that coalescence principally involves multiple mineral occurrences within individual pf particles that are brought into close proximity as the coal char burns out. This is discussed further in the section on the mechanism of slagging.

Entrained Flow Reactor (Imperial College)

The Entrained Flow Reactor (EFR) at Imperial College is a unique facility within the UK that enables ash deposition behaviour to be studied at small scale, but under realistic time/temperature conditions. The reactor consists of four vertically linked furnaces that can be independently controlled to give the desired temperatures. Pulverised coal samples are fed in at the top and deposit and particulate samples can be taken at the sample points between the furnaces. A cross-sectional diagram of the reactor is shown in Fig 2. By adjusting the furnace temperature settings, ash deposition behaviour can be studied over a wide range of gas temperatures from about 1400°C downwards.

In addition to the 3 coals fired at full-scale, an additional 8 coals were studied on the EFR to give results for a much wider range of coal ash compositions (Table 1). Under the chosen experimental conditions, both deposition rate and deposit type, ie degree of fusion, appeared to be relatively insensitive to coal type. However, a novel method for assessing slagging potential based on the detailed porosity and chemical composition of the collected deposits appears promising.

Sample Characterisation

It can be seen from the diagrammatic representation of the Project (Fig 1) that the detailed characterisation of the various samples produced was seen as a key factor in providing an improved understanding of coal ash slagging. These expectations have been fully realised and the Computer Controlled Scanning Electron Microscope (CCSEM) has provided a wealth of data on the distribution of mineral matter in coal and its subsequent transformation through flyash to a consolidated deposit. This will become clear from the summary of the findings, most of which are based on CCSEM results.

SUMMARY OF FINDINGS

The Mechanism of Slagging

The slagging process can be broken down into three stages:

·   Deposit initiation

·   Bulk ash deposition

·   Deposit consolidation

As a result of the work undertaken during this Project the controlling factors in each of these stages and their relative importances are much more clearly understood.

Examination of ash deposits collected on an air-cooled probe during the Ratcliffe trials clearly shows an inner deposit layer that is highly enriched in iron and, to a lesser extent, calcium. This phenomenon has been reported before (Cunningham et al, 1991) and is due primarily to preferential deposition of low melting, dense iron-rich particles derived from pyrite. The work at Babcock Energy has shown that only relatively large pyrite particles will survive long enough to be sticky at the furnace walls as a partially oxidised Fe-O-S phase. Calcium is rapidly assimilated into the aluminosilicate phases in the flyash (see below) and the enrichment in this element is assumed to be due to the relatively low viscosity of calcium-rich aluminosilicate glass. It is reasonable to assume that the concentration of pyrite and calcite (the prime sources of iron and calcium in UK coals) will have some effect on the rate of deposit initiation, but observations of operational boilers indicate that this initiation phase will inevitably occur and, therefore, plays no significant part in determining overall slagging behaviour.

Once the initial layer attains a certain thickness, about 2mm, the insulation from the boiler tube will be sufficient for the temperature of the outer surface to reach a level at which bulk ash can be retained and the bulk ash deposition phase begins. This transformation is quite sharp and is characterised by a change from selective deposition that is reliant on the stickiness of the impacting particles to a situation where the substrate is sufficiently sticky to collect impacting particles, irrespective of their stickiness. The ash deposition rate is then a function of (1) the ash flux which is dependent on the ash content of the coal and flow patterns through the furnace and, (2) the collection efficiency which depends on the stickiness of the substrate. This then explains the fact that the bulk deposit composition is invariably very similar to the composition of the coal ash from which it is derived. The slight, but consistent, enrichment of the bulk deposits in iron relative to the flyash that was noted is assumed to have been due to the greater density (inertia) of iron-rich particles, but this is probably a second order effect as far as ash deposition is concerned.

It follows that the key to furnace slagging is, therefore, the rate of ash deposition and, most importantly, the longer term behaviour of the deposits, ie the consolidation process (Phase 3). It is well known that deposit consolidation occurs by a viscous flow sintering mechanism and is dependent on both temperature and the composition of the individual ash particles that make up that deposit. The CCSEM characterisation of the samples collected during this programme provides a new insight into the relationship between mineral distribution in pf and flyash and deposit composition, and hence provides a basis for predicting the nature of consolidated deposits formed from individual coals.

Flyash and, therefore, ash deposits are principally composed of aluminosilicate glass and it is the composition of this glass that determines the viscosity and hence rate of deposit consolidation. The aluminosilicate glass is derived from the major aluminosilicate mineral constituents in the parent coal, ie kaolin (a pure aluminosilicate) and the illitic clays (degraded mica), but may also incorporate iron and calcium from the pyrite and calcite in the coal. It is the degree of assimilation of these fluxing oxides that ultimately determines deposit behaviour. As discussed earlier, this occurs by coalescence of different minerals, eg pyrite and clay, within individual coal particles and will be influenced by the proportions of the minerals that are included or excluded.

The changes that occur are illustrated by the ternary FAS diagrams (iron oxide/alumina /silica) obtained by CCSEM for pf, flyash, and slag samples from the 3 coal trials (Figs 3-5). Bentinck coal (Fig 3) shows the greatest degree of tightening of the composition distribution from pf to slag. The FAS diagram for the pf shows fairly distinct peaks along the aluminosilicate axis and at the SiO_2 (corresponding to quartz) and Fe_2O_3 (pyrite) apexes. The flyash diagram shows that much of the pyrite and some of the quartz has been assimilated to produce mainly aluminosilicate glass of variable composition. This composition range is greatly reduced in the

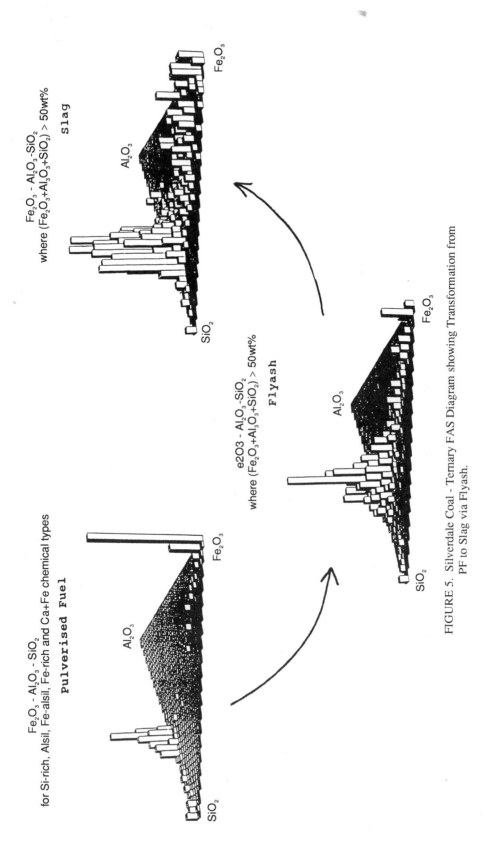

FIGURE 5. Silverdale Coal - Ternary FAS Diagram showing Transformation from PF to Slag via Flyash.

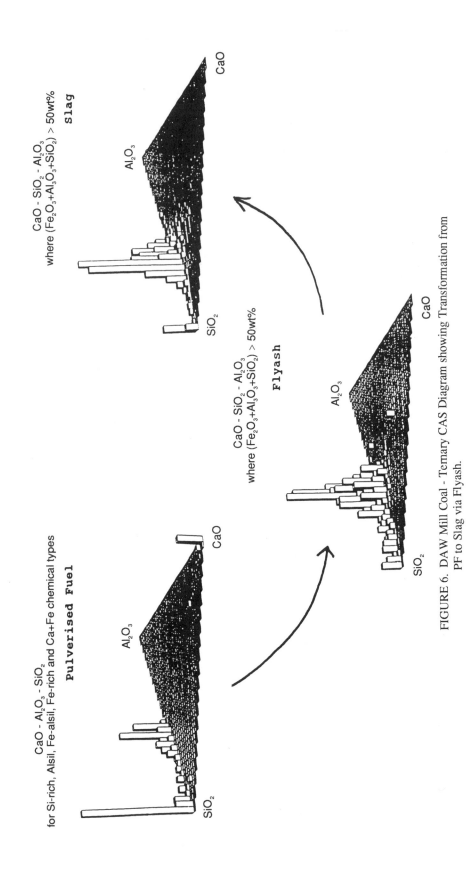

FIGURE 6. DAW Mill Coal - Ternary CAS Diagram showing Transformation from PF to Slag via Flyash.

slag sample which shows an extremely narrow chemical distribution. It should be noted that the "slag" samples were obtained from Unit 2 ash hopper during the trials and have an unknown time/temperature history. Accordingly, the degree of consolidation apparent from the FAS diagrams will be variable depending on sample history.

For the other coals, a similar pattern can be seen, though there are differences in the relative abundance of the individual minerals and the degree of homogenisation. For Daw Mill (Fig 4) a larger quartz peak can be seen in the pf diagram and, for the flyash, the degree of assimilation of the iron is less than for Bentinck. The slag shows virtually complete assimilation of the iron-rich phases, but some quartz remains unassimilated. For Silverdale (Fig 5) the high pyrite content is apparent from the pf FAS diagram and both the flyash and slag diagrams show that assimilation of the resultant iron-rich phases is far from complete.

The behaviour of the other major fluxing element calcium is illustrated by the ternary CAS diagram (calcia/alumina/silica) for Daw Mill pf, flyash, and slag (Fig 6). The pf diagram has a large peak at the CaO apex corresponding to calcite in the coal, but this is completely lost in the flyash and assimilated into the aluminosilicate phase. A similar pattern was observed for Bentinck and Silverdale coals.

As all the calcium is assimilated, it follows that the degree of assimilation of iron into the aluminosilicate glass is a key factor in predicting slagging behaviour. As the iron concentration of aluminosilicate glass increases, up to a certain level, the viscosity will tend to decrease and the rate of deposit sintering is enhanced. However, at higher concentrations the solubility limit is reached and iron will crystallise out in the form of spinels. Alternatively, unassimilated iron may simply occur as crystalline magnetite within the glass matrix. Either way, the presence of crystalline phases will increase the viscosity and inhibit sintering. This is illustrated by the link between observations of deposit type made during the slagging trials and CCSEM characterisation of the slag samples taken at the time.

Considering now the types of deposits observed during the trials, Daw Mill coal produced the most fluid deposits due to the combined fluxing effect of the calcium and iron present and the absence of significant crystallisation within the glass melt. In contrast, the Silverdale coal gave fairly massive deposits, but they were less fluid due to extensive magnetite crystallisation in the melt phase. For Bentinck coal the slagging was least troublesome due to the relatively low levels of iron and calcium in that coal and the consequent higher viscosity of the glass melt.

This then begs the question of what type of deposit is most troublesome from an operational point of view. In practice, highly fluid deposits such as those formed when firing Daw Mill coal are traditionally thought of as ones to be avoided. However, for such deposits thickness is self-limiting and they may have little effect on heat transfer. More massive, thermally resistant deposits such as those formed by the Silverdale coal could potentially cause greater heat transfer problems. Returning to the description of the slagging mechanism at the start of this Section, the other factor that must be taken into account is the rate of ash deposition. For the same reasons as outlined above, the Daw Mill deposits will be stickier over a wider temperature range than those formed when firing either Silverdale or Bentinck coal. The overall level of slagging will, therefore, be greater resulting in the potential fouling of a greater proportion of the furnace and an increased risk of slag accumulation at secondary sites, eg the ash hopper slope. This illustrates the complexity of the slagging process and the difficulty of deriving a slagging index that accurately reflects the potential for operational problems.

Coal Assessment: The Prediction of Slagging Characteristics

By Coal Analysis

As stated earlier, the primary objective of the project was to develop an improved method for predicting the slagging potential of coals, most probably based on a knowledge of the mineral distribution within the coal. The need for this information is demonstrated by the above discussion of the mechanism of slagging. It has been shown that, as far as coal quality is concerned, the nature of slag deposits depends on the occurrence and distribution of the aluminosilicate, pyrite, and calcite minerals in the coal.

This determines the viscosity of the deposited ash particles and, hence, the ultimate degree of consolidation of the deposit. This in turn will determine the nature of that deposit, its effect on boiler heat transfer, and ease of removal by sootblowing.

A new slagging index is proposed based on CCSEM analysis that takes account of mineral occurrences in the coal and, thereby, the degree of assimilation of the fluxing oxides, CaO and Fe_2O_3. The proposed index is based on CCSEM analysis of the coal to provide data on the composition and size distribution of the individual mineral matter particles present. These may be pure minerals (eg quartz, clay, pyrite) or mixtures.

It has been shown that all the calcium in the coal is assimilated into the aluminosilicate glass and it, therefore, has an "effectiveness" of 1.0 as a fluxing agent. For iron, it has been shown that the degree of assimilation and "effectiveness" is potentially variable and will depend on size and distribution of the pyrite within the coal. Any pyrite that occurs with the clay as a mixed mineral particle has more chance of coalescing to form an iron-rich glass than a pure pyrite particle on its own. For calculating the index it is assumed that pure pyrite particles have an "effectiveness" of 0.5 and any clay/pyrite mix an "effectiveness" between 0.5 and 1.0 in line with the amount of clay present in the mix. The CCSEM-based index is then defined as follows:

$$\text{Index} = \sum_{\text{All mineral occurences}} \left[\text{mass fraction} \times \left(CaO + Fe_2O_3 \times \left(1 - 0.5 \times \frac{Fe_2O_3}{Fe_2O_3 + Al_2O_3 + SiO_2} \right) \right) \right]$$

The results obtained for the 11 coals that have been analysed are shown in Table 1, together with their full ash compositions, base/acid ratios, and the derived slagging indices, Rs. This ash composition-based index is the most widely used and is defined as the product of base/acid ratio and sulphur content (dry basis), where the basic oxides are Fe_2O_3, CaO, MgO, K_2O, and Na_2O, and the acidic oxides SiO_2, Al_2O_3, and TiO_2. Base/acid ratio alone is used as a guide to slagging propensity and, as the key basic oxides are Fe_2O_3 and CaO, is similar in concept to the proposed CCSEM-based index. Indeed, a plot of base/acid ratio against the new index shows a reasonable trend for many of the coals examined - see Fig 7. The outliers are due to differences in Ca:Fe ratio and the proportion of the pyrite present in the coal as mixed pyrite/clay occurrences. For example, the CCSEM-based index predicts that the high calcium content and high Ca:Fe ratio coals Asfordby and Daw Mill have higher slagging propensities than indicated by base/acid ratio. Conversely the CCSEM-based index for Oxcroft is lower than indicated by Base/Acid ratio due to the low Ca:Fe ratio and relatively small proportion of pyrite present as mixed pyrite/clay occurrences.

The CCSEM-based indices for the 3 main test coals place the coals in order of increasing severity as follows:

Bentinck (9.6) < Silverdale (15.2) < Daw Mill (17.3)

This prediction agrees well with the nature of the slagging observed during the trials at Ratcliffe PS where Daw Mill gave the most fluid deposits and Bentinck the least. In contrast, based on either base/acid ratio or Rs (see Table 1) Silverdale would be considered much worse than Daw Mill. The slagging index Rs emphasises the importance of pyrite as this will influence both the base and sulphur terms, ie it is effectively proportional to $(Fe)^2$. It has long been recognised that this index underestimates the effect of calcium as exemplified for example by the virtually identical values for Bentinck and Daw Mill. In the absence of CCSEM data, based on this study, Base/Acid ratio is probably a more reliable ash composition-based index than Rs. However, based on both theory and experience, the proposed CCSEM-based index will provide a more reliable prediction.

Under this project, just 11 coals have been studied, 10 UK coals and a single US coal from the Pittsburgh No 8 seam. However, the principles on which the CCSEM index is based should be applicable to any bituminous coal. Pyrite is the main mineral contributing to slagging in such coals and the "effectiveness" factor built into the index should be generally applicable. Pending studies of a wider range of coals, there is more uncertainty concerning the effectiveness of fluxes occurring as carbonates. Based on the coals studied here calcium derived from calcite appears to be totally effective, ie is totally assimilated into the aluminoaticate glass. It must be assumed, therefore, that where iron is mainly present as the carbonate mineral siderite, it is also likley to be totally effective. Further work is planned to test this assumption.

By Laboratory-Scale Testing (the EFR)

The Entrained Flow Reactor was designed to enable accurate predictions of slagging behaviour to be made from laboratory-scale testing with a few kilograms of coal. Though small in scale, it has been shown that deposits with a similar morphology to those found in real plant can be collected. The predictions based on EFR testing should, therefore, be more reliable than those based on coal analysis, however comprehensive. At the chosen test conditions little variation in deposit type was seen, but a novel method of ranking coals based on CCSEM analysis of the microstructure of the deposits looks promising.

The deposit sample is mounted in resin and a polished cross-section prepared. An area of deposit is then selected for point analysis on a 16x16 square grid and the results plotted as FAS and CAS ternary diagrams. The purpose of this analysis is to quantify the proportion of the deposit having a composition range that is likely to promote sintering by viscous flow, and thereby, deposit consolidation. Based on liquidus temperatures, the composition ranges for the two ternary systems that were chosen are given below:

FAS System $Fe_2O_3 > 10\%$ and $< 50\%$

CAS System $CaO > 5\%$ and $< 40\%$

The proposed EFR-derived index is then taken as the sum of the numbers of point analyses falling within these composition ranges and the results obtained for deposits collected on an uncooled ceramic probe at 1200°C (Port 2) are included in Table 1. The use of an uncooled ceramic probe, rather than a cooled metal probe, was preferred as this accelerates the critical deposit growth and consolidation process.

The relationship between this index and the proposed CCSEM-based index is shown in Fig 8 and for most of the coals tested agreement is good. The outliers above the line are Dalquhandy and Pittsburgh No 8, ie the EFR indicates a worse slagging behaviour than the CCSEM analysis. In the case of Dalquhandy this is certainly surprising as this coal would appear to be particularly inert. The other clear outlier (below the line) is Silverdale with an EFR index of just 38 compared to 41 for Bentinck and 68 for Daw Mill.

The apparently unrealistic predictions for Dalquhandy and Silverdale may be related to test conditions. The type of deposit obtained depends on gas temperature at the sampling port as this will directly affect the degree of assimilation of the iron phases into the aluminosilicate melt. It is possible, therefore, that the ranking order obtained from the EFR index will vary with gas temperature. For example, at a higher temperature, for Silverdale, the degree of assimilation may be higher relative to the other coals. Similarly, for Dalquhandy it may be lower. It follows that the relative full-scale behaviour of different coals may vary depending on which area of the furnace is being considered; to obtain a full characterisation it will be necessary to undertake EFR tests over a range of gas temperatures. This further illustrates the complexity of the slagging process and the inherent difficulties in attempting to apply any single index.

By Rig-Scale Testing

For coal assessment, the next step up in scale is to go to rig testing. This obviously requires special facilities that are considerably more expensive to operate. However, a rig provides a real combustion environment and, therefore, much more realistic test conditions. The $50kW_{th}$ Ash Deposition Rig at CRE probably represents the smallest scale at which this type of testing can be carried out and the results obtained within this Project have shown that this facility can be used to reproduce the slagging behaviour observed at full scale.

The most realistic prediction of slagging behaviour was obtained from the nature of the deposits found on the refractory walls of the combustion chamber, rather than the air-cooled metal deposition panel and probe. It is concluded that, for relatively short term tests, the best approach is to use an uncooled ceramic probe for slagging assessment. This greatly accelerates phases 1 and 2 of slagging (initiation and deposition - see Section 4.1) and enables the critical consolidation phase to be realistically simulated. This comment applies to short term testing at any scale from plant tests to laboratory-scale tests such as those performed on the EFR at Imperial College.

The Interpretation of Slagging Indices

Completely novel methods of assessing slagging propensity from CCSEM analysis and a unique laboratory-scale furnace, the EFR, are proposed. The principle of both of these single number "indices" is that the type of deposit obtained depends on the mineral occurrences within the coal and the degree of assimilation of the fluxing oxides into the aluminosilicate glass. More work is needed to optimise the EFR methodology and fully assess the reliability of both indices. However, it is believed that both these assessment techniques will prove superior to existing indices based on bulk chemistry or bulk properties such as viscosity and ash fusion temperatures. In both cases, the higher the index, the more fluid the deposit. It should be noted though, that a more fluid deposit does not necessarily imply greater operational problems due to slagging.

During the plant trials, it was clear that Daw Mill produced the most fluid deposits in line with the CCSEM-based index. Because of the fluidity, the deposit thickness was limited by "dripping" of low viscosity slag and the effect on heat transfer would be small. In contrast, the Silverdale slag was generally less fused, again as predicted, more massive and, therefore, a more effective thermal barrier. These observations were made on the rear wall at top burner level, ie at a high gas temperature. It is likely that at a lower gas temperature and/or radiation flux level a different ranking might be obtained in terms of effect on heat transfer. For example, in the upper furnace Daw Mill might produce less fused, more massive deposits due to the different conditions. As indicated earlier, the influence of gas temperature on deposit type can be investigated on the EFR.

Returning to the single number index concept, this is a valid approach for the initial screening of coals. However, the above arguments suggests that, rather than ranking coals on the basis that the higher the index the worse the coal, there is a range of values indicative of potential problem coals. Coals with a high index that is linked to deposit fluidity, eg Daw Mill, would appear to be unlikely to cause major operational problems. Coals with a low index, eg Bentinck, do not cause problems either. Looking at the CCSEM-based index, the coals that potentially might cause the greatest problems may be those with indices in the mid-range, say 12 to 16.

It must be stressed that all of this Section has been concerned with the assessment of the inherent properties of coals with respect to slagging potential. There is a need for numerical indices to assist coal buyers and users in assessing the suitability of coals. However, apart from the properties of the coal, that suitability is very strongly dependent on the design and operation of individual boilers. It is only by experience that the range of coals that can be tolerated is established. For example, the appropriate CCSEM index range will vary from station to station.

Boiler Design and Operation

A major driving force for the project was a concern that the retrofitting of low-NOx burners might lead to an increase in slagging problems. In the UK, the first boilers to be converted were tangentially-fired and experience at the time this Project was initiated had been good. The Low-NOx Concentric Firing System (LNCFS) tends to blanket the furnace walls with

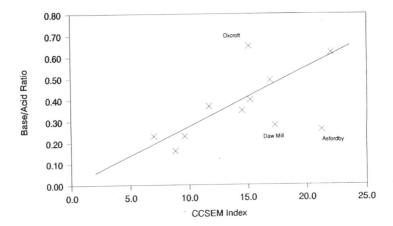

FIGURE 7. Slagging Indices: The Relationship bewteen Base/Acid Ration and the Proposed CCSEM-Derived Index.

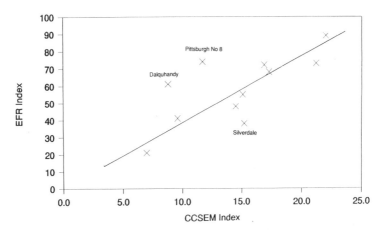

FIGURE 8. Slagging Indices: The Relationship between the EFR-Derived Index and the Proposed CCSEM-Derived Index.

offset air and would be expected to inhibit slagging. However, at that time there was no experience with wall-fired boilers where it was believed the low-NOx burners might exacerbate slagging due to an increase in local reducing conditions and flame impingement on the rear wall.

The results obtained during this Project have clearly demonstrated that the retrofitting of low-NOx burners on Ratcliffe Unit 2 has not had any detrimental effect on boiler slagging. Indeed, the reverse appears to be true. Both Daw Mill and coals similar to Silverdale are known to have caused slagging problems at Ratcliffe in the past, but no measurable effect on boiler operation was detected during the trials. This finding was novel at the time of the trials, but is now supported by findings worldwide as experience with low-NOx systems increases.

The greater tolerance towards slagging of the boilers fitted with low-NOx burners may be due in part to the lower heat flux in the burner belt that is characteristic of low-NOx firing systems. Although the same heat is released in the same overall furnace volume, in the low-NOx system the peak flame temperature is lower and the heat is released over an increased volume in the burner belt. This leads to lower local heat fluxes.

There are two other coincident benefits of low-NOx burner retrofits. First, they are normally preceded by improvements in the fuel distribution to the burners giving more uniform combustion conditions and a consequent reduction in slagging. Second, the retrofit also involves burner quarl modifications which tend to reduce slagging in that specific location and elsewhere and secondary deposition sites, eg the ash hopper slopes.

It should be noted that for any new boiler design the control of NOx emissions is a prime requirement. Apart from the actual burner design, the overall boiler design factors used, in particular a return to relatively large furnaces with a low heat release per plan area, will tend to reduce the risk of slagging anyway.

The observations made during the trials and subsequent slag mapping exercise have confirmed the view that the presence of keying points which encourage slag growth are design features which should be avoided as far as possible. The most important example of this is burner quarl slagging which is localised enough to have no great effect on overall heat transfer, but which can have a serious effect on burner performance and potentially lead to major slag falls that can block the ash hopper.

Regarding boiler operation, the results also indicate that, because of the increased tolerance of the low-NOx boiler, slagging is also less sensitive to excess air level. It should, therefore, be possible to operate at a lower level thereby increasing boiler efficiency and reducing NOx emissions. There is an optimum level, at low O_2 concentrations the level of unburnt carbon in the flyash will increase and may reach an unacceptable level from either an efficiency or an ash sales point of view.

SUMMARY

The slagging process can be broken down into three phases - initiation, ash deposition, and consolidation. The results of this comprehensive study show that the overall key to the long-term behaviour of ash deposits is the consolidation phase. Consolidation occurs by a viscous flow sintering mechanism and is dependent on both temperature and the composition of the individual ash particles that make up the deposit. Flyash and, therefore, deposits are mainly composed of aluminosilicate glass and it is the composition of this glass that determines the viscosity and rate of consolidation. The aluminosilicate glass is mainly derived from the clay minerals in the coal, but may also incorporate iron and calcium from the pyrite and calcite in the coal. These elements are the main fluxes in the glass and their concentration and distribution in the coal determine their degree of assimilation into the glass and, hence, the slagging characteristics of the coal.

The value of CCSEM for quantifying the mineral matter distribution in coals has been demonstrated and a novel CCSEM-based slagging index is proposed. This index is considered superior to conventional indices based on bulk ash composition as it takes account of the potential degree of assimilation of the fluxing oxides. It is believed that the basic approach should be generally applicable to all bituminous coals.

A methodology for studying the ash deposition characteristics of coals using a laboratory-scale Entrained Flow Reactor has also been established. CCSEM analysis of the ash deposits formed in the EFR shows the potential for coalescence and provides the basis for an alternative slagging index which can include the effect of deposition temperature.

Finally, it has been clearly demonstrated at plant scale that the retrofitting of low-NOx burners makes wall-fired boilers more tolerant towards potentially slagging coals. This is believed to be due to the less intense heat release in the near-burner region that is an integral feature of low-NOx burner design.

ACKNOWLEDGMENT

As noted earlier, this Project has been a collaborative effort and the support of all concerned is gratefully acknowledged. In addition, the work could not have been undertaken without the financial support of the UK Department of Trade and Industry. Finally, this paper is published by permission of PowerGen PLC.

REFERENCES

Gibb, W.H., Jones, A.R. and Wigley, F.(1993). "The UK Collaborative Research Programme on Slagging in Pulverised Coal-Fired Furnaces: Results of Full-Scale Plant Trials." In J.Williamson and F.Wigley (Eds), *The Impact of Ash Deposition on Coal Fired Plants*, Washington: Taylor and Francis, 1994.

Cunningham, A.T.S., Gibb, W.H., Jones, A.R., Wigley, F. and Williamson, J.(1991). "The Effect of Mineral Doping of a Coal on Ash Deposition Behaviour". *Proceedings of the Engineering Foundation Conference on Inorganic Transformations and Ash Deposition during Combustion*. March, 10-15, 1991, Palm Coast, Florida.

Hutchings, I.,West, S.S. and Willamson, J.(1995). *An Assessment of Coal Ash Slagging Propensity Using an Entrained Flow Reactor*. This conference.

Lee, F.F., Riley, G.S. and Lockwood, F.C.(1995). *The UK Collaborative Research on Slagging and Fouling in PF-Fired Furnaces - Mathematical Modelling of Ash Deposition Behaviour*. This conference.

Wigley, F. and Williamson, J.(1995). *Modelling Flyash Generation for UK Power Station Coals*. This conference.

Williamson, J. and Wigley, F.(1994). "The Impact of Ash Deposition on Coal-Fired Plants". *Proceedings of the Engineering Foundation Conference*, Solihull, England, June 20-25, 1993. Washington: Taylor and Francis, 1994.

ASH DEPOSIT PROPERTIES AND RADIATIVE TRANSFER IN COAL FIRED PLANT - CURRENT UNDERSTANDING AND NEW DEVELOPMENTS

T.F. Wall[1], S.P. Bhattacharya[1], L.L. Baxter[2], B.C. Young[3], and A.A. Grisanti[3]

[1]Department of Chemical Engineering, University of Newcastle
Callaghan, NSW 2308, Australia
[2]Combustion Research Facility, Sandia National Laboratories
Livermore, Calif. 94550
[3]Energy and Environmental Research Center, University of North Dakota
P.O. Box 9018, Grand Forks, North Dakota 58202-9018

ABSTRACT

Previous research has offered speculations relating the physical and chemical character of ash deposits and their radiative properties. Recent work has confirmed that the physical character - whether the deposit is particulate, sintered or fused, or slag - has the dominant effect. Both on-line measurements and theoretical predictions suggest that measurements of emissive or reflective spectra from deposits may be used to identify this physical character and possibly the chemical species comprising the deposit, as an aid to on-line cleaning.

1. INTRODUCTION

Recent reviews[1,2] have presented the current understanding of the properties and thermal effects of ash deposits in coal-fired furnaces. The properties influencing heat transfer were identified as both radiative and conductive and therefore both property types were considered. The reviews presented new insights into the relationships between the chemical and physical character of deposits and radiative properties while offering few insights into such relationships for conductive properties. More recent theoretical and experimental research has extended the work relating deposit character to radiative properties, and is detailed here.

2. DEPOSIT CHARACTERISTICS AND FURNACE HEAT TRANSFER

2.1 Definitions

In a furnace the thermal energy transfer from flame gases (at temperature T_f) to the exterior surface of the ash deposits (at temperature T_s) is by radiative transfer and convection (with convection usually of secondary importance in the furnace regions but dominant in the superheater region). The transfer through the deposit and tube is by conduction and then by convection to the contained steam. Figure 1 illustrates the process. Equating the net radiative

Applications of Advanced Technology to Ash-Related Problems in Boilers
Edited by L. Baxter and R. DeSollar, Plenum Press, New York, 1996

transfer at the surface (per unit area of surface) to the heat conducted through the deposit, neglecting the thermal resistance of the tube and steam film gives (using notation in Fig. 1):

$$Q = H - R - \varepsilon E = \alpha H - \varepsilon E \qquad (1)$$

so that, for a large furnace, taking a flame emissivity of unity

$$Q = \frac{k}{x}(T_s - T_t) = \alpha \sigma T_f^4 - \varepsilon \sigma T_s^4 \qquad (2)$$

The properties determining heat transfer and therefore the deposit thermal conductivity (k), thickness (x), absorptivity (α) and emissivity (ε). In many cases, the deposit layer is sufficiently thin so that $T_f^4 \gg T_s^4$ and the surface emission (the last term in Eq.(2)) is small. The only radiative property of importance is then the absorbance of the deposit surface.

Deposit spectral emissivities vary with wavelength from a low of approximately 0.2 to greater than 0.9. A Planck-weighted average is used to obtain a total emissivity appropriate for many engineering calculations. This total emissivity (ε) depends on both the spectral emissivity and the Plank function (P) evaluated at the deposit surface temperature and is given by

$$\varepsilon (T_s) = \frac{1}{\sigma T_s^4} \int_{\lambda=0}^{\infty} \varepsilon_\lambda (\lambda, T_s) P (\lambda, T_s) d\lambda \qquad (3)$$

where ε_λ is the spectral emissivity, T is absolute temperature, λ is wavelength and σ is the Stefen-Boltzmann constant. Subscripts s and λ represent surface and spectral respectively.

In general, the spectral emissivity, ε_λ and spectral absorptivity, α_λ are equal according to Kirchoff's law under the assumption of diffuse radiation. Therefore, the Planck-weighted total absorptivity can be determined as follows:

$$\alpha (T_s,T_f) = \frac{1}{\sigma T_s^4} \int_{\lambda=0}^{\infty} \alpha_\lambda (\lambda, T_s) P (\lambda, T_f) d\lambda = \frac{1}{\sigma T_s^4} \int_{\lambda=0}^{\infty} \varepsilon_\lambda (\lambda, T_s) P (\lambda, T_f) d\lambda$$

$$(4)$$

In contrast to the total emissivity, the total absorptivity depends on both the surface temperature and the flame temperature[1]. Planck-weighted radiative properties are used here to approximate total radiative properties applicable to furnace calculations. For a large variety of materials relevant to combustion problems, the Planck-weighted properties decrease with increasing temperature[1,3,4], this being due to the low spectral emissivity of such materials at low wavelengths.

2.2   Charges in furnace heat transfer as deposits grow and mature

Suggested trends in α and k/x during deposit growth are given[1] on Fig. 2. For a deposit-free wall (or tube), the absorbance is high, say 0.8. The initial deposit layer (b) is likely to be fine condensible salts or fine ash transported by a thermophorosis mechanism. For a layer of 2μm particles the absorbance was speculated to be less than 0.3. As the fine layer thickens and results in the collection of larger particles (which rebound from a thin layer) a layer of coarse ash (c) develops with an absorbance of, perhaps, 0.4-0.5. As this particulate layer builds to a greater thickness (d), the absorbance will not change but the conductance coefficient (k/x) will continue to reduce. During the development of this particulate layer, the temperature gradient

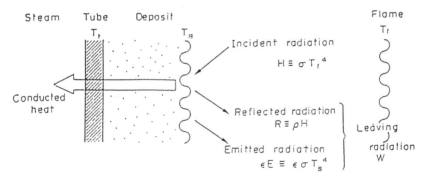

Fig. 1.   An illustration of the heat transfer mechanisms to and through
           deposits[1].

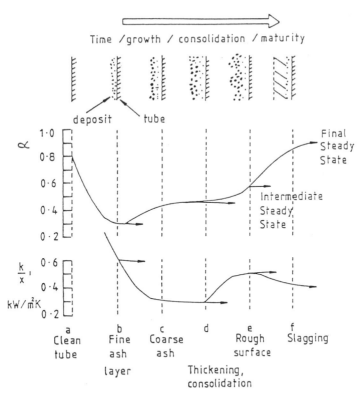

Fig. 2.   Speculations on the trends in deposit properties during their growth[1].

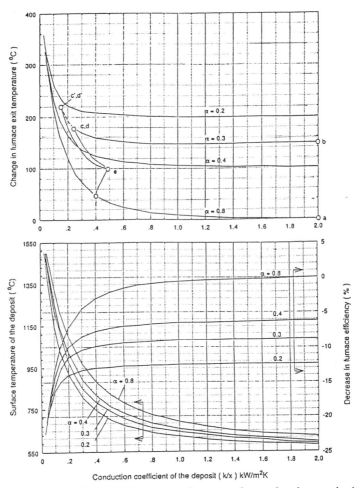

Fig. 3. Predicted changes in furnace performance due to the changes in deposit properties given on Fig. 2.

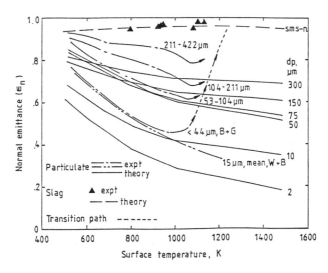

Fig. 4. The effect of particle size and slag character, comparison of measurements[6,7] with theory[1].

through the deposit, and therefore the temperature of its surface, will continuously increase. Sintering and fusion will result. Also a thick deposit is likely to leave an irregular (rough) surface. Therefore (e) both α and k/x will increase. Once the thickness builds, and the deposit temperature increases further, a liquid slag will develop. α and k will increase (f), and k/x may increase or decrease compared to its value for the sintered deposits. As shown by the horizontal arrows in Fig. 2, an intermediate steady state may develop during the deposit growth. For example, if the deposit does not sinter appreciably, the erosion rate of the deposit by large particles may balance the deposition rate and a permanent particulate character will result.

The approximate changes in furnace exit temperature (FET), furnace efficiency and surface temperature of the deposit may be estimated by a well-mixed furnace model (4) which has also been used by Richter et al[5]. The model assumes a uniform build-up of ash deposits over the complete furnace walls. The effects are given on Fig. 3, and are related to the physical and property changes identified on Fig. 2. The initial reflective layer (b) results in an increase in FET of 150°C, with a build-up of deposit (c,d) the increase exceeds 200°C. At this condition, the surface temperature of the deposit exceeds 1000°C, and melting will therefore occur as this temperature increases further. The property changes associated with slag formation (it is predicted) result in a reduction of FET (e,f).

2.3 The significance of radiative properties

The greatest predicted effect of the radiative properties is the change a-b on Fig. 3 which is associated with an initial fine layer of particles, if the absorptivity of clean tubes is 0.8 then FET will increase by 100°C, 150°C and 200°C if the deposit absorptivity reduces to 0.4, 0.3 and 0.2 respectively. Such ash layers are called 'reflective' and the evidence for such reflective layers is a major aspect of recent research. As deposits sinter and slag, the absorptivity is expected to increase with the change in FET (relative to a clean furnace) being then determined by the conduction coefficient as is again illustrated on Fig. 3.

3. IDENTIFIED RELATIONSHIP BETWEEN THE CHEMICAL AND PHYSICAL CHARACTER OF DEPOSITS AND RADIATIVE PROPERTIES

3.1 Measurements

For *total emissivity* the often-quoted results for emissivity of Boow and Goard[6] are given on Fig. 4. The results are for particles derived by crushing synthetic slag prepared by melting oxides. All particles therefore have the same chemical composition removing the uncertain effect of the non-homogeneous chemical nature of coal ash. Data for several size fractions is also reported and compared with calculations by Wall et al[1] for particles. The predicted trends with size and temperature agree with the measurements. On melting the particles sinter and eventually form slag with resulting emissivity values from measurements and theory both being about 0.9-0.95. The path on Fig. 4 followed by the ash as it sinters will depend on its melting temperature, with a typical transition path being indicated.

The *spectral emissivity* changes associated with the above changes in total emissivity are given in the results of Markham et al[7] given on Figure 5. For particles of fly ash a region of low spectral emissivity is observed at wavelengths below 6μm. This results in the decrease of emissivity with temperature observed in the total emissivity results of Fig. 4. On sintering of

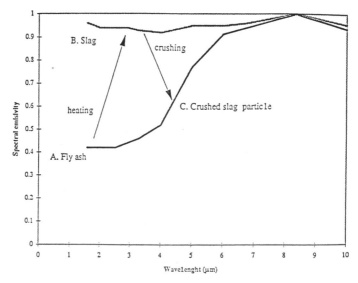

Fig. 5. Experimental evidence for the spectral character of particulate layers (either fly ash or crushed slag) and the high emissivity of slag, as provided by Markam et al[7].

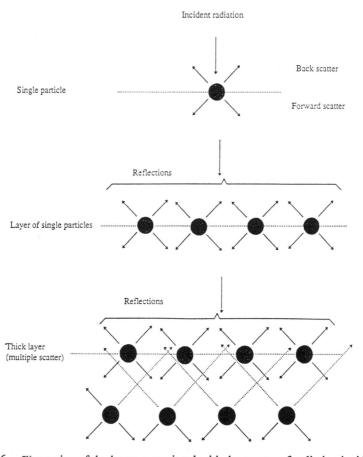

Fig. 6. Illustration of the beams associated with the scatter of radiation incident onto particles.

the same samples of fly ash to a slag the spectral character is destroyed and the emissivity increases to values from 0.9 to 0.95, again agreeing with the slag data on Fig. 4.

3.2   Theoretical predictions

The predictions presented for particle and slag layers on Fig. 2 are based on the following theories and concepts:

For *particle layers* the theory is based on the well known Mie theory for scatter of electromagnetic radiation by single particles. As illustrated on Fig. 6 for a single particle some incident radiation is scattered in the hemisphere of the incident beam, this is seen by an observer as reflected radiation. For mono-layer of particles, the reflectivity is the same as that for a single particle. For a thick layer of particles, radiation scattered in the forward direction due to the interaction with the initial particles can eventually emerge from the surface of the layer following interaction with other particles, increasing the reflectivity above that of the mono-layer. This process is called *multiple* scatter, and is the basis of the theoretical predictions of Fig. 4, which are based also on the discrete-ordinate (DO) method of radiative transfer calculation. Spectral emissivity predictions using the same technique are given on Fig. 7 for opaque particle layers based on optical data for slags of ($SiO_2/A\ell_2O_3 \sim 2$. $CaO/A\ell_2O_3 \sim 0.6$. $Fe_2O_3 \sim 5\%$) and show a region of low emissivity at wavelengths $< 5\mu m$ where particle size influences the predicted results[8]. The features indicated between 4 and $6\mu m$ are believed to be associated with gas-phase carbon dioxide (4-$5\mu m$) and water (5-$6\mu m$) absorption that was not completely eliminated from the literature values used for the optical constants. There are no known absorption bands for this type of material in this region. These residual water emission features tend to exaggerate the particle size effect. The true particle size effect, as indicated in the data from 3 to $4\mu m$ is largely a result of particle scattering outlined earlier, as small particles tend to back scatter radiation (leading to lower emissivities and higher reflectivities) while larger particles primarily scatter in the forward direction. By contrast, the spectral emissivity generally exceeds 0.9 for a smooth iron-containing slag layer of the same optical properties (chemical composition).

The results illustrate some important qualitative trends which may be expected in real systems[9]. First, the spectral nature of the emissivity is clearly evident. In particular, the spectral emissivity is quite low at the short wavelengths, the spectral region where much of the radiation in industrial furnaces are exchanged. Second, the particulate nature of the deposit has measurable influence on its radiative properties. This, again, is especially evident at shorter wavelengths where the ash is relatively transparent and most of the radiation is back scattered by multiple interactions with particles in the deposit. In contrast, no scattering occurs for a smooth opaque slag layer and any incident radiation is either reflected or absorbed.

A theoretical complexity must be considered for layers occur when particles are in close proximity when the interaction of the particle with the radiation field is influenced by neighbouring particles. The scattering is then said to be *dependant,* and scatter will be less than that estimated using multiple scatter alone, and the emissivity greater.

Recent measurements to estimate the significance of dependant effects have been made by Bhattacharya[10]. Transmission (τ) and reflection measurements (ρ) through, and of, particle layers were measured using an FTIR spectrometer with an integrating sphere at the Physikalisches Institut der Universitat. Particles prepared from crushed slag, of $5.2\mu m$ mean size, were used with two iron levels (1% and 5% Fe_2O_3). The emissivity was then estimated as

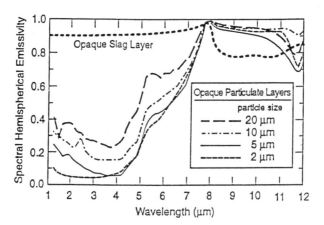

Fig. 7.   Theoretical spectral emissivity estimates for particulate and slag layers, for deposits of 5% Fe_2O_3[8].

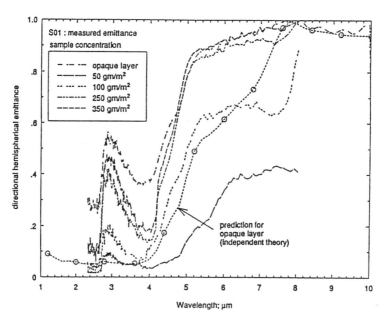

Fig. 8.   Measured directional hemispherical emittance for particulate deposits of varying levels of ash coverage and comparison with predictions for opaque particulate layer (considering independent radiative effects and multiple scatter). Sample material : S01 (1% Fe_2O_3)[10].

$$\epsilon = 1 - \rho - \tau \qquad\qquad\qquad (5)$$

The results, given on Fig. 8 are also compared with predictions based on multiple (independent) scatter theory. The measurements confirm the region of high reflectivities and low emissivity at wavelengths lower than 6μm evident on Fig. 5 and give emittance estimates increasing with sample thickness (indicating transparent layers). However, for the opaque layer, the measured spectral emissivity values may be double that predicted. The difference is an estimate of the effect of dependent scatter and errors in the use of a multiple non-dependant scatter theory. A comparison of total emissivity estimates from Eq.(3), given on Fig. 9, indicate that multiple (independent) scatter theory may be in considerable error.

For *slag layers,* the emissivity can be simply approximated by an optically smooth layer which is opaque[1]. Such predictions, given on Fig.7, give values of greater than 0.9 which agree with measurements such as those of Fig. 5.

3.3   On-line measurements

The above measurements on collected samples of ash and slag, supported by the theoretical predictions, indicate that the emission spectra from a deposit will be primarily related to the physical character of the deposit. Spectra from growing deposits measured on-line provide the best way of evaluating this conclusion.

3.3.1. Measurements at the Sandia National Laboratories

A diagnostic tool based on emission Fourier transform infrared (FTIR) spectroscopy[8] has been developed to measure *in situ* properties of ash deposits, in real time, on a stainless steel probe in Sandia's Multifuel Combustor (MFC). The MFC is a down-fired facility with an open test section at the furnace exit to allow optical and probe access to the flow of combustion products. The ceramic walls of the combustor are electrically heated to provide the desired temperature profile. Deposits were formed on an air-cooled cylindrical probe (1.6 cm O.D.) in cross flow. The probe was rotated continuously about its axis during each three hour test to form a nominally uniform deposit around the tube circumference. The radiative emission from a region approximately 5 mm in diameter on the bottom half of the probe was collected by a set of off-axis paraboloidal mirrors (effective f# = 4.3) and focused into the aperture of the FTIR spectrometer. A commercial, rapid-scanning interferometer (Biorad FTS-40/60) was used with a liquid nitrogen-cooled, broad-band mercury-cadmium-telluride (MCT) detector and a germanium-based beam splitter. The MCT detector is sensitive over the wavelength range of 2.8-22 microns. Emitted radiation from the bottom of the probe was collected to minimise the contribution of reflected furnace radiation to the total signal. While this may be possible in a measurement on a research rig we will see that reflected radiation may dominate the energy from deposits in practical situations. The radiative emission from the probe was compared to that of a blackbody calibration source to determine the spectral emissivity of the deposit surface. The surface temperature of the probe was held constant, typically between 450°C and 600°C whereas the deposit surface temperature varied with time as the deposit grew.

Figure 10 illustrates the time resolved spectral emissivity measurements. These spectra were obtained at a bulk gas temperature of 900°C and have been digitally filtered to remove some of the narrow peaks attributed to the gas-phase interference of infrared active components in the vitiated flow. Residual interference appears as features at 4.2-4.7μm and 14.4-15.2μm (CO_2 interference) and at 2.9-3.0μm and 5.5-7.5μm (H_2O interference). These infrared-active gases are opaque over the pathlength of this experiment in these regions. Their temperatures differ from that of the surface by large amounts. The gases in the combustor effluent are at 900 to 1300°C, depending on the experiment, whereas those in the laboratory air are at room temperature. By contrast, the deposit surface temperature varied from 300 to 600°C, depending on the experiment. Hot band emission from the hot gases produced emissivities greater than one when normalised by the Planck function evaluated at the probe temperature. However, gas interference does not always increase the computed emissivity. Radiative absorption by room temperature gases in the laboratory environment prevented probe (or gas) radiation from reaching the detector, producing very low values of computed emissivities. For these reasons, gas interferences in the regions cited above produced emissivities that are artefacts that can be either much too high or much too low. The data in the remaining regions do not suffer from gas-phase interferences.

The initial emissivity shown in Fig. 10 is that of the clean stainless steel probe taken at temperature prior to the initiation of coal flow. The emissivity appears relatively constant across the spectrum at approximately 0.2, except for the regions of gas phase interference. An emissivity less than 0.3 is consistent with literature values for polished stainless steel. There is little change in the spectra with time at wavelengths less than 7 μm as ash is deposited on the probe. Ash deposits from western coals rarely contain chemical species with strong absorption features in this region. Therefore, the measured emissivity of initial deposits from these coals will be influenced significantly by the emissivity of the deposition probe and the scattering of small particles. The influence of the deposition probe was seen in a similar experiment performed with a heavily oxidised mild steel deposition probe which had an emissivity of 0.8 at wavelengths less than 7μm. In that experiment, the measured emissivity (< 7 μm) was approximately 0.75, significantly higher than the measured values as a result of the higher tube emissivity. Calculations indicate that a deposit of 5 μm ash particles (Fe_2O_3 ~ 5%) must be approximately 50 μm thick to be opaque at a wavelength of 4 μm.

Changes in emissivities with time are greater at wavelengths exceeding 7 μm are evident on Fig. 10 as chemical species with strong absorption features in this region are deposited on the probe surface. The spectral structure derives from the presence of sulfate, silica and silicate species in the deposit. Some progress[11] has been made in the identification of aspects of the deposit chemistry by location of peaks (or features) in the spectra, such as those observed on Fig. 11. These features are observed at wavelengths exceeding 7 μm. For example, the spectra given on Fig. 11 indicates the presence of calcium sulfate, in the form of anhydrite. Two of the five peaks for this species are obvious in this spectrum, other peaks are either obscured by CO_2 emission from the gas or spectral features associated with other species.

3.3.2 Measurements at the Energy and Environmental Research Center

The EERC is investigating in situ monitoring of infrared emissions from combustion deposits during their development in the convective section of a pilot-scale combustor under full-load conditions. The goal of the work is to assess the feasibility of applying Fourier transform infrared (FTIR) spectroscopy to determine the mineral species responsible for fouling and slagging deposits in a pilot-scale coal-fired combustor. A specially designed accessory, denoted as the Infrared Emission Sampling Probe (IESP), allows the radiation from deposits to be collected by an FTIR spectrometer. The probe, consisting of an optical section and a protective outer jacket, fits into a standard optical access of a furnace duct but still enables operator viewing through the port. The optical section of the IESP incorporates a cadmium telluride (CdTe) objective plano-convex lens of 1.25 mm in diameter and an effective focal length of 125 mm. This lens focuses the FTIR spectrometer optics onto the heat transfer tube

TABLE 1. Suggested character - spectra relationships

Deposit property	Illustrative properties monitored	Analysis of spectra which is necessary	Relevant wavelength region (μm)
1.Chemical character	Salts (SO_4^{--}, CO_3^{--}), aluminosilicates	Identified by location (μm) of bands in spectra	8 - 20
2.Physical character	Transformation of powdery deposits to sintered/fused deposits and molten slags	The magnitude, and trends with time, at spectral region outside bands of 1	< 6
3.Radiative properties	Emissivity, absorptivity, reflectivity, emissive power, temperature	Spectral radiation is integrated w.r.t black body function to obtain properties	Full thermal spectrum required, say, 1.5-15 μm

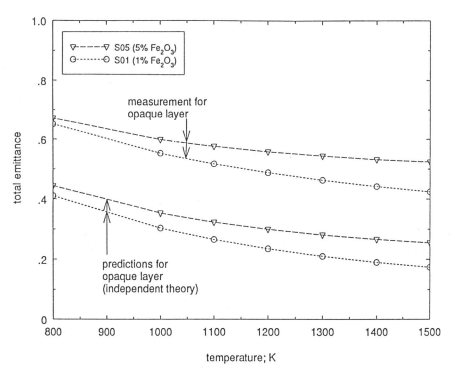

Fig. 9. Planck weighted emittance for opaque particulate layer, measured and predicted. Based on the data of Fig. 8, gas phase interference removed from the measured emittance[10].

where ash and slag deposits are formed. The face of the lens, exposed to the furnace duct, is cooled by a continuous nitrogen flow. The lens and lens holder are mounted on an 8mm diameter gold-coated light pipe, 1 meter in length, whose opposite end is fixed to a 1:4 beam expander. The expanded 32mm end of the light pipe connects to the FTIR spectrometer port. Enclosed in a protective outer jacket consisting of two concentric tubes, the entire light pipe assembly is purged continually with nitrogen (inner tube) and cooling water (outer tube). The nitrogen flow prevents deposit formation on the lens, and the water flow maintains an acceptable temperature at the instrument optics. The temperature of both the lens and cooling water is monitored with thermocouples.

The viewing geometry of the IESP when used with the EERC pilot-scale combustor is shown in Fig. 12. This diagram depicts an overhead view of the convective pass of the combustor. Deposition tubes are located in the approximate centre of the figure. To the left of the probes is the fireside region; combustion gases and flame radiation propagate from left to right. The IESP views the centre deposition tube. The diagram reveals that the IESP will collect reflected as well as emitted radiation from the deposition tubes as well as the refractory wall. Theoretical estimates indicate that the bulk of the energy will be from reflected radiation, with the deposit emissions making up only a fraction of the total energy. This will also be the situation in industrial furnaces. Therefore, the IESP will be most useful in measuring reflected radiation from the deposits rather than deposit emissions. To reduce the viewing area of the IESP, it has been equipped with a plano-convex objective lens which will limit the reflected radiation collected from refractory walls.

The IESP in combination with the FTIR spectrometer has been designed to accomplish monitoring of combustion deposits in a viewing geometry similar to that pertaining to conventional combustion systems. Preliminary tests using the IESP have been conducted involving the collection of emission spectra of the furnace duct and deposition probes under gas-fired operating conditions of the combustor. For these tests, the plano-convex CdTe lens had not been fabricated, but, instead, a CdTe optical flat, which gave the IESP an acceptance cone half angle of 1.5°, was employed. These baseline tests demonstrated the level of background infrared radiation as well as the efficiency of the protective housing of the IESP optics. The combustor (fireside) temperature during gas firing was approximately 1200°C, and the gas temperature in the probe region was 550°C. Figure 13 shows two emission spectra resulting from the FTIR spectrometer/IESP probe monitoring. The upper spectrum was taken with the IESP 203 mm from the deposition probe, whereas the lower spectrum was recorded with the IESP 51 mm from the deposition probe. The spectra have the general shape of the Planck blackbody curve. Riding on top is the emitting and absorbing carbon dioxide gas peaks in the 2200-2300 cm^{-1} region and the water vapor peaks in the regions of 3100-4000 cm^{-1} and 1100-2000 cm^{-1}. Note that as the IESP is moved further from the deposition probe, more of the furnace duct is viewed (more radiant energy) and the overall intensity increases. Additionally, as more duct-emanating gases come into view, the intensity of the CO_2 and water vapor peaks increases.

The FTIR spectrometer/IESP is currently being set up for measurements in the pilot plant combustor using coal at full load conditions. Spectral data will be analyzed to determine changes in deposit character as a function of time. In addition, selected deposit samples will be later analyzed by computer-controlled scanning electron microscopy and an electron probe to identify chemical species. The electron microanalysis data will be correlated with FTIR/IESP spectra to identify IR spectral bands associated with major chemical species in a similar way as for the Sandia measurements. The success of this correlation will determine the viability of

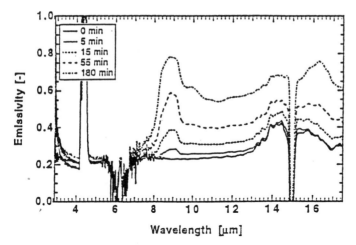

Fig.10. On-line spectral emissivity measurements, during the early stages of ash deposition, for times up to 180 minutes[8,9].

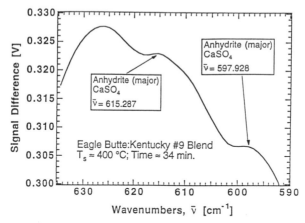

Fig.11. Spectral peaks associated with calcium sulfate in a deposit. The peaks agree most closely with anhydrite and not hydrated forms of calcium sulfate (gypsum). The anhydrite emission peak at 672 cm^{-1} is completely obscured by CO_2 emission from the gas.

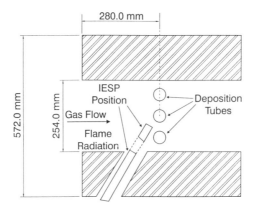

Fig.12. Overhead view of the IESP sighted onto deposits in the EERC furnace.

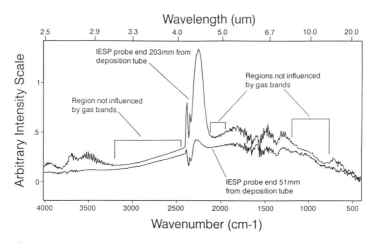

Fig.13. FTIR spectra of radiation as a function of distance from the deposition
tube in the EERC pilot-scale combustor under gas-firing.

developing FTIR spectroscopic monitoring as a control technology for minimising ash buildup in furnace ducts of full-scale coal-fired boilers at power generation utilities.

4.   DISCUSSION AND IMPLICATIONS OF RECENT WORK

Recent work has supported the speculation[1] that *ash deposits of fine particles will be reflective* and that the physical character of the deposit (that is, whether it comprises particles, is sintered or has formed a slag) has a greater effect on radiative properties than chemical character. Increases in furnace exit temperatures of from 100 to 150°C are expected when clean tubes in a boiler are covered with an opaque layer of fine dust.

Recent work has indicated that *three types of properties might be identified from measurements of the emissive or reflective spectra* from deposits, these being:

- Identification of aspects of *chemical character* (for example, the presence of salt and silicates).

- Identification of *physical character* (for example, whether the deposit is powdered, sintered, fused or molten).

- Measurement of *radiative properties* (emittance, absorbance or reflectance), and possibly temperature.

Information may be obtained for the elements and compounds which are preferentially deposited, the identification of the physical character may monitor the changes associated with the development of strength in deposits. Both aspects may be used in strategies for optimising cleaning (soot blowing) cycles in practical plant. On-line measurement of radiative properties will therefore assist in the monitoring of the thermal performance of furnaces, and identifying coals associated with performance problems. The properties monitored will clearly be those of the outer layers of the deposit, as determined by its transparency, but probably less than a depth of 0.5mm.

It is not clear, at present, if the banded spectra suitable for extracting chemical information will be evident once deposits assume an irregular surface and reaction between the species have occurred. It is believed that the above understanding is sufficient to allow spectra to be interpreted even though chemical and physical properties influences may overlap, or occur together during deposit growth. Changes in spectra observed as deposits grow will clearly be easier to interpret than a single spectra.

A summary of the properties which may be monitored and the relevant spectral regions is given on the Table

Related to *the estimation of radiative properties from theory*, recent work has indicated that the theories for particulate layers which consider multiple scatter but not dependent scatter will underestimate the emissivity, particularly in the spectral regions below a wavelength of 8 μm. Previous speculations on the emissivity of such layers, such as those given on Fig. 2, should therefore be increased (by about 0.1).

The conclusions related to the association of physical character to radiative properties emphasise the importance of characterising the surface of deposits when radiative measurements are reported. Such characteristic could include roughness, porosity and the size of particles and would be most important for the outer 1 mm thick layer which determines the radiative properties.

ACKNOWLEDGMENTS

We would like to thank the organisations supporting the research area of the paper at our institutions, these being the Australian Research Council (UN), the US Department of Energy, Pittsburg Energy Technology Centre (Sandia and EERC) and the Electric Power Research Institute (EERC).

REFERENCES

1. Wall, T.F., Bhattacharya S.P., Zhang, D.K., Gupta, R.P., and He, X, The properties and thermal effects of ash deposits in coal fired furnaces, Prog. Energy Combust. Sci., 19: 487-504, 1993.

2. Wall, T.F., Mai-Viet, T., Becker, H.B. and Gupta, R.P., Fireside deposits and their effect in pf boilers, Proceedings of Conference on pulverised coal firing - The effect of mineral matter, Department of Chemical Engineering, University of Newcastle, pp L8.1-16, 1979.

3. Wall, T.F., Lowe, A., Wibberley, L.J. and Stewart, I.McC, Mineral matter in coal and thermal performance of large boilers, Prog. Energy Combust. Sci., 5: 1-29, 1979.

4. Hottel, H.C. and Sarofim, A.F., Radiative transfer. McGraw-Hill, New York, 1967.

5. Richter, W., Payne, P. and Heap, M.P., Influence of thermal properties of wall deposits on performance of pulverised fuel fired boiler combustion chambers. In: Karl Vorres (Ed.), Mineral Matter and Ash in Coal, American Chemical Society, Washington, DC, 375-383, 1986.

6. Boow, J. and Goard, P.R.C., J.Inst. Fuel, 42, 412-419, 1969.

7. Markam, J.R., Best, P.E., Soloman, P.R., Yu, Z.Z., J. Heat Transfer, 114, 450-467, 1992.

8. Richards, G.H., Harb J.N., Baxter, L.L., Bhattacharya, S.P., Gupta, R.P., Wall, T.F., Radiative heat transfer in pulverised-coal-fired furnaces - Development of the absorptive/reflective character of initial ash deposits, 25th (Int) Symp. on Combustion, Proceedings, The Combustion Institute, 511-518, 1994.

9. Wall, T.F., Bhattacharya, S.P., Baxter, L.L., Richards, G.H., Harb, J.N., The character of ash deposits and the thermal performance of furnaces, Fuel Processing Technology (in press, 1995).

10. Bhattacharya, S.P., PhD thesis, University of Newcastle, NSW, Australia, 1995.

11. Baxter, L.L., Fate of mineral matter in coal, Combustion Science Quarterly Progress Report, April-June, SAND94-8231, 1993.

RATES AND MECHANISMS OF STRENGTH DEVELOPMENT IN LOW-TEMPERATURE ASH DEPOSITS

John P. Hurley, Jan W. Nowok, Tina M. Strobel, Cathy A. O'Keefe,
Jay A. Bieber, and Bruce A. Dockter

Energy & Environmental Research Center
University of North Dakota
PO Box 9018
Grand Forks, ND 58202-9018

ABSTRACT

At temperatures below approximately 1900°F, ash particles formed in coal-fired energy systems are relatively hard and not prone to sticking to system surfaces. However, if the ash collects on a surface not exposed to a shearing gas flow such as the downstream side of a heat exchanger or the surface of a hot-gas filter, the deposit can develop enough strength over periods of minutes to days so that it becomes difficult to remove, in some cases growing to sizes that impede the flow of gas. This paper presents data from ongoing measurements of the significance of ash and gas composition, deposit temperature, and time on the rates of strength development in simulated low-temperature ash deposits. Preliminary results of surface composition and particle-size distribution analyses of the ash, including submicron material, are also presented to explain the possible mechanisms of strength development.

INTRODUCTION

Ash formed during combustion of pulverized low-rank coal in utility boilers causes a variety of problems, including adhesion to heat-transfer surfaces, erosion, and corrosion. Approximately 20% to 40% of the ash produced in a pulverized coal-fired utility boiler deposits within the boiler, primarily in the hoppers at the bottom of the furnace, but also on interior surfaces as slag. The remaining 60% to 80% of ash leaves the boiler with the flue gas and can deposit on steam tubes as ash-fouling deposits or continue through the system to become a collection problem for particulate control devices. It is common to deal with many of these ash-related problems in an *a priori* fashion, that is, by predicting the problems that might occur when burning a particular coal and designing the boiler specifically to ameliorate the problems. However, the rates and positions at which deposits form depend on the size distribution and the chemical composition of the ash, which is often times not well known before the coal is fired. Also, in order to reduce operating costs or plant emissions, coal is increasingly being purchased from fields that may be very different from those for which the system was designed. These coals can have significantly different ash-related problems than

Applications of Advanced Technology to Ash-Related Problems in Boilers
Edited by L. Baxter and R. DeSollar, Plenum Press, New York, 1996

83

those of the specification coal. In addition, unforseen problems often arise in new ash-handling systems, such as hot-gas cleanup systems, for which little operating experience is available. In order to better predict these problems, basic information about the mechanisms leading to the formation of the deposits must be developed.

In the following discussion, we address the development of strength in ash deposits formed at relatively low temperatures in systems where the ash is enriched in calcium. These low-temperature deposits have received relatively little attention in the literature because they have only been observed in U.S. boiler systems since the mid-1980s. The discussion centers on work performed at the University of North Dakota Energy & Environmental Research Center through funding from three government–industry consortiums. Two types of low-temperature depositions are addressed: those that occur on steam tubes toward the rear of the convective pass of pulverized coal-fired boilers burning high- calcium coal and the ash bridges between candle filters in hot-gas cleanup systems operating in conjunction with pulverized fluidized-bed combustors.

BACKGROUND

The inorganic constituents of western U.S. coals occur as discrete mineral particles and, in low-rank coals, as cations associated with organic acid groups or other organic complexation sites. Minerals abundant enough to form a significant portion of the ash include quartz, pyrite, gypsum, calcite, dolomite, and the clays kaolinite, illite, and montmorillonite [Raask, 1985]. Organically associated elements, which can form as much as half of the ash, include primarily sodium, potassium, magnesium, calcium, and sulfur. During combustion, the inorganic constituents undergo a variety of physical and chemical changes that depend on their original mode of occurrence, their time-temperature history, and interactions among the constituents. Usually the mineralogy and size distribution of the ash is quite different from that of the inorganic particles originally in the coal. For example, x-ray diffraction of fly ash samples from utilities burning western U.S. low-rank coals identified seventeen mineral species, only one of which, quartz, commonly exists in the coals [McCarthy et al., 1984]. The others formed through interactions among the inorganics during combustion.

As the ash particles are carried through the combustor system by the flowing gas, they may separate from the gas and deposit on boiler surfaces. Ash deposits that form in the radiant section of the boiler are known as slag deposits, whereas those that form in the convective pass on steam tubes are known as ash-fouling deposits. Deposits that form at gas temperatures over 2000°F are known as high-temperature or conventional fouling deposits. These deposits are usually formed from glassy silicate-based ash that is sticky at these higher temperatures [Crossley, 1952; Wibberley and Wall, 1982; Benson et al., 1993; Baxter and DeSollar, 1993; Walsh et al., 1990; Sondreal et al., 1977]. At temperatures below 2000°F, most silicates are relatively hard and do not stick to boiler surfaces. However, experience with high-calcium low-rank coals has shown that deposition at lower temperatures can still be significant. Below 1900°F, the sticky material is sulfate-based [Walsh et al., 1992; Osborn, 1992; Skrifvars, et al., 1991], although the mechanism and rate of formation of the sulfates have not been clear.

Extensive sampling and analysis of coal, ash, and deposits from a variety of boilers burning high-calcium western U.S. coals have revealed significant similarities in the properties of ash deposits that form at low temperatures in utility boilers. Four types of low-temperature deposits were observed: three types forming on the upstream side of the tubes and one type forming on the downstream side [Hurley and Benson, 1995]. Upstream deposits include massive ones that form toward the back of the secondary superheater and into the reheater at gas temperatures between approximately 1700° and 1900°F, enamel-like deposits that form at lower temperatures on primary superheaters and economizers, and double-crested deposits that form in the primary superheater and economizer tubes. The deposits forming on the downstream sides of the tube occur from the reheater section to the economizer. The common and defining characteristic of the four types of low-temperature deposits is their high concentration of sulfur that is fixed after the ash has deposited [Hurley and Benson, 1995].

In addition to fouling of steam tubes at relatively low temperatures in boilers burning high-calcium coals at temperatures below 1900°F, significant deposition has been observed in hot-gas cleaning systems designed to remove ash in direct coal-fired turbine systems. The problems have been more extensively described in pressurized fluidized-bed combustors where the ash deposits can grow between candle filters, forcing them apart and causing breakage of the filters at temperatures as low as 1475°F.

In general, two phenomena are important in the formation of low-temperature ash deposits. The first is the stickiness of the particles that allows them to stay in place on surfaces against the pull of gravity. The second is the slow development of much greater strength through sintering, leading to the formation of massive deposits that can block the flow of gas through the system or cause damage to low-strength objects such as candle filters.

RESULTS AND DISCUSSION

Particle Sticking

The cohesive forces which cause powder cakes to support themselves against gravity can be of varying types, including van der Waals, surface tension, electrostatic, crystallization at contact points, and interlocking of particles. These forces cause cakes made of more densely packed particles [Bortzmeyer, 1992] or smaller particles to have greater tensile strengths [Koch et al., 1992], although finer particle sizes generally result in a lower bed density which somewhat counteracts the increase in strength caused by the smaller particles.

The effect of particle-size distribution in the development of thick ash cakes on hot-gas cleanup filters can be seen by comparing the size distributions of ashes that caused bridging problems versus those that did not. Figure 1 shows the size distributions of ash collected from the Westinghouse advanced particle filter (APF) hot-gas particulate control system operating on the American Electric Power Tidd pressurized fluidized-bed combustor. The ash was sized with a Malvern Model 2600 laser particle-size analyzer. This instrument determines particle size by laser diffraction in the range of 0.5–564 microns. A low-power visible laser transmitter produces a parallel monochromatic beam of light which illuminates the particles. An assumption when using a laser particle analyzer is that all particles are spherically shaped. The results from the laser particle analyzer are expressed as volume percentages of the ash in each diameter bin. The samples were submerged in a sonic bath in attempts to break apart agglomerated particles.

The ash collected in the APF in May of 1994 was stickier and formed thicker residual cakes and more extensive ash bridges than the ash collected in February of 1995. The May 1994 ash was also much finer than the February 1995 ash, with 75% less than 10 microns in diameter for the May ash, versus only 25% less than 10 microns for the February ash.

In addition to the size distribution of the ash, the composition distribution of the ash can affect its stickiness by affecting the magnitude of the van der Waals forces between the particles or by forming thin liquid layers on the surface of the ash. Determining the surface composition

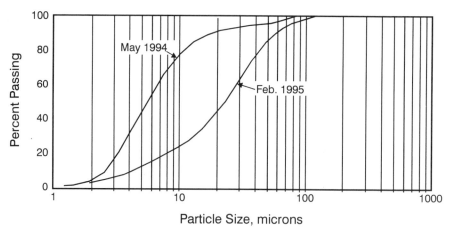

Figure 1. The particle-size distribution of the Tidd APF ashes collected in May 1994 and February 1995.

of ash is difficult because the x-rays used for elemental analysis in scanning electron microscopy (SEM) can originate from a depth of several microns within a particle, which may be the entire volume of the smaller, stickier ash. Instead, Auger electron spectroscopy can be used because Auger electrons with energies of between 0 and 2000 eV emerge from only a depth of 5 to 20 angstroms in the particles.

To determine the variations in composition with depth in the May 1994 Tidd APF ash, ion sputtering was used. In ion sputtering, a beam of ions, in this case Ar^+, is focused on the area to be analyzed. This ion beam etches away the molecular layers of compounds on the surface of the sample on the atomic scale. The sputter rate is dependent upon many factors, including ion-beam energy, beam current density, and the structure and composition of the surface to be sputtered. Ion-beam conditions of 2 keV, 0.3 microamp, over a 3-mm × 3-mm area of ash particles smeared on an indium foil gave a sputter rate of about 6 angstroms per minute. This sputter rate was chosen because it gave a resolution of approximately 1 atomic layer per minute on SiO_2.

Figure 2 shows the average depth profile of the May 1995 Tidd APF ash. Data for iron, sodium, and chlorine are left out of the figure because their concentrations were below the detectability limit under these conditions. This profile shows atomic concentration as a function of sputtering time in minutes. The actual depth this represents is difficult to estimate, because the structure and composition of the ash particles change with depth, which will affect the sputter rate, but we may assume by rough approximation, based on the use of a sputter standard, that each minute should represent approximately one atomic layer. Phosphorus, potassium, and sulfur decrease on average with depth, while oxygen, aluminum, and silicon remain relatively constant. Calcium increased throughout the profile, while magnesium initially increased up to 15 minutes, then decreased thereafter. The most pronounced effect was for phosphorus, which decreased by a factor of 3, from roughly 5% on the surface to 1.5% after only 10 minutes of sputtering. This profile indicates a small average surface enrichment

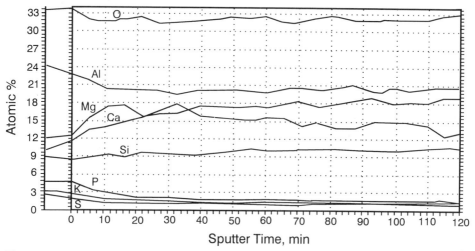

Figure 2. A depth profile of the concentration of elements near the surface of particles of ash collected from the Tidd APF in May 1994.

of compounds that may include a potassium phosphate–sulfate compound. The eutectic temperature for the K_2SO_4–KPO_3 system is 1324°F, indicating that the surface atomic layers of the ash may have been molten in the APF, which could lead to sticking of the ash if it was well packed. Further surface analysis is currently being performed in an attempt to determine the specific chemical compounds present by measuring the binding energies of the Auger electrons through x-ray photoelectron spectroscopy (XPS). In addition, Auger electron maps of the particles will be prepared, along with elemental depth profiles on a particle-by-particle basis.

Particle Sintering

In order for low-temperature deposits to block the flow of gas through a system or cause damage to system components, they must develop much more strength than just that necessary to hold the ash in place against gravity. Sintering is most often pointed to as the way in which deposits develop strength, but the mechanisms and rates of sintering in low-temperature deposits are not well understood.

Figure 3 illustrates the importance of sulfation in the development of strength in low-temperature deposits. It shows the compressive strength versus sintering time of 1-cm-diameter by 2-cm-tall pellets of 5- to 15-micron-aerodynamic-diameter high-calcium ash formed from coal from the Eagle Butte mine, Wyoming, and sintered in air and simulated flue gas containing 1000-ppm SO_2. The ash was sized aerodynamically and does not include submicron ash which would increase sintering rates but would also mask the effects of particle composition which was the main focus of this effort. Also, note that the rate of gas penetration into the pellets does not simulate the rates in a real deposit, so this method is used only for determining the relative importance of sintering mechanisms, not the actual rate of strength

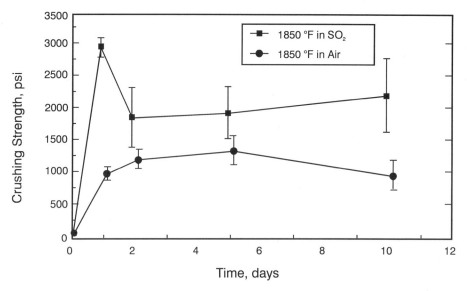

Figure 3. Compressive strength versus sintering times for pellets of 5- to 15-micron-aerodynamic-diameter Eagle Butte ash sintered in air and simulated flue gas containing 1000-ppm SO_2.

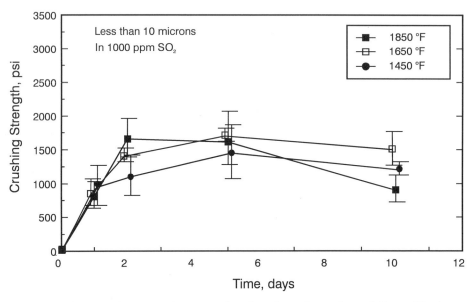

Figure 4. Compressive strength versus sintering times for pellets of 5- to 15-micron-aerodynamic-diameter Shoshone ash sintered in simulated flue gas containing 1000-ppm SO_2 at 1450°, 1650°, and 1850°F.

development. The plotted points are the average strengths of four pellets prepared at each condition. The vertical lines through each point are the standard deviations for the measured strengths. The points have been displaced horizontally, somewhat, from the actual lengths of the tests to allow the standard deviations for specific length tests to be seen more clearly. For each sintering time, the pellets prepared in SO_2 are stronger than those prepared in air. In both cases, pellet strength develops rapidly over the first day. For the pellets sintered in SO_2, strength dropped considerably after 1 day, but then remained relatively constant. The reason for the drop in strength is not clear, but is believed to be due to the formation of crystalline calcium sulfate from amorphous calcium-containing phases present at the start of the tests. Most of the coal ashes tested exhibited similar trends, but the loss of strength was not of the magnitude seen in the Eagle Butte pellets. In contrast, the pellets sintered in air do not lose significant strength, although they are significantly weaker than those sintered in SO_2, indicating that sulfur definitely plays a role in the development of strength in low-temperature deposits.

Figure 4 illustrates the importance of sintering time and temperature to the development of strength in low-temperature deposits. The figure shows the compressive strength versus sintering times for pellets of 5- to 15-micron-aerodynamic-diameter high-calcium ash formed from coal from the Shoshone mine, Wyoming, sintered in simulated flue gas containing 1000-ppm SO_2 at 1450°, 1650°, and 1850°F. The pellets were prepared and sintered under similar conditions as the Eagle Butte pellets. The shapes of the strength curves are similar for the three temperatures, indicating the phenomena related to strength development does not change, but the rates of strength increases are higher at higher temperatures. For the Shoshone ash, peak strength was reached at 1850°F in two days, whereas it took closer to 5 days at lower temperatures. Maximum strength was similar for the two higher temperatures and lower for the lowest temperature. The rate of loss of strength was decreased at lower temperatures as well, probably because crystalline phase formation is reduced at lower temperatures.

The Shoshone and Eagle Butte ash pellets showed spectacular crystal-like regions of nearly complete sintering. Figure 5 shows the edge of the Eagle Butte pellet sintered for 5 days at 1850°F in 1000-ppm SO_2. Figure 6 shows a high-magnification view of a crystal formed in the 10-day pellet. It shows that the crystalline areas have very low porosity, typically 5%, whereas the initial ash pellet had a porosity of 50%. The elemental x-ray maps shown in Fig. 7 indicate the composition of the crystalline regions. They are formed primarily of a calcium sulfate matrix in which silicate and aluminosilicate particles are embedded.

In general, the amount of crystalline material in the pellets increased with time and temperature, but there did not appear to be a relationship between crystal size and sintering time. Overall, the pellets with the highest strength and greatest alkaline-to-silicon ratio had the most crystalline material. The formation of crystals is not believed to be related to the increase of strength in the pellets since the crystals are not continuous. In fact, their formation is likely the cause of the measured weakening of the pellets. However, the magnified appearance of the crystals as shown in Fig. 5 is very similar to that of some low-temperature deposits collected from boilers. It is possible that as boiler deposits form, the ash sulfates immediately after deposition, leading to the highly sintered, high-strength nature of the deposits. In other words, the upstream massive deposits are essentially large crystals, such as those shown in Fig. 5. The downstream deposits never become as hard as the upstream deposits because their rate of sulfation is low compared to the rate of deposition, so crystals may not form or are certainly not continuous.

The propensity of the deposits to shed from the tubes under their own weight can be described through the concept of the shedding index, a dimensionless number which is the ratio of the rate of deposit growth divided by the rate of strength development. For reference it is given as:

$$SI = DR/SDR \qquad (Eq. 1)$$

where
SI = shedding index
DR = ash deposition rate (weight/[(time)(cross-sectioned area of tube)])
SDR = Strength development rate (crushing weight/[(time)(cross-sectioned area of deposit)])

A high shedding index implies deposits will tend to shed under their own weight, whereas a low shedding index implies that deposits will form slowly in relation to the strength and will be less likely to shed. Figure 8 shows the comparative shedding indices of downstream ash deposits formed in a pilot-scale combustor during testing of blends of an eastern U.S. bituminous coal and a western U.S. subbituminous coal. The ash deposition rates were determined by weighing deposits formed in the pilot-scale tests, whereas strength development rates were determined in laboratory tests with ash pellets, as described previously. As can be seen from the figure, the blends had the lowest shedding indices, indicating that they will likely cause the greater low-temperature deposition problem. However, the figure also indicates that the subbituminous coal has a higher shedding index than the bituminous coal, which is not consistent with utility experience. We believe that the inconsistency develops because in the laboratory strength test the gas must diffuse into the pellets rather than be in intimate contact during deposition. This reduces the rate of strength development due to sulfation in the laboratory tests as compared to the rate in an actual boiler.

Activation Energy

Sulfation of low-temperature deposits occurs through chemical vapor deposition of sulfur by converting calcium present as calcium oxide or calcium aluminosilicate glass to calcium sulfate. The kinetics of sulfation of calcium oxide are well understood, but the sulfation of calcium present in an aluminosilicate glass was not well understood before the present work. Thermogravimetric analysis (TGA) was used to determine the kinetics of the sulfation of calcium magnesium aluminosilicate glass in simulated combustion gas. The glass was made by melting reagent-grade single-element oxides in a platinum crucible at 1600°C and holding for 15 minutes at temperature to homogenize the mixture. The composition was chosen to reflect the normalized composition of 1–3-micron aerodynamic diameter Powder River Basin coal ash (Hurley and Benson, 1995). It consists of 28% SiO_2, 25% Al_2O_3, 38% CaO, and 9%MgO. The molten glass was quenched on a brass plate, ground, and sieved to minus 20 microns. The sulfation reaction conditions were SO_2 (5000 ppmv), NaCl (10 ppmv), CO_2 (15 vol%), O_2 (3 vol%), H_2O (15 vol%), and N_2 balance at temperatures ranging from 700° to 1000°C. Surface area for samples prior to and after the sulfation reaction was determined by the BET method. During sintering, the surface area for all samples was reduced from about 2.45 m^2/g to below 1 m^2/g, with the lowest of 0.58 m^2/g at 1000°C as a result of sintering. Since the surface area of all tested samples continually changed with the reaction time, we have introduced a method of calculating surface area versus time based on the following equation:

$$S = S_o - \gamma (\Delta m) \quad \text{at } T = \text{const.} \qquad [Eq. 2]$$

where S_o and S correspond to surface areas prior to sulfation and during sulfation at a designated time, and γ is a coefficient that depends on reaction temperature. In this equation, we have assumed that changes of surface area are caused by the plugging of pores and micropores by the calcium sulfate formed during the reaction. We have also assumed that the plugging is proportional to the weight gain of the sample.

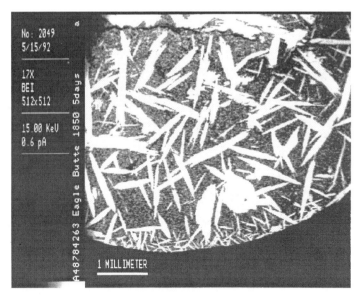

Figure 5. Backscattered electron micrograph of the edge of the Eagle Butte coal ash pellet sintered at 1850°F for 5 days in 1000-ppm SO_2.

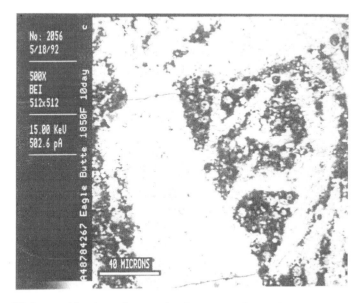

Figure 6. High-magnification micrograph of a crystal-like region in the Eagle Butte ash pellet sintered at 1850°F for 5 days.

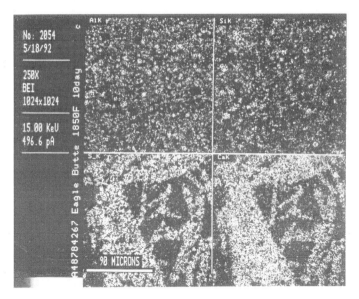

Figure 7.   X-ray map of Fig. 6 showing the relative concentrations of aluminum, silicon, calcium, and sulfur.

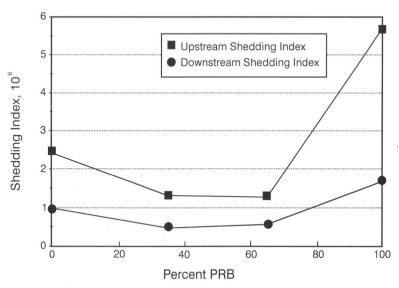

Figure 8.   The shedding indices of ash deposited on the downstream sides of simulated steam tubes during pilot-scale testing of an eastern U.S. bituminous and a western U.S. high-calcium subbituminous coal.

The kinetics of sulfation reactions are governed by the magnitudes of their activation energies, usually designated by the symbol E. The rates of chemical reactions have been found to depend very strongly on the composition of the reaction mixture and the system temperature. Many chemical reactions follow the Arrhenius rate law that is given by the following equation:

$$k_p = A \exp(-E/kT) \qquad \text{[Eq. 3]}$$

where k_p represents the rate constant, A is a preexponential factor independent of temperature, and k is Boltzman's constant. The implication of this observation is that the temperature of the reaction mixture must be held as constant as possible throughout the course of the reaction.

The rate of weight gain per unit surface area is a function of the rate constant and time as given by:

$$(\Delta m/S)^n = k_p t \qquad \text{[Eq. 4]}$$

where n represents a constant coefficient, depending on the mechanism of the sulfation reaction. Usually, when the sulfate layer is very porous, the sulfation process is expected to proceed linearly and is governed by the surface area of the material. When the sulfate forms a dense layer on the surface and n is equal to 2, then the reaction is diffusion-controlled (otherwise known as a parabolic relationship). The exponent is larger than 2 if other reactions occur. As seen in Fig. 9, the sulfation reaction obeys a nonparabolic relationship. The exponent n is about 1.3–1.5. Also, there is a transient period at very early stages of sulfation at which the mechanism differs from that recorded in longer time. An analysis of the slopes of the linear portion of the curves was used to determine the rate constant and coefficient n. The rate constants for the sulfation of the glass in a combustion atmosphere are listed in Table 1.

The logarithm of the rate constants as a function of the reciprocal of the absolute temperature (shown in Fig. 10) yields an activation energy of 22.4 kcal/mol for the sulfation reaction of calcium magnesium aluminosilicate glass. The activation energy for the sulfation of CaO in SO_2 is much higher at 39 kcal/mol [Gopalakrishnan and Seehra, 1990]. We speculate that the higher reaction rate of the calcium in aluminosilicate glass is due to the higher mobility of the calcium in the glass than in the crystalline oxide. Also, the sudden decrease of $\ln k_p$ for the glass at 1000°C is likely caused by additional reactions at that temperature or densification out of proportion to the weight gain as would occur in viscous flow sintering. These hypotheses are currently being tested.

After reaction, the particles were sintered into a mass that could be crushed with light finger pressure. SEM analysis showed that a reaction layer typically less than 3 microns thick formed on the particles in 24 hours, which is thin given the relatively high reaction order. Energy-dispersive x-ray data show that the layer is composed of essentially stoichiometric calcium sulfate with a small amount of the other elements, a portion of which may be interference in the x-ray signal. X-ray diffraction confirms the formation of crystalline calcium sulfate along with minor amounts of calcium magnesium sulfate. The unreacted cores of the particles are depleted in calcium and enriched in magnesium, aluminum, and silicon.

CONCLUSIONS

Low-temperature ash deposits form at temperatures below 1900°F and are most prevalent when high-calcium ashes are present. In order to grow and become hard, the ash must be initially sticky enough to stay in place on vertical surfaces, then harden sufficiently as the deposit grows to prevent shedding, erosion, or breakage from gas pressure and particle

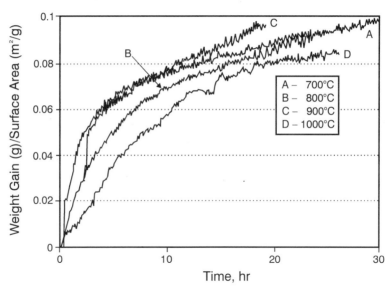

Figure 9. TGA curves obtained from calcium magnesium aluminosilicate glass powder at different isothermal temperatures.

TABLE 1

Rate Constants for Sulfation of Calcium Magnesium Aluminosilicate Glass

Temperature, °C	Rate Constant, k_p, g/h·S[1] $n = 1.3–1.5$
700	1.5×10^{-3}
800	5.0×10^{-3}
900	12.0×10^{-3}
1000	1.5×10^{-3}

[1] S corresponds to the surface area of the silicate.

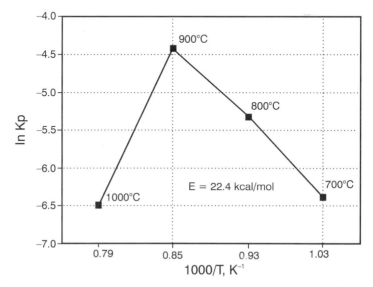

Figure 10. Arrhenius plot of the reaction of calcium magnesium aluminosilicate glass in a simulated pressurized fluidized-bed combustor atmosphere. The slope of the line gives an activation energy of 22.4 kcal/mol.

erosion. Initial sticking of ash particles is controlled by the ash size distribution, which affects contact area/weight ratios, and ash surface compositions. Particles with diameters below 10 microns and with thin liquid surfaces, such as are possible with Na–K–P–S–O eutectics, are most sticky. Long-term sintering is controlled by gas–solid reactions including nonstoichiometric sulfur substitution in silicates to decrease glass viscosity and the sulfation of calcium through chemical vapor deposition leading to molar volume increases that cement adjacent particles. The activation energy for sulfation of calcium in a Ca–Mg–Al–Si oxide glass is 22.4 kcal/mol, whereas for calcium in calcium oxide it is only 39 kcal/mol. Energy-dispersive x-ray data show that the layer is composed of essentially stoichimetric calcium sulfate with a small amount of the other elements, a portion of which may be interference in the x-ray signal. X-ray diffraction confirms the formation of crystalline calcium sulfate along with minor amounts of calcium magnesium sulfate. The unreacted cores of the particles are depleted in calcium and enriched in magnesium, aluminum, and silicon.

We speculate that the higher reaction rate of the calcium in aluminosilicate glass is due to the higher mobility of the calcium in the glass than in the crystalline oxide. Also, the sudden decrease of $\ln k_p$ for the glass at 1000°C is likely caused by additional reactions at that temperature or densification out of proportion to the weight gain as would occur in viscous flow sintering. These hypotheses are currently being tested.

REFERENCES

Baxter, L.L. and DeSollar, R.W. (1993). *Fuel, 72,* 1411.

Benson, S.A., Jones, M.L., and Harb, J.H. (1993). "Ash Formation and Deposition." In L.D. Smoot (Ed.), *Fundamentals of Coal Combustion for Clean and Efficient Use.* Amsterdam: Elsevier, 299.

Bortzmeyer, D. (1992). "Tensile Strength of Ceramic Powders." *J. of Materials Science, 27,* 3305–3308.

Crossley, H.E. (1952). *J. Inst. Fuel, 24,* 222–225.

Gopalakrishnan, R. and Seehra, M.S. (1990). "Kinetics of the High-Temperature Reaction of SO_2 with CaO Particles Using Gas-Phase Fourier Transform Infrared Spectroscopy." *Energy and Fuels, 4,* 226–230.

Hurley, J.P. and Benson, S.A. (1995). "Ash Deposition at Low Temperatures in Boilers Burning High-Calcium Coals 1. Problem Definition." *Energy and Fuels.*

Koch, D., Cheung, W., Seville, J.P.K., and Clift R. (1992). "Effects of Dust Properties on Gas Cleaning using Rigid Ceramic Filters." *Filtration and Separation, 29*(4), 337–341.

McCarthy, G.J., Swanson, K.D., Keller, L.P., and Blatter, W.C. (1984). *Cement Concrete Res., 14,* 471.

Osborn, G.A. (1992). *Fuel, 71,* 131.

Raask, E. (1985). *Mineral Impurities in Coal Combustion.* Washington: Hemisphere.

Skrifvars, B.J., Hupa, M., and Hyoty, P. (1991). *J. Inst. Energy, 64,* 196.

Sondreal, E.A., Tufte, P.H., and Beckering, W. (1977). *Combustion Science and Technology*, *16*, 95–110.

Walsh, P.M., Sarofim, A.F., and Beér, J.M. (1992). *Energy and Fuels*, *6*, 709.

Walsh, P.M., Sayre, A.N., Loehden, D.O., Monroe, L.S., Beér, J.M., and Sarofim, A.F. (1990). *Prog. Energy Combust. Sci.*, *16*, 327–346.

Wibberley, L.J. and Wall, T.F. (1982). *Fuel*, *61*, 93–99.

DEPOSIT FORMATION DURING THE
CO-COMBUSTION OF COAL-BIOMASS BLENDS

Hein, K.R.G., Heinzel, T., Kicherer, A., Spliethoff, H.

Institute of Process Engineering
and Power Plant Technology (IVD)
University of Stuttgart
Pfaffenwaldring 23
D-70569 Stuttgart, Germany
Fax +49 711 685 3491

ABSTRACT

In order to minimise the negative effect that energy conversion has upon the global climate, the utilisation of CO_2-emitting fossil fuels has to be reduced by applying conversion processes with higher efficiencies or by changing the fuel base. One of the options under investigation in large scale electricity production is to partly replace coal with biomass.

During recent years, there has been extensive research as well as demonstrations concerning combustion of biomass as a single fuel or combined with coal in various countries. Especially noteworthy is in this connection an EU-funded international project carried out in 1993/94. In this project it was shown that the utilisation of biomass may lead to operational problems through fouling and slagging of heat transfer surfaces due to the change in fuel composition. In particular, the high and variable alkali content of different types of biomass was found to be most troublesome under certain conditions.

In the paper the major results of the research group at the University of Stuttgart (IVD) will be presented; the appearance of deposits during biomass co-combustion will be

Applications of Advanced Technology to Ash-Related Problems in Boilers
Edited by L. Baxter and R. DeSollar, Plenum Press, New York, 1996

97

described and their dependence on fuel properties and on parameters of operation explained. Also some examples of experience with process alternations for improvements will be given.

1. INTRODUCTION

1.1   Co-combustion of biomass in existing power plants

The increasing concentration of CO_2 in the atmosphere is regarded as one of the main causes for global warming. Improved efficiency in fossil-fired power stations, the substitution of low-carbon fuels for high-carbon fuels, the application of renewable energies and the use of nuclear energy are at present potential techniques to reduce the CO_2 emissions in electricity production. One option in using renewable energies is the thermal use of biomass.

About 40% of the European electricity production is presently generated in pulverised-coal power plants. It can be expected that this percentage will not be substantially reduced in the next decades. However, the partial substitution of biomass for coal would reduce the emissions of the greenhouse gas CO_2 and other noxious gases, too. Combining biomass such as wood, straw, and energy crop with coal in full-scale boilers for electrical generation (co-combustion) could lead to a high energy-efficient utilisation at comparatively low investment costs. In recent years, various countries have done extensive research on biomass combustion both as separate fuel and in combination with coal. One event in this context was an international multi-partner project carried out in 1993/94 with financial support by the European Union (APAS Clean Coal Programme) [1].

Because co-combustion in existing power plants will have consequences on combustion behaviour, corrosion, emissions, and ash-related problems, these effects were studied in this project. It has been shown that the potential for fireside slagging and fouling due to the changed fuel composition may become one of the major operational problems.

1.2   Biomass fuel characteristics

There are various types and sources of biomass considered for co-combustion:

- by-products and wastes from agriculture, e.g. different types of straw (cereal, rape), grass, residuals from forestry and from the wood production industry (e.g. sawdust).

- energy crops, which are specially grown for thermal use on surplus agricultural areas, e.g. cereals, and alternative plants such as the C4-plant miscanthus sinensis, or short-rotation coppices of willow, poplar, and other fast growing trees.

In general combustion behaviour basically depends on the fuel properties. Biomass as a natural resource has variable characteristics different from coal. As an example, principal properties are shown in Table 1 for the fuels used in the IVD experiments. In general, the ash content of energy crops and straw is in the range of coals, the one of miscanthus sinensis is between 2% and 6%, and the ash content of wood is very low, usually below 1% [Figure 1].

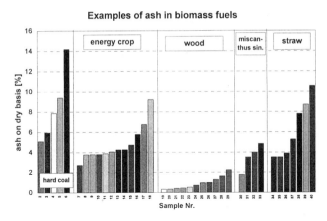

FIGURE 1. Ash content of coal and biomass

The moisture content in fresh wood is about 40-50%; straw, crops, miscanthus, and aged wood contain 5-25% of water. Net calorific values on dry basis are about 15 to 20 MJ/kg and usually about 13-15 MJ/kg on delivery basis. The sulphur and nitrogen contents are low, but the chlorine content in cereals and straw is sometimes very high in comparison with coal. Biomass has a high amount of volatiles, corresponding to the low fixed-carbon content.

For practical application in pulverised-fuel co-combustion, biomass can be injected in a larger particle size than coal. Combined with the high water content this leads to a delayed combustion of biomass in co-utilisation resulting in lower peak flame temperatures [6].

TABLE 1. Main Properties of Fuels (IVD Experiments)

	coal	straw	misc. sinen.	beech	oats
net caloric value (dry basis) [MJ/kg]	32.4	17.32	18.4	18.7	18.1
ash content (dry basis) [%]	7.9	8.71	4.85	0.73	5.79
moisture [%]	2.23	8.04	22.5	10.8	6.7
volatiles (dry basis) [%]	32.5	78.2	75.5	84.2	75.6
C (dry basis) [%]	82.0	40.4	42.7	44.2	42.6
H (dry basis) [%]	2.62	5.77	5.73	5.98	6.06
N (dry basis) [%]	1.33	0.79	0.56	0.23	1.18
S (dry basis) [%]	1.11	0.31	0.35	0.09	0.36
Cl (dry basis) [%]	0.23	0.72	0.36	0.05	1.57

The composition of elements is different from those found in coal, as exemplified by the fuels used in the deposit tests. Figure 2 shows that the main element oxides found in the German high rank coal are SiO_2, Al_2O_3 and Fe_2O_3. A dominance of SiO_2 with fairly large amounts of K_2O and CaO were found for straw, miscanthus sinensis, and oats. This is also known for wood (not shown). Noteworthy, too, is that the wood material analysed by the IVD had an unusual high iron content. The contents of sulphur-, titanium- and sodium-oxides in the examined biomass are low and generally more P_2O_5 than in coal was found.

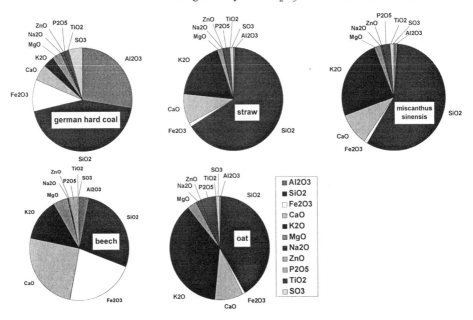

FIGURE 2. Ash elements of coal and biomass

A remarkable but well-known fact is that biomass exhibits low ash fusion temperatures in contrast to coal due to its high alkali contents [fig. 3]. In ash fusion tests according to the German

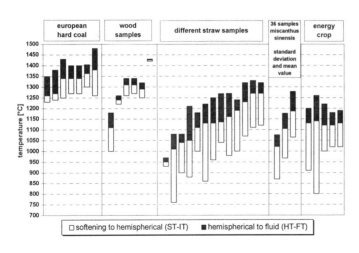

FIGURE 3. Ash fusion temperatures of coal and biomass (German standard)

standard[1] high rank coal ashes soften at ST 1250-1300° C and fluidize at FT 1350-1450° C. Wood ash mainly softens at ST 1200-1250° C and fluidizes at FT 1250-1350° C, often within a small temperature difference between ST and FT. However, some types of wood show a deviation from these values. Straw, energy crop and miscanthus sinensis samples show a wider scattering in their ash softening temperatures ST between 750-1100°C (mean values 950-1000°C) and fluid temperatures FT between 1000° C and 1350° C (mean values 1150-1200° C).

Thus the comparison of the ash melting data of coals and biomass reveals that sticky particles and slags in a much wider temperature range are expected for biomass. In consequence, a high risk of slagging and subsequent corrosion depending on the fuel characteristics can be expected when firing biomass in existing pulverised fuel boilers which are designed for coal only.

---

[1] The German method is shown for an example in [Appendix 1]. On this standard three temperature points are observed at a cubic probe. In this paper they will be named, corresponding to the international usage, Softening Temperature (ST), Hemispherical Temperature (HT) and Fluid Temperature (FT). HT and FT are more or less comparable to ASTM standard. Softening, though defined differently (by lower temperatures) in German usage, is nevertheless referred to as ST.

2. EXPERIMENTS AT THE IVD TEST FACILITY

To obtain information about the fuel-dependent characteristic phenomena involved in ash deposition during biomass co-combustion, tests were performed at the 0.5 MW semi-industrial pulverised-fuel combustion test facility of the IVD (Figure 4).

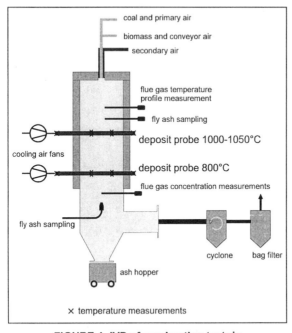

FIGURE 4. IVD pf combustion test rig

Six tests were performed during the experiments: one reference combustion with German bituminous coal, and five tests with a thermal input share of 25% each of various biomass types (sewage sludge, straw, the Asian grass type of miscanthus sinensis, beech, and oats). The sewage sludge experiments are described elsewhere [4]. For the investigations two air-cooled probes simulating the heat exchangers in the convective pass were inserted in different zones of the chamber. The probes remained inserted over a period of several hours in the combustion chamber at flue gas temperatures of 1000°C to 1050°C and of 800°C, corresponding to the section of a superheater 2 and a reheater, in the following referred to as SH2-probe and RH-probe, respectively [table 2].

All trials were run with a constant thermal input power of 300 kW. There was only a low variation of O_2 in the flue gas; mean values ranged between 2.9% and 3.5% in the trials. SO_2 was reduced in the flue gas in co-combustion of biomass due to the lower sulphur content in biomass and the sulphur capture in ash. Flue gas temperatures at the SH2-probe were between 1000°C and 1050°C. At the entrance of the cooling air the temperature of the probe tube walls was 500°C, at the cooling air outlet the temperature had risen to 650°C. These temperatures remained almost constant in each test.

In addition to the regular measurements at the test rig the probes were visually inspected and video-recorded. The time-dependent changes of the temperatures and the heat flux at the probe tubes were recorded by continuous in-flame registration of the

TABLE 2. Experiment parameters

biomass addition	100% coal	25% straw	25% misc. sinensis	25% beech	25% oats
total thermal input	300 kW				
flue gas O_2 [%]	3.26	3.4	3.3	3.5	2.9
flue gas SO_2 [mg/m³]	2007	1388	1400	1496	1457
probe test time [min.]	460	420	265	415	355
flue gas temperature at SH2-probe [°C]	1010	1040	1040	1000	1015
probe temperature pattern SH2-probe	500°C - 650°C from cooling air inlet to cooling air outlet				
flue gas temperature at RH-probe [°C]	780	825	825	820	810
probe temperature pattern RH [°C]	500°C - 650°C. from cooling air inlet to cooling air outlet				

temperatures of flame, tube walls, and cooling air. Fly ash samples of the test flames were collected and analysed. At the end of the experiments, the different deposit fractions were sampled to determine the rate of dry deposition.

Ash fusion tests of laboratory furnace ash, both of plain fuels and fuel blends, and of fly ash samples and deposit samples were carried out and compared with regard to their oxide contents.

3. RESULTS AND DISCUSSION

3.1 Ash Fusion Temperatures

Hot stage microscopy is applied world-wide to determine the slagging and fouling potential of a solid fuel. This technique was used to measure the characteristic deformation temperatures (HT, ST, FT) for laboratory furnace ash samples of selected biomass materials, their 25%-blend (on thermal basis) with high rank coal and compared with the data of the pure coal (fig. 5). It is obvious that in comparison with coal the non-fossil biomass materials - straw, miscanthus, and oats - show a much lower deformation temperature and a wider range between softening and fluid temperature, which seems to increase with a decreasing absolute temperature. This tendency is expected in line with the alkali content of the materials as shown in fig. 2 and the tendencies already mentioned in the interpretation of fig. 3 (page 5)

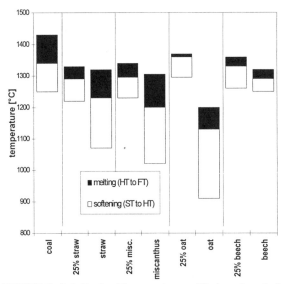

FIGURE 5. Ash Fusion Temperatures of Laboratory Ash

However, if blended with coal up to the investigated mass fractions corresponding to 25 % of thermal input, the lower critical temperatures are almost identical to those of the pure coal sample, while the upper levels are considerably lower than the upper pure coal temperatures. With regard to industrial application these results may be interpreted in such a way that, although the initial deformation temperature of coal as main fuel is hardly affected by the biomass addition, fusion - and thus slagging - is strongly influenced by the fluxing properties of the biomass components. Therefore, the biomass addition might not alter the initial deposit temperatures of coal ash but slag forming will occur at considerably lower temperatures than with coal only.

Somewhat surprising is that the beech sample does not show the expected low deformation temperature and the wide temperature range. For this case, an explanation is offered by the chemical analysis given in fig. 2. The fluxing component potassium here is considerably lower than in the other biomass kinds. Also, calcium and iron are much more pronounced in the beech sample, components which are known to reduce the initial deposit potential in the low temperature range. However, slag formation may be promoted by the iron content (by calcium only at much higher temperatures) leading to the small temperature range ("short slag") observed in the case of pure beech and this superimposes the coal ash characteristics in the blend.

During the preparation for ash fusion tests, alkali species may be volatilized through incineration and hence be lost in the sample. This might affect the obtained results of the biomass samples. For this reason, the ash melting behaviour of the samples incinerated in the laboratory furnace might be different from that of the composition of the inorganic matter in the fuel. Still, because of the good corresponding of both the data on the alkali shares, obtained from the chemical analysis of the laboratory samples, with the analysis data of the furnace deposits (chapter 3.3), and the ash fusion temperatures of these samples (following chapter), it can be marked that the laboratory ash samples in general give a good description of the ash and deposition particles found in the combustion trials.

In figures 6-8 the softening temperature ST and the fluid temperature FT of blends are shown depending on the thermal input share of biomass. The black line represents the calculated linear ash mixture temperature of the laboratory ashes, the grey shaded area shows the range between minimum and maximum values of plain fuel samples.

In principle, the fly ash and deposit samples behave in the same way as the laboratory-made blends. Only one major exception was found for straw co-combustion [fig. 8]. For this reason the melting behaviour of ash and dust deposits in the furnace can be estimated by testing laboratory furnace ashes.

Under normal conditions, neither the laboratory-made blends nor the fly ash and deposit samples show the low softening and fluid temperatures of plain biomass. But the fluid temperature FT is lower in contrast to the expected linear blend behaviour in the case of blends of coal with miscanthus and straw, a feature which may be due to

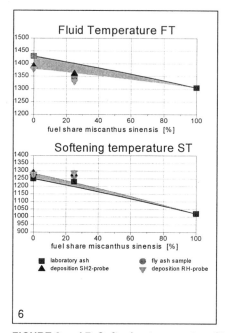

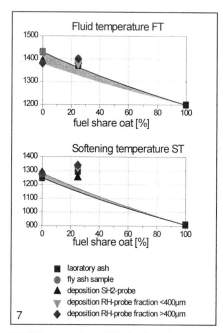

FIGURE 6 and 7. Softening temperature ST and fluid temperature FT of ashes from coal-miscanthus blends and coal-oats blends

low-melting minerals or eutectics. However, the softening temperature is higher than expected. The range between softening and fluid temperatures decreases for the blends. In principle, this effect is the same for the coal-oats blend, but shifted to a higher level, so that the fluid temperature of the blend is still higher than the linear mixture temperature.

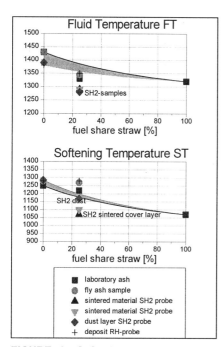

FIGURE 8. Softening temperature ST and fluid temperature FT of ashes from coal-straw blends

When co-firing straw a sintered upper layer of the deposits, originating from the still glowing large sticky straw particles, softens and fluidizes much more like plain straw ash, which corresponds to the chemical analysis and the observations, where the glowing particles were sticking to the surface of the probes and burning out there.

3.2 Ash Content and Deposit Characteristics

While producing the same thermal output, the biomass blends caused no or only a slight increase of the total ash flow. Co-combustion of beech reduced the amount of ash in the flue gas, due to the low ash content typical for wood. Only 0.03 to 0.8 % of the ash flow through the projected area of the probes can be found as deposit on the probes, the main fraction is moving on with the gas flow. The total amount of deposits depended strongly on experimental conditions. In fig. 9 results of accumulated deposit mass dependent on position and, thus, on temperature are presented. The left bar in every picture shows the deposit amount on the probe near the furnace wall at low probe temperatures of about 500°C (entrance of the cooling air). The second bar shows the main part of the probes in the middle of the combustion chamber, and the right bar shows the deposit amounts at the cooling air outlet, where the highest probe temperatures of 650°C were measured. Loose dusts on the probes were mainly accumulated near the furnace wall where the flow velocity is reduced. Generally, the amount of deposits increased with the probe temperature.

The total amount of deposit was lowest in beech co-combustion on both probes [fig. 9 and table 3]. This can be explained by the effect of solid beech coke particles (>1mm) hitting the probes. The deposit amounts of .pure coal combustion were low on the centre part of the probes. The deposit amounts of the straw blend combustion were

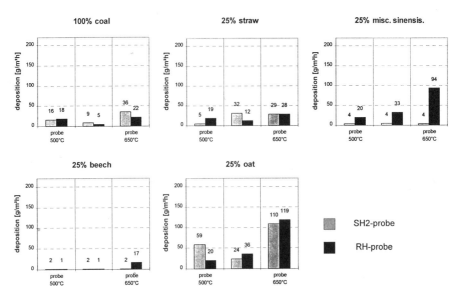

FIGURE 9. Deposition rates on the SH2-probe (grey) and on the RH-probe (black)

TABLE 3. Ash content and deposit characteristics

thermal fraction	100% coal	25% straw	25% miscanthus sinensis	25% beech	25% oats
biomass mass fraction [%]	0	39	40	38	38
total ash content [g/MJ]	2.4	3.0	2.3	1.9	2.5
biomass ash of total [%]	0.0	39.2	26.1	5.1	30.1
ash deposition on SH2-probe [%]	0.26	0.29	0.07	0.03	0.76
deposition rate SH2 [g/m²h]	17.5	24.2	4.3	1.7	53.9
ash recovery on RH probe [%]	0.19	0.22	0.7	0.1	0.74
deposition rate RH-probe [g/m²h]	12.8	17.9	44.9	5.1	52.6

All depositions were either fine dry ash dust layers or loose dry ash dust. The only exception was a cover layer of slightly sintered material on the SH2-probe on straw co-combustion. This layer originated from large straw particles which were still glowing and sticking on the probes.

low, too, but the highest amounts were found on the SH2-probe in the main flue gas stream. Low deposition rates on the SH2-probe but very high amounts of loose dusts on the RH-probe were found when co-firing miscanthus sinensis. Highest amounts on both probes were sampled when co-firing oats.

The deposits found on the probes were always easily removable. They either appeared as a fine layer of dry ash dust or loosely bonded ash agglomerations. When co-firing straw, an upper layer of slightly sintered material partly covering the dry ash dust was found on the SH2-probe.

3.3   Chemical Composition of Ash

Besides the fuel ashes, the fly ashes and deposits on the probes were analysed with regard to their composition, and compared with the ash element input in the combustion chamber. In Figure 10 important components of the ashes and deposits are shown for pure coal combustion as well as for the test blends with 25% straw and 25% miscanthus sinensis.

The percentage of Al_2O_3 is much lower in the deposit samples than in laboratory furnace fuel-blend ash for all test cases. In co-combustion tests Al_2O_3 is also reduced in fly ash samples. Calcium oxide remains on input level when firing pure coal and miscanthus sinensis but increases in depositions when co-firing straw. A higher share of potassium can be found in deposit samples of straw co-combustion. The chemical composition of the deposit samples from straw co-combustion in comparison to the other samples can be interpreted as an accumulation of straw ash elements in the deposits.

Silicon, the main component of ash, shows a heterogeneous behaviour. In coal combustion its share in the flame samples increases, making up for the reduction of aluminium. Silicon in straw co-combustion is reduced in the fly ash and on the RH probe, however, it shifts towards the amounts found in plain straw ash on the SH2 probe. In the case of miscanthus, the Si share slightly increases in the fly ash, decreasing again in the samples from the probes where sulphur is bound into the ash.

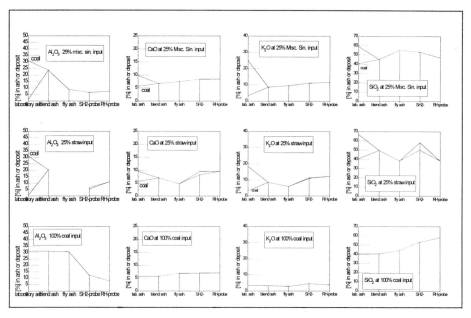

FIGURE 10. Ash elements in fuel ash, blend ash, fly ash and depositions

3.4 Sulphur binding in ash

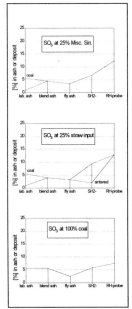

Sulphur accumulates in the deposits. Sulphur-binding in ash is higher when co-firing biomass in comparison to pure coal combustion, although the percentage in fuel blend ash is reduced due to the low sulphur content of biomass. The highest sulphur content was measured in the probes at 800°C flue gas temperature. In fly ash these high shares could not be found. A dependence of the sulphur content on the time during which the particles have contact to the flue gas is expected. The sintered particles in the upper layer of the SH2-probe when co-firing straw did not bind the sulphur due to the higher temperatures there.

FIGURE 11. Sulphur in fuel ash, blend ash, fly ash and depositions

3.5   Slagging and Fouling Prediction Indices

Indices found in the literature were calculated both for pure fuels and blends. However, it remains to be determined whether these indices are applicable to co-combustion. Additionally, the determining procedures were applied to the fly ash and deposit samples in order to identify severe shifts of the characteristics. Table 4 shows a list of the calculated indices for laboratory-made ash selected from literature, which seems to be applicable to the fuel blends, with their estimated slagging and fouling tendency. The numerical values can usually be classified as "low", "medium", "high", or "severe".

TABLE 4. Slagging and fouling indices

slagging indices	coal	25% straw	25% misc.	25% beech	25% oats
B/A-ratio [15] et al.	low	low	low	low	low
R_S [15] et al.	low	low	low	low	low

fouling indices	coal	25% straw	25% misc.	25% beech	25% oats
F_U [15] et al.	medium	medium	medium	medium	medium
K_V [16]	high	high	high	low	high
Cl- content [15] et al.	low	medium	low	low	high

The base-acid ratio of coal ash and all blend ashes is low. The slagging factor RS is low as well because of the low sulphur content in biomass. The fouling index FU, which takes into account the alkali content, is medium. The chlorine content of the blends is low with the exception of straw blend (medium) and oats blend (high). The lignite fouling index KV, taking into account alkaline and sulphur content, is high for pure coal and the blends, with the exception of the beech blend. For this index, the results are classified as "low" or "high".

In general, only minor differences can be found between coal ashes and the blend ashes. Beech is an exception and performs best in co-combustion, followed by pure coal, miscanthus, and straw blends. Oats blend performs worst among the evaluated biomass types. The coal/biomass blends assessment by those indices is comparable to German bituminous coal.

In summary it can be stated that the predictions are not necessarily in agreement with the observations and even contradictions can be found. This underlines the well known fact that such indices have only a very limited value for practical purposes and must be applied with extreme care to large scale processes or avoided completely. The major reason is the process dependent temperature regimes in operation which are not accounted for by the considered indices.

4. CONCLUSIONS

The investigations presented here brought about a first estimation of the deposit behaviour of the different coal/biomass combinations.

The experiments and evaluations at the IVD's test facility reveal that the co-combustion tests with a 25% share of biomass, given a satisfactory burnout of the biomass even at such low flue gas temperatures ranging from 1000° C to 1050° C, did not cause slagging and fouling problems at the heat exchange surfaces. The higher temperatures of the uncooled walls in the IVD's test facility lead to higher fly ash and deposit particle temperatures compared with the corresponding areas where the equivalent power plant heat surfaces are arranged.

In straw co-combustion the build-ups on the SH2 probe show a melting behaviour which is distinctly decreased by the biomass. Particles of the coarsely ground straw were still glowing and sticky at the entrance of the convective section. These straw particles bonded selectively on the SH2 probe. The observation of this phenomenon was confirmed by the melting behaviour and the chemical analyses of the samples which reveal a distinct shift towards the ash composition of plain straw.

The obtained results at the IVD-test rig agree well with the findings from co-combustion experiments in the facility tests from the APAS-project [1]. There, too, the fouling assessed at the heat exchange surfaces showed either no or only a minor increase, thus being always uncritical.

The tests point out that ash deposition and fouling of bituminous coal-biomass fuel blends is in the scale of bituminous coal ash deposition. Deposit quantities and qualities on heat transfer surfaces in the temperature range from 800°C up to higher than 1050°C were never severe. One reason is the softening behaviour, which remains at high temperature levels for the blends, although the melting behaviour of most pure biomass ashes is very low due to the high alkaline content. The softening temperatures of the blends are not reduced from the level of the coal as much as expected, but the melting temperatures of the blends are reduced in comparison to the coal ash. In consequence there is a higher slagging risk in the furnace, but a low fouling risk at the heat exchanger surfaces at the examined flue gas temperatures. If particles hit the surface in a glowing state, however, they can adhere to and be taken up by it selectively. This may cause slagging which should be avoided by a sufficient grinding of fuel. Although the biomass has a low sulphur content, biomass co-combustion leads to a high SO_3 accumulation in ash and deposits, but not to an accumulation of alkali components in the determined depositions. The coal slagging and fouling indices indicate only a slight increase in fouling and slagging in comparison to the coal for the examined biomass types. Some indices generally do not seem to be practicable for biomass co-combustion and the differences between the biomass types seem to be not correctly represented by most indices.

Straw, miscanthus sinensis and cereals in common have an ash amount referring to the calorific values comparable to German hard coal. Sulphur content is very low but chlorine content is sometimes very high (up to 2% in dry fuel). Due to a high alkaline and silicon content as plain fuel the ash has very low ash fusion temperatures. Therefore the input share of biomass must be restricted in power plants constructed for coal. When co-firing biomass at low input shares no selective accumulation of biomass ash was found on the probes and the ash behaviour was dominated by the ash behaviour of the main coal ash.

Based on the above specific results it can be stated that biomass co-combustion - apart from the positive effect on NO_x emission reduction reported elsewhere, e.g. [17] - is advantageous with pure biomass utilisation provided certain temperature ranges are not exceeded. Blends of coal with biomass containing devoted alkali contents will not necessarily influence the initial deposit formation but may promote slagging at considerably lower temperatures than for coal only.

Thus, slagging problems at the entrance of the convective pass might occur if low melting biomass particles are available and, in particular, if not completely burned out at this position.

ACKNOWLEDGMENTS

The authors wish to acknowledge the involved APAS-partners and the financial support of the European Union in the APAS Clean Coal Technology Programme, APAS-Contract COAL-CT92-0002.

REFERENCES

[1]   *APAS Clean Coal Technology Programme 1992-1994, "Combined Combustion of Biomass/Sewage Sludge and Coals"*. Volume II: Final reports. Edited by the Institute for Process Engineering and Power Plant Technology, University of Stuttgart, for the European Commission, DG XII: Science, Research and Development, Stuttgart 1995 (ISBN 3-928123-16-5).

[2]   Baldacci, Bianchi, De Robertis, Paolicchi, Rognini, Tani: "Fireside Fouling During Caol/Biomass Co-Firing in a Circulating Fluidized Bed Combustor and a Pulverized Coal Power Plant". In: *APAS Clean Coal Technology Programme 1992-1994, "Combined Combustion of Biomass/Sewage Sludge and Coals"*. Volume II: Final reports [1].

[3]   DIN 51370: "Bestimmung des Ascheschmelzverhaltens", 5/1984

[4]   Heinzel, T.: "Ermittlung der Verschlackungs- und Verschmutzungsneigung von Kohle-/Biomasse-Mischungen bei der Verbrennung in Staubfeuerungen". Diplomarbeit Nr. 2530. Institute for Process Engineering and Power Plant Technology (IVD), University of Stuttgart, 1994.

[5]   Juniper, Lindsay: "Applicability of Ash Slagging Indices" Australian Combustion Technology Center ACTC: Combustion News, Feb. 1995,

[6]   Kicherer, A.: „Untersuchungen zur Biomasseverbrennung in Staubfeuerungen: Technische Möglichkeiten und Schadstoffemissionen", Preliminary report, Institute for Process Engineering and Power Plant Technology (IVD), University of Stuttgart, 1995 (will be published soon)

[7]   Kicherer, A., Gerhardt, T., Spliethoff, H., Hein, K.R.G.: "Co-Combustion of Biomass/Sewage Sludge with Hard Coal in a Pulverized Fuel Semi-Industrial Test Rig". In: *APAS Clean Coal Technology Programme 1992-1994, "Combined Combustion of Biomass/Sewage Sludge and Coals"*. Volume II: Final reports [1].

[8]   Petersen, K. Hansen, K.: "Co-Firing of Straw in a Pulverized Coal-Fired Boiler
        I/S Vestcraft, Unit 1". In: *APAS Clean Coal Technology Programme 1992-
        1994, "Combined Combustion of Biomass/Sewage Sludge and Coals"*. Volume
        II: Final reports [1].

[9]   Rasmussen, I. Poulsen, P. Wieck-Hansen. K.: "Grenaa Demo-Test at Enhanced
        Straw Input". In: *APAS Clean Coal Technology Programme 1992-1994,
        "Combined Combustion of Biomass/Sewage Sludge and Coals"*. Volume II:
        Final reports [1].

[10]   Reisinger, Klaus.: „Bestimmung des Ascheschmelzverhaltens verschiedener
        Biomassearten". Diplomarbeit. Institut für Verfahrenstechnik,
        Brennstofftechnik und Umwelttechnik, TU Wien, 4/1993

[11]   Williamson, J.; Wigley, F. (eds.): *The Impact of Ash Deposition on Coal Fired
        Plants.* Proceedings of the Engineering Foundation Conference, Solihull,
        England, 1993. Taylor & Francis, Bristol, UK, 1993 (ISBN 1-56032-293-4)

[12]   Zelkowski, J.: *Kohleverbrennung.*VGB-Kraftwerkstechnik, volume 8. VGB,
        Essen 1986

[13]   Meschgbiz, A., Krumbeck, M.: "Combined Combustion of Biomass and Brown
        Coal in a Pulverized fuel and Fluidized Bed Combustion Plant". In: *APAS
        Clean Coal Technology Programme 1992-1994, "Combined Combustion of
        Biomass/Sewage Sludge and Coals"*. Volume II: Final reports [1].

[14]   ASTM Standard D1857-87: "Test Method for Fusibility of Coal and Coke
        Ash"

[15]   Couch, G.: *Understanding slagging and fouling in pf combustion*, IEA coal
        research, London, 8/1994, IEACR/72, ISBN 92-9029-240-7

[16]   Reidick, H., Schuhmacher, D.: „ *Verschmutzungsverhalten bei der Verfeuerung
        von Braunkohlen.* " International VGB conference „Slagging, Fouling and
        Corrosion in Thermal Power Plants", VGB, Essen 1984

[17]   Siegle, V., Kicherer, A., Spliethoff, H., Hein, K.R.G.: *Biomass Co-combustion
        for Polluant Control in Pulverized Coal Units.* 3rd Int. Conference on
        Combustion Technologies for a Clean Enviroment, 3.-6.July, Lisbon, Portugal

APPENDIX
Ash fusion temperatures according to German DIN standard tests

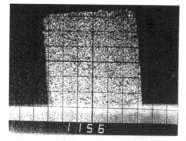

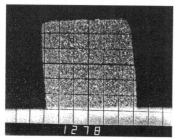

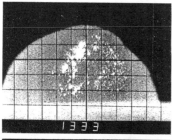

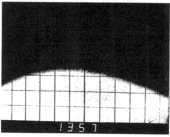

A cubic sample with an edge length of 3mm is heated with 10 K/min. The sample can be either observed through microscopic lens, photographed or video recorded. The pictures presented here are video prints. The overlayed grid distance is 0.5mm.

The DIN standard method uses ash from a laboratory furnace, but the method can be applied as well to deposit and fly ash samples.

The sample shown on the left is fly ash from the IVD 0.5 MW test facility pf co-combustion of German hard coal with 25% miscanthus sinensis. The ash sample was prepared for the fusion test according to the standard method.

Up to 1156°C no deformation is visible. Using a higher microscopic magnification, changes in surface structure can be observed (not shown here). In the temperature range between 1156°C and 1229°C the sample is shrinking without changes in geometry. Since a fixed temperature for these continuous changes cannot be declared, no sintering point is defined. The temperature, at which first deformations can be detected (usually at the corners of the cube), is defined as softening temperature ST. The three temperature points referring to the standard are presented as shown for this sample, while the temperature points are given in steps of 5 or 10 K:

1280°C	softening	ST
1335°C	hemispherical	HT
1360°C	fluid temperature	FT

QUANTIFICATION OF DEPOSIT FORMATION RATES AS FUNCTIONS OF OPERATING CONDITIONS AND FRACTION OF BIOMASS FUEL USED IN A CONVERTED PC BOILER (100 MW)

Anders Nordin
Energy Technology Centre in Piteå
Department of Inorganic Chemistry
University of Umeå, Sweden

Bengt-Johan Skrifvars
Department of Chemical Engineering
Åbo Akademi University, Turku, Finland

ABSTRACT

Statistically designed experiments were carried out to study the potential effects of operating parameters on the extent of slagging, fouling and sulphur retention in a 100 MW PC boiler burning a combination of wood powder and oil. The influence of primary air, burner air distribution (secondary/tertiary air), secondary air (upper and lower burners), over fire air (east, west and rear), flue gas recirculation, fuel distribution, load and fraction of biomass fuel used were evaluated using first a screening design and then a face-centred composite design (CCF). Polynomial models were deducted from statistical analysis of the experiments. The general results from the study are that wood powder in combination with oil can be used in a safe way with relatively small amounts of deposits formed. The dominating influential factor was the fraction of biomass fuel used, where an increase resulted in somewhat higher sulphur retention but a slightly greater fouling. However, chemical analysis of samples collected during the tests did not indicate any severe corrosion tendencies and the fuel combination has been successfully used for more than two years now, without any ash related operational problems.

INTRODUCTION

An increasing number of pulverised coal (PC) boilers are currently being converted to facilitate use of more biomass fuels due to the environmental benefits associated with these fuels. In addition to the insignificant atmospheric CO_2 contribution and low sulphur emissions, most biomasses contain relatively large amounts of potassium, calcium and sodium in reactive forms [Nordin 1994]. These metals may be useful for

Applications of Advanced Technology to Ash-Related Problems in Boilers
Edited by L. Baxter and R. DeSollar, Plenum Press, New York, 1996

117

an increased sulphur retention in the ash when co-combusting biomass and sulphur containing fuels. However, they have also been reported to cause severe slagging and fouling. Both the degrees of sulphur retention and deposit formation are strongly dependent on the temperature distribution and gas composition throughout the furnace and boilers. Thus, primary measures could be valuable for reducing the sulphur emissions as well as decreasing the formation of deposits. Previous studies reported by Hupa [1980, 1989] showed that the extent of deposit formation varies considerably depending on both fuel combinations and operating conditions. Several publications can be found in the literature where the influence of separate factors on the deposit formation rates have been studied. However, to our knowledge, no previous studies have been reported where the effects of all potential operating variables have been evaluated in a systematic way, utilising statistical experimental designs. In the present work, the effects of operating conditions and fraction of biomass fuel used, on both SO_2 emissions and deposit formation rates, were estimated by the use of empirical models. This was accomplished by performing statistically designed experiments in a full scale boiler originally intended for pulverised-coal.

PROCESS STUDIED

Boiler and fuels

The measurements were carried out over a two week period in November 1994 and a two week period in March 1995 on a 100 MW PC boiler located in Hässelby, Stockholm. A schematic view of the boiler is shown in Figure 1.

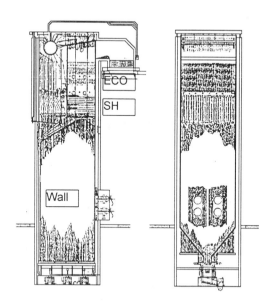

Figure 1. The PC Boiler and Sampling Locations.

The characteristics of the fuels are presented in Table 1. Theoretical sulphur emissions without any retention in the ash would be 67 and 22 mgS/MJ for oil and wood respectively. Total intrinsic "alkali" to sulphur ratio for the wood fuel is 1.8.

Experimental designs

In process evaluation there has always been a strong belief that to determine how one factor influences a response, all other factors must be held constant while only varying that particular factor. As a result, many investigations are still carried out using the inefficient "Changing One Separate factor at a Time" (COST) approach (Fig. 2, left). The COST approach might lead to incomplete mapping of the behaviour of the system, often resulting in poor understanding as well as incorrect conclusions. Instead, both process evaluation and optimisation benefit from the use of statistical designs [Box et al. 1987, Nordin et al. 1995a and Nordin 1995b]. These are informationally optimal

Table 1. Fuel Characteristics (wt%).

Wood		Fuel oil	
Dry substance	93	HV ((MJ/kg$_{Fuel}$)	44.6
Ash[a]	0.3	Ash	0.04
HV (MJ/kg$_{Fuel}$)	17.2	C	87.3
C[a]	43.7	H	12.3
H[a]	6.8	N	0.2
N[a]	0.2	S	0.3
S[a]	0.04	Cl	n.a.
Cl[a]	< 0.01	O	0.3
O[a]	41.9		
SiO_2[b]	6.1	V_2O_5	4ug/g
Al_2O_3[b]	1.3	NiO	4.2 ug/g
Fe_2O_3[b]	1.6	Na_2O	4 ug/g
MgO[b]	5.6		
CaO[b]	39		
K_2O[b]	12.2		
Na_2O[b]	1.4		
P_2O_5[b]	2.3		

[a]wt% dry matter
[b]wt% of ash

mathematical schemes in which all important factors are changed systematically, thereby facilitating the identification of process relations as well as the location of the real process optimum (Fig. 2, right). Normally 2^k experiments are required for evaluating a process with k factors. For a more complete investigation, main effects, as well as quadratic and interaction terms have to be determined. However, as a process may be influenced by a multitude of factors, the number of experiments rapidly grows, making it impractical to carry out all experimental runs. It is therefore desirable to reduce the number of experiments while still retrieving the most information. This is accomplished by first performing a screening study, to identify the most influential (of a number of potentially important) factors and those that may be regarded as inert. Once the most important parameters are identified the process may be optimised using some kind of extended experimental design.

In the present work, a large number of operating factors were considered in a screening study, common in an extensive evaluation study, concerning the effect of primary measures on nitrogen oxide emissions [Österlund, 1995]. The effects of air distribution (lower/higher burner planes), over fire air (OFA), flue gas recirculation (FGR), burner air (secondary/tertiary air) and fuel distribution (upper/lower burner planes) were evaluated by the use of a design based on the algorithms described by Morris and Mitchel [1983]. Thereby, the analysis of both emissions and deposition rates was accomplished utilising the same underlying screening design. In Table 2, the factors and their design levels are given. All these experiments were performed during full load operation, using 70 %$_{energy}$ biomass fuel and 30 %$_{energy}$ oil.

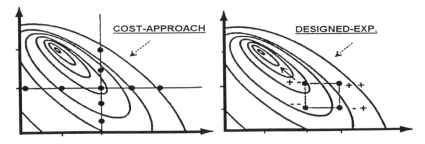

Figure 2. Illustration of the Benefits of Using Statistical Experimental Designs.

Based on the results from the screening study, our subsequent optimisation study could be restricted to two factors; load and fraction of biomass fuel used. Here, a face-centred composite design (CCF) according to Fig. 3 was utilised.

To minimise the effect of uncontrolled factors and time trends, the experiments were performed in randomised order. At each experimental point, the conditions were kept constant for three hours, after allowing one hour for equilibration to stable conditions. Analytical data were collected every two minutes during the three hours, and the average values were used in the evaluation. This procedure resulted in a total experimental time of about 100 hours.

Analysis

To quantify the deposit formation rates, short-term deposit samples were collected on air-cooled sampling probes as described by Hupa [1980]. Probes were inserted at three locations in the furnace/boiler, corresponding to the wall (Wall), superheater (SH) and economiser (ECO) tubes (Fig. 1). The probes were 2.5 m long, consisting of a double tube, with a detachable sampling ring (28 cm^2) of normal wall tube material. Eurotherm temperature controllers connected to electrical ball valves were used to keep the probe temperatures at the same surface temperature as the local tubes (~350°C). The probe rings and deposits were further analysed using SEM EDS and X-ray diffraction analysis.

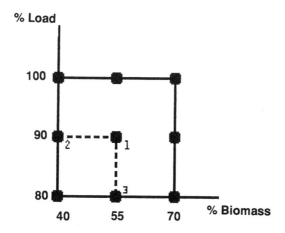

Figure 3. Illustration of the CCF Design Used.

The gases NO, NO_2, SO_2, O_2, CO_2 and CO were continuously determined by conventional instruments. Evaluation of the data has been restricted to relative changes, such that the influence of errors is minimised. As unit for the emission of SO_2, mgS/MJ_{fuel} was used to facilitate comparisons of the results with other investigations and legislation's, as well as to avoid dilution effects. To facilitate construction of the optimisation design, modelling and analysis procedures, the computer programs MODDE and SIMCA-P were used.

Table 2. Experimental Design and Corresponding Results for the Screening Study.

Exp. run	Air high kNm³/h	Air low kNm³/h	OFA west	OFA east	OFA back	FGR kNm³	Burner air sec/tert.	Fuel distribution upper/lower	SO_2 mgS/MJ	Flame mg/h[*]	SH mg/h[*]	ECO mg/h
3	47 (45)	59 (47)	open	closed	closed	0	100/0	min/max	14.2	17.1	8.5	2.8
17	49(52.2)	47(43.2)	open	open	open	5	100/0	max/min	12.0	41.1	10	3.4
16	46	43	closed	closed	closed	5	100/0	max/min	12.0	21.9	15.3	3.3
20	41(42)	51(52)	closed	open	open	0	100/0	min/max	12.0	21.6	10.2	3.2
6	59(52)	47(41.5)	open	open	open	5	100/0	max/min	12.0	34.2	10.2	3.1
4	50(49)	50(49)	open	open	open	5	50/50	50/50	12.0	25.1	9.7	3.2
5	51(50.7)	41	open	open	closed	0	0/100	max/min	11.0	21.3	24.0	3.4
11	51(50.8)	41(40.8)	closed	open	closed	5	0/100	min/max	12.0	89.7**	11.1	2.8
18	59(58.3)	47	closed	closed	open	0	0/100	max/min	11.9	20.0	11.3	4.2
22	41	51	open	closed	open	5	0/100	min/max	11.8	17.9	9	3.6
10	51	41	open	open	closed	0	0/100	max/min	12.0	29.5	19.5	3.1

[*] Area of the sampling rings: 28 cm²   ** Identified as outlier (measurement error)

122

RESULTS

The results from the screening study is summarised in Table 2. Relatively low deposit formation rates and very little variation in the measured responses were obtained.

Some corrective actions (within parenthesis) had to be performed to secure a safe operation of the plant. Therefore, the multivariate evaluation was performed by Principal component analysis (PCA) [Wold, 1987] and Partial Least Squares Projection to Latent Structures (PLS) [Wold et al. 1983, Dunn et al. 1984] instead of multiple regression. The results showed that none of the studied variables exhibited a significant effect on any of the ash related responses. The sulphur retention was not much influenced and relatively constant at 30-40 %. The evaluation of the deposit formation rates indicated strong coincidental variation (noise), independently of the studied factors. Comparisons between the replicates also show a significant influence of noise.

Table 3. Experimental Design (CCF) for the Optimisation Study.

Exp. run	Biomass % of load	Load MW	SO_2 mgS/MJ	SH mg/h	ECO mg/h
10	40	80	31.6	10.4	4.2
11	70	80	8.1	41.8	7.5
5	40	100	32.1	15.1	6.9
9	70	100	12.4	51.5	7.0
2	40	90	33.2	13.2	4.0
8	70	90	10.6	30.7[*]	64.7[*]
3	55	80	20.8	12.2	3.6
4	55	100	21.4	20.6	4.0
1	55	90	18.1	21.3	10.8
7	55	90	21.7	35.8	3.4
6	55	90	20.9	10.0	5.3
12	55	90	19.4	41.7	7.6

[*]uncertain values

According to these results, all screening factors may therefore be used exclusively to minimise nitrogen oxide emissions, and the results from the parallel NO_x project indicated a relatively high potential for such measures. The optimisation study could therefore be restricted to the two operating parameters that would be expected to be most important for both sulphur retention and deposit formation; load and fraction of biomass fuel used. The results are summarised in Table 3.

The responses SO$_2$ and deposit formation rates in the super heater area were evaluated as described in the screening study. The effects of the two factors on these responses are presented in Fig. 4 and 5. Only the fraction of biomass used have significant effects. An increased use of biomass fuel will reduce the emissions of SO$_2$ but also somewhat increase the fouling. The amounts of deposits formed in the economiser were to small and showed to little variation to be evaluated. In addition, the probe close to the flame front suffered from a severe leakage and could not be used at all during the optimisation study.

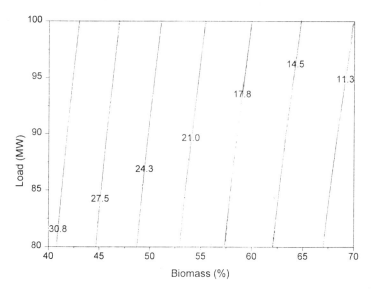

Figure 4. Effect of Fraction of Biomass Used and Load on the Emissions of SO$_2$ (mgS/M).

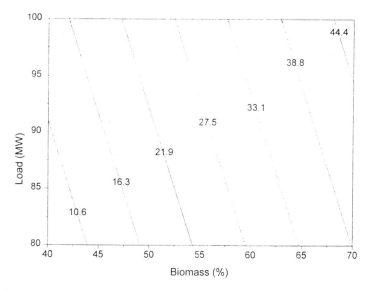

Figure 5. Effect of Fraction of Biomass Used and Load on Depositions Formed on the Sample Ring (28 cm2) Placed in the Superheater Area (mg/h).

Results from cross sectional SEM EDS analysis of the sampling rings indicated relatively harmless deposits, consisting mainly of K, S and Ca. No signs of initial severe corrosion could be found.

The calculated sulphur retention, based on the measured emission of SO2, as function of the fraction of biomass used is shown in Fig. 6.

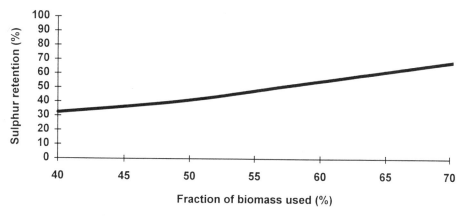

Figure 6. Sulphur Retention versus Fraction of Biomass Used.

In Figure 7 the sulphur capture efficiency from the present study have been compared with the normal efficiencies of conventional Ca-based absorbents for in-furnace sulphur dioxide reduction in PF-boilers. The very high utilization rate for biomass fuels can be explained by the highly avaliable and reactive form of Ca and K in biomass fuels [Nordin 1993].

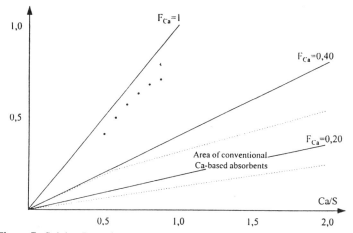

Figure 7. Sulphur Retention versus Total "Alkali" to Sulphur Ration (••••; Results from the Present Study).

CONCLUSIONS

- No severe slagging or fouling were experienced in the plant during the present study. In addition, two years of documented normal operation using the same fuel combination have not resulted in any disturbing ash related operational problems.

- Of all the operational parameters varied, the degree of ash deposition was significantly influenced only by the fraction of biomass used, where an increase resulted in a slightly greater fouling.

- A sulphur retention of 30-70 % was obtained, and influenced only by the fraction of biomass fuel used.

ACKNOWLEDGMENTS

Thanks are due to Per Levén, ÅF-Energikonsult AB, for initiating the work, Ulf Hagström, Ingemar Goldkuhl and Karin Hedin, ÅF-Energikonsult AB, for the experimental work. We are also grateful to the personal at Stockholm Energi AB, Hässelbyverket for their help, as well as the permission to publish the results of this study. Financial support given by the Thermal Engineering Foundation is gratefully acknowledged.

REFERENCES

Box, G. E. P. Hunter, W. G. and Hunter, J. S. Statistics for experimenters; An introduction to design, data analysis and model building. Wiley, NY, 1978

Dunn. III, W. J., Wold, S., Edlund, U., Hellberg, S. and Gasteiger, J. (1984). "Multivariate structure-activity relationships between data from a battery of biological tests and ensemble of structure descriptors: The PLS method.", *Quant. Struct.-Act. Relat.,3*, 131-137.

Hupa, M. (1980). " En undersökning av rökgassidans beläggningar i ångpannor." Thesis, Åbo Akademi University.

Hupa. M. (1989) "Predicting slagging and fouling tendency." Symposium on Low-Grade Fuels, Helsinki, June 12-16.

Morris. M. D. and Mitchell, T. J. (1983). "Two-level multifactor designs for detecting the presence of interactions." *Technometrics, 25*, 345-355 and ref. herein.

Nordin. A. (1993). "On the chemistry of combustion and gasification of biomass fuels. peat and waste - environmental aspects." Thesis, Dept. of Inorganic Chemistry, Umeå University

Nordin. A., Öhman, M. and Eriksson, L. (1995a). "NO reduction in a fluidized bed combustor with primary measures and selective non-catalytic reduction; A screening study using statistical experimental designs." *FUEL, 74*, 128-135.

Nordin, A. (1995b). "Optimization of sulphur retention in ash when cocombusting high sulphur fuels and biomass fuels in a small pilot scale fluidized bed." *FUEL, 74,* 615-622.

Wold, S. (1987). "Principal component analysis.", *Chemometrics and Intelligent Laboratory Systems, 2,* 37-52

Wold, S., Albano, C., Dumm, III, W. J., Espensen, K., Hellberg, S., Johansson, E. and Sjöström, M. (1983). "Pattern recognition: Finding and using regularities in multivariate data." In: Martens, H. and Russwurm, H. (Eds.), *Food research and data analysis,* Applied Science Publishers, London 147-188

Österlund, U. and Andersson, K. (1995). Internal report, Stockholm Energi. SEU 210.

IN SITU MEASUREMENTS OF BOILER ASH DEPOSIT EMISSIVITY AND TEMPERATURE IN A PILOT-SCALE COMBUSTION FACILITY

David W.Shaw[1] and Scott M. Smouse

U.S. DOE, Pittsburgh Energy Technology Center
Pittsburgh, Pennsylvania 15236-0940 USA

ABSTRACT

An instrument for in situ determination of ash deposit temperature and emissivity has been designed, assembled, calibrated, and used to characterize deposits in a coal-fired pilot-scale combustor. The temperature information has also been used to estimate the effective thermal conductivities of ash deposits. These measurements were made in the radiant section of the Pittsburgh Energy Technology Center's pilot-scale combustion facility, the Combustion and Environmental Research Facility, which is nominally rated at 500,000 Btu/hr (145kW), for ash deposits from a Pittsburgh seam coal, an Alaskan coal, a Russian coal, and a blend of the latter two. The estimated thermal conductivities fall within the range of values reported by others; however, the deposit formed by the coal blend had a substantially higher thermal conductivity than either of the two parent coals. The in situ measurement system, which is called INSITE for In Situ Temperature and Emissivity, has been successfully operated in adverse conditions and without the need of specialized training.

INTRODUCTION

Knowledge of the thermal properties of ash deposits is critical to the realistic modeling of boiler heat transfer [Wall et al., 1979; Mulcahy et al., 1966 a, b]. These properties will depend on the composition and temperature of the deposit, on its thermal history, and possibly on local flow conditions. A good database of such properties at realistic conditions is one approach to improving the predictive capability of boiler models. Other approaches include (1) detailed physical modeling of deposit structure and subsequent calculation of thermal properties based on this structure, (2) measuring properties of simulated deposits, and (3) measuring properties of deposits removed from the boiler

---

[1] Assistant Professor of Mechanical Engineering, Geneva College, Beaver Falls, PA 15010 and ORISE part-time faculty research participant at PETC.

Applications of Advanced Technology to Ash-Related Problems in Boilers
Edited by L. Baxter and R. DeSollar, Plenum Press, New York, 1996

129

environment. A significant advantage in making in situ measurements is that the deposits can be observed throughout their development, thereby providing information that is not available by most other approaches. Another advantage is that the measurement system can be proven initially at a relatively small scale, e.g., in a pilot-scale combustor, but in principle can be used in full-scale combustors such as utility boilers. Thus, pilot-scale and full-scale data obtained with the same or a similar system can be compared. This information is required for verifying/calibrating/refining existing or new physical models of deposits. Knowledge of ash thermal properties could also be used to improve boiler operation. For example, soot blowing could be optimized for different regions of a boiler if information on the in situ thermal properties of ash deposits were known on a time scale compatible with soot blower operation.

Thermal properties of ash deposits have been measured in the laboratory by several groups [Goetz et al., 1979; Wall and Becker, 1984; Mulcahy et al., 1966a; Boow and Goard, 1969; Anderson et al., 1987]. However, few in situ measurements have been reported for pilot-scale and full-scale combustion systems. For example, a limited number of in situ measurements were made in a 150,000 Btu/hr (44 kW) and a 5-million Btu/hr (1.45 MW) pilot-scale combustor at the Babcock and Wilcox Company's Alliance Research Center, as reported by Wagoner et al. (1984), and Clark et al. (1989). The major limitations of the B&W work involved dealing with the background radiation from hot walls in the refractory-lined combustor and insufficient signal from the relatively cool deposits. Data acquisition time would also limit the application of the technique of Clark et al.(1989), because the wide range of wavelengths scanned required several detectors and order-sorting filter changes.

These problems are not present in the optical probe system developed and used in this work, since the instrument was designed to operate at several fixed wavelengths rather than scanning, and the geometric aspects of the reflected radiation are being used to deal quantitatively with background radiation.

The In Situ Temperature and Emissivity (INSITE) system was described in detail by Shaw and Smouse, (1994). Figure 1 illustrates the essential physics. A probe viewing a deposit at some temperature that is lower than its surroundings will receive both directly emitted energy, which is a function of the deposit temperature and emissivity, and reflected energy, which is a function of the temperature of the surroundings and the deposit reflectivity. By using a probe that blocks some known fraction of the energy from the surroundings and making measurements of intensity at several wavelengths, the temperature of the surroundings, the temperature of the deposit, and the deposit emissivity can be determined, since varying blockage will produce different changes in sensor signal at the different wavelengths. The wavelengths were selected to avoid gas absorption/emission bands and, at the location and depth of field viewed, particle emissions will have only a second-order effect [Shaw and Smouse, 1994]. The diameter of the cooled probe and the distance from the deposit to the probe control the fraction of background radiation blocked. This is described by a geometric view factor.

The INSITE probe is shown in Fig. 2. A water cooled stainless steel probe is used to block the background radiation and provide a cold collimating tube for direction of infrared radiation through the filters and onto the sensor. A mechanical filter wheel is used to select the wavelength of interest.

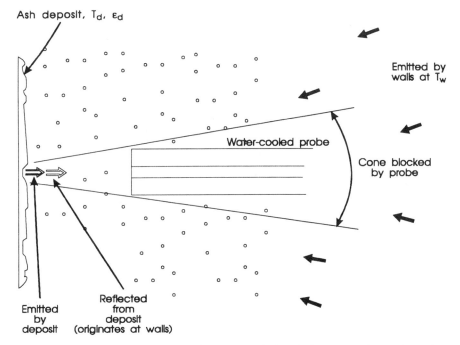

FIGURE 1. Sources of thermal radiation and their relation to probe and combustor geometry.

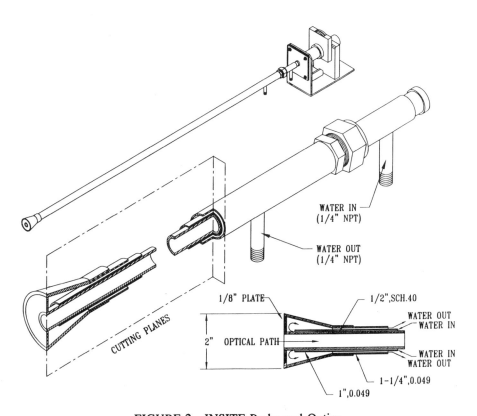

FIGURE 2. INSITE Probe and Optics.

Table 1. Coal Analyses

Proximate Analysis (Dry Basis)	Alaskan	Russian	90/10 Blend
Volatile Matter	46.7%	29.2%	46.2%
Fixed Carbon	40.0%	37.8%	38.2%
Ash	13.2%	33.0%	15.6%

Ash Fusion Temperature (Reducing) °F	Alaskan	Russian	90/10 Blend
Initial Deformation	2120	2745	2081
Softening	2170	over 2800	2140
Hemispherical	2240	over 2800	2162
Fluid	2350	over 2800	2250

Ash Mineral Composition (wt%)	Alaskan	Russian	90/10 Blend
Silicon Dioxide	43.1	62.2	57.2
Aluminum Oxide	17.0	23.3	20.1
Ferric Oxide	8.7	4.2	4.3
Titanium Dioxide	0.8	1.1	0.8
Phosphorus Pentoxide	0.1	0.1	0
Calcium Oxide	20.6	3.3	8.8
Magnesium Oxide	2.6	1.0	1.2
Sodium Oxide	0.6	0.6	0.6
Potassium Oxide	1.2	1.9	2.2
Sulfur Trioxide	5.6	1.3	0

The measurements described here were made during a series of combustion tests of an Alaskan coal, a Russian coal (13% and 33% ash respectively on a dry basis), and a blend of these coals (90% Alaskan/10% Russian). The laboratory analyses of the individual coals and the mixture are shown in Table 1. This report focuses on the first pilot-scale temperature and emissivity measurements using the INSITE system. The full results of the effects of blending will be reported elsewhere. Deposit characteristics for tests using a baseline Pittsburgh seam coal are also reported here for comparison.

EXPERIMENTAL

The measurements reported here were made during the course of regular testing in the 500,000 Btu/hr (145 kW) Combustion and Environmental Research Facility (CERF) at the Pittsburgh Energy Technology Center (PETC). This combustor, described previously by Smouse and Walbert (1990), simulates the radiant and convective sections of a boiler and has ports at 18-inch intervals for sampling and for insertion of various probes (see Fig. 3). It is routinely used for comparison of combustibility, emissions, and slagging/fouling characteristics of various fuels [Smouse, 1994] as well as for studies of in-flame NOx reduction techniques. The deposits for which properties were measured in these tests were formed at port #4 in the radiant section.

Deposits were formed on air-cooled deposition probes with a flat carbon steel deposition surface mounted flush with the furnace wall. These probes were instrumented with surface and depth thermocouples (Fig. 4) so the surface temperature could be controlled and heat flux could be measured. The heat flux was determined by applying Fourier's Law

$$(q/A) = k\frac{\Delta T}{\Delta x}$$

to the heat flow through the steel face of the probe. The thermocouples are placed near the centerline of the probe face, as shown in Fig. 4. Finite difference heat transfer calculations predict the maximum error in heat flux caused by two-dimensional heat transfer effects to be approximately 10%.

The deposition probes differ from the slagging panels described by Smouse and Walbert (1990) in several important ways. Configuration as a probe, rather than as a panel, allows probes to be inserted easily at any port at any time during a combustion test. The probes can also be removed at any time to inspect or remove the deposit. The multiple depth thermocouples make it possible to calculate heat flux even if one or more thermocouples fail. The smaller exposed area also makes probe temperature control at high firing rates easier to accomplish.

Data were collected using the procedure described previously [Shaw and Smouse, 1994]; although some modifications to the operating procedure were used because potential sources of error had become apparent. One important modification was made to better deal with changes in the absolute calibration of the system. The INSITE probe/optics assembly is shown in Fig. 2. The optics are mechanically coupled to the probe to prevent misalignment problems and to minimize the possibility of damage to components. There is, however, a potential for non-repeatable alignment in this system. Small changes or distortions in the optical path will change the **absolute** signal, but not the **relative** signals at the several wavelengths. This is because a filter wheel is used to change wavelengths. The absolute drift caused by alignment changes is easily dealt with by making an initial measurement with the INSITE probe flush with the near wall of the combustor (Fig. 5a) so that the measurement is essentially a measurement of the refractory temperature.

133

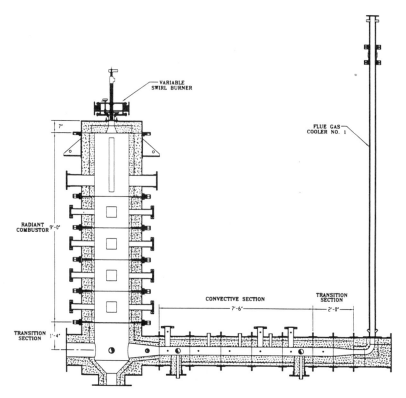

FIGURE 3.   CERF 500,000 Btu/hr combustor.

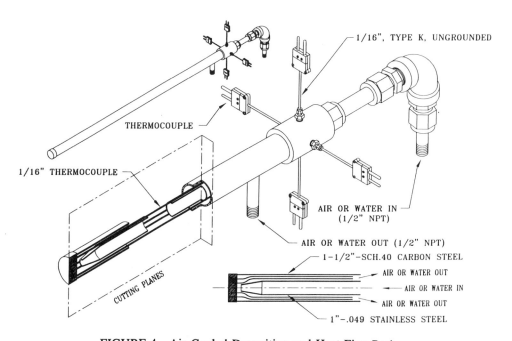

FIGURE 4.   Air-Cooled Deposition and Heat Flux Probe.

The INSITE system can be treated as a simple multi-wavelength pyrometer to determine a wall temperature and this measurement can be used to rescale the calibration constants for all the wavelengths. Knowing this wall temperature, the absolute scaling of voltages can be adjusted to match previous calibration data from a blackbody furnace

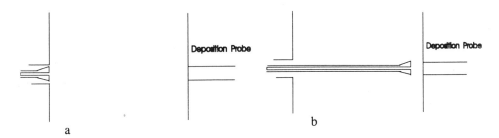

Figure 5. Insertion Depths for absolute calibration check (a) and for data taking (b).

Following the absolute calibration check, the probe was then moved to regions near the deposition probe (Fig. 5b). Data were taken at several intervals (typically 1.5, 2.0, 2.5, and 3.5 inches from the deposit). Nonlinear regression techniques were then used to determine the deposit and wall temperatures and the deposit emissivity based on the equations and methods presented by Shaw and Smouse, 1994. The voltage at each wavelength/position combination can be expressed as:

$$V_{\lambda,x} = a_\lambda [\varepsilon_d I(\lambda, T_d) + F_b(x)(1 - \varepsilon_d) I(\lambda, T_w)]$$

where: a_λ is the instrument calibration factor at wavelength λ, determined using a blackbody furnace
λ is the wavelength
x is the distance from the deposit surface to the INSITE probe
ε_d is the deposit emissivity
T_d is the deposit surface temperature
T_w is the wall (surroundings) temperature
$I(\lambda,T)$ is intensity as given by Planck's law at the appropriate wavelength and temperature
$F_b(x)$ is the fraction of background radiation blocked by the probe, expressed as a function of distance, x

With measurements taken at several values of λ and x, non-linear regression can be used to fit for ε_d, T_d, and T_w.

Additional data were collected to estimate the effective thermal conductivities of the deposits. These included heat flux measurements using the deposition probe, deposition probe surface temperature, and measurements of the masses of deposits on the probe at the end of the test, from which deposit thicknesses were estimated. Effective thermal conductivities were then calculated using Fourier's Law

$$k = (q/A) \frac{\Delta x}{\Delta T}.$$

RESULTS

Table 2 summarizes the results for the Alaskan and Russian coals as well as some baseline measurements for Pittsburgh seam coal. The conductivities are also plotted as a function of temperature in Fig. 6 and compared to results from laboratory measurements on ash [as summarized by Wall et al, 1994].

Table 2. Conductivities and Emissivities of Deposits

Coal	Deposit thickness (mm)	Deposit mean temperature (°C)	Effective thermal conductivity (W/m °C)	Emissivity
Pitt 3/28/95	0.225	505	0.31	0.63
AK 11/15/94	0.665	844	0.30	0.68
Russ 12/15/94	0.178	510	0.19	0.81
90/10 2/1/95	0.872	538	1.01	0.64
Pitt 8/11/94	*	515	**	0.73
Pitt 7/13/94	*	519	**	0.65

* - Not measured
** - Not able to determine without thickness data

The deposit mean temperature listed above is the average of the deposit surface temperature and the deposition probe surface temperature.

All tests were conducted at a firing rate of 500,000 Btu/hr (145 kW) using 20% excess air. The furnace exit gas temperatures were between 980 and 1100°C, measured using a suction pyrometer. All the deposits tested to date have been powdery and loose, and their steady state thickness has been less than 1 mm. These deposits were collected on carbon steel surfaces maintained at temperatures between 425 and 485°C in the radiant section of the CERF, and typical local gas temperatures were 1315-1375°C.

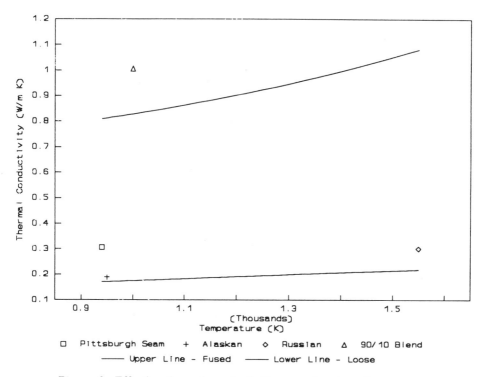

Figure 6. Effective thermal conductivities as a function of temperature.

Estimation of deposit thickness required knowledge of the apparent density of the deposit layer. Because the deposit was disturbed when it was scraped off the probe and weighed, the density could not be determined directly. Instead, typical values for the solid volume fraction and true ash density were used to estimate a thickness for each deposit. The values used were $\varepsilon=0.25$ and $\rho=2.7$ g/cm^3. These values are appropriate for loose deposits of coal ash as observed in these and previous tests [Ramer and Martello, 1994; Ramer, 1995].

Although the data can be fitted for both of the temperatures (deposit and surroundings) and the emissivity, it was found that convergence of the regression analysis was enhanced by imposing a value of the temperature of the surroundings that was an average value of the local refractory temperature. The modified operating procedure for checking for absolute calibration drift using a partially inserted INSITE probe also provides another measurement of the temperature of the surroundings by ratio pyrometry. Such an approach should work even in a system without wall thermocouples as long as the temperature of the surroundings is changing slowly.

DISCUSSION

The results in Table 1 and Fig. 6 show an interesting effect when the Alaskan and Russian coals were blended. The deposit thickness increased, but the effective thermal conductivity of the deposit was significantly higher. This means that even with the thicker deposit, the heat flux to the deposition probe was actually higher for the blend than for the Alaskan coal alone, and not much lower than the heat flux with the Russian coal alone. This interaction of the coal ashes could lead to some interesting results when using blended coals, with soot-blowing requirements significantly less than might have been expected. The blended ash deposit has a thermal conductivity that is more characteristic of fused deposits than of loose deposits [Wall et al., 1994]. Because its average temperature was substantially below the softening temperatures for either of the ashes, it seems unlikely that substantial fusion occurred. The deposit also did not appear to be fused when it was scraped from the probe.

OPERATIONAL CONCERNS/INSTRUMENT IMPROVEMENTS

The tests described here have led to some changes in operation of the INSITE system to help achieve more consistent and useful results. The most important of these is the additional data point taken to adjust for absolute calibration drift. This is a fairly simple solution to a common problem, not requiring any hardware modifications, and it also gives an independent indication of the temperature of the surroundings.

It was also found during one period of testing that the seal between the probe and the optics had failed, resulting in a flow of dusty gas from the furnace into the optical train instead of a positive flow of purge gas. A thin coating of particulates developed on the focusing lens, which caused not only an absolute calibration change, but also a change of relative signals between wavelength channels. This was probably because the fine particles on the lens scattered different wavelengths of light to different extents. After this failure, new gasket material was added to the system. Between runs, the probe is tested for such problems by viewing a standard quartz photographic lamp (600 W) that maintains a nearly constant color temperature over its lifetime. By comparing the relative magnitudes of the output for the wavelengths, any relative (wavelength-to-wavelength) variations will be seen.

CONCLUSIONS

These initial tests show the utility of in situ measurements of temperature and emissivity, and extension to the determination of thermal conductivity. A thermal conductivity significantly higher than that measured for deposits from either of the parent coals was measured for a deposit formed by an Alaskan/Russian coal blend. Continued in situ measurements combined with modeling efforts and evaluation of the structure of ash deposits should provide a better predictive ability to aid in fuel switching and blending. This could also be very useful in situations where ash deposits can vary dramatically in structure and composition. Particular possibilities include the combustion of biomass and refuse derived fuel.

The INSITE system would also be useful in other process industries in any situation where the surface temperature of a reflective body must be measured while it is being heated or cooled. The temperature of reflective materials during heat treating could be monitored continuously with such a system, and even materials with varying reflectivity could be measured.

ACKNOWLEDGMENTS

This work was carried out while the first author was a U.S. DOE Fossil Energy part-time faculty research participant at the Pittsburgh Energy Technology Center in a program administered by the Oak Ridge Institute for Science and Education.

Thanks also to Clement Lacher and Dennis Stanko for instrument operation and data collection as well as useful input on modified operating procedures and to Mark Freeman for assistance in organizing furnace temperature and heat flux data and suggestions for deposition probe modifications.

REFERENCES

Anderson, D.W., Viskanta, R. and Incropera, F.P. (1987). "Effective Thermal Conductivity of Coal Ash Deposits at Moderate to High Temperatures," *Eng. for Gas Turbines Power*, Vol. 109, pp. 215-221.

Boow, J. and Goard, P.R.C. (1969). "Fireside Deposits and their Effect on Heat Transfer in a Pulverized-fuel-fired Boiler, Part III: The Influence of the Physical Characteristics of the Deposit on Its Radiant Emittance and Effective Thermal Conductance," *J. Inst. Fuel*, Vol. 42, pp. 412-418.

Clark, G.A., Holmes, M.J., Vecci, S.J. and Bailey, R.T. (1989). "In Situ Measurements of Coal Ash Deposit Optical Properties," Final Report, DOE/PC/79902-T1 (DE90000641), August.

Goetz, G.J., Nsakala, N.Y. and Borio, R.W. (1979). "Development of Method for Determining Emissivities and Absorptivities of Coal Ash Deposits," *J. Eng. Power*, Vol. 101, No. 4, pp. 607-614.

Mulcahy, M.F.R., Boow, J. and Goard, P.R.C. (1966a). "Fireside Deposits and their Effect on Heat Transfer in a Pulverized-fuel-fired Boiler, Part I: The Radiant Emittance and Effective Thermal Conductance of the Deposits," *J. Inst. Fuel*, Vol. 39, pp. 388-396.

Mulcahy, M.F.R., Boow, J. and Goard, P.R.C., (1966b). "Fireside Deposits and their Effect on Heat Transfer in a Pulverized-fuel-fired Boiler, Part II: The Effect of the Deposit on Heat Transfer from the Combustion Chamber Considered as a Continuous Well-stirred Reactor," *J. Inst. Fuel*, Vol. 39, pp 394-398.

Ramer, E.R. and Martello, D.V. (1994). "Quantitative Microstructural Characterization of Ash Deposits from Pulverized Coal-Fired Boilers," *The Impact of Ash Deposition on Coal Fired Plants*, J. Williamson and F. Wigley (eds.), Taylor and Francis, Ltd., London, pp. 487-498.

Ramer, E.R. (1995). Private Communication

Shaw, D.W. and Smouse, S.M. (1994). "An Instrument for In Situ Temperature and Emissivity Measurements (INSITE) on Boiler Ash Deposits," *The Impact of Ash Deposition on Coal Fired Plants*, J. Williamson and F. Wigley (eds.), Taylor and Francis, Ltd., London, pp. 539-551.

Smouse, S.M. (1994). "To Clean or not to Clean? A Technical and Economic Analysis of Cleaning Pittsburgh Seam Coal," *The Impact of Ash Deposition on Coal Fired Plants*, J. Williamson and F. Wigley (eds.), Taylor and Francis, Ltd., London, pp. 189-218.

Smouse, S.M. and Walbert, G. (1990). "Design of a Pilot-Scale Fuels Evaluation Facility," *Mineral Matter and Ash Deposition from Coal*, R. W. Bryers and K.S. Vorres (eds.), United Engineering Trustees Inc., New York, pp. 585-600.

Wagoner, C.L., Haider, G., Berthold, J.W. and Wessel, R.A. (1984). "Measurement of Fundamental Properties Characterizing Coal Minerals and Fire-Side Deposits," Final Report, DOE/PC/40266-3 (ARC RDD:84:4696-01-01:03), October.

Wall, T.F., Lowe, A., Wibberley, L.J. and Stewart, I. McC. (1979). "Mineral Matter in Coal and the Thermal Performance of Large Boilers," *Prog. Energy Combus. Sci.*, Vol. 5, pp. 1-29.

Wall, T.F. and Becker, H.B. (1984). "Total Absorptivities and Emissivities of Particulate Coal Ash from Spectral Band Emissivity Measurements," *J. Eng. Gas Turbine Power*, Vol. 106, pp. 771-776.

Wall, T.F., Bhattacharya, S.P., Zhang, D.K., Gupta, R.P., and He, X. (1994). "The Properties and Thermal Effects of Ash Depositions in Coal Fired Furnaces: A Review," *The Impact of Ash Deposition on Coal Fired Plants*, J. Williamson and F. Wigley (eds.), Taylor and Francis, Ltd., London, pp. 463-478.

EROSION-OXIDATION OF CARBON STEEL
AND DEPOSITION OF PARTICLES ON A TUBE DURING
COMBUSTION OF COAL-BASED FUELS IN AN INDUSTRIAL BOILER

James J. Xie*† and Peter M. Walsh§

*Department of Materials Science and Engineering
Pennsylvania State University
University Park, PA 16802

§Energy and Environmental Research Corporation
c/o Sandia National Laboratories
Livermore, CA 94550

ABSTRACT

An *in situ* test was used to assess the erosivity of particles in the convection section of an industrial boiler during combustion of micronized coal and while cofiring coal-water fuel with natural gas. Erosion was accelerated using a small jet of clean gas to increase the velocities of ash and unburned char particles at the surface of a carbon steel coupon. Because the jet alters the velocity distribution of the particles, a simulation of the system including particle behavior in the jet, erosion of both tube metal and oxide scale, and scale formation was needed to estimate erosion rates at lower velocities. Although the particles formed from coal-water fuel were more erosive than the particles from micronized coal in the accelerated test, calculations showed that the opposite would be the case on an isolated 51 mm diameter tube at typical convection section velocities. The change in relative erosion rates was due to differences in the size distributions and impaction efficiencies of the particles (effects of gas velocity and size of target). An accurate model is therefore essential to meaningful application of the results of the accelerated test. The calculations indicated that erosion of tube material at 550 K would be slower than 0.05 μm/hour at convection section velocities less than 8 m/s while firing coal-water fuel with natural gas and at velocities less than 10 m/s during combustion of micronized coal. At these velocities, under the conditions of gas and particle composition investigated, erosion is expected to be most rapid on the upstream stagnation line of an isolated tube and to remove material only from the oxide layer, not from the underlying carbon steel.

Deposition of particles, rather than erosion by particles, was the dominant process observed at the low velocities (3 to 4 m/s) present in the convection section of the boiler, in the absence of the accelerating jet. Deposition of particles from micronized coal decreased with increasing tube temperature, suggesting that thermophoresis made an important contribution to particle transport. At a tube temperature of 450 K, deposition of particles from micronized coal on the upstream stagnation line of a tube was 300 times greater than deposition of the particles from coal-water fuel. Erosion of the deposits by

---

†Present address: National Semiconductor Corporation, 2900 Semiconductor Drive, Santa Clara, CA 95052.

Applications of Advanced Technology to Ash-Related Problems in Boilers
Edited by L. Baxter and R. DeSollar, Plenum Press, New York, 1996

large particles was the explanation proposed for this behavior. The particles from coal-water fuel were larger (mean size of 65 μm vs. 43 μm for the particles from micronized coal) and had higher approach velocity (4.0 m/s vs. 3.1 m/s). The deposits were easily removed, but the sootblowing required to control flue gas temperature and maintain steam flowrate might lead to tube erosion. Comparison of the two fuels, which had nearly identical ash compositions, demonstrated the importance of fuel preparation and combustion to erosion and deposition, through their effects on carbon burnout and particle size distribution. The interaction between erosion and deposition is one of the areas most in need of additional work.

INTRODUCTION

A study of erosion of heat exchanger surfaces by fly ash is part of an investigation of the prospects for increased utilization of coal as fuel for industrial boilers, being conducted in the demonstration boiler at the Penn State Energy and Fuels Research Center [Miller et al., 1995a, 1995b, 1995c]. The objective of the erosion work was to estimate, from measurements during tests lasting 1000 hours or less, the dependence of the rate of tube metal loss in the convection section of a boiler on fuel properties, combustion conditions, gas velocity, and metal temperature.

Long tube life is a requirement for reliable and economic operation of steam generators. Raask [1985, 1988] suggested that 0.05 μm/hour is a useful figure for the maximum acceptable rate of tube wastage, corresponding to loss of roughly 4 mm of wall thickness during 10 years of continuous service. For the purposes of an investigation of the influences of several variables, much more rapid erosion is needed. A method was devised to accelerate erosion of a test coupon mounted on a probe in the convection section of the boiler, a technique which permitted measurements to be made during tests lasting 2 hours, and allowed for variation of metal temperature and the composition of gas adjacent to the sample surface. The test coupons were cut from carbon steel tube, the material of choice for convection heat exchangers in packaged boilers. Measurements were made during combustion of coal-water fuel with natural gas and during combustion of micronized (finely ground) coal.

EXPERIMENTAL MEASUREMENTS

Combustor and Fuels

The probes were mounted in a superheater cavity in a 6760 kg steam/hour (at 481 K and 1.7 MPa) D-frame watertube boiler manufactured by Tampella Keeler (Williamsport, Pennsylvania), as shown in Fig. 1. The convection section of the boiler is generously-sized, has relatively low gas velocities, and is equipped with a sootblower. The low ash, low sulfur coal-water fuel and micronized coal were prepared using high volatile A bituminous coal from the Brookville seam in Lawrence County, Pennsylvania. Properties of the fuels are given in Table 1. The coal-water fuel was cofired with natural gas to improve flame stability. The liquid fuel spray burner (Faber Burners) was original equipment on the boiler [Miller et al., 1995c]. The micronized coal burner was retrofit by ABB Combustion Engineering (Windsor, Connecticut) [Jennings et al., 1994; Miller et al., 1995a; Sharifi and Scaroni, 1995]. Typical conditions during the tests are given in Table 2.

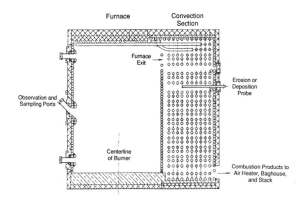

Figure 1. Plan of the boiler, showing the furnace and convection sections and the location of the erosion and deposition probes in a superheater cavity [Tampella Keeler, Williamsport, PA]. The length and width of the furnace (inside) are 2.5 and 1.8 m, respectively.

Size distributions of samples of particles separated from flue gas in the baghouse, downstream from the convection section, are shown in Fig. 2. The high combustible content of the particles, high silica and alumina in the ash, and low ratio of silica to alumina (Table 2), suggest that unburned char and aluminosilicates are the principal constituents of the particles, with char accounting for most of the particle volume. The higher unburned combustible content and larger particles produced during combustion of coal-water fuel are consequences of the agglomeration of coal particles during drying of the coal slurry spray droplets in the boiler flame.

Erosion of Carbon Steel

The test specimens were cut from 25.4 mm diameter, 2.8 mm wall, ASTM A179 carbon steel tube having the following composition (weight basis): C, 0.08%; Mn, 0.42%; P, 0.027%; S, 0.017%; Si, 0.030%; Al, 0.048%; and Fe, balance. A portion of the outer surface of each piece was ground and polished to produce a flat area 10 mm x 38 mm having roughness less than 0.05 μm. The specimens were mounted in a notch in an air-cooled support made from the same tubing. A small jet of gas was directed at the polished surface of the sample from a nozzle mounted on a sidearm attached to the tube. The jet gases were mixtures of oxygen and nitrogen from compressed gas cylinders, fed through a passage inside the probe. The axis of the jet was horizontal, parallel to the direction of flow in the convection section of the boiler. The clean gas issuing from the nozzle entrained the surrounding flue gas and particles, accelerating them toward the metal coupon. The arrangement of the nozzle and test specimen is shown in Fig. 3. Additional information about the erosion probe may be found in a separate publication [Walsh et al., 1994].

The erosion measurements reported here were obtained using nitrogen and air as the accelerating jet gases. The nitrogen contained 3 vol ppm oxygen. The initial average gas velocity at the nozzle was 200 m/s. Metal temperature, measured using thermocouples placed in the sample (Fig. 3) was set at 450, 550, or 650 K by adjusting the cooling air flowrate. Each test coupon was exposed to the jet and entrained combustion products for 2 hours. After removal from the convection section the shape of the surface was measured using a surface profiler (Rank Taylor Hobson, Leicester, UK; Talysurf 10) [Snaith et al., 1987]. An example of a profile is shown in Fig. 4a. Complete sets of profiles and electron micrographs of the samples may be found in previous publications [Walsh et al., 1994, Xie and Walsh, 1995]. The average of the two lowest

Table 1. Properties of the Coal-Based Fuels.

Fuel	Coal-Water Fuel	Micronized Coal
Coal rank	hvAb	hvAb
Proximate analysis (wt%, as received)		
Moisture	41.6	6.4
Ash	2.1	3.3
Volatile matter	21.1	33.5
Fixed carbon (difference)	35.2	56.8
Ultimate analysis (wt%, as received)		
Moisture	41.6	6.4
Ash	2.1	3.3
Carbon	46.6	74.5
Hydrogen	3.2	5.0
Nitrogen	0.9	1.5
Sulfur	0.5	0.7
Oxygen (difference)	5.1	8.6
Higher heating value (MJ kg^{-1}, as received)	19.3	31.0
Particle size distribution (volume fraction smaller than size)		
10 %	4 μm	6 μm
50 %	23 μm	22 μm
90 %	78 μm	62 μm

points (greatest material loss) in the profile, on opposite sides of the jet axis, was divided by the exposure time to obtain a time-averaged erosion rate. The dependence of the erosion rate on metal temperature for four combinations of fuel and the oxygen content of the jet are shown in Fig. 5. The erosion rate (1) increased with increasing temperature when the jet contained 3 mol ppm oxygen, but showed little dependence on temperature in the presence of the air jet; (2) decreased with increasing oxygen partial pressure at 550 and 650 K, but had negligible dependence on oxygen at 450 K; and (3) was greater during combustion of micronized coal than during coal-water fuel/natural gas cofiring, except at 550 K in the presence of the air jet, when erosion by particles formed from coal-water fuel was greater than erosion by the particles from micronized coal.

Deposition of Particles

In a separate set of measurements, a 25.4 mm diameter air-cooled carbon steel tube, without any sidearm and nozzle for accelerating particles, was placed across the convection section at the same location as the erosion probe. During periods of 2 to 6 hours, deposits of ash accumulated on the tube surface. Local average deposition rates

Table 2. Experimental Conditions.

Fuels	Coal-Water Fuel and Natural Gas	Micronized Coal
Firing rate (MW)		
Coal	3.8	4.4
Natural gas	1.7	0
Total	5.5	4.4
Carbon conversion (%)		
Coal	87	94
Natural gas	100 [a]	-
Overall	91	94
Gas composition at the boiler exit (volume fraction) [b]		
N_2 [b]	68.3 %	75.4 %
H_2O [b]	14.3 %	6.9 %
CO_2	13.6 %	14.2 %
O_2	3.7 %	3.4 %
CO	190 ppm	480 ppm
SO_2	430 ppm	430 ppm
NO_x	570 ppm	410 ppm
Conditions at the erosion probe		
Gas temperature (K)	850	890
Gas velocity [b] (m s^{-1})	4.0	3.1
Particle concentration [b] (kg m^{-3})	0.0027	0.0030
Particle apparent density (kg m^{-3})	870	835
Volume-based mean particle size (μm)	65	43
Analysis of particles from the baghouse (wt%, dry)		
Ash	23	37
Combustible (difference)	77	63
Composition of ash in particles from the baghouse (wt%, dry, SO_3-free)		
SiO_2	49.4	50.3
Al_2O_3	35.7	35.2
TiO_2	1.2	1.4
Fe_2O_3	7.3	6.6
CaO	2.7	3.2
MgO	0.6	0.7
Na_2O	0.5	0.4
K_2O	1.3	1.5
other	1.3	0.7

a. assumed
b. calculated

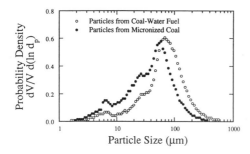

Figure 2. Volume-based size distributions of particles separated from flue gas in the baghouse, determined using laser diffraction (Malvern droplet and particle sizer). The mean sizes were 65 μm during coal-water fuel/natural gas firing and 43 μm during micronized coal firing.

were determined by scraping off and weighing the particles. At a tube temperature of 450 K, as shown in Fig. 6a, the distributions of deposit around the circumference of the tube were different for the two fuels, and heavier deposits were formed during micronized coal combustion than during coal-water fuel/natural gas combustion. Scanning electron micrographs showed that most of the deposits collected during combustion of coal-water fuel were ash particles having sizes between 0.5 and 10 μm, while the deposits collected during combustion of micronized coal contained ash and char particles as large as 20 μm. The deposition rate decreased with increasing tube temperature during micronized coal combustion (Fig. 6b), suggesting that thermophoresis contributed to transport of the particles. However, the measurements at 450 K show a minimum in deposition rate near the upstream stagnation line, where deposition by thermophoresis is expected to be greatest. This suggests that deposition was offset by erosion of the deposits, and that the observed deposit growth is the difference between the addition and removal rates. The competition between deposition and erosion is apparently quite sensitive to particle properties and convection section conditions. Attempts to make measurements of deposition at 550 and 650 K during coal-water fuel/natural gas combustion were not successful. The deposits under these conditions were uneven in thickness, apparently due to their spontaneous sloughing off.

EROSION-OXIDATION MODEL AND CALCULATIONS

The model is based on mechanisms of erosion-oxidation described by Levy [1982], Levy et al. [1985], Raask [1985, 1988], Wright et al. [1986, 1991], Kang et al. [1987], Sethi et al. [1987], Sethi and Wright [1989], Rishel et al. [1991], Sundararajan [1991a, 1991b], and Kosel [1992]. The oxidation of carbon steel and formation of oxide scale were assumed to occur at a rate inversely proportional to the thickness of the scale. An estimate of the parabolic rate coefficient was derived from published work [Davies et al., 1951; Schmahl et al., 1958; Caplan and Cohen, 1966; Jansson and Vannerberg, 1971]. However, the oxidation rate of carbon steel exposed to jets of oxygen/nitrogen mixtures at 450 to 650 K in the presence of erosion is uncertain because: (1) the rate coefficient was extrapolated from measurements at higher temperatures, (2) it is derived from measurements on iron in oxygen, (3) it is based on measurements during growth of relatively thick scale, (4) the rate of oxidation may be influenced by the effects of collisions and erosion on properties of the scale, and (5) the actual composition of the gas adjacent to the sample surface, after some mixing with flue gas, is not known. (If particles are able to diffuse into the interior of the jet, as shown in Fig. 3, then gaseous combustion products diffuse there as well.) Allowance was made for an oxygen concentration dependence of the rate, which was required to fit the measurements, by including a power of the oxygen partial pressure as a factor in the rate expression. The oxygen content of the jet nozzle fluid was used as an approximation to the oxygen

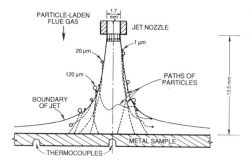

Figure 3.   Arrangement of the carbon steel specimen, thermocouples, and the gas jet used to accelerate the particles. Examples of calculated trajectories are shown for 1, 20, and 120 µm particles having the average particle density.

content of the gas adjacent to the surface at the point where particle collisions and erosion were simulated, 2 mm from the jet axis, near the minima in the erosion profile (Fig. 4a).

The composite erosion resistance of metal and scale was assumed to be a linear combination of the resistances of metal and oxide alone, weighted by their volume fractions in eroded fragments [Levy, 1982; Hovis et al., 1986; Conrad, 1987]. The contribution from metal was described by the relations for ductile erosion [Finnie 1958, 1960, 1980; Bitter, 1963a, 1963b; Finnie et al., 1967; Neilson and Gilchrist, 1968; Engel, 1978], greatest at a collision angle of 30°, having a rate at normal incidence equal to 0.3 times the maximum [Shida and Fujikawa, 1985], and increasing with increasing temperature. The contribution from oxide was described by an expression for brittle erosion derived by Xie [1995], from the work of Sheldon [1970] and Evans et al. [1976]. The rate of brittle erosion is proportional to the 3.4 th power of the normal component of particle velocity, greatest at a collision angle of 90°, and independent of temperature [Levy, 1982; Conrad, 1987]. Contributions to erosion resistance from deposits or isolated particles attached to the surface were neglected.

The polished carbon steel specimens were stored in ambient air and used without pretreatment. The rate of oxidation of a fresh specimen increases on exposure to oxygen in the jet gas and combustion products at high temperature, but decreases as the oxide layer grows. During this transient period, the scale approaches the steady average thickness at which its rates of formation from metal and removal by erosion are equal. The volume fractions of metal and scale in eroded fragments were determined from a material balance at steady state, in which the product of the collision frequency of the volume-weighted average size of particles and average volume of oxide removed per collision is equated to the rate of oxide formation. When particles penetrate the oxide scale to metal it is necessary to specify the area of scale removed by a single impact.

There were five adjustable parameters in the erosion-oxidation model: (1) the effective order of the parabolic oxidation rate with respect to oxygen, (2) the preexponential factor in the coefficient for the erosivity of the particles toward metal, (3) the activation temperature describing the temperature dependence of the erosivity of particles toward metal, (4) the erosivity of the particles toward oxide, and (5) the ratio of the area of oxide removed per collision to the cross section area of the mean size of particles. Kinetic energies and collision angles of the impacting particles were needed to extract values for these parameters from the measurements of maximum depth in the erosion profiles.

The trajectories of particles after entering the accelerating jet from the free stream were found from the equations of motion for the particles and gas in an axisymmetric, incompressible stagnation point flow, with the boundary condition of no slip at the target surface [Laitone, 1979a, 1979b; Dosanjh and Humphrey, 1985; Schuh et al., 1989, Humphrey, 1990]. Particles were subjected to mean and fluctuating components of gas

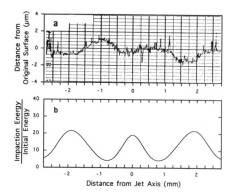

Figure 4. a. Profile of erosion, deposition, and scale formed on the surface of carbon steel during 2 hours of exposure to particles from micronized coal, accelerated by a small jet of air, as shown in Fig. 3. The metal temperature was 450 K. Raised portions of the surface were formed by oxidation or particle deposition; depressed portions of the profile show the effect of erosion. b. Particle kinetic energy at the surface of the sample as a function of distance from the jet axis. The energy is shown as a multiple of the kinetic energy of particles in the free stream. The eroded regions in Fig. 4a correspond approximately to the peaks in the kinetic energy distribution.

velocity, with the standard deviations of the normally distributed fluctuating components proportional to the local mean axial and radial velocities. Some examples of particle trajectories are shown in Fig. 3. Additional information about the calculations may be found in other publications [Xie, 1995; Xie and Walsh, 1993]. An alternative boundary condition (no radial flow at the nozzle) yields a distinctly different shape of streamlines [Xie, 1995], expected to increase the calculated frequency of collisions in the vicinity of the jet axis.

The highest velocity at the target was approximately 20 m/s, for 20 µm particles. Larger particles are less strongly influenced by the jet; smaller particles follow the gas flow more closely. Very large particles arrive at the surface having velocities near the approach velocity, while very small particles have velocities approaching zero at the surface, the boundary condition for the gas flow. Therefore, although the initial velocity of the jet at the nozzle is very high, the velocities of particles at the metal sample surface were not excessive. However, the distribution of velocity with respect to particle size produced by the jet is different from the velocity-size distribution of particles colliding with a tube under normal convection section conditions, in the absence of the jet. The spatial distribution of the kinetic energy of particles at the sample surface is shown in Fig. 4b. The average size and velocity of particles colliding with the target were calculated at points 2 mm from the stagnation point of the jet, where the total kinetic energy of particles at the surface was a maximum. These particle properties were used in conjunction with the erosion-oxidation model to simulate the processes occurring at the locations on the surface where the rate of erosion was a maximum. The effects of differences in density, shape, size, and erosivity of char and ash particles were neglected.

The particle properties and erosion parameters used to fit the experimental measurements are given in Table 3. The value of 25 for the ratio of area of oxide removed to particle cross section area is unrealistically large. This is especially so, considering that the calculated steady average thickness of oxide on carbon steel at 550 K in the presence of the air jet, subject to erosion by particles formed from coal-water fuel, was only 0.06 nm, less than the thickness of a monolayer of magnetite. The ratio of areas is expected to be of order 1. An alternative approach is to set the ratio of area of oxide removed to particle cross section area equal to 1, then adjust the preexponential factor in the parabolic oxidation rate coefficient to fit the data. When this was done, the oxidation rate was a factor of 25 faster than the rate estimated from the published work on iron. The result is independent of the choice of area ratio and oxidation rate, as long as their product is kept

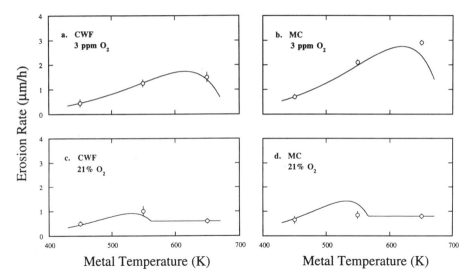

Figure 5. Examples of the temperature dependence of the measured maximum rates of erosion of carbon steel under the influence of the accelerating gas jet and entrained particles, as functions of the metal temperature. Left: coal-water fuel/natural gas, Right: micronized coal; Top: 3 vol ppm oxygen in the jet gas, Bottom: 21 vol % oxygen in the jet gas. The curves were obtained using the erosion-oxidation model, after adjusting the order of the oxygen dependence in the parabolic oxidation rate, the erosivities of the particles, and the average area of the surface from which oxide was removed by a single particle impact, when particles are able to penetrate the scale to metal.

the same. The extent to which the model and empirically determined parameters are able to reproduce the experimental measurements is shown in Fig. 5.

According to the calculations, over most of the temperature range investigated (above 475 K when firing coal-water fuel with natural gas, above 460 K when firing micronized coal), oxide scale is more resistant to erosion than carbon steel. The contribution of scale to the combined erosion resistance of scale and metal increases with scale thickness, reaching its maximum when particles are not able to penetrate the scale to underlying metal. The steady average thickness of scale increases with increasing oxygen concentration or temperature. At low temperature and low oxygen partial pressure, when the oxidation rate coefficient is low, erosion is controlled by the ductile metal, whose rate of erosion increases with increasing temperature, the behavior shown over most of the range of temperatures in Figs. 5a and 5b. At high temperature and high oxygen partial pressure, erosion is controlled by brittle oxide, whose erosion is independent of temperature as shown on the right in Figs. 5c and 5d. The portions of the curves in Fig. 5 having negative slope are the regions in which increasing temperature causes a transition from thin scale to thick, changing the erosion mechanism from largely ductile to completely brittle.

The approach velocity dependence of erosion on the upstream stagnation line of an isolated tube was estimated using the average particle size and average impaction efficiency on a tube in crossflow [Serafini, 1954; Israel and Rosner, 1983; Walsh et al., 1988]. The results for coal-water fuel/natural gas and micronized coal are shown in Fig.

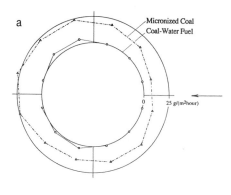

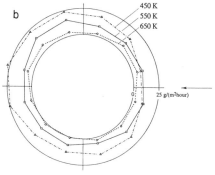

Figure 6. Measurements of particle deposition on an air-cooled tube exposed to the convection section flow during coal-water fuel/natural gas and micronized coal combustion. a. Comparison of net deposition rates during coal-water fuel/natural gas and micronized coal combustion with tube temperature at 450 K. The average gas velocities were 4.0 m/s (coal-water fuel/natural gas) and 3.1 m/s (micronized coal). b. The dependence of deposition rate on tube temperature during micronized coal combustion. The average gas velocity was 3.1 m/s.

7. At low velocity, where erosion is slow, the steady thickness of scale is large enough that particles remove only oxide. At a critical velocity, 12 m/s for coal-water fuel/natural gas and 16 m/s for micronized coal, particles begin to penetrate the scale to metal, and the erosion rate increases rapidly with further increase in velocity, approaching the rate characteristic of metal at high velocities as the thickness of scale approaches zero. Erosion on the upstream stagnation line of an isolated tube, by the particles formed from coal-water fuel and micronized coal, is expected to be slower than 0.05 μm/hour at convection section velocities below 8 and 10 m/s, respectively. According to the calculation, this rate falls in the oxide-controlled regime, in which erosion is greatest at normal incidence (brittle mechanism), on the upstream stagnation line of the tube.

Local impaction efficiencies, kinetic energies, and collision angles were used to estimate local rates of wastage around the circumference of an isolated tube, as shown in Fig. 8. The calculations of particle properties here are more accurate than those used to obtain the results shown in Fig. 7, in which particles were described by properties averaged over the circumference of the tube. For the calculations shown in Fig. 8 (and Fig. 9) the ratio of the area of oxide removed to particle cross section area was set equal to 25 and the metal oxidation rate set equal to the value derived from the published work on iron. Effects of changes in the area ratio and parabolic rate coefficient were not examined in this case. When the approach velocity is low and the oxide layer relatively thick, erosion is most rapid at normal incidence, as described above. As approach velocity is increased, the penetration of scale to metal occurs first on the stagnation line, where particles are most effective in removing oxide. With further increase in velocity the erosion pattern near the stagnation line first broadens, then the location of maximum erosion moves away from the stagnation line, producing two maxima, as the contribution from ductile erosion increases. At high velocity, when the scale is very thin, the highest rates of wastage occur where the collision angle of the average size of impacting particles is 30° [Bauver et al., 1984]. Other patterns appear when the approach velocity and concentration of particles are not uniform [Fan et al., 1990].

Calculations of the tube cross sections expected under conditions at which material loss occurs by the brittle, ductile, and combined mechanisms are shown in Fig. 9. This calculation translates the results of the accelerated erosion measurements into a form which can be tested by comparison with observations in the field.

Table 3. Values of the Parameters Used to Fit the Accelerated Erosion Measurements.

Fuels	Coal-Water Fuel and Natural Gas	Micronized Coal
Distance from jet axis (mm)	2	2
Particle kinetic energy flux ($J\ m^{-2}\ s^{-1}$)	0.57	0.55
Average collision angle (degree)	83	81
Mean size of impacting particles (μm)	78	51
Impaction efficiency	0.64	0.59
Collision frequency ($m^{-2}\ s^{-1}$)	32×10^6	95×10^6
Rate coefficient for erosion of metal ($m^3\ J^{-1}$)	2.0×10^{-7} $\cdot exp(-2530/T)$ [a]	2.9×10^{-7} $\cdot exp(-2530/T)$ [a]
Rate coefficient for erosion of oxide ($m^2\ kg^{0.5}\ J^{-1.7}$)	3.5×10^{-5}	7.7×10^{-5}
Ratio of area of oxide removed to particle cross section area	25 [b]	25 [b]
Parabolic rate coefficient for oxidation of metal ($m^2\ s^{-1}$)	$1.6 \times 10^{-5}\ P_{O_2}^{1/2}$ $\cdot exp(-19240/T)$ [b,c]	$1.6 \times 10^{-5}\ P_{O_2}^{1/2}$ $\cdot exp(-19240/T)$ [b,c]

a. T is the temperature in Kelvin.
b. If the area ratio is assumed to be 1, the same results (Figs. 5 and 7) are obtained by increasing the oxidation rate coefficient by a factor of 25.
c. T is the temperature in Kelvin, P_{O_2} is the partial pressure of oxygen in atmospheres.

DISCUSSION

Comparison of Fuels and Importance of Combustion Conditions

In the accelerated erosion test, the erosivities of the particles formed during micronized coal combustion were greater than the erosivities of the particles derived from coal-water fuel. As shown in Figs. 5a and 5b, in the metal-controlled erosion regime, at low oxygen and low temperature, the ratio of the erosion rates for the two types of particles is approximately equal to the ratio of their ductile erosion rate coefficients. This is not the case in the oxide-controlled regime, at high oxygen and high temperature (Figs. 5c and 5d), because the rate of brittle erosion depends upon particle size. The higher erosion rate coefficient of the particles produced during micronized coal combustion was partly offset by the effect of their smaller size.

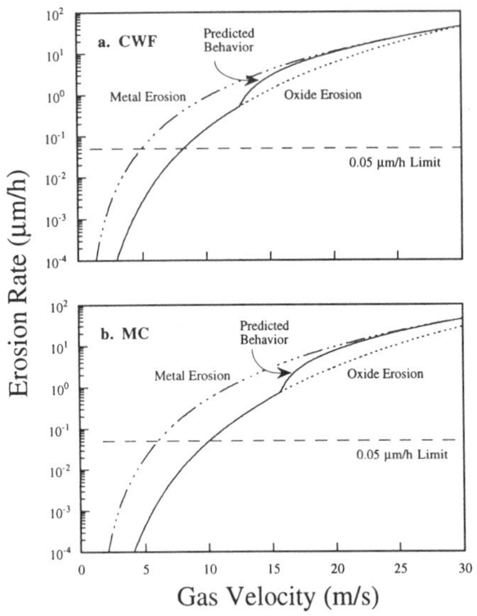

Figure 7. Estimate, using the model, of the velocity dependence of the erosion rate on the upstream stagnation line of an isolated carbon steel tube by particles suspended in the products from coal-water fuel/natural gas and micronized coal combustion, under the conditions specified in Table 2. The limit of 0.05 μm/h is the maximum acceptable erosion rate suggested by Raask [1985, 1988]. Particle deposition was not included in the model. Deposition rather than erosion was the dominant process occurring at velocities of 3 to 4 m/s, as shown in Fig. 6.

According to the calculations, the relative rates of erosion by the two types of particles on a 51 mm diameter tube are reversed, as shown in Fig. 7. In this case, where particle Stokes numbers and impaction efficiencies are lower than in the accelerated erosion test, the smaller size of the particles produced from micronized coal outweighs their higher erosion rate coefficients in both the metal and oxide-controlled regimes. At a velocity of 10 m/s, erosion on the upstream stagnation line of an isolated tube by the micronized coal-derived particles was estimated to be a factor of 3 slower than erosion by particles produced from the coal-water fuel.

Because the particles entrained in the flue gas contain high proportions of unburned carbon (Table 2) and because the apparent density of char is much lower than the apparent density of ash, the volume-based size distributions of the particles in flue gas (Fig. 2) are determined, to a large extent, by the size distributions of the micronized coal feed and coal-water fuel spray droplets and the extent of burnout of their char residues. Therefore, erosion of the convective heat exchanger is expected to depend upon the aspects of fuel preparation, burner design, and combustion conditions which influence carbon conversion [Hargrove et al., 1985, 1987; Chow et al., 1992].

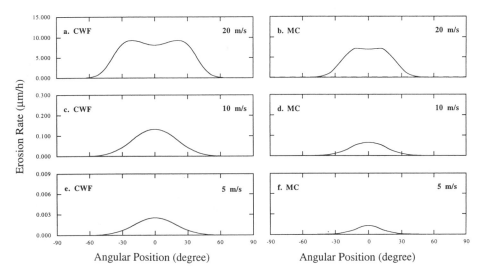

Figure 8. Calculated local rates of metal wastage as a function of position around the circumference of a carbon steel tube under conditions in the convection section of the boiler. Left: during combustion of the coal-water fuel with natural gas; Right: during combustion of the micronized coal. The effect of particle deposition (Fig. 6) was not included in these calculations.

Sources of Uncertainty and Importance of Deposition

The fact that the calculations indicate that the relative erosivities of two types of particles are reversed on extrapolation from the accelerated erosion test to practical conditions highlights the problems inherent in using the results of accelerated tests as indicators of performance in the field. If erosion must be accelerated to be studied, then an accurate model is required to interpret and apply the results, especially when the situation is as complicated as that in a coal-fired boiler. Some sources of error and uncertainty have been mentioned: the description of particle trajectories in the jet; the carbon steel oxidation rate; extent of mixing of flue gas with the jet; the need to account for joint distributions of particle size, density, shape, hardness, and velocity; the existence of other erosion mechanisms, such as spalling; the structure and composition of the surface and their temperature dependence; the distribution of particles over the convection section flow; and the competition between erosion and deposition.

In the absence of the accelerating jet, the dominant process observed at the tube surface was ash deposition. However, the deposition rate was so dependent upon particle size distribution and gas velocity that an increase in mean particle size from 43 µm (particles from micronized coal) to 65 µm (particles from coal-water fuel) and simultaneous increase in velocity from 3.1 to 4.0 m/s, reduced deposition on the upstream stagnation line by a factor of 300. In the case of coal-water fuel/natural gas, the thickness of the deposit near the upstream stagnation line was approximately 0.5 µm, less than a monolayer of the mean size of particles, so erosion of tube material might occur at this location. The effect of a layer of particles in inhibiting tube erosion was not considered in the extrapolation of the accelerated erosion measurements to lower velocities. In

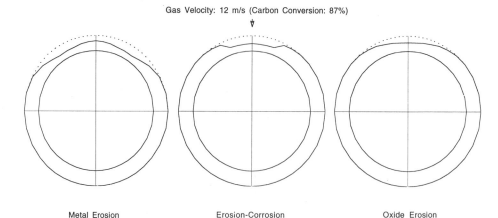

Gas Velocity: 12 m/s (Carbon Conversion: 87%)

| Metal Erosion | Erosion-Corrosion | Oxide Erosion |
| (1500 hours) | (3000 hours) | (6000 hours) |

Figure 9. Tube cross sections expected after long-term exposure to products from coal-water fuel/natural gas combustion at a convection section velocity of 12 m/s. The profiles on the left and right were calculated under the assumptions that conditions are weakly oxidizing (erosion of bare metal) and strongly oxidizing (erosion of thick oxide), respectively. The profile in the middle was obtained for intermediate conditions, under which particles are able to remove both oxide and metal simultaneously.

practice, the high deposition rates of particles on the convective tubes required the operators of the boiler to use the sootblower to control flue gas temperature and maintain steam flowrate. Erosion of tubes by particles entrained in the high velocity steam jets from the sootblower [Ots, 1994] may be more important than erosion during normal operation.

CONCLUSIONS

Formation of deposits and erosion of tube material by fly ash and unburned char were measured in the convection section of an industrial boiler while cofiring coal-water fuel with natural gas and while firing micronized coal. Particle deposition was the dominant process observed at the low gas velocities (3 to 4 m/s) in the convection section of the boiler under normal conditions, but erosion of tube material by particles is expected at higher gas velocities. Erosion of carbon steel by particles suspended in the flue gas in the convection section was measured during tests lasting 2 hours, using a small jet of clean gas to increase the velocity of particles colliding with the surface of a test specimen. Because the jet alters the distribution of velocity with respect to particle size, a model including the impaction, erosion, and metal oxidation processes was needed to extrapolate the results to lower velocities.

The transition from oxide erosion to metal erosion, with decreasing temperature, decreasing oxygen concentration, or increasing velocity, was accompanied by a marked increase in the erosion rate. Using experimentally determined values for the erosion rate coefficients of metal and oxide, the model for simultaneous erosion and oxidation provided estimates of the erosion rate as a function of gas velocity. The sensitivity to particle size and velocity is such that the calculated relative rates of erosion by particles formed from coal-water fuel and micronized coal under normal conditions were the opposite of the relative rates observed in the accelerated test. Under the conditions of metal temperature, flue gas oxygen concentration, particle composition, particle size distribution, and particle concentration investigated, erosion of carbon steel is expected to be slower than 0.05 µm/hour when gas velocity in the convection section is less than approximately 8 m/s during coal-water fuel/natural gas combustion and less than approximately 10 m/s during micronized coal combustion.

The best use of the model, at present, is not to provide estimates of erosion, because the uncertainty in the calculations is very large; but to identify limitations on our ability to make reliable estimates. Among the principal problems are: (1) the difficulty of accounting rigorously for the effects of joint distributions of particle size and velocity, (2) the particles must be separated into individual components (quartz, aluminosilicates, iron oxides, char, etc.), with separate size distributions, densities, and erosion rate coefficients for each, (3) uncertainty in the rate of oxide formation on carbon steel in the presence of flue gas at low temperature, under the influence of particle collisions, when the scale is very thin, (4) more detailed analysis of the erosion of oxide on metal is needed, with extension to other erosion regimes, such as spalling, (5) particles are not uniformly distributed across the flow, after turning from the furnace into the convection section and passage through one or more rows of tubes, (6) the structure of the surface is complex, it consists of more than just two materials ("metal" and "oxide"), and is temperature-dependent, and (7) interaction between erosion and deposition.

ACKNOWLEDGMENTS

This work was supported by the Commonwealth of Pennsylvania/Penn State Coal Utilization Program, and by the U.S. Department of Energy/Pittsburgh Energy Technology Center, under Cooperative Agreement No. DE-FC22-92PC92162. The Project Officer of the US DOE/PETC program is John C. Winslow. The authors thank the Director of the Penn State Energy and Fuels Research Center, Alan W. Scaroni; Associate Directors Bruce G. Miller and Sarma V. Pisupati; and staff members David A. Bartley, Scott A. Britton, Carl J. Martin, Roger L. Poe, Ronald S. Wasco, and Ronald T. Wincek for valuable contributions to the work.

REFERENCES

Bauver, W. P., Bianca, J. D., Fishburn, J. D., and McGowan, J. G. (1984). "Characterization of Erosion of Heat Transfer Tubes in Coal Fired Power Plants." Paper No. 84-JPGC-FU-3, New York: American Society of Mechanical Engineers.

Bitter, J. G. A. (1963a). "A Study of Erosion Phenomena, Part I." *Wear*, *6*, 5-21.

Bitter, J. G. A. (1963b). "A Study of Erosion Phenomena, Part II." *Wear*, *6*, 169-190.

Caplan, D., and Cohen, M. (1966). "Effect of Cold Work on the Oxidation of Iron from 400-650°C." *Corrosion Science*, *6*, 321-335.

Chow, O. K., Nsakala, N., Hargrove, M. J., and Levasseur, A. A. (1992). "Fireside Combustion Performance Evaluation of Beneficiated Coal-Based Fuels." Paper No. 92-JPGC-FACT-1, New York: American Society of Mechanical Engineers.

Conrad, H. (1987). "Erosion of Ceramics." A. V. Levy (Ed.), *Proceedings of the Conference on Corrosion-Erosion-Wear of Materials at Elevated Temperatures*. Houston: National Association of Corrosion Engineers, pp. 77-94.

Davies, M. H., Simnad, M. T., and Birchenall, C. E. (1951). *Transactions of the American Institute of Mining and Metallurgical Engineers*, *191*, 889-896.

Dosanjh, S., and Humphrey, J. A. C. (1985). "The Influence of Turbulence on Erosion by a Particle-Laden Fluid Jet." *Wear*, *102*, 309-330.

Engel, P. A. (1978). *Impact Wear of Materials*. Amsterdam: Elsevier.

Evans, A. G., Gulden, M. E., Eggum, G. E., and Rosenblatt, M. (1976). "Impact Damage in Brittle Materials in the Plastic Response Regime." Report No. SC5023.9TR, Thousand Oaks, CA: Rockwell International Science Center.

Fan, J., Zhou, D., Cen, K., and Jin, J. (1990). "Numerical Prediction of Tube Row Erosion by Coal Ash Impaction." *Chemical Engineering Communications*, *95*, 75-88.

Finnie, I. (1958). "The Mechanism of Erosion of Ductile Metals." R. M. Haythornthwaite (Ed.), *Proceedings of the Third U.S. National Congress of Applied Mechanics*. New York: American Society of Mechanical Engineers, pp. 527-532.

Finnie, I. (1960). "Erosion of Surfaces by Solid Particles." *Wear*, *3*, 87-103.

Finnie, I. (1980). "The Mechanisms of Erosive Wear in Ductile Metals." K. Natesan (Ed.), *Corrosion-Erosion Behavior of Materials*. New York: The American Institute of Mining, Metallurgical, and Petroleum Engineers, pp. 118-126.

Finnie, I., Wolak, J., and Kabil, Y. (1967). "Erosion of Metals by Solid Particles." *Journal of Materials*, *2*, 682-700.

Hargrove, M. J., Levasseur, A. A., Miemiec, L. S., Griffith, B. F., and Chow, O. K. (1985). "Performance Characteristics of Coal-Water Fuels and their Impact on Boiler Operation." *Proceedings of the Seventh International Symposium on Coal Slurry Fuels Preparation and Utilization*. Washington: Coal & Slurry Technology Association, pp. 855-872.

Hargrove, M., Kwasnik, A., Grinzi, F., and Romani, G. (1987). "Boiler Performance Predictions and Field Test Results Firing CWF in an Industrial Boiler." B. A. Sakkestad (Ed.), *Proceedings of the Twelfth International Conference on Slurry Technology*. Washington: Coal & Slurry Technology Association, pp. 393-399.

Hovis, S. K., Talia J. E., and Scattergood, R. O. (1986). "Erosion in Multiphase Systems." *Wear*, *108*, 139-155.

Humphrey, J. A. C. (1990). "Fundamentals of Fluid Motion in Erosion by Solid Particle Impact." *International Journal of Heat and Fluid Flow*, *11*, 170-195.

Israel, R., and Rosner, D.E. (1983). "Use of a Generalized Stokes Number to Determine the Aerodynamic Capture Efficiency of Non-Stokesian Particles from a Compressible Gas Flow." *Aerosol Science and Technology*, *2*, 45-51.

Jansson, L., and Vannerberg, N.-G. (1971). "The Effect of the Oxygen Pressure and the Growth of Whiskers on the Oxidation of Pure Fe." *Oxidation of Metals*, *3*, 453-461.

Jennings, P. L., Borio, R. W., Miller, B. G., Scaroni, A. W., and McGowan, J. G. (1994). "Installation and Intial Testing of Micronized Coal in a Gas/Oil-Designed Package Boiler." *Proceedings of the 19th International Technical Conference on Coal Utilization & Fuel Systems*. Washington: Coal and Slurry Technology Association, p. 63.

Kang, C. T., Chang, S. L., Pettit, F. S., and Birks, N. (1987). "Synergism in the Degradation of Metals Exposed to Erosive High Temperature Oxidizing Atmospheres." A. V. Levy (Ed.), *Proceedings of the Conference on Corrosion-Erosion-Wear of Materials at Elevated Temperatures*. Houston: National Association of Corrosion Engineers, pp. 61-76.

Kosel, T. H. (1992). "Solid Particle Erosion." P. J. Blau, S. D. Henry, G. M. Davidson, T. B. Zorc, and D. R. Levicki (Eds.), *ASM Handbook, Volume 18, Friction, Lubrication, and Wear Technology*. ASM International, pp. 199-213.

Laitone, J. A. (1979a). "Aerodynamic Effects in the Erosion Process." *Wear*, *56*, 239-246.

Laitone, J. A. (1979b). "Erosion Prediction near a Stagnation Point Resulting from Aerodynamically Entrained Solid Particles." *Journal of Aircraft*, *16*, 809-814.

Levy, A. V. (1982). "The Erosion of Metal Alloys and their Scales." A. V. Levy (Ed.), *Proceedings of the Conference on Corrosion-Erosion-Wear of Materials in Emerging Fossil Energy Systems*. Houston: National Association of Corrosion Engineers, pp. 298-376.

Levy, A. V., Yan, J., and Patterson, J. (1985). "Elevated Temperature Erosion of Steels." K. Ludema (Ed.), *Proceedings of the International Conference on Wear of Materials*. New York: American Society of Mechanical Engineers, pp. 708-716.

Miller, B. G., Bartley, D. A., Poe, R. L., and Scaroni, A. W. (1995a). "A Comparison between Firing Coal-Water Slurry Fuel and Dry, Micronized Coal in an Oil-Designed Industrial Watertube Boiler." *Proceedings of the 20th International Technical Conference on Coal Utilization & Fuel Systems*. Washington: Coal & Slurry Technology Association, pp. 267-278.

Miller, B. G., Bartley, D. A., Poe, R. L., and Scaroni, A. W. (1995b). "The Development of Coal-Based Technologies for Department of Defense Facilities." *Proceedings of the Eleventh Annual Coal Preparation, Utilization, and Environmental Control Contractors Conference*, Pittsburgh, PA, July 12-14.

Miller, B. G., et al. (1995c). "The Development of Coal-Based Technologies for Department of Defense Facilities." Phase I Final Report, prepared for the U. S. Department of Energy, Pittsburgh Energy Technology Center under Cooperative Agreement No. DE-FC22-92PC92162, draft submitted.

Neilson, J. H., and Gilchrist, A. (1968). "Erosion by a Stream of Solid Particles." *Wear*, *11*, 111-143.

Ots, A. (1994). "The Influence of Cleaning on Corrosive-Erosive Wear of Steam Boiler Heating Surface Tubes." J. Williamson and F. Wigley (Eds.), *The Impact of Ash Deposition on Coal-Fired Plant*. London: Taylor and Francis, pp. 759-766.

Raask, E. (1985). *Mineral Impurities in Coal Combustion*. Washington: Hemisphere Publishing Corporation, pp. 361-396.

Raask, E. (1988). *Erosion Wear in Coal Utilization*. Washington: Hemisphere Publishing Corporation, pp. 393-456 and 547-554.

Rishel, D. M., Pettit, F. S., and Birks, N. (1991). "Some Principal Mechanisms in the Simultaneous Erosion and Corrosion Attack of Metals at High Temperature." A. V. Levy (Ed.), *Proceedings of the Conference on Corrosion-Erosion-Wear of Materials at Elevated Temperatures*. Houston: National Association of Corrosion Engineers, pp. 16-1 to 16-23.

Schmahl, N. G., Baumann, H., and Schenck, H. (1958). "Die Temperaturabhängigkeit der Verzunderung von reinem Eisen in Sauerstoff." *Archiv für das Eisenhüttenwesen, 29*, 83-88.

Schuh, M. J., Schuler, C. A., and Humphrey, J. A. C. (1989). "Numerical Calculation of Particle-Laden Gas Flows Past Tubes." *AIChE Journal, 35*, 466-480.

Serafini, J. S. (1954). "Impingement of Water Droplets on Wedges and Double-Wedge Airfoils at Supersonic Speeds." Report No. 1159, Washington: National Advisory Committee for Aeronautics.

Sethi, V. K., Corey, R. G., and Spencer, D. K. (1987). "Wear Corrosion Synergism Studies." A. V. Levy (Ed.), *Proceedings of the Conference on Corrosion-Erosion-Wear of Materials at Elevated Temperatures*. Houston: National Association of Corrosion Engineers, pp. 329-344.

Sethi, V. K., and Wright, I. G. (1989). "Observations on the Erosion-Oxidation Behavior of Alloys." V. Srinivasan and K. Vedula (Eds.), *Corrosion & Particle Erosion at High Temperatures*. The Minerals, Metals & Materials Society, pp. 245-263.

Sharifi, R., and Scaroni, A. W. (1995). "Comparative Analysis of Two Commercial, Low-NO_x Pulverized Coal Swirl Burners." Joint Technical Meeting, Central States/Western States/Mexican National Sections of the Combustion Institute and the American Flame Research Committee, San Antonio, TX, April 23-26, Paper No. 95S-005.

Sheldon, G. L. (1970). "Similarities and Differences in the Erosion Behavior of Materials." Transactions of the ASME, *Journal of Basic Engineering*, Series D, *92*, 619-626.

Shida, Y., and Fujikawa, H. (1985). "Particle Erosion Behavior of Boiler Tube Materials at Elevated Temperature." *Wear, 103*, 281-296.

Snaith, B., Witton, J. J., and Wright, I. G. (1987). "Progress in Surface Topographical Analysis of Erosion Craters." A. V. Levy (Ed.), *Proceedings of the Conference on Corrosion-Erosion-Wear of Materials at Elevated Temperatures*. Houston: National Association of Corrosion Engineers, pp. 359-366.

Sundararajan, G. (1991a). "The Solid Particle Erosion of Metallic Materials at Elevated Temperatures." A. V. Levy (Ed.), *Proceedings of the Conference on Corrosion-Erosion-Wear of Materials at Elevated Temperatures*. Houston: National Association of Corrosion Engineers, pp. 11-1 to 11-33.

Sundararajan, G. (1991b). "An Analysis of the Erosion-Oxidation Interaction Mechanisms." *Wear*, *145*, 251-282.

Walsh, P. M., Beér, J. M., and Sarofim, A. F. (1988). "Estimation of Aerodynamic Effects on Erosion of a Tube by Ash." *Proceedings: Effects of Coal Quality on Power Plants*. EPRI CS-5936-SR, Palo Alto: Electric Power Research Institute, pp. 2-19 to 2-34.

Walsh, P. M., Xie, J., Poe, R. L., Miller, B. G., and Scaroni, A. W. (1994). "Erosion by Char and Ash and Deposition of Ash on Carbon Steel in the Convective Section of an Industrial Boiler." *Corrosion*, *50*, 82-88.

Wright, I. G., Nagarajan, V., and Stringer, J. (1986). "Observations on the Role of Oxide Scales in High-Temperature Erosion-Corrosion of Alloys." *Oxidation of Metals*, *25*(3/4), 175-199.

Wright, I. G., Sethi, V. K., and Nagarajan, V. (1991). "An Approach to Describing the Simultaneous Erosion and High-Temperature Oxidation of Alloys." Transactions of the ASME, *Journal of Engineering for Gas Turbines and Power*, *113*, 616-620.

Xie, J. (1995). "Erosion of Heat Exchanger Tubes in the Convective Section of an Industrial Boiler by Products of Coal Combustion." Ph.D. Dissertation, Department of Materials Science and Engineering, Pennsylvania State University, University Park.

Xie, J., and Walsh, P. M. (1993). "Erosion-Corrosion of Carbon Steel in the Convection Section of an Industrial Boiler Cofiring Coal-Water Fuel and Natural Gas." Paper No. 93-JPGC-PWR-23, New York: American Society of Mechanical Engineers.

Xie, J., and Walsh, P. M. (1994). "Erosion-Corrosion of Carbon Steel in the Convection Section of an Industrial Boiler." *The Impact of Ash Deposition on Coal-Fired Plants*. J. Williamson and F. Wigley (Eds.), London: Taylor and Francis, pp. 735-746.

Xie, J., Walsh, P. M., Miller, B. G., and Scaroni, A. W. (1994). "Evaluation of Erosion-Oxidation and Ash Deposition in the Convective Section of an Industrial Watertube Boiler Retrofitted to Fire Coal-Water Fuel." *Proceedings of the 19th International Technical Conference on Coal Utilization & Fuel Systems*. Washington: Coal & Slurry Technology Association, pp. 721-732.

Xie, J., and Walsh, P. M. (1995). "Erosion-Corrosion of Carbon Steel by Products of Coal Combustion." *Wear*, *186-187*, 256-265.

LABORATORY MEASUREMENTS OF ALKALI METAL CONTAINING VAPORS RELEASED DURING BIOMASS COMBUSTION

David C. Dayton and Thomas A. Milne

National Renewable Energy Laboratory
Industrial Technologies Division
1617 Cole Boulevard
Golden, CO 80401-3393
Phone: (303) 384-6216 FAX: (303) 384-6103
e-mail: daytond@tcplink.nrel.gov

ABSTRACT

Alkali metals, in particular potassium, have been implicated as key ingredients for enhancing fouling and slagging of heat transfer surfaces in power generating facilities that convert biomass to electricity. When biomass is used as a fuel in boilers, the deposits formed reduce efficiency, and in the worst case lead to unscheduled plant downtime. Blending biomass with other fuels is often used as a strategy to control fouling and slagging problems. Depending on the combustor, sorbents can be added to the fuel mixture to sequester alkali metals. Another possibility is to develop methods of hot gas cleanup that reduce the amount of alkali vapor to acceptable levels. These solutions to fouling and slagging, however, would greatly benefit from a detailed understanding of the mechanisms of alkali release during biomass combustion. Identifying these alkali vapor species and understanding how these vapors enhance deposit formation would also be beneficial.

Our approach is to directly sample the hot gases liberated from the combustion of small biomass samples in a variable-temperature quartz-tube reactor employing a molecular beam mass spectrometer (MBMS) system. We have successfully used this experimental technique to identify alkali species released during the combustion of selected biomass feedstocks used in larger scale combustion facilities. Multiple combustion conditions have been investigated to target those conditions that minimize alkali metal release. The results of these laboratory studies indicate that initial feedstock composition has the most pronounced effect on alkali metal released during combustion. Four mechanisms of alkali metal release have been identified depending on the feedstock composition. Primary alkali metal release in the combustion of relatively low alkali metal containing woody feedstocks is through the vaporization or decomposition of alkali sulfates. Alkali metal chlorides are the primary alkali metal species released during combustion of herbaceous feedstocks, grasses, and straws with high alkali metal and chlorine contents. For feedstocks with high alkali metal and low chlorine content, alkali metal hydroxides are the most abundant alkali vapor released. If a high alkali content is coupled with high levels of fuel-bound nitrogen, the dominant form of alkali metal vapor is the alkali cyanate. In general, the chlorine content of biomass has been identified as an important parameter that facilitates alkali release.

Applications of Advanced Technology to Ash-Related Problems in Boilers
Edited by L. Baxter and R. DeSollar, Plenum Press, New York, 1996

INTRODUCTION

Ash deposition and high temperature corrosion in power generating facilities have been problems for many years and as a result have received considerable attention [Bryers, 1978 and references therein; Reid, 1981; and Benson, 1992]. Deposits that form on heat transfer surfaces in combustion facilities lead to reductions in heat transfer rates and unscheduled downtime. Past studies have concentrated on the transformations of mineral matter during coal combustion and how the inorganic material in coal leads to ash deposition [Raask, 1985; Vorres, 1986; Bryers and Vorres, 1990; and Benson et al, 1993]. Alkali metals in conjunction with sulfur oxides were identified as the materials that had the most important role in deposit formation.

Recently, the concern about global warming and the finite supply of fossil fuels has made biomass an attractive alternative energy resource. Biomass is a domestic, renewable, and sustainable energy resource, and the increased use of biomass for power generation could have significant environmental, economic, and social impacts on the United States [Bain, 1993]. Current biomass resources are comprised primarily of waste materials such as: sawdust or pulp process wastes; hog fuel; forest residues; clean landfilled wood waste; and agricultural prunings and residues [Turnbull, 1993]. Next generation biomass power facilities, however, are being designed in conjunction with dedicated feedstock supply systems that are projected to provide low-cost energy crops for power generation.

Biomass combustion facilities have benefited from the technologies developed for coal fired power plants. Coal and biomass, however, are very different fuels in terms of composition, in particular the inorganic content. In general, biomass tends to have less sulfur, fixed carbon, and fuel-bound nitrogen compared to coal. Biomass also has a higher oxygen content. The ash content of biomass is typically less than coal but the elemental compositions of biomass and coal ash differ. Coal ash tends to be composed of mineralogical material while biomass ash tends to reflect the inorganic material required for plant growth. As a result, the alkali metals in coal tend to be less volatile compared to biomass. Deposit formation from biomass combustion, however, has posed a significant challenge that has limited the increased use of biomass in power generating facilities.

Deposit formation is a complex process which involves mechanisms of alkali metal release during biomass combustion, thermodynamics and chemical reactions in terms of gas composition, particle transport, gas and surface temperatures, fluid dynamics, and surface interactions and reactions. The release of alkali metal vapor accelerates fouling and slagging. Thermodynamics favors the release of alkali chlorides under most combustion conditions if chlorine is available in the gas [Scandrett, 1984]. In the absence of chlorine, hydroxides are the next most likely alkali species to be released, and in the absence of hydrogen, alkali oxides form.

The form of the alkali vapor species is also a function of temperature. At lower temperatures the alkali sulfates are stable. Deposit formation, therefore, is strongly dependent on the alkali species composition in the gas phase as well as the temperature of the gas and the surface on which the deposits form. Many of the agricultural residues targeted as low cost biomass feedstocks, such as grasses and straws, pose the greatest threat for deposit formation, not only because of the high alkali metal and chlorine content, but because of the high ash content as well. These feedstocks tend to be 5% to 18% (rice straw) ash which tends to be predominantly silica, SiO_2, that reacts with the gas phase alkali metal on surfaces.

Baxter [Baxter, 1993] has described four modes of deposit formation: inertial impaction, thermophoresis, condensation, and chemical reaction. In a typical boiler, ash particles become entrained in the hot gas flow and stick to heat transfer surfaces by inertial impaction. In the case of biomass fuels, the ash can have a high silica content, but the silica alone does not pose too great a threat for deposit formation because it has a high melting point ($1700°C$) [Levin, Robbins, and McMurdie, 1964]. Fouling and slagging is accelerated by the alkali vapors, primarily potassium, that condense on or react with the silica on the surfaces to form potassium silicates. The melting point of this potassium/silica mixture is dramatically reduced thus enhancing the tenacity and fluidity of the deposit, even on relatively hot surfaces. Alkali metal condensation on cooler surfaces is more rapid which increases the rate of accumulation of the deposit. Given this mechanism for deposit formation, it is clear that even a small amount of alkali vapor can dramatically alter the rate of deposit formation and the nature of the deposit that is formed.

The ultimate goal of these alkali screening studies is to provide information that could be used to develop a predictive model relating feedstock composition to the fouling and slagging potential of that feedstock when used in an industrial combustor to generate electricity. Knowing the major gas phase alkali metal containing species should improve the understanding for the condensation and chemical reaction pathways in the deposit formation mechanism described above. Once these concepts are further refined it could be possible to start with the temperature profile of a given combustor and a feedstock composition and identify where and how long it will take for a deposit to form. This model will require details of the gas phase chemistry and an in depth study of the condensed phase products and reactions that occur. A detailed understanding of the chemistry of deposit formation would allow one to seek ways of altering the chemistry, through additives or hot gas cleanup, to control deposit formation.

Our approach to studying gas phase alkali metal released during biomass combustion is to directly sample the hot gases liberated from the combustion of small biomass samples in a variable-temperature quartz-tube reactor employing a molecular beam mass spectrometer (MBMS) system. This experimental technique has successfully been used to identify alkali metal containing species released during the combustion of selected biomass feedstocks used in larger scale combustion facilities. Multiple combustion conditions have been investigated to understand the details of the chemistry of alkali metal release and to target those conditions which minimize alkali metal release.

EXPERIMENTAL

Alkali metal vapor released during the combustion of 23 different feedstocks was investigated using a direct sampling, molecular beam mass spectrometer (MBMS) system [Evans and Milne, 1987a,b] in conjunction with a high temperature quartz-tube reactor which has been described in detail in the literature [French, Dayton, and Milne, 1994 and Dayton, French and Milne, 1995]. The following is a list of the feedstocks investigated: planer shavings of lodgepole pine (*Pinus contarta*); eucalyptus (*Eucalyptus saligna Sm.*); poplar (*Populus deltoides x nigra var. Caudina*); corn stover (*Zea mays L.*); switchgrass (*Panicum virgatum L.*); wheat straw (*Triticum aestivum.*); rice straw; (Sandia) switchgrass (*Panicum virgatum L.*); pistachio shells (*Pistacia vera*); almond shells (*Prunnus amygdalus*); almond hulls (*Prunnus amygdalus*); wood waste #1; wood waste #2; waste paper; Danish wheat straws (*Triticum aestivum*) from Slagelse and Haslev (SLAG001, SLAG002, and HAS001); alfalfa stems (*medicago sativa L.*) (IGT001 and IGT002); summer switchgrass (*Panicum virgatum L.*); Dakota switchgrass (*Panicum virgatum L.*); and two willows (*Salix veminalus*, *Salix alba*, tops only). The proximate, ultimate, and ash analyses of all 23 feedstocks are listed in Appendix A.

This set of feedstocks can be divided into various classes of biomass identified as woody feedstocks, herbaceous feedstocks, grasses, agricultural residues, and waste feedstocks. It will become apparent that the varying nature of the feedstock compositions results in unique differences in terms of alkali metal release during combustion.

Most of the biomass samples were ground in a mill to a size of +20/-80 mesh. The particle size of some of the samples was smaller such that the material was flour-like. In the following experiments, the biomass particle size does not have much of an effect on the types of alkali species released during combustion but it is expected to have an effect on the rate of release of alkali metal. The investigation of particle size effects are reserved for future studies.

Twenty to sixty milligrams of sample was loaded into hemi-capsular quartz boats that were placed in a platinum mesh basket attached to the end of a quarter inch diameter quartz rod. This quartz rod can be translated into a heated quartz-tube reactor enclosed in a two-zone variable temperature furnace. Furnace temperatures were maintained at 1100°C and 800°C, respectively. A mixture of helium and oxygen (5%, 10% or 20%) was flowed through the reactor at a total flow rate of 4.4 standard liters per minute. For several experiments 20% steam was also added to the gas mixture by injecting water into the rear of the reactor through a needle fed by a syringe pump. Gas temperatures near the quartz boat were measured with a type-K thermocouple inserted through the quartz rod. The actual boat temperature and the flame temperature of the combustion event were not measured.

The molecular beam sampling system consists of a three stage, differentially pumped vacuum chamber. A conical, stainless steel molecular beam sampling orifice was positioned at the downstream end of the quartz-tube reactor to sample the high temperature, ambient pressure biomass combustion gases. Sampled gases underwent a free jet expansion into the first stage of the vacuum system. The high temperature combustion gases were rapidly cooled during the expansion such that chemical reactions were quenched and condensation was inhibited. As a result, the integrity of the sampled gases was preserved, and reactive and condensable alkali metal containing species remained in the gas phase at temperatures far below their condensation point for long periods of time in comparison to reaction rates.

A molecular beam was formed by collimating the gas stream in the free-jet expansion with a conical skimmer located at the entrance to the second stage of the vacuum system. The molecular beam was directed into the ionization region of the mass spectrometer located in the third stage of the vacuum system. Electron impact ionization (25 eV) of the species in the molecular beam yielded ions that were filtered by a triple quadrupole mass analyzer and detected with an off axis electron multiplier. The mass spectrometer was scanned continuously at a rate of approximately 100 amu/sec to record complete mass spectra during the combustion event every 1.0 second to 1.5 seconds.

The biomass feedstocks listed above were screened for alkali release and speciation at four different combustion conditions: $1100°C$ in $He/O_2(20\%)$; $800°C$ in $He/O_2(20\%)$; $1100°C$ in $He/O_2(5\%)$; and $1100°C$ in $He/O_2(10\%)/Steam(20\%)$. These conditions were chosen to study the effect of temperature, oxygen concentration, and excess steam on alkali release and speciation.

RESULTS AND DISCUSSION

Baseline Screening of Alkali Release at 1100°C in 20% O₂ in Helium

Alkali release and speciation during biomass combustion was investigated by examining the mass spectral results recorded during biomass combustion in a 20% O_2 in helium atmosphere at $1100°C$. As discussed in the literature [French, Dayton, and Milne, 1994; Dayton, French, and Milne, 1995], biomass combustion occurred in three distinct phases: the devolatilization phase, the char combustion phase, and the ash cooking phase.

Typical combustion products such as CO, CO_2, and H_2O, as well as SO_2 and NO, were produced during the devolatilization phase while O_2 was consumed. Most of the alkali metal was released into the gas phase during the char combustion phase and it was possible to identify the alkali species in the mass spectrum averaged over the char combustion phase. Consequently, the char combustion phase contains the most useful information concerning alkali metal release and speciation. Most of the volatile matter has been liberated by the beginning of the ash cooking phase, however, the remaining ash was left in the high temperature reactor to insure that all of the volatile matter had been released.

Masses 18 (H_2O^+), 32 (O_2^+), and 44 (CO_2^+) were not routinely scanned during these experiments to avoid saturating the detector. The ion intensities in all subsequent mass spectra have been normalized to the $^{34}O_2^+$ signal before the sample was inserted into the hot zone of the reactor. Relative ion intensities are, therefore, semi-quantitative and it is reasonable to compare the relative ion signal intensities between spectra. Uncertainties of a factor of two, however, can arise because of day-to-day variations in mass spectrometer tuning, varying time intervals of spectral averaging, and different sample weights. The insets in the figures contain the compositions of the feedstock of interest as determined in the ultimate and ash analysis of the sample on a dry basis.

Combustion of Woody Feedstocks

For the most part, the woody samples, planer shavings of lodgepole pine, poplar, eucalyptus, and the two willow samples have a very low chlorine content and a low alkali metal content. As a result, very little alkali metal was released into the gas phase during the combustion of these feedstocks. For example, the mass spectrum averaged over the char phase of the willow (SV13YR) combustion shown in Figure 1 exhibits a large ion intensity at masses 45 and 46 which are the isotopes of CO_2. The peak at $m/z = 28$, corresponding to CO^+, is also intense. The peak at $m/z = 30$ assigned to NO^+ is considerably weaker. No parent alkali metal species can be identified in Figure 1, however, the K^+ ion can be assigned to the small peak at $m/z = 39$. It is not possible to determine whether the potassium was released as the free metal or whether K^+ is a fragment ion from some other parent alkali metal containing species.

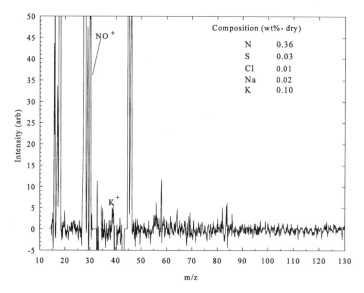

FIGURE 1. Mass spectrum averaged over the char phase of willow combustion at 1100°C in 20% oxygen in helium. The signals have been normalized to the $^{34}O_2^+$ signal measured before sample insertion. The composition in the inset is from the ultimate and ash analyses of the feedstock on a moisture free basis.

The mass spectrum averaged over the char phase during lodgepole pine combustion is similar to the willow char phase spectrum in Figure 1. The carbon dioxide and carbon monoxide peaks dominated the spectrum. Based on the relative intensities of the NO^+ peaks at m/z = 30, less NO was released during willow combustion compared to lodgepole pine combustion. Small HCl^+ peaks were observed at m/z = 36 and m/z = 38. There was some evidence of alkali metal release because K^+ was observed at m/z = 39 and Na^+ was observed at m/z = 23. Similar trends were observed in the char phase mass spectrum of eucalyptus. According to the ultimate and ash analysis of poplar, there is an order of magnitude more potassium and less sodium in this sample than in the planer shavings and eucalyptus. This was reflected in the char phase mass spectrum of poplar which displayed a very intense K^+ peak at m/z = 39. In fact the potassium isotope at m/z = 41 (6.7% natural abundance [Weast, 1985]) was also observed.

The willow tops (SA223YR-tops only) sample has a higher potassium content compared to the other woody feedstocks making it possible to identify K^+ (m/z = 39) in the corresponding char phase spectrum. SO_2^+ was also identified in the char phase spectrum recorded during willow tops combustion. Intense peaks at m/z = 56 and m/z = 81 were assigned to KOH^+ and $KOCN^+$, respectively. The identity of the peak at m/z = 81 was confirmed by recent collision induced dissociation experiments [Dayton and Wang, 1995]. This assignment is reasonable considering the relatively high nitrogen content (0.89 weight %) of this feedstock and the large amount of excess oxygen in the reactor atmosphere. In support of this assignment, the temperature at which this species was released is consistent with the melting/decomposition temperature (700 - 900°C [Weast, 1985]) of potassium cyanate. The formation mechanism of this species is currently not known. It is unclear whether KOCN formed in the condensed phase and then vaporized or potassium cyanide formed in the condensed phase, vaporized, and then reacted with the excess oxygen to form the cyanate.

Combustion of Grasses and Straws

Compared to the woody feedstocks, the grasses and straws have higher alkali metal and chlorine contents. In fact, the rice straw sample has the highest chlorine content of all 23 feedstocks studied. This affects the amount and form of alkali metal released during combustion as observed in the corresponding char phase mass spectra. For example, the mass spectrum averaged over the char phase during rice straw combustion shown in Figure 2 exhibits large signals resulting from CO and CO_2, but the K^+ signal at m/z = 39 is considerably more intense than the NO^+ and SO_2^+ peaks. There are also many features in the spectrum that can be attributed to alkali chlorides. Two peaks at m/z = 74 and m/z = 76 are assigned to KCl^+ while the two peaks at m/z = 113 and m/z = 115 are assigned to K_2Cl^+, a fragment ion of the KCl dimer [Milne and Klein, 1960; and Hastie, Zmbov, and Bonnell, 1984]. Sodium chloride was also released during rice straw combustion as indicated by the peaks at m/z = 58 and m/z = 60 corresponding to $NaCl^+$.

Although an abundance of alkali chlorides appeared to be released during the combustion of this high alkali metal containing feedstock, there was no evidence of other alkali containing species being released during the char combustion phase. As a result, the increased intensity of the K^+ and Na^+ signals is mostly due to fragmentation of the parent chloride species. One curious feature was observed, however, in the char phase mass spectrum of rice straw shown in Figure 2. Two unique peaks were evident at m/z = 97 and m/z = 99 that appeared to have the characteristic intensity pattern of a species containing one chlorine atom. Given the high potassium and sodium content of this feedstock, these peaks are tentatively assigned to $KNaCl^+$, a fragment of the mixed dimer of sodium and potassium chloride.

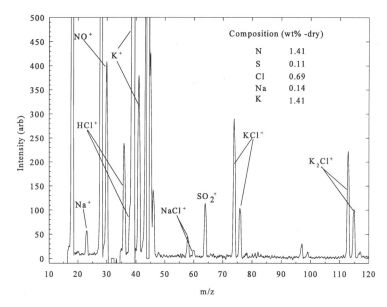

FIGURE 2. Mass spectrum averaged over the char phase of rice straw combustion at 1100°C in 20% oxygen in helium. The signals have been normalized to the $^{34}O_2^+$ signal measured before sample insertion. The composition in the inset is from the ultimate and ash analyses of the feedstock on a moisture free basis.

Four varieties of switchgrass were screened for alkali metal release at 1100°C in 20% O_2 in helium. The alkali metal released during the combustion of switchgrass has been discussed in detail in the literature [Dayton, French, and Milne, 1995]. This switchgrass sample has the highest potassium and chlorine content of the four switchgrasses studied. Therefore, potassium chloride was the dominant alkali metal species released during switchgrass combustion. The mass spectrum averaged over the char combustion phase of Sandia switchgrass combustion in 20% O_2 in helium at 1100°C displayed signals corresponding to the typical combustion products. According to the ultimate analysis of this switchgrass sample, it contains much less chlorine (by a factor of 7) and three times less potassium compared to the switchgrass sample discussed in a previous study [Dayton, French, and Milne, 1995]. This is reflected in the char phase mass spectrum by the absence of the HCl^+ (m/z=36 and m/z = 38) and KCl^+ (m/z=74 and m/z = 76) peaks which were prominent in the char phase spectra recorded during the combustion of switchgrass. The only indication of alkali release was the K^+ peak at m/z = 39.

Summer switchgrass (MM001) and Dakota switchgrass (MM002), contain less alkali and chlorine compared to other two switchgrass samples. The mass spectra averaged over the char phase of summer switchgrass and Dakota switchgrass combustion indicated that little alkali vapor was released. K^+ was identified and SO_2^+ was also observed.

Four different wheat straw samples were also studied. These wheat straws have relatively high chlorine contents and moderately high alkali contents. The mass spectrum recorded during the char phase of wheat straw combustion contains peaks assigned to K^+ (m/z = 39) and KCl^+ (m/z = 74 and m/z = 76) as expected from a high chlorine containing feedstock. It was also possible to identify the KCl dimer fragment (K_2Cl^+ at m/z = 113 and m/z = 115). Very little HCl was detected in the char combustion phase, however, there was a significant amount of HCl released during the devolatilization phase. Peaks in the char phase spectrum of Haslev wheat straw were also assigned to SO_2^+ at m/z = 64 and $KOCN^+$ at m/z = 81. Similar species were also identified in the char phase spectra recorded during the combustion of the two Slagelse wheat straw samples (SLAG001 and SLAG002).

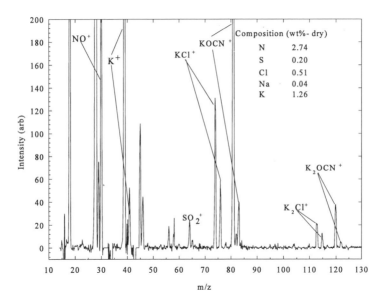

FIGURE 3. Mass spectrum averaged over the char phase of alfalfa stems combustion at 1100°C in 20% oxygen in helium. The signals have been normalized to the $^{34}O_2^+$ signal measured before sample insertion. The composition in the inset is from the ultimate and ash analyses of the feedstock on a moisture free basis.

The two alfalfa stems samples (IGT001 and IGT002) have similar compositions according to the ultimate analysis of each sample. Alfalfa stems are very high in potassium, nitrogen, chlorine, and sulfur. Consequently, the mass spectrum averaged over the char phase of alfalfa stems combustion reveals substantial alkali metal and nitrogen release as displayed in Figure 3. The K^+ peak at m/z = 39 is very intense and potassium chloride is identified by the peaks at m/z = 74 and m/z = 76. The potassium chloride dimer fragment was also assigned to the peaks at m/z = 113 and m/z = 115. One of the most intense peaks in the spectrum is at m/z = 81, assigned to $KOCN^+$, the potassium cyanate ion. The peak at m/z = 120 is assigned to the KOCN dimer fragment ion, K_2OCN^+. It is also possible to identify KOH^+ at m/z = 56. Alfalfa stems combustion is an example of how multiple alkali metal release mechanisms can occur simultaneously.

Combustion of Agricultural Residues

The char phase mass spectra of the agricultural residues, almond hulls, almond shells, pistachio shells, and corn stover, were all different as a result of the varying compositions of the feedstocks. The almond hulls have the highest potassium content of the 23 feedstocks studied. The almond hulls are also high in nitrogen and very low in chlorine. These factors play an important role in the alkali release mechanism observed during almond hulls combustion. The mass spectrum averaged over the char phase during almond hulls combustion is shown in Figure 4. As observed in the previous char phase mass spectra, there are intense peaks at m/z = 28 and m/z = 45 corresponding to CO^+ and $^{13}CO_2^+$. A substantial amount of NO was also released given the intensity of the m/z = 30 peak. The dominant alkali release mechanism during almond hulls combustion was the release of potassium cyanate as observed by the intense peak at m/z = 81, corresponding to $KOCN^+$. The potassium cyanate dimer fragment ion, K_2OCN^+ at m/z = 120 was also observed. Potassium hydroxide release was also observed based on the peak at m/z = 56 assigned to KOH^+. The release of these alkali metal containing species also resulted in substantial intensity at m/z = 39, the K^+ ion which was formed by the fragmentation of these parent alkali metal containing species. The release of potassium chloride was not a dominant alkali metal release mechanism during almond hulls combustion because of the low level of chlorine in the almond hulls.

The almond shells have substantially less potassium and fuel-bound nitrogen compared to the almond hulls. Comparing the mass spectral results for these two feedstocks indicates that twice as much NO was released during almond hulls combustion compared to almond shells combustion. The intensity of the K^+ peak (m/z =39) was also twice as large in the almond hulls char phase mass spectrum compared to the almond shells char phase mass spectrum. Less potassium cyanate was released during the almond shells combustion compared to the almond hulls combustion. The $KOCN^+$ peak at m/z = 81 was an order of magnitude less intense in the almond shells char phase mass spectrum because the almond shells contain less potassium and nitrogen compared to the almond hulls. Potassium hydroxide release was still evident during the almond shells combustion.

The pistachio shells have less potassium and nitrogen compared to the almond hulls and almond shells. In contrast, the pistachio shells have a higher sulfur content than the other agricultural residues. Surprisingly, given the high sulfur content of the pistachio shells, no sulfur containing species were observed in the mass spectrum averaged over the char phase of pistachio shells combustion. Alkali metal release was evident based on the peak observed at $m/z = 39$ assigned to the K^+ ion, however, peaks corresponding to individual alkali containing species were not observed.

The corn stover sample has higher potassium and nitrogen levels comparable to the almond shells. The main difference between the corn stover and the other agricultural residues is that corn stover has a higher chlorine content. The higher chlorine content affects the form of the alkali metal that was released during corn stover combustion. The high potassium and chlorine content of the corn stover results in alkali metal release as potassium chloride. Hydrogen chloride was also released during corn stover combustion.

Combustion of Waste Feedstocks

The final three feedstocks screened were categorized as wastes. The waste paper exhibits a unique combustion behavior compared to the other feedstocks screened to date. A distinct devolatilization phase was observed during which CO_2, H_2O, CO, HCl, and SO_2 were liberated. After this phase, the total ion current (TIC) recorded with the MBMS decayed very rapidly indicating a comparatively short char combustion phase. In terms, of alkali metal release, however, the sample is relatively clean. Only K^+ at $m/z = 39$ was observed and no definitive alkali species were identified. This is not surprising considering the ultimate analysis of the waste paper.

The two wood wastes are blends of wheat straw and almond shells, respectively with urban wood waste. Comparing the char phase mass spectra of wood waste #1 and wood waste #2, more K^+ and NO was released during the combustion of wood waste #1 compared to the combustion of wood waste #2. There was also an indication of KCl^+ in the wood waste #1 char phase mass spectrum although individual alkali metal containing species could not be identified in the wood waste #2 char phase mass spectrum. These two feedstocks have a similar alkali content according to the ultimate analyses, but wood waste #1 has four times more chlorine because of the wheat straw blended in the sample. This explains the greater release of alkali metal into the gas phase during combustion of wood waste #1 and was further evidence that chlorine facilitates the release of alkali. Considering that these feedstocks are classified as wastes, the char phase mass spectra indicate that these samples are relatively clean in terms of alkali metal released into the gas phase.

The Effect of Temperature

Investigating the combustion behavior of the 23 feedstocks screened above at $800°C$ in 20% O_2 in helium revealed the effect of temperature on the release of alkali metal during biomass combustion. Despite which feedstock was combusted, the devolatilization phase was dominated by the release of organic hydrocarbons at the lower furnace temperature. The mass spectrum recorded over this phase for a given feedstock was complicated and congested by the fragmentation of these organic hydrocarbons. The identification of alkali species liberated during this phase was obscured by the dominant mass spectral fingerprints of the hydrocarbon species.

In general, the amount of alkali released into the gas phase appeared to be less at the lower furnace temperature. This was confirmed in a previous study on switchgrass combustion where the alkali metal release was quantified at various combustion conditions [Dayton, French, and Milne, 1995]. For the case of switchgrass combustion, the K^+ signal intensity was reduced and the peak at m/z = 64, corresponding to SO_2^+, was absent from lower temperature char phase mass spectrum. These observations revealed the alkali release mechanism that involved the vaporization of condensed phase potassium sulfate. Indeed, the lower furnace temperature was below the melting temperature of K_2SO_4. This alkali release mechanism involving potassium sulfate has also been observed during coal combustion [Hastie, Plante, and Bonnell, 1982].

A similar reduction in the K^+ signal intensity and the absence of the SO_2^+ ion was observed in the char phase mass spectra recorded during the combustion of the woody feedstocks. These observations were used to determine that the primary alkali metal release mechanism during the combustion of the woody feedstocks, with low alkali metal and chlorine contents, occurs via the vaporization of potassium sulfate.

The effect of furnace temperature on the combustion characteristics of the remaining feedstocks was minimal. As stated above, less alkali metal was released at the lower furnace temperature because of the decreased vapor pressures of the alkali metal containing species at the lower temperatures. This was most noticeable in the case of KOCN release. The signal intensity of the m/z = 81 peak in the char phase spectra recorded during the combustion of almond shells, almond hulls, the Danish wheat straws, and the alfalfa stems was less intense compared to the higher temperature results. This is consistent with a thermodynamic investigation of solid KOCN where the calculated vapor pressure of KOCN is two orders of magnitude lower at 800° C compared to 1100° C [Miller and Skudlarski, 1983], assuming the relationship can be extrapolated to higher temperatures.

The Effect of Oxygen Concentration

The oxygen concentration in the reactor atmosphere was reduced to 5% O_2 in helium to investigate any possible effect this might have on alkali metal release during biomass combustion. The furnace temperature was maintained at 1100° C to compare these results with the initial screening of the feedstocks at 20% O_2 in helium. Reducing the oxygen concentration in the reactor atmosphere had a major effect on the devolatilization phase of biomass combustion. The reduced oxygen concentration prevented complete conversion of the released volatile materials to combustion products. As a result, the mass spectra averaged over the devolatilization phase were dominated by the presence of organic hydrocarbons, mostly organic tars such as benzene and naphthalene. Few differences were observed, however, in the spectra averaged over the char combustion phase. For all of the feedstocks, similar species were identified in the char phase spectra recorded during biomass combustion in a reduced oxygen atmosphere compared to the higher oxygen concentration results.

The Effect of Excess Steam

Excess steam was added to the reactor atmosphere to simulate an environment in which water vapor is continuously being supplied by the random combustion of multiple biomass particles. Such an environment is one that would occur in an industrial boiler or combustor in which biomass is constantly fed. Only 14 of the 23 feedstocks screened for alkali metal release were subjected to combustion in an excess steam environment. The combustion conditions established to study the effect of added steam were 10% O_2 and 20% steam in helium at a furnace temperature of 1100°C.

The added steam had little effect on the combustion of the clean, woody feedstocks. No new species were identified in the char phase mass spectra recorded during the combustion of planer shavings of lodgepole pine and eucalyptus as a result of the added steam. Poplar contains considerably more potassium than both planer shavings and eucalyptus. As a result, a feature at $m/z = 56$ assigned to KOH^+ was clearly observed in the mass spectrum recorded during the char combustion phase of poplar with steam added indicating that some of the alkali was being released as the hydroxide in the presence of water vapor. Combustion of the willow samples was not investigated in the excess steam atmosphere.

The release of potassium hydroxide was also observed for switchgrass combustion in the presence of steam [Dayton, French, and Milne, 1995]. Even though there is a considerable amount of chlorine in the switchgrass, alkali release was partially shifted from the chloride to the hydroxide during switchgrass combustion in excess steam. This release is not completely shifted because potassium chloride is thermodynamically more stable than potassium hydroxide [Scandrett, 1984]. The excess water vapor in the reactor atmosphere also resulted in an enhanced the formation of HCl.

The char phase spectra recorded during the combustion of rice straw and wheat straw in excess steam showed that the K^+/HCl^+ signal intensity was dramatically reduced when steam was added to the reactor atmosphere and indicated that more HCl was being released. It also appeared that less alkali metal was released as potassium chloride. The intensity of the KCl^+ peaks at $m/z = 74$ and $m/z = 76$ were considerably reduced and the K_2Cl^+ peaks that were observed in the char phase spectra recorded without steam were not observed. Given the high potassium content of these two feedstocks, it is not surprising that a prominent peak at $m/z = 57$ (KH_2O^+ see Dayton, French, and Milne, 1995 for details) was observed in char phase spectra of rice and wheat straw combustion in excess steam. KOH was not observed during rice straw combustion suggesting that the primary alkali species released was potassium chloride, even in an atmosphere with excess water vapor. Some potassium hydroxide was detected during the combustion of wheat straw indicating that alkali metal release was shifted in the excess steam atmosphere. The char phase spectrum recorded during combustion of Sandia switchgrass with steam at 1100°C suggested that the amount of alkali released decreased when steam was added. Combustion of the remaining switchgrass samples was not investigated in the excess steam atmosphere.

Excess steam does not have much of an effect on the char combustion phase of pistachio shells, however, adding steam appears to have a substantial effect on the combustion of almond shells and almond hulls. KOCN release was dramatically affected by the presence of excess water vapor. The features at m/z = 81, assigned to $KOCN^+$, and m/z = 120, assigned to K_2OCN^+, that were so prominent in the absence of steam (see Figure 4) were absent from the char phase spectrum of almond shells and there was only a hint of a $KOCN^+$ peak in the almond hulls char phase spectrum. The amount of KOH released during almond shells and almond hulls combustion was not significantly affected by adding steam, however, KH_2O^+ (m/z = 57) was detected in the char phase spectra.

Adding 20% steam to the reactor atmosphere has little effect on the combustion of the waste feedstocks. More HCl was liberated during waste paper combustion in excess steam compared to dry combustion. There also appeared to be less K^+ signal intensity in the char combustion phase in excess steam. Similar conclusions can be drawn for combustion of the two wood wastes in the presence of excess steam.

Linear Relationships Between the Ultimate Analysis and Mass Spectral Data

Mass spectra averaged during the char combustion phases of all 23 feedstocks screened for alkali metal release qualitatively reflect the feedstock composition as determined from the ultimate and ash analyses. Although the ultimate analysis can be used to determine the total amount of alkali metal in a given feedstock, it does not reflect how much alkali metal is released into the gas phase nor the form of the alkali metal released.

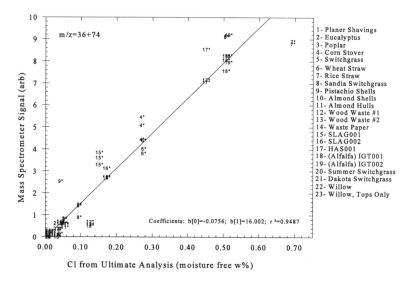

FIGURE 4. Correlation between feedstock chlorine (Cl) content, on a moisture free basis, as determined in the ultimate analysis and the cumulative mass spectrometer signals for HCl^+ (m/z=36) and KCl^+ (m/z=74).

Once the major biomass combustion products were identified, including the major alkali metal containing species released, linear correlations were sought between the mass spectral data and the ultimate analysis data for the 23 feedstocks screened at $1100°C$ in 20% O_2 in helium. Elemental compositions determined in the ultimate analyses, on a dry basis were plotted versus the areas under various ion time profiles (normalized to sample weight and background $^{34}O_2^+$ signals). The chlorine content determined in the ultimate analyses correlated well with the mass spectral intensities of identifiable chlorine containing species, HCl and KCl, monitored at m/z=36 and 74 (HCl$^+$ and KCl$^+$, respectively) as shown in Figure 5. The plotted numbers in Figure 5 correspond to the 23 feedstocks according to the legend in the figure. The straight line in this figure represents a least squares fit of the data. The fitted parameters and the correlation coefficient are indicated in the figure. Rice straw (7*) has the highest chlorine content and as a result, the combustion of this feedstock released the most chlorine into the gas phase as HCl and KCl. Switchgrass (5*) and the alfalfa stem samples (18* and 19*) have the next highest chlorine levels as confirmed by the mass spectral data. Most of the woody feedstocks and the waste feedstocks have low chlorine levels and are grouped tightly around the origin.

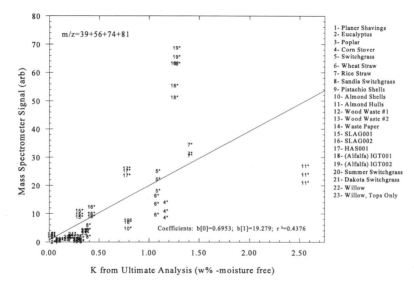

FIGURE 5. Correlation between feedstock potassium (K) content, on a moisture free basis, as determined in the ultimate analysis and the cumulative mass spectrometer signals for K$^+$ (m/z = 39), KOH$^+$ (m/z = 56), KCl$^+$ (m/z=74) and KOCN$^+$ (m/z=81).

Figure 6 attempts to correlate the potassium content of a given feedstock with the parent and fragment ions; KOH$^+$, KOCN$^+$, KCl$^+$, and K$^+$, identified in the char phase mass spectra that result from the dominant gas phase potassium species. The straight line is a least squares fit to the data with the coefficients and correlation coefficient indicated in the figure. This plot and the corresponding correlation coefficient suggest that the overall potassium content of a given feedstock does not necessarily correspond to a high level of alkali metal vapor released during biomass combustion. For example, the almond hulls (11*) have the highest potassium content of the 23 feedstocks, yet combustion of almond hulls does not yield the most potassium vapor, either as KCl, KOH, or KOCN. Rice straw (7*) has a high potassium content, however, it also has the highest chlorine content of any of the 23 feedstocks. As a result, significant alkali metal vapor was released as KCl during rice straw combustion. Alfalfa stems (18* and 19*) and switchgrass (5*) also have a high potassium and chlorine content that accounts for the above average (compared to the least squares fit) alkali metal vapor released during combustion. The Haslev (Danish) wheat straw (17*) has a moderately high potassium content, however, the high chlorine content of this feedstock results in an above average release of alkali metal vapor. It appears that those feedstocks with high potassium and the higher chlorine levels tend to lie above the least squares fitted line in the figure.

The form of the alkali metal containing species present and the mechanism of alkali metal release play an important role in determining the level of gas phase alkali species released during biomass combustion. Of particular importance is the synergistic effect the chlorine content of a given feedstock has on alkali metal release. Clearly, the chlorine content determined from the ultimate analysis is accurately accounted for in the mass spectral results because most of the chlorine was released into the gas phase as HCl and KCl.

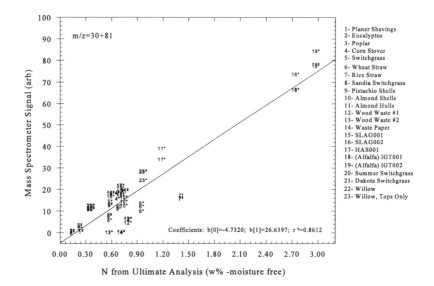

FIGURE 6. Correlation between feedstock nitrogen (N) content, on a moisture free basis, as determined in the ultimate analysis and the cumulative mass spectrometer signals for NO^+ (m/z = 30) and $KOCN^+$ (m/z=81).

The laboratory combustion atmosphere comprised only helium, oxygen, and steam in certain instances. Because air was not used as the oxidizer, the conversion of fuel-bound nitrogen to NO_x could be directly studied without interferences from thermal NO_x formed during combustion. NO_x formation as a function of fuel-bound nitrogen was investigated by plotting the mass spectral intensity for NO^+ (m/z = 30) versus the nitrogen content of the feedstocks. Given the wide range of fuel bound nitrogen in the various feedstocks there is a direct correlation between the amount of gas phase NO and the amount of fuel bound nitrogen. In an attempt to account for all gas phase nitrogen, the signal intensity from m/z = 81 ($KOCN^+$) was added to the NO^+ signals and plotted versus the amount of fuel-bound nitrogen in Figure 7. The correlation coefficient (0.8612) indicates a linear relationship between the gas phase nitrogen oxides released during biomass combustion and the amount of fuel-bound nitrogen. The balance of the nitrogen released during combustion of these feedstocks was most likely nitrogen gas. Unfortunately, a gas phase nitrogen balance cannot be achieved. The molecular nitrogen levels were not quantified because carbon monoxide (a major combustion product) and nitrogen have the same mass and both species contribute to the mass spectral intensity at m/z = 28.

Similar correlations were sought that connect the sulfur content of a given feedstock with the sulfur species observed in the mass spectra recorded during biomass combustion. The mass spectral intensity of SO_2^+ (m/z = 64) was plotted against the sulfur content of the feedstocks. There was a weak trend suggesting that more SO_2 was released during combustion of the higher sulfur containing feedstocks. A possible relationship between alkali and sulfur species was probed by plotting the mass spectral intensities of KCl^+ and K^+ versus the sulfur content of the feedstocks. There was little correlation between gas phase alkali and sulfur content.

CONCLUSIONS

Of the multiple phases that occur during biomass combustion, the char combustion phase is the most important in terms of studying alkali release. The mass spectrum averaged over a given char phase of biomass combustion qualitatively reflects the feedstock composition as determined in the ultimate analysis. Although the ultimate analysis is important for determining the total amount of alkali in a given feedstock, it does not reflect how much alkali metal is released into the gas phase nor the form of the alkali metal containing species released. For this reason, the MBMS technique continues to be valuable for directly studying alkali metal released during biomass combustion.

To date, 23 biomass feedstocks have been screened for alkali release and speciation under various combustion conditions. The conditions were chosen to investigate the effect of temperature, oxygen concentration, and excess water vapor on the release of alkali vapor during biomass combustion. The focus of the alkali screening studies has been to identify alkali metal containing species released during biomass combustion, and conditions, if any, which reduce the amount of alkali released.

Changing the combustion conditions has little effect on the release of alkali vapor species. The largest effects on alkali metal release were observed when excess steam was added to the reactor atmosphere. The alkali released during the combustion of the high chlorine containing species was shifted partly from potassium chloride to potassium hydroxide. The presence of excess steam during combustion of high chlorine containing feedstocks also increases the amount of HCl released. This has important implications in terms of high temperature corrosion in boilers, turbines, and industrial combustors. Excess steam in the reactor atmosphere also dramatically affects the release of KOCN during the combustion of almond shells and almond hulls.

The most significant parameter that affects alkali vapor release is the feedstock being combusted. Linear relationships were uncovered between the amount of gas phase chlorine and nitrogen released during biomass combustion. The higher the chlorine content and fuel bound nitrogen of a given feedstock, the more gas phase chlorine and NO_x that is released during combustion. The combustion of a feedstock with a high potassium content does not necessarily translate into high levels of gas phase potassium.

Although each individual feedstock appears to have it's own unique combustion properties, these feedstocks can often be grouped together into classes of feedstocks. Woody feedstocks have comparatively little alkali content and low levels of chlorine. Consequently, combustion of woody feedstocks leads to very little alkali vapor release.

Herbaceous feedstocks, grasses and straws contain very high levels of alkali and chlorine compared to the woody feedstocks. Large amounts of alkali are released into the gas phase during combustion of herbaceous feedstocks that results in a high fouling and slagging potential for these feedstocks. This prevails even when excess steam is present during combustion, however, the alkali release under these conditions was shifted partly, but not completely, to the hydroxide. The results of the alkali screening studies implicate the chlorine content of a given feedstock as a facilitator of alkali metal release.

Although chlorine content has been linked to substantial alkali vapor release, it is not essential. The agricultural residues still release significant amounts of alkali vapor during combustion, however, the alkali metal is present as the hydroxide and potassium cyanate instead of the chloride. The present studies identify a new mechanism for alkali metal release that involves formation of potassium cyanate. This mechanism of alkali transport seems to dominate during the combustion of feedstocks that have high potassium and nitrogen contents. This pathway for alkali metal release can be minimized or eliminated by reducing the furnace temperature below 700°C. Lower furnace temperatures, however, translate into reduced efficiencies in power generating facilities.

Investigating biomass combustion has revealed several dominant mechanisms for alkali metal release depending on the composition of the feedstock being combusted as summarized below:

Low alkali metal and chlorine feedstocks: vaporization or decomposition of potassium sulfate

| High alkali and low chlorine feedstocks: | potassium hydroxide, and if also high in nitrogen, potassium cyanate |
| High alkali and high chlorine feedstocks: | potassium chloride |

The challenge is to relate the effects of alkali metal release during biomass combustion on how deposits form in industrial combustors. The results of the alkali screening studies are relevant to the fouling and slagging observed in larger scale combustion facilities and industrial power generating facilities. Clearly the herbaceous feedstocks, with high potassium and chlorine content, represent the highest probability for severe fouling and slagging in industrial boilers. Herbaceous feedstocks also release too much alkali by three orders of magnitude for these feedstocks to be considered useful in direct fired turbine applications [Moses and Bernstein, 1994].

ACKNOWLEDGMENTS

The authors are grateful for the support of the Solar Thermal and Biomass Power Division of the Department of Energy, Office of Energy Efficiency and Renewable Energy. Special thanks go to Richard L. Bain and Ralph P. Overend for both programmatic and technical support and guidance. Richard J. French is also acknowledged for developing some of the experimental techniques. We would like to thank Dr. Larry Baxter, Sandia National Laboratories, for supplying some of the biomass samples and their analyses. Gerald Cunningham of Hazen Research, Inc. is acknowledged for providing feedstock analyses. We would also like to acknowledge the efforts of Thomas R. Miles, Sr. and Thomas R. Miles, Jr. of Thomas R. Miles Consulting Engineers.

REFERENCES

Bain, R.L., (1993). "Electricity From Biomass in the United States: Status and Future Direction," *Bioresource Technology*, 46, 86-93.

Baxter, L.L. (1993). "Ash Deposition During Biomass and Coal Combustion: A Mechanistic Approach," **Biomass and Bioenergy** 4, pp. 85-102.

Benson, S.A., Ed. (1992). *Inorganic Transformations and Ash Deposition During Combustion*, Engineering Foundation Conference March 10-15, 1991, Palm Beach, FL Published on behalf of the Engineering Foundation by the American Society of Mechanical Engineers, New York, NY.

Benson, S.A., Jones, M.L., and Harb, J.N. (1993). "Ash Formation and Deposition." In L.D. Smoot, Ed. *Fundamentals of Coal Combustion for Clean and Efficient Use* Elsevier, New York, NY.

Bryers, R.W., (1978). <u>Ash Deposits and Corrosion Due to Impurities in Combustion Gases</u>. Hemisphere Publishing Corporation, New York, NY.

Bryers, R.W. and Vorres, K.S., Eds. (1990). *Mineral Matter and Ash Deposition From Coal,* Engineering Foundation Conference February 22-26, 1988 Santa Barbara, CA

Dayton, D.C., French, R.J., and Milne, T.A. (1995). "The Direct Observation of Alkali Vapor Release During Biomass Combustion and Gasification. I. The Application of Molecular Beam/Mass Spectrometry to Switchgrass Combustion." *Energy&Fuels*, 9, pp. 855-865.

Dayton, D.C. and Wang, D., (1995). "CID Studies of Inorganic Species Released During Biomass Combustion." 43rd ASMS Conference of Mass Spectrometry and Allied Topics, May 21-26, 1995, Atlanta, GA. Paper TPA 034.

Evans, R.J. and Milne, T.A. (1987a). "Molecular Characterization of the Pyrolysis of Biomass: I. Fundamentals," **Energy and Fuels** 1, pp. 123-127.

Evans, R.J. and Milne, T.A. (1987b). "Molecular Characterization of the Pyrolysis of Biomass: II. Applications," **Energy and Fuels** 1, pp. 311-319.

French, R.J., Dayton, D.C., and Milne, T.A. (1994). "The Direct Observation of Alkali Vapor Species in Biomass Combustion and Gasification," NREL Technical Report (NREL/TP-430-5597). January 1994.

Hastie, J.W.; Plante, E.R.; Bonnell, D.W. "Alkali Vapor Transport in Coal Conversion and Combustion Systems," in <u>Metal Bonding and Interaction in High Temperature Systems</u>. Gole; Stwalley; Eds.; ACS Symposium Series #179, 1982. Chapter 34.

Hastie, J.W., Zmbov, K.F., and Bonnell, D.W. (1984). "Transpiration Mass Spectrometric Analysis of Liquid KCl and KOH Vaporization." *High Temperature Science*, **17**, 333-364.

Levin, E., Robbins, C.R., and McMurdie, H.F. (1964). *Phase Diagrams for Ceramists.* Columbus, OH: American Chemical Society.

Miller, M. and Skudlarski, K. (1983). "Mass Spectrometric Study of Potassium Cyanate and Potassium Cyanate-Potassium Cyanide System at High Temperatures." *Int. Journal of Mass Spec. and Ion Phys.*, **47**, 243-246.

Milne, T.A. and Klein, H.M., (1960). "Mass Spectrometric Study of Heats of Formation of Alkali Chlorides," *J. Chem. Phys.*, **33**, 1628-1637.

Moses, C.A. and Bernstein, H. (1994). "Impact Study on the Use of Biomass-Derived Fuels in Gas Turbines for Power Generation," NREL Technical Report (NREL/TP-430-6085). January 1994.

Raask, E. (1985). *Mineral Impurities in Coal Combustion* Hemisphere Publishing Corporation, Washington, DC.

Reid, W.T., (1981). "Coal Ash-Its Effect on Combustion Systems," Chapter 21 in <u>Chemistry of Coal Utilization</u>, Elliot, M.A.; Ed; New York, NY.

Scandrett, L.A. (1984). "The Thermodynamics of Alkali Removal From Coal-Derived Gases," *Journal of the Institute of Energy*, December 1984, 391-397.

Turnbull, J.H., (1993). "Use of Biomass in Electric Power Generation: The California Experience." *Biomass and Bioenergy*, **4**, 75-84.

Vorres, K.S., Ed. (1986). *Mineral Matter and Ash in Coal* ACS Symposium Series 301 American Chemical Society, Washington, DC.

Weast, R.C., Ed. (1985). "66th Edition of the CRC Handbook of Chemistry and Physics," CRC Press, Inc., Boca Raton, FL.

Appendix A

Compositional Analyses of The Biomass Feedstocks

	Planer Shavings	Eucalyptus	Poplar	Switchgrass	Rice Straw	Wheat Straw	Corn Stover	(Sandia) Switchgrass	Pistachio Shells	Almond Shells	Almond Hulls	Waste Paper
Proximate (Weight % as received)												
Ash		0.48	1.16	4.22	17.79	7.48	4.75	5.64	1.30	2.80	5.67	7.51
Volatile		78.52	80.99	72.73	57.92	68.60	75.96	66.49	75.49	70.13	67.22	78.61
Fixed Carbon		11.66	13.05	14.89	12.56	14.73	13.23	13.68	15.67	19.22	18.93	8.71
Moisture		9.34	4.80	8.16	11.73	9.19	6.06	13.68	7.53	7.85	8.18	5.17
Heating Value, dry (HHV) Btu/lb	8760	8262	8382	8012	6339	7599	7782	8126	8469	8189	8069	9126
Ultimate (Weight % as received)												
Moisture	34.5	9.34	4.80	8.16	11.73	9.19	6.06	13.68	7.53	7.85	8.18	5.17
C	34.5	44.89	47.05	43.04	34.64	40.88	43.98	40.9	46.41	46.2	42.7	46.76
H	3.76	5.21	5.71	5.37	4.39	5.14	5.39	4.90	5.83	5.48	5.48	6.7
O*	26.65	39.92	41.01	38.58	29.70	36.06	39.10	33.78	38.05	36.94	36.83	33.01
N	0.16	0.13	0.22	0.53	1.12	0.83	0.62	0.48	0.63	0.68	1.06	0.66
S	0.01	0.03	0.05	0.10	0.09	0.17	0.10	0.05	0.20	0.03	0.04	0.14
Cl**	0.02	0.05	<0.01	0.46	0.55	0.24	0.25	0.08	0.04	0.01	0.05	

* Oxygen by difference   ** Not usually reported as part of the ultimate analysis

Appendix A continued

Ash Composition (Weight % as received)	Planer Shavings	Eucalyptus	Poplar	Switchgrass	Rice Straw	Wheat Straw	Corn Stover	(Sandia) Switchgrass	Pistachio Shells	Almond Shells	Almond Hulls	Waste Paper
Si	0.024	0.04	0.05	0.94	6.71	2.29	1.20	1.72	0.05	0.16	0.19	0.88
Fe	0.014	---	---	---	0.11	0.054	---	0.14	0.32	0.07	0.04	0.03
Al	0.021	0.02	0.02	0.03	0.14	0.077	0.05	0.21	0.01	0.04	0.04	3.11
Na	0.005	0.02	0.001	0.007	0.11	0.116	0.006	0.04	0.04	0.05	0.04	0.07
K	0.057	0.038	0.311	0.989	1.12	0.938	1.08	0.33	0.2	0.72	2.29	0.02
Ca	0.116	0.091	0.392	0.223	0.25	0.149	0.294	0.29	0.09	0.34	0.34	0.54
Mg	0.013	0.021	0.081	0.117	0.22	0.118	0.175	0.11	0.03	0.11	0.17	0.1
P	---	0.061	0.075	0.284	0.13	0.044	0.180	0.07	0.07	0.12	0.2	0.01

Appendix A continued

	Wood Waste #1	Wood Waste #2	Wheat Straw SLAG001	Wheat Straw SLAG002	Wheat Straw HAS001	Alfalfa Stems IGT001	Alfalfa Stems IGT002	Summer Switchgrass MM001	Dakota Switchgrass MM002	Willow SV13YR	Willow Tops SA223YR
Proximate (Weight % as received)											
Ash	6.21	6.58	5.25	3.21	4.96	4.78	4.83	2.33	3.15	0.85	2.17
Volatile	67.31	67.61	71.05	67.09	66.44	71.59	73.77	71.93	71.07	76.52	73.92
Fixed Carbon	16.26	16.53	13.96	12.29	12.78	14.34	13.8	12.47	13.14	12.40	16.70
Moisture	10.22	9.28	9.74	17.41	15.82	9.29	7.6	13.27	12.64	10.23	7.21
Heating Value, dry (HHV) Btu/lb	8164	7868	7988	7905	7847	8025	8014	7979	8014	8330	8424
Ultimate (Weight % as received)											
Moisture	10.22	9.28	9.74	17.41	15.82	9.29	7.6	13.27	12.64	10.23	7.21
C	44	42.61	42.19	39.27	39.07	42.79	43.23	41.21	41.45	44.07	45.86
H	5.19	5.06	5.17	4.84	4.77	5.44	5.5	5.03	5.02	5.29	5.47
O*	33.5	35.91	36.91	34.71	34.72	35.09	36	37.81	37.02	39.21	38.28
N	0.69	0.5	0.64	0.48	0.58	2.43	2.66	0.31	0.65	0.32	0.89
S	0.07	0.03	0.1	0.08	0.08	0.18	0.18	0.04	0.07	0.03	0.12
Cl**	0.11	0.03	0.23	0.14	0.37	0.45	0.46	<0.01	0.03	<0.01	<0.01

Appendix A continued

Ash Composition (Weight % as received)	Wood Waste #1	Wood Waste #2	Wheat Straw SLAG001	Wheat Straw SLAG002	Wheat Straw HAS001	Alfalfa Stems IGT001	Alfalfa Stems IGT002	Summer Switchgrass MM001	Dakota Switchgrass MM002	Willow SV13YR	Willow Tops SA223YR
Si	1.67	1.71	1.55	0.82	1.34	0.13	0.07	0.67	0.90	0.03	0.02
Fe	0.22	0.22	0.05	0.02	0.02	0.01	0.01	0.02	0.02	0.004	0.01
Al	0.34	0.37	0.06	0.01	0.01	0.001	0.001	0.02	0.01	0.01	0.02
Na	0.12	0.11	0.04	0.04	0.03	0.04	0.05	0.01	0.01	0.02	0.04
K	0.25	0.24	0.46	0.34	0.64	1.11	1.14	0.16	0.20	0.09	0.33
Ca	0.46	0.66	0.35	0.28	0.27	0.63	0.67	0.18	0.27	0.28	0.53
Mg	0.11	0.12	0.05	0.05	0.05	0.3	0.27	0.07	0.10	0.01	0.04
P	0.004	0.04	0.08	0.06	0.06	0.16	0.16	0.03	0.05	0.04	0.07

AN IMPROVED ASH FUSION TEST

Charles D.A. Coin, Hakan Kahraman, and Adrian P. Peifenstein
ACIRL Ltd.
P.O. Box 242, Booval QLD, 4304 Australia

ABSTRACT

A new method of measurement of ash fusion temperatures has been developed using essentially the same equipment as is used for measurement of ash fusibility under Standards such as AS1038.15-87 and ASTM D1857-87. However, unlike the standard method the new method produces quantitative results of progressive dimensional changes during heating and melting of the ash. Further, the new method has much improved precision in determination of the temperatures at which these changes take place. Repeatability and reproducibility of the results are much improved and have scope for further improvement. Correspondence between the index points of current ash fusion tests [Initial Deformation, Sphere, Hemisphere and Flow] with reference points in the new method is poor, particularly in relation to ID temperatures and initial dimensional changes. The temperatures of significant movement in the new test appear to be systematic and therefore are likely to correspond to mineralogical melting points. As slagging and fouling mechanisms depend on relative melting of mineral phases, the new test should provide a significant improvement on the current ash-fusion method.

INTRODUCTION

The current standard ash fusion test [ASTM D1857, 1987; AS1038.15, 1987] is somewhat subjective, being based on the operator's judgement of the degree of deformation of the ash sample as the temperature rises at a specified rate in a controlled atmosphere. The specified repeatability limits [single operator and apparatus] of 30°C and reproducibility limits [different operator and apparatus] of 50-80°C reflect the subjectivity of the current test. The change from the original form of the sample pellet, through all the stages until complete melting, is a continuous process. In cases of congruent melting, the entire transformation can occur over a very narrow temperature range. However, in cases of incongruent melting, the transformation can be protracted, minor and gradual.

The main criticism of the standard ash fusion tests has been for unreliability in predicting the behaviour of coal ashes in real combustion processes. It has been emphasised that initial deformation temperature is not the temperature at which ash melting begins and most ashes were found to have commenced melting at

Applications of Advanced Technology to Ash-Related Problems in Boilers
Edited by L. Baxter and R. DeSollar, Plenum Press, New York, 1996

187

temperatures far below the initial deformation temperatures [Gerald *et al.*, 1981; Huggins *et al.*, 1981]. These authors also found that all of the ash fusion temperatures including the initial deformation temperatures were dependent on ash composition in a manner which mirrored liquidus curves and not the solidus curves of ternary oxide phase diagrams. The degree of partial ash melting is significant at temperatures as much as 200°C to 400°C below the initial deformation temperature and the initial deformation temperature values would make little distinction among the coals analysed in the ash fusion test whereas the Mössbauer measurement of iron in glass can give a fairly clear rating of the samples.

A number of recent studies have made use of the technique of multiple linear regression analysis whereby any fusion temperature such as initial softening temperature is estimated by combining and weighting the effects of several compositional variables [Rees, 1964; Riley *et al.*, 1989; Winegartner & Rhodes, 1975; Sondreal & Ellman, 1975; Vorres, 1979; Gray, 1987; Sleiger *et al.*, 1988; Lloyd *et al.*, 1989a; Lloyd *et al.*, 1989b; Lloyd *et al.*, 1993]. However, attempts by the present authors to use such regressions have not been successful and are unlikely to be so, considering the complex mathematics that would be needed to describe even the SiO_2-Al_2O_3-CaO system, even without consideration of incongruent melting and the concept that the melting of the ash can be a gradual process and not a discrete event.

Alternative methods have also been proposed. An ash fusion test based on the electrical resistivity of coal was suggested [Sanyal & Mehta, 1993]. In this test the electrical resistance and the current flowing through an ash are monitored up to 1350°C when the sample is heated at the same rate and in the same environment as that specified for the ASTM ash fusion test. The duplicate results for the temperatures at which the major electrical resistance begins to occur, showed a repeatability of ± 15°C for 17 out of 20 cases compared with 15 out of 20 cases of ID temperatures within ± 15°C. However, it has been claimed that the technique is superior to the standard ASTM test. Although it has been stated that the apparatus and its components are inexpensive to build and are based on commercially available material, it is still new equipment and requires a capital investment. Similar techniques based on the measurement of electrical resistance of ash also have been proposed [Raask, 1979; Sanyal & Cumming, 1981; Gibson & Livingston, 1991].

Electrical conductance methods for detection of the onset of fusion in coal ash and deposits have also been proposed [Cumming & Sanyal, 1981; Conn & Austin, 1984; Cumming *et al.*, 1985]. It was claimed that these methods provide a better measure of likely performance of an ash in a boiler than the aforementioned methods, particularly the standard ash fusion temperature. However, another study showed that these methods would be difficult because satisfactory contact between the ash and the electrodes is hard to achieve and maintain, particularly in oxidising atmospheres [Wall *et al.*, 1989].

A thermo-mechanical analysis device has been also constructed to characterise ash and deposit samples from both experimental-scale combustion test facilities and full-scale operating boilers [Ellis *et al.*, 1987]. It was indicated that the results obtained from this thermo-mechanical analysis device can be used to broadly assess the temperatures at which sintering and fusion occur.

There are many variants to the conventional ash fusion test and these are described in detail in their respective standards [ISO 540-1981; ASTM D1857-1987; AS1038.15-1987; BS 1016.15-1970; DIN 51730-1984; GOST 2057-1982; SABS 932; GE 219-1974] and summarised by Hough [1990] and this is indicative of the recent interest in the subject and the limitations of the existing methods. This paper does not seek to address the consequential questions raised by Hough regarding the mechanisms of ash fusion and the application of the method to actual ash fusion during coal combustion. These matters are currently being researched and indications are that the results from the new ash fusion method are enabling resolutions of the pathways and mechanisms of coal combustion ash fusion.

Another criticism of ash fusion tests in general is the relevance of the ash sample which is subjected to the test. The temperature at which the ash is derived is much lower than that experienced in most combustion situations. It is not the purpose of this paper to examine this question. However, research work is currently being undertaken by the authors using samples derived from slagging panels from a pilot scale combustion facility.

EXPERIMENTAL

Samples

A total of 30 coal ash samples were subjected to the new ash fusion test. To ensure a wide variety of response, coal ash samples were selected on the basis of their ash chemistry data available in publications [Queensland Coal Board, 1990; Australian Coal Report, 1994]. An inert sample of alumina refractory material also was tested.

Apparatus

The new test requires two standard ash fusion ceramic tiles [25*25*6 mm, Figure 1] and a moulding device as cited in Figure 4 of AS1038.15. The mould produced four cylinders of ash of 2.0 mm diameter and 2.0 mm height. Ash fusion tests were performed in a standard Carbolite ash fusion furnace.

Test method

The new ash fusion test uses the existing test set-up. Performing the new ash fusion method is easier than the standard methods. The preparation of the ash pellets in the cylindrical mould is easier and faster. The four pellets after being prepared in the mould are disposed toward the corners of an ash fusion tile and a second tile is then placed on top of the pellets [Figure 1]. As the temperature increases at 5°C/minute, photographs of the assemblage were taken at 20 °C intervals. The separation of the tiles was measured from the photographic negatives using a transmitted light microscope at 25x magnification with a graticule in one eyepiece.

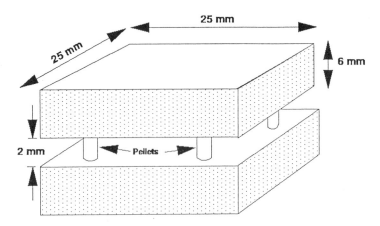

OBLIQUE VIEW

Figure 1 Diagrammatic representation of the arrangement of the ash pellets and the ceramic tiles for the improved ash fusion test. The separation of the tiles is measured from photographs.

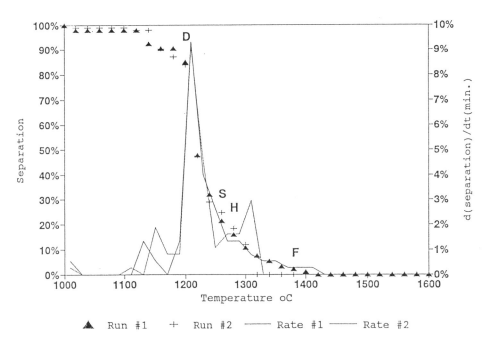

▲ Run #1 + Run #2 ——— Rate #1 ——— Rate #2

Figure 2 An improved ash fusion test result in which an ash sample displayed periods of significant movement over several narrow temperature ranges. The current ash fusion test results are also indicated.

The results are presented as the primary curves and the rate of change of separation with respect to temperature [e.g. Figure 2]. The x-axis is the temperature range of the test. The separation of the tiles is given as percentage [left y-axis] and plotted as marked points. Sets of points that are horizontal indicate no movement in the sample over the associated temperature range. Sets of points exhibiting slopes indicate some contraction of the ash cylinders, probably due to melting. The rate of change of separation (first derivative) is calculated as the function of the average change in separation per minute i.e.

$$y = \frac{(x_1 - x_2)}{\Delta t}$$

where y is the first derivative of tile separation per minute, x_1 is the reading of tile separation (%) from a previous photograph, x_2 is the reading of tile separation (%) from the following photograph, and Δt is the time interval between the two photographs (minutes). These results refer to the right y-axis [Figure 2]. The rate of change allows easy identification of periods of maximum contraction.

The results from testing according to AS1038.15, i.e. initial deformation [D], spherical [S], hemispherical [H] and flow [F] temperatures are also marked in the graphs.

RESULTS

Coal ash samples

The coal ash samples exhibited a range of responses to the heating regime. Some samples displayed one or more periods of significant movement over narrow temperature ranges [Figure 2]. Other samples exhibited no change or very slow change in the new ash fusion test [Figure 3] and these corresponded to ashes with S > 1600°C in the conventional test. Other samples exhibited more gradual collapse over an extended temperature range [Figure 4]. A gradation of responses was also observed between these three groups.

Inert sample

An inert sample of alumina refractory material was used to confirm that the system measured the proportional differences in height of the ash pillars through the gradual temperature increases and to give an indication of any background correction that might be needed. This sample did not display any movement in the test indicating that the new ash fusion test did not require base-line correction.

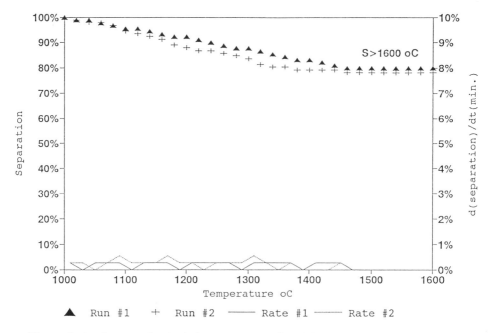

Figure 3 An improved ash fusion test result in which an ash sample exhibited protracted change. The current ash fusion test [AS1038.15] temperatures are also indicated.

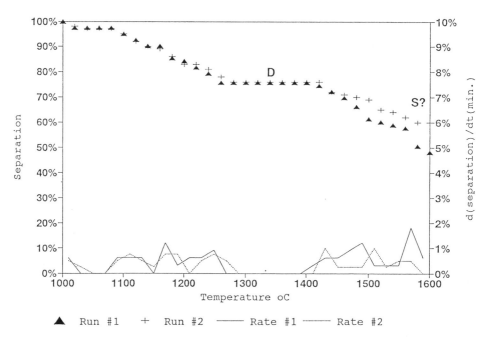

Figure 4 An improved ash fusion test result in which an ash sample exhibited gradual collapse over an extended temperature range. The current ash fusion test [AS1038.15] temperatures are also indicated.

Repeatability of results

The repeatability of the results can be viewed as the repeatability of the *magnitude of the response* and separately as the repeatability of the *temperature at points of critical movement* [analogous to the current ash fusion test]. All the ash samples were tested in duplicate to check the repeatability of the new method with respect to the allocation of temperature at the points of critical movement. A statistical analysis of the differences of the magnitude of response for two consecutive runs at the same temperature level revealed a standard deviation of 11.92% for each test at the 95% confidence level.

Points of critical movement were arbitrarily regarded as significant when the first derivative curve exceeded 1.5%/minute. The majority of results [94%] were identical within the interval of the photography [±20°C]. This is substantially better than the ±50°C repeatability for the AS1038.15 test. It is thought that with increased frequency of photographic recording or continuous video-recording the interval could be reduced to much less than 20°C and would be only limited by the variation in the proximity of the samples to the thermocouple, the inherent precision of the thermocouple and the homogeneity of the sample.

Reproducibility of results

Reproducibility due to operator variation or bias was determined by using two operators to measure the tile separation from the photographs using 15 randomly selected samples. A statistical analysis from the differences of two operators revealed that at the same temperature, the degree of deformation had a standard deviation of 10.30% for each test at the 95% confidence level.

DISCUSSION

Correspondence of the improved test with AS1038.15 ash fusion test

The major changes observed from the primary and first derivative curves of some samples in the new method were broadly coincident with only a few of the temperatures indicated by the AS1038.15 ash fusion test [Table 1]. These coincidences were generally observed when the new method exhibited significant tile movement over a narrow temperature range. However, the general correspondence between the results of the two methods was poor. The temperature of the first movement or initial deformation in the new test was generally observed well before the AS1038.15 Initial Deformation temperature. In some cases the cylinders had reduced to approximately 35% of their height prior to the Initial Deformation of the AS1038.15 test. This supports previous criticisms of the standard ash fusion tests [Gerald *et al.*, 1981; Huggins *et al.*, 1981] by explaining why conventional ash fusion temperature determinations have had such poor reproducibility.

Table 1   Correspondence between the first major peak of the new test on Category 4 samples with the index temperatures of the AS1038.15 ash fusion test

Sample No.	Initial Deformation (°C)	Sphere (°C)	Hemi-sphere (°C)	Flow (°C)	Improved Test First Major Peak (°C)
2	**1200**	1260	1280	1380	1210
7	**1280**	1420	1440	1480	1300
8	**1160**	1320	1360	1400	1130
9	1340	**1580**	1600		1540
14	1200	**1460**	1480	1540	1450
15	1180	1560	1580	1600	~1500
16	1220	1480	1500	1540	1300
18	1180	**1380**	1420	1480	~**1360**
20	1360	1480	1500	1540	1290
21	1280	1370	1390	1450	1110
22	1320	**1400**	1420	1460	**1370**
23	**1440**	1540	1560	1580	1490
25	1340	1600<			1140
26	1300	1540	1580	1600	1450
27	**1180**	1320	1360	1440	**1170**
28	1220	1340	1360	1420	1180

Approximate coincidences of the conventional ash fusion temperatures with the temperatures of the first major point of change in the improved test are outline in **bold**.

The new method was compared with twelve results from an Australian inter-laboratory exercise that used the standard ash fusion method (AS1038.15-87). The new method allowed interpretation of the variability in reproducibility between current ash fusion results. In the new method, samples which displayed gradual collapse over an extended temperature range had significantly poorer reproducibility in the standard ash fusion test [Figure 5]. This is consistent with the cognitive difficulty of marking a point of change in a system which is changing very slowly. Furthermore, the new method revealed more details in these samples than the standard ash fusion test.

In comparison to the previously described case, in the new test, samples that exhibited low ash fusion temperatures and high rates of movement over a narrow temperature range, were found to display improved reproducibility in the standard ash fusion test. Further, the Initial Deformation temperature on the current ash fusion tests were found to correlate more closely with initial significant movement in the new ash fusion test [e.g. Figure 6]. However, it should be noted that the Initial Deformation temperature can just as frequently occur up to 200°C prior to any significant movement taking place [Table 1].

Limitations to further improvement of the test

The sources of statistical variation are predominantly in the recording of the temperature [i.e. variability in the proximity of the cylindrical pillars to the thermocouple] and the frequency of the photographic recording. The second source

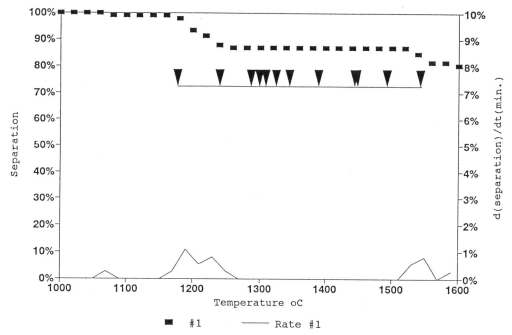

Figure 5 Comparison of the new method with an inter-laboratory exercise in which Initial Deformation temperatures exhibited poor reproducibility. Initial Deformation temperatures are indicated with ▼.

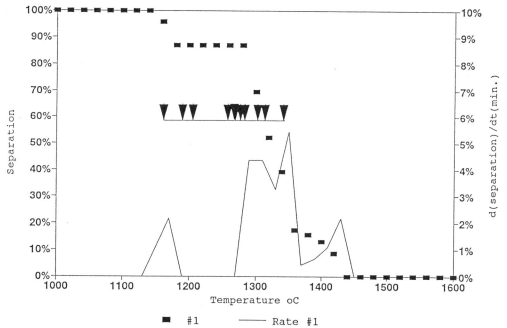

Figure 6 Comparison of the new method with an inter-laboratory exercise in which the Initial Deformation temperatures exhibited generally acceptable reproducibility. Initial Deformation temperatures are indicated with ▼.

of variation may be readily addressed as the new method is particularly suited to automation by either video recording and image digitising or by laser scanning. Future development work is being continued in the areas of automation and optimisation of the test.

Correlation of improved ash fusion test results with mineralogical changes

Combining the results of all the samples indicates that some temperatures of significant movement occur frequently [Figure 7]. The first derivative peaks might be coincident with melting points of known mineral phases which would indicate that the ash was melting incongruently. This is consistent with mechanisms of melting and fractionation of the ash in coal combustion [Creelman, 1994]. Correlation of melting profiles with phase changes in coal derived and synthetic ashes is currently being undertaken, as is direct testing of slag deposits in commercial and experimental boilers. It is apparent that there is a need for this type of information to correlate the mineralogical data with the ash fusion and relate this to real slagging and fouling behaviour.

CONCLUSIONS

An improved ash fusion test has been developed. The apparatus is essentially the same as that used in current standard ash fusion tests with the exception of the

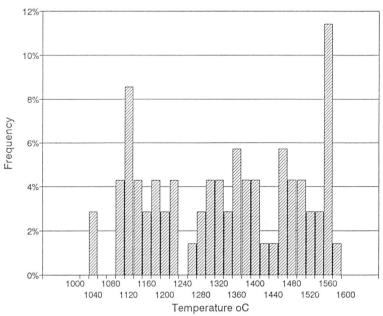

Figure 7 Temperatures of significant movement in the improved ash fusion test for the samples in this study.

arrangement of the coal ash sample. In the new test, four pellets of ash are disposed towards the corners of an ash fusion tile and a second tile is then placed on the top of the pellets. As heating progresses, the separation of the tiles is measured.

The improved ash fusion test is superior to the existing AS1038.15 ash fusion test. The information gained from the new ash fusion test is objective rather than subjective and contains much more information than the current AS1038.15 test. The method gives superior repeatability and reproducibility and could be further improved by increasing the recording frequency. The changes in tile separation are easy to determine and measuring requires very little skill.

The correspondence between the temperatures of significant movement of the new test and the index points of the current ash fusion test [Initial Deformation, Sphere, Hemisphere & Flow] is poor, with the best correlation occurring with Initial Deformation temperatures in samples that exhibit significant tile movement over a narrow temperature range.

The temperatures of significant movement in the new test appear to be systematic and correspond to mineralogical melting points. As such they should provide much more useful information on slagging and fouling than the current ash fusion method which is already known to be deficient in this regard.

There is a necessity for further investigations into the relationship between mineralogical and chemical compositional and ash fusion behaviour.

ACKNOWLEDGMENTS

This project was sponsored and funded by Australian Coal Association Research Program [Project No. 3092][Coin *et al.*, 1994]. Samples were provided by ACIRL Ipswich Laboratory, ACIRL Maitland Laboratory, R.A. Creelman and Associates, CSIRO - Queensland Centre for Advanced Technologies, Callide Coalfields, South Blackwater Coal Ltd., BHPAC and Yarrabee Coal Company Pty Ltd.

REFERENCES

AS 1038.15 (1987). "Methods for the analysis and testing of coal and coke - fusibility of higher rank coal ash and coke ash". Australian Standards Association.

ASTM D1857-87 (1987). "Standard test method for fusibility of coal and coke ash". American Society for Testing and Materials.

Australian Coal Report (1994). "Coal 1994". Published by Australian Coal Report, Sydney, 168.

BS 1016, Part 15 (1970). "Fusibility of Coal Ash and Coke Ash". British Standard.

Coin C.D.A., Kahraman H. & Reifenstein A.P. (1994). "Improved Ash Fusion Test". Australian Coal Association Research Project 3092, 54.

Conn R.E. and Austin L.G. (1984). "Studies of sintering of coal ash relevant to pulverised coal utility boilers. 1:Examination of the Raask shrinkage-electrical resistance method". *Fuel*, 63, 1664.

Creelman, R.A. (1994). "Using mineralogy as a guide to understanding slagging: a case study". In proceedings of Practical Workshop on Impact of Coal Quality on Power Plant Performance, May 15-18, Brisbane, Paper 14, 12.

Cumming I.W., Joyce W.I. and Kyle J.H. (1985). "Advanced techniques for the assessment of slagging and fouling propensity in pulverised coal fired boiler plant" *J. Inst. E.*, 58.

Cumming I.W. and Sanyal A. (1981). "An electrical conductance method for predicting the onset of fusion in coal ash". Proceedings of the Engineering Foundation Conference, 12-17 July, Henniker, New Hampshire. Richard W Bryers (Ed), 1983, Engineering Foundation.

DIN 51730 (1984). "Determination of Fusibility of Fuel Ash". German Standard.

Ellis G.C., Ledger R.C. and Ottrey A.L. (1987). "Thermo-mechanical analysis as a technique for characterisation of ashes and deposits formed during combustion". Steaming coal testing and characterisation workshop, Institute of Coal Research, Univ. of Newcastle, NSW, Australia, 1987, 17.

GE 219-74. "Coal Ash Fusibility Determination Method". Peoples' Republic of China National Standard.

Gerald P.H., Huggins F.E. and Dunmyre G.R. (1981). "Investigation of the high-temperature behaviour of coal ash in reducing and oxidising atmospheres". *Fuel*, 60, 585.

Gibson J.R. and Livingston W.R. (1991). "The sintering and fusion of bituminous coal ashes". Engineering Foundation Conference on Inorganic Transformations and Ash Deposition During Combustion, Palm Coast, Florida, 425-447.

GOST 2057 (1982). "Methods of Determining Ash Fusibility". State Standard of the USSR.

Gray V.R. (1987). "Prediction of ash fusion temperatures from ash composition for some New Zealand coals". *Fuel*, 66, 1230.

Hough, D.C. (1990). "ASME Ash Fusion Research Project". Amer. Soc Mech. Eng. Report No. FT/88/01, 7.

Huggins F.E., Deborah A.K. and Gerald P.H. (1981). "Correlation between ash-fusion temperatures and ternary equilibrium phase diagrams". *Fuel*, 60, 577.

ISO 540 (1981). "Determination of Fusibility of Ash". International Standard Organisation.

Lloyd W.G., Riley J.T., Risen M.A., Gilleland S.R. and Tibbitts R.L. (1989)a. "Estimation of ash softening temperatures using cross terms and partial factor analysis". *Energy & Fuels*, 4, 360.

Lloyd W.G., Riley J.T., Risen M.A., Gilleland S.R. and Tibbitts R.L. (1989)b. "Estimation of ash fusion temperatures from elemental composition: a strategy for regressor selection". Amer. Chem. Soc. Divn. Fuel Chem. Preprints, 36, 325.

Lloyd W.G., Riley J.T., Risen M.A., Gilleland S.R. and Tibbitts R.L. (1993). "Estimation of ash fusion temperatures". *J. Coal Quality*, 12(1), 30-36.

Queensland Coal Board (1990). "Queensland coals: physical and chemical properties, colliery and company information". 8th Edition. Published by The Queensland Coal Board, Brisbane.

Raask E. J. (1979). "Sintering characteristics of coal ashes by simultaneous dilatometry-electrical conductance measurements". *J. Thermal Analysis*, 16, 91.

Rees O.W. (1964). "Composition of the ash of Illinois coals". Illinois State Geological Survey, Urbana, Circular 386.

Riley J.T., Lloyd W.G., Risen M.A., Gilleland S.R. and Tibbitts R.L. (1989). "Predicting ash fusion temperatures from elemental analysis". In Proceedings of the 6th International Coal Testing Conference, 1989, 58.

SABS Method 932. "Fusibility of Coal Ash". South African Bureau of Standards.

Sanyal A and Cumming I.W. (1981). "An electrical resistivity method for detecting the onset of fusion in coal ash". US Engineering Foundation Conference on Slagging and Fouling from Combustion Gases, Henniker, 329-341.

Sanyal A. and Mehta A.K. (1993). "Development of an electrical resistance method based on ash fusion test". Engineering Foundation Conference on Impact of Ash Deposition on Coal Fired Plant, June 20-25, Sollihul, England.

Sleiger W.A., Singletary J.H. and Kohut J.F. (1988). "Application of a microcomputer to the determination of coal ash fusibility characteristics". *J. Coal Quality*, 7, 48.

Sondreal E.A and Ellman R.C. (1975). "Fusibility of ash from lignite and its correlation with ash composition", U.S. Bureau of Mines Rept., GFREC/RI-75-1, Pittsburgh.

Vorres K.W. (1979). "Effect of composition on melting behaviour of coal ash". *J. Eng. Power*, 101, 497.

Wall T.F., Gupta R.P., Polychroniadis P., Ellis G.C., Ledger R.C. and Lindner E.R. (1989). "The strength, sintering, electrical conductance and chemical character of coal ash deposits", NERDDC Project No. 1181 - Final Report, Vol. 1, Summary Report.

Winegartner E.C. and Rhodes B.T. (1975). "An empirical study of the relation of chemical properties to ash fusion temperatures". *J. Eng. Power*, 97, 395.

AN ASSESSMENT OF COAL-ASH SLAGGING PROPENSITY USING AN ENTRAINED FLOW REACTOR

I. S. Hutchings, S. S. West[T] and J. Williamson

Department of Materials, Imperial College, London SW7 2BP, UK
[T]ETSU, Harwell, Didcot, Oxfordshire, OX11 0RA, UK

ABSTRACT

This paper describes the design and operation of an entrained flow reactor to assess the slagging propensity of a coal-ash. Temperatures and residence times have been chosen to closely simulate those experienced by pulverised fuel (pf) particles in a full-size utility boiler. Ash deposits have been collected on ceramic coupons at 1500°C and 1200°C and on an air-cooled metal probe at 830°C.

Ten UK coals and one US coal were selected to give a wide range of coal-ash compositions, a range similar to that found at many power stations. Deposits ranged from dense, highly fused material collected at 1500°C, to lightly sintered ash-particles collected at 830°C. A visual inspection of the deposits allowed a provisional ranking of the slagging propensity to be made. A computer-controlled scanning electron microscope (CCSEM) technique has been developed to provide a quantitative characterisation of each microstructure, thus providing the basis for a more rigorous assessment of the slagging propensity.

The technique described provides the basis for a reliable assessment of coal-ash slagging propensity to be made from a few kgs of coal. It removes many of the uncertainties associated with conventional indices and the previous subjectively based laboratory techniques.

INTRODUCTION

The efficient operation of modern coal-fired pf plant is dependant on fuel quality, high conversion efficiencies and high availability. The nature and quantity of the mineral matter present in a coal can have a profound effect on the operation and performance of the boiler. The formation of slags and deposits not only reduces the thermal efficiency of a boiler, but can give rise to increased exit gas temperatures and fouling of the convective heat exchangers. Enforced outages due to blocked ash hoppers or fractured boiler-tubes are costly occurrences. A reliable method for predicting the

Applications of Advanced Technology to Ash-Related Problems in Boilers
Edited by L. Baxter and R. DeSollar, Plenum Press, New York, 1996

201

slagging propensity of a coal has long been a goal set by fuel-technologists and boiler-operators.

The UK collaborative project "Minimising the Effects of High Temperature Coal-Ash Deposition in Pulverised Coal-Fired Boilers", was designed to address these issues, with particular reference to the problems that can be caused when burning UK bituminous coals. The project was described in some detail at the previous Engineering Foundation conference by Gibb (1994), and some of the major findings by Gibb (1995) at the present conference.

The project involved extensive collaboration between the UK electricity generators, manufacturing industry and academic institutions, with tests ranging from a full-size utility boiler (PowerGen's Ratcliffe Power Station), to combustion rigs and laboratory testing. This paper presents in more detail one specific activity, namely the design and construction of a scaled reactor that would simulate pulverised-coal combustion conditions, thus enabling representative deposits from small quantities of coal to be produced from which an assessment of the slagging propensity of the coal-ash could be established.

A key element of the UK project was the series of trials on a unit at the Ratcliffe Power Station, with the same coals then being burnt on rigs at smaller scale. Each trial provided a series of deposits and ashes that could be used to compare with slags and ashes obtained from the laboratory facilities.

As coal passes through a pf boiler, many varied reactions take place at the high temperatures to the adventitious and inherent mineral matter. The changes from pf mineral occurrences to pulverised-fuel ash (pfa) for the three main coals used during the collaborative project have already been described (Wigley and Williamson, 1994). Only with the most advanced CCSEM techniques can these changes be characterised in any detail, to provide the potential to establish a reliable slagging predictor.

Since the 1950's, when pf combustion became the widely accepted practice for coal-fired power generation, attempts to predict the slagging propensity of a coal have taxed the coal community. Laboratory based techniques, such as ash-fusion temperatures, high-temperature microscopy, electrical conductivity of ash pellets and strengths of sintered ash pellets have all been based on the properties of the high-temperature coal-ash and fail to reflect the complex mineral interactions that occur during the combustion processes. Empirical indices based on the chemical composition of the ash, such as the base:acid ratio, Rs value and the temperature of critical ash viscosity all assume that the ash behaves as a homogeneous material. The chemical inhomogeneity of most boiler deposits is sufficient to indicate that such approaches are likely to have little chance of success. Inevitably the most reliable predictions come from the use of combustion rigs that closely replicate the combustion processes; both of which are costly and may be difficult to arrange.

It has been clear for a number of years that no advance in laboratory based techniques was likely unless an assessment could be made on an ash/slag that had been produced under conditions that closely simulated those in a large boiler. Drop-tube furnaces have been used in coal research for many years. They have provided valuable information on volatile matter release, N_2 evolution and carbon burn-out. Typically these furnaces operated at temperatures up to 1200°C, with particle residence times of about one second. The temperature gradients seen by pf particles in these furnaces were different to those in large boilers. Thus, the mineral matter transformations were quite different to those that occur at full-scale.

This paper describes the design and operation of a reactor in which pf particles would experience a temperature gradient of 1650°C to 1200°C, with a residence time of 2-3 seconds. Thus, the conditions would closely match those in a large boiler as a pf coal particle passes from the burner to the superheaters. It was also desirable that all particles, ie ash and char, should have the same time-temperature history, thus all particles should be entrained in the gas flow. Facilities for sampling ash particles at various temperatures and times would enable:

- mineral transformations to be followed;

- deposits to be collected on a probe;

- deposits to be aged.

Both particulate material and deposits would be needed for characterisation and this would require the latest CCSEM techniques from which a measure of ash particle interactions could be quantified to enable the slagging propensity of the ash to be assessed.

DESIGN OF THE ENTRAINED FLOW REACTOR

The entrained flow reactor (EFR) consists of a vertical multi-zone furnace approximately 5m in length, with an internal diameter of 100mm. The assembly of four high temperature furnaces to form a continuous unit was designed to allow maximum flexibility in setting up a temperature gradient from 1650°C at the top to 1200°C or less at the base, see Figure 1. The joint between each furnace accommodates two diametrically opposed sight or sampling tubes, each tube being fitted with either a silica glass window or a blanking plate.

Coal particles are introduced into the top of the reactor via a vibrating hopper system, and are entrained in the primary gas flow of approximately 70 l/min (STP), equivalent to approximately 450 l/min at 1650°C. With laminar gas flow, gas velocities in the central region of the worktube would be approximately twice the average velocity. Calculations have shown that the flow rate was sufficient to entrain all particles <80μm. Particles would have a residence time of approximately three seconds from the top to bottom of the reactor, spending approximately 0.5 seconds in the 1650°C zone. It must be recognised that a laboratory based combustion facility will never approach the levels that flux a real boiler. However, establishing particle residence times equivalent to those for particles between the burner and convective passes in a boiler, will at least enable mineral transformations to proceed to the same degree. The specification for each of the four furnaces is shown in Table 1.

Much effort was given to the design and construction of a reliable coal feed system. The objective was to be able to deliver a whole pf feed at a rate of 50-200gh^{-1}. A screw feed system was found to operate without blocking only if pf particles of <38μm were removed, thus changing the composition of the ash. Fluidised bed systems were investigated, but concerns over density and size segregation were too large to ignore. The system that was finally adopted is based on a vibratory hopper with precise control

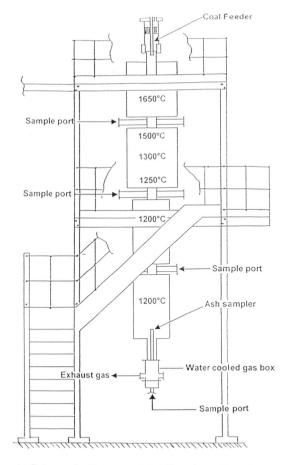

Figure 1. Schematic diagram of the Entrained Flow Reactor with supporting structure

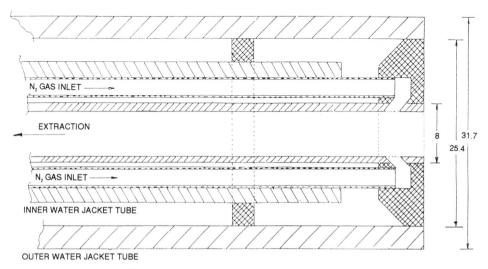

Figure 2. Water-cooled particle extractor

of the vibration amplitude (Coote Vibratory Co Ltd). The vibratory hopper is enclosed in a gas-tight Perspex box, into which the primary air is introduced. Pf particles leaving the hopper mix with the air stream and pass through a water-cooled stainless steel delivery tube into the top of the reactor. Typically the hopper is charged with 200-300g of pf. Tests have shown that the feed rate remains constant over a 1-2 hour period and that no detectable size separation occurs in the hopper during a normal run. Only coals with an abnormally small particle size distribution have given any difficulty in providing a continuous feed.

The top of the reactor is completed with a water-cooled stainless steel gas-box with a total volume of 10 l, into which the secondary air is introduced. Adjustments to the primary and secondary gas-flows enabled a well defined column of pf particles to be established in the central region of the work-tube. The use of flow straighteners was found to cause more, not less, perturbations of the gas flow, it also created unacceptably high backpressures.

The bottom of the reactor is fitted with a water-cooled stainless steel jacket to cool the exit gases and particulate material before they pass into a cyclone. Ash collected from the cyclone was found to closely resemble the pfa collected from electrostatic precipitators at a power station. A small percentage of the finest ash particles escape the primary cyclone, but may be collected with a second smaller cyclone. The second cyclone was found to operate with too high a pressure drop, causing the reactor to operate over-pressure, an undesirable condition for operation. Gases leaving the cyclone pass through a water-cooled heat exchanger to drop the gas temperature to <100°C, and then through a bag-filter before being vented to atmosphere.

All essential gas temperatures, probe temperatures and gas flow rates were monitored continuously and logged on a PC, allowing any changes to the conditions during a run to be monitored.

Ash particle collection probe

To remove ash particles from regions of the furnace where the gas and wall temperatures could be as high as 1550°C it was necessary to use a water-cooled probe, connected to a vacuum line to extract the gas stream with its burden of ash particles. However, in order to prevent further oxidation and mineral transformations, it was also necessary to quench the particles in N_2 and to remove them from the gaseous environment as quickly as possible, as shown in Figure 2. The probe was constructed from a set of concentric stainless steel tubes. The central tube was 6mm ID, 8mm OD. Near the end which was inserted into the furnace, eight 3mm holes were aligned with a rebated slot in the plug into which the central tube and the most external tube (31.7mm in diameter) were welded. Around the diameter of the central 8mm tube, four one-eighth inch OD tubes were placed and welded into the end plug. These are used to bring N_2 to the end of the probe, where it exits through the holes in the 8mm tube. By angling the stream away from the entrance, the flow of N_2 actually helps the removal of particles rather than causing a baffling at the entrance. The extraction of particles from the gas stream was accomplished using the BCURA cyclone.

Table 1. Specifications of furnaces forming the Entrained Flow Reactor

	Furnace number			
	1	**2**	**3**	**4**
Furnace temperature (°C)	1700	1500	1200	1200
Heating elements	$MoSi_2$	SiC	Kanthal AF wire	Kanthal AF wire
Work tube	Recrystallised alumina	Impervious mullite	Kanthal APM	Kanthal APM
Length (mm)	1250	1200	1150	1150
Internal diameter (mm)	100	100	104	104
Heated length (mm)	650	750	800	800
Temperature controller	Eurotherm 902P	Eurotherm 808	Eurotherm 808	Eurotherm 808
Temperature sensor	Pt20%Rh-Pt40%Rh	Pt-Pt13%Rh	Pt-Pt13%Rh	Pt-Pt13%Rh
Thermal insulation	High alumina hot-face brick	Ceramic fibre board	Ceramic fibre board	Ceramic fibre board

Table 2. Typical operating conditions for Entrained Flow Reactor

	Settings (a)	Settings (b)
Primary air (l/min)	4	4
Secondary air (l/min)	70	60-65
Top pressure (mm H_2O)	-2	-1
Bottom Pressure (mm H_2O)	-8	-6
Furnace 1 (°C)	1650	1650
Furnace 2 (°C)		
Zone 1	1500	1500
Zone 2	1400	1450
Zone 3	1300	1400
Furnace 3 (°C)	1200	1200
Furnace 4 (°C)	1200	1200
Coal feed rate (g/hr)	100	100

Settings (a) and (b). An increase in the temperatures of Furnace 2 gave higher temperatures at Port 2, with more fused deposits formed

Air-cooled metal deposition probe

This was based on earlier designs of deposition probes used by the CEGB and by the Coal Research Establishment. Early trials were conducted with mild steel (MS) coupons which had thermocouple location points drilled along their lengths. However, MS proved unsatisfactory for applications over 600°C, as considerable oxide formation and scaling occurred. It was thought that the use of P22 low alloy (2¼% Cr-1% Mo) boiler tubing would overcome this problem. However, this also scaled at 800°C. Therefore, the metal deposition work was conducted using grade 316 stainless steel coupons, the same stainless steel used in the construction of the probe itself. A separate SS holder for the thermocouple was used because of the difficulty of drilling small holes along the coupon lengths with these materials (Figure 3). As long as the holder was placed well into the heated zone of the furnace, there was little evidence of loss of accuracy of temperatures recorded in the heated zone.

Ceramic deposition probe

The use of an uncooled ceramic deposition probe has several advantages. It provides a hot oxide surface with a surface temperature similar to that of a mature boiler deposit. Thus arriving ash particles behave in a similar way to those in a boiler which already shows some deposition. The choice of mullite provides a solid surface with a composition similar to the coal ash. The mullite is also resistant to thermal shock on removal of the hot probe.

The probe used was of simple construction, consisting of a mullite tube or coupon 16 mm OD and +12.5 mm ID, over a metre length of 12 mm OD mullite tubing, see Figure 4. The coupon was originally 150 mm in length but was later increased to 200 mm length for ease of working. The coupon was held in place on the 12 mm tube with a layer of high alumina cement mixed with a sodium silicate. The cement was also used to block the end of the 12 mm tubing to prevent gas leakage when it was inserted into the furnace. An aluminium holder with an "O" ring assembly was used to hold the probe in the furnace.

OPERATING CONDITIONS FOR THE EFR

Furnace temperature and gas flow rates were adjusted to give conditions that would closely simulate the passage of pf particles through a boiler. A Stokes Law calculation showed that for a mean gas temperature of 1400°C, a gas velocity of $1.5 ms^{-1}$ would be required to entrain the heaviest particles (pyrite and its derivatives) for particles with diameters of up to 80μm, see Figure 5. This would be achieved with gas flow rates of 150 l/min. However, with near laminar flow conditions down the reactor, gas velocities in the central region would be approximately twice the average gas velocity. Thus, a gas flow rate of 70 l/min at STP would be sufficient to establish conditions for entrained flow of most particles, see Figure 6. A typical set of operating conditions for the EFR is shown in Table 2.

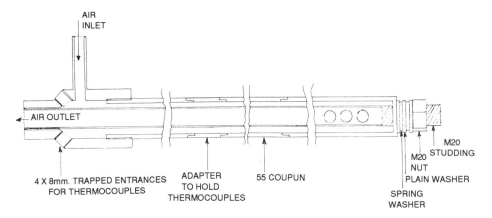

Figure 3. Air-cooled metal deposition probe

AIR
INLET

AIR OUTLET

4 X 8mm. TRAPPED ENTRANCES
FOR THERMOCOUPLES

ADAPTER
TO HOLD
THERMOCOUPLES

55 COUPUN

M20
NUT

M20
STUDDING

PLAIN WASHER

SPRING
WASHER

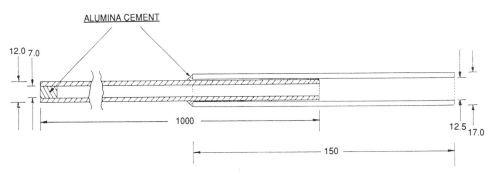

Figure 4. Ceramic deposition probe

ALUMINA CEMENT

12.0 7.0

1000

150

12.5 17.0

Choice of coals

Ten UK and one US bituminous coals were selected to standardise and characterise the reactor. The coals included the three UK coals used for the power station and other large scale trials as part of the UK slagging project (Gibb, 1995). The additional coals, all of which are used in UK power stations, were selected to extend the range of known slagging propensity, which ranged from innocuous to severe. The coals and their ash analyses are shown in Table 3.

Production of ash and deposits

The ceramic probe was used to collect deposits at Ports 1 and 2, where the gas temperatures were approximately 1500°C and 1200°C respectively. The deposits collected at 1500°C were all of a highly fused nature, with the most fused deposits having flowed around the coupon representing mature deposit growth. Deposits collected at 1200°C were of a lightly sintered nature, with distinctive visual differences in the degree of fusion depending on which coal had been used. Deposits collected at Port 2 on the air-cooled metal probe, where the metal temperature was cooled to 830°C, were of a dusty to lightly sintered nature. Temperatures which were below 830°C, and which could have been lowered to 370°C to represent a water wall, would have given little or no deposition under realistic collection times. Again differences could be noted by eye depending on which coal had been used. From each run a representative bottom ash was collected from the cyclone for chemical analysis and carbon content determination.

CHARACTERISATION OF DEPOSITS

Following the removal of samples (deposit and coupon) from the EFR, the amount of material collected was calculated, the samples were then photographed and a small sample taken for x-ray diffraction analysis. Each deposit was then coated with a thin layer of epoxy resin to hold the fragile material together. Sections through the centre of the deposit were then cut with a diamond saw, normal to the coupon, to give a sample 5 mm in length. This sample was then set in resin so that a cross section through the deposit, from the interface with the coupon to the surface, could be examined. Set samples were ground and polished to a 0.25 μm finish and carbon coated for SEM examination. At low magnification (10×), three back scattered images (BSI) were obtained to give an overall view of the sample. A more detailed examination was then made of two separate areas at a higher magnification, typically 50-200×. Quantitative data and image analysis was then performed on the BSI using a 16×16 grid, giving 256 analysis points from which quantitative EDS data was collected using PROSA corrections. The porosity of each deposit was determined by two methods. The first being from a grey scale image, using this to produce a binary image where only the deposit was shown, and the second method being from the fraction of analysis points reporting as deposit out of the total number of analysis points.

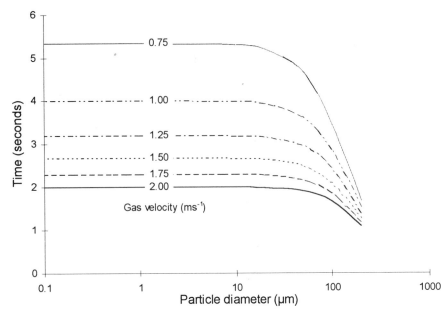

Figure 5. Residence times for pyrite particles, neglecting turbulence, buoyancy, edge effects, and initial acceleration. An average temperature of 1400°C has been used to calculate density and viscosity of the gas phase

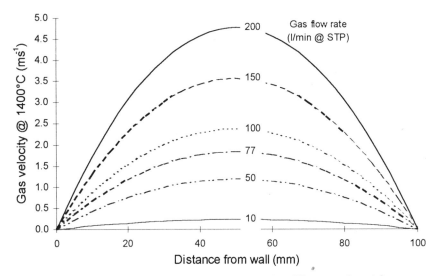

Figure 6. Velocity profile across the reactor for different volumetric gas flow rates

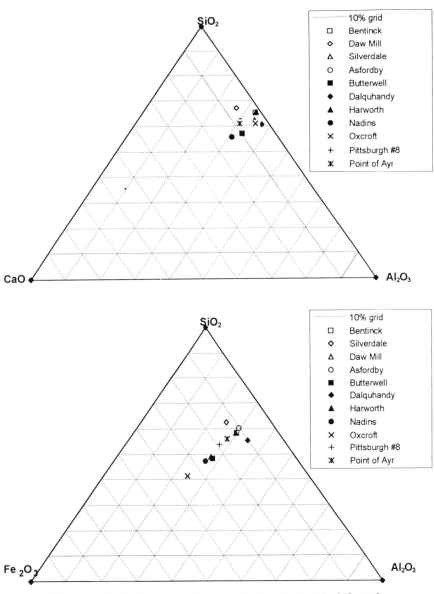

Figure 7. Coal ash compositions plotted on CaO-Al₂O₃-SiO₂ and Fe₂O₃-Al₂O₃-SiO₂ ternary diagrams

The composition of each point reporting as ash was first stored for subsequent processing. The bulk composition of each sample was calculated and the results shown graphically on either $CaO-Al_2O_3-SiO_2$ or $Fe_2O_3-Al_2O_3-SiO_2$ ternary diagrams. For subsequent analysis of data only compositions where $CaO+Al_2O_3+SiO_2$ or $Fe_2O_3+Al_2O_3+SiO_2$ >50wt% were plotted on ternary plots. This was because other compositions could be normalised to 100% wt but could give erroneous results. Each of the ternary plots shows not only the range of individual analyses, i.e. the degree of inhomogeneity of the sample, but also the degree of interaction between the clay residues with either iron- rich or lime- rich particles from the coal.

The EFR slagging index now proposed is based on a microstructure analysis, in which a comparison of the microstructure from each of the deposits showed a range of particle-particle interactions with varying degrees of fusion. Ash samples from the coals with the lowest slagging propensity would be expected to give rise to few agglomerates. At the other end of the scale, large agglomerates, formed from interactions between many ash particles, would be expected with a high degree of fusion observed.

In order to provide an estimate of the potential fusion behaviour of an ash, composition limits were placed on each analysis. Both CaO and FeO/Fe_2O_3 oxides act as strong fluxing agents, lowering both the liquidus temperatures and the ash viscosity at any temperature. However each oxide is itself quite refractory, so that with increasing concentration of either CaO or FeO/Fe_2O_3 a point is reached when the liquidus temperatures start to increase (West et al, 1993). Limits for the concentration of CaO at 5-40wt% and for Fe_2O_3 at 10-50 wt% were chosen to determine the fraction of particles which were most likely to act as potentially sticky particles, or substrates, with the sum of both sub-sets likely to more closely represent the real situation.

RESULTS AND DISCUSSION

Characterisation of the coals

Bentinck, Daw Mill and Silverdale coals were the UK coals chosen for the full-size boiler trials. Their slagging behaviour during these trials has been described in detail by Gibb (1995). Bentinck is a coal with a relatively low iron and calcium content, with an Rs value of 0.44, ash shown in Table 3. This is a coal that causes few , if any, problems at a power station. On the other hand, both Daw Mill and Silverdale coals, the first with a high Fe_2O_3 and CaO, and the second a very high Fe_2O_3 ash, have Rs values of 0.46 and 1.35 respectively. Both coals are known to cause boiler slagging. Of the remaining coals that were used to calibrate the EFR, Oxcroft, an ash with an Fe_2O_3 content of 32wt% had the highest Rs value at 2.40, while Dalquhandy, a low Fe_2O_3 coal had the lowest Rs value of 0.17. Ash compositions, when normalised to either $CaO-Al_2O_3-SiO_2$ or $Fe_2O_3-Al_2O_3-SiO_2$ are shown in Figure 7. The coal ash compositions tend to lie in a narrow band that stretches from a point on the SiO_2-Al_2O_3 binary to either the CaO or Fe_2O_3 apex. This suggests that the coal-ash compositions are derived from mixtures of quartz (SiO_2) and clays (aluminosilicates) to which with limestone (CaO) or pyrite ("Fe_2O_3") had been added. In fact x-ray diffraction analysis of LTA from each coal showed that the major phases were kaolinite, illite and quartz, with smaller amounts of pryite, calcite, dolomite, ankerite, magnetite and hematite.

Nature of the bottom ash

Ash collected from the cyclone at the bottom of the reactor closely resembled a pfa from a power station burning the same coals, in terms of both ash composition and phases present. The entrained flow reactor ashes typically contained 2wt% or less of unburnt carbon, indicating a relatively high level of burn-out in the reactor. Much of the ash was glassy with quartz, mullite and iron oxides, including spinels from the solid solution series FeO. Al_2O_3-FeO.Fe_2O_3, as the crystalline phases. A Malvern particle size analysis of the bottom ashes showed only a few weight percent to be less the $10\mu m$ in size. Thus a small proportion of the fly ash remained in the exhaust gases after passage through the cyclone.

Depositions formed at 1500°C

Deposits were collected on ceramic coupons at Port 1, where the gas temperatures were close to 1500°C. All the coals gave deposits that were fused to the ceramic substrate. Ash from Point of Ayr coal had been so fluid that it had remained largely glassy on cooling. The Dalquhandy ash gave a fused deposit with a rough surface and some internal porosity and less homogeneity, suggesting a much higher ash fusion temperature. These observation were later confirmed by CCSEM analysis that showed a wide scatter of chemical compositions for the Dalquhandy deposit compared to a much tighter distribution for the other ashes.

Each deposit was ranked visually on the Jones Index (Jones and Riley, 1987), which ranks deposits on a scale 0-10, with 0 for dusty deposits and 10 for well fused slags with a high degree of mechanical strength. Deposits collected from Port 1 at 1500°C ranged from 8 for Dalquhandy to 10 for Daw Mill, Silverdale, Asfordby, Butterwell, Nadins, Oxcroft, Point of Ayr and Pittsburgh #8. Since a very high degree of interaction and move to homogenisation had taken place at this temperature, these deposits were considered to be too well fused for a sensible slagging assessment. Reactor conditions that displayed a wider range of deposition phenomena were considered to be more desirable.

Deposits formed at 1200°C

Gas temperatures at Port 2 were typically in the range 1200-1250°C. Both ceramic and metal coupons were used to collect deposits at this point; this section deals with the deposits collected on an uncooled ceramic probe that was assumed to be at the gas temperature. The deposits were of a friable to dusty nature, with a Jones Index of 0 to 2. Deposits were impregnated with an epoxy resin before mounting and polishing a cut section. The microstructures showed high porosities, typically in the range 65-80%, as found in real deposits (Wain et al, 1991). SEM backscattered images showed a wide range of particle-particle interactions, with noticeable differences between each coal. The more fused deposits consisted of aggregates of particles up to several

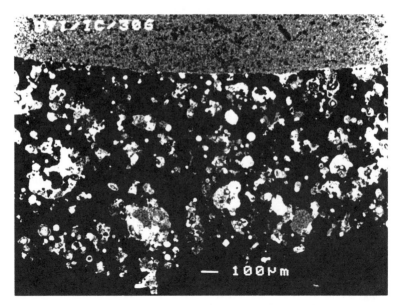

Figure 8. Section through Butterwell coal-ash deposit collected on a
ceramic cupon at 1200°C, backscattered image

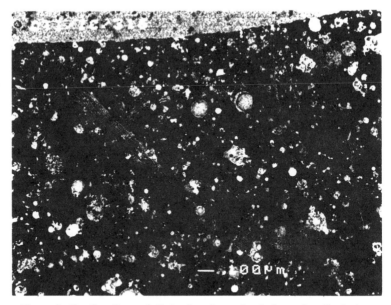

Figure 9. Section through Dalquhandy coal-ash deposit collected on
a ceramic cupon at 1200°C, backscattered image

hundred micrometers in size, with few discrete ash particles visible. Figure 8 shows a section through the Butterworth ash deposit, showing extensive particle interactions. As an example of an ash showing minimal particle interactions, Figure 9 shows a section through the Dalquhandy deposit.

A CCSEM analysis of each microstructure was made using a 16X16 grid, giving 256 point analyses. Data was acquired from at least two areas, one close to the surface of the deposit and the other adjacent to the interface with the coupon, from each microstructure to give a total of 512 analyses. Ash compositions were found to vary widely from point to point. The majority of the analysis were of aluminosilicates, with varying amount of Fe_2O_3 or CaO. The analysis data obtained from sections shown in Figures 8 and 9 is shown graphically in Figures 10 and 11. All deposits gave a high density of analysis points close to Al_2O_3-SiO_2 binary. These are clearly ash particles of clay origins. A significant number of the aluminosilicate particles have reacted with either CaO or Fe_2O_3, or both, giving a distribution that extends to either the CaO or the Fe_2O_3 apex. Pure aluminosilicate ash particles would be considered to be relatively refractory, with a very high viscosity at the deposition temperature. However, CaO and Fe_2O_3 are both effective fluxing oxides lowering both the liquidus temperature and the viscosity at a given temperature. Additions of CaO will initially lower the liquidus temperature of the coal-ashes, and a minimum is reached at 25-30wt% CaO, after which further additions increase the liquidus temperatures (West et al, 1994). Similarly a minimum is reached in the viscosity-composition curve with approximately the same amount of added CaO. Additions of Fe_2O_3 to an aluminosilicate melt show somewhat different behaviour. Initially both liquidus temperatures and viscosities fall, but Fe_2O_3 has a limited solubility in the melt and once exceeded, precipitation of dendritic iron oxides effectively increase the bulk viscosity.

With these effects in mind, limits were placed on the concentrations of both CaO and Fe_2O_3 to determine a range of ash compositions that would contribute to or enhance, the slagging propensity of an ash. If ash particles with <5wt% but >40wt% CaO, and <10wt% but >50wt% are excluded from the CCSEM analysis, then the remaining points would be expected to contribute significantly to the agglomeration and sintering processes. The fraction of data points that lay between these limits is shown in Table 4 for each coal. Thus deposits showing the highest number of points within these limits contain the highest proportion of particles which at some stage would have been well above the solidus temperature, and these coal-ashes would therefore be ranked as having the highest slagging propensity. Conversely, deposits showing the fewest data points within these limits would show a much reduced tendency to agglomerate. This method for ranking the slagging propensity of a coal-ash has the advantage of producing a numerical index that is based on the actual microstructure of a deposit that has been produced under conditions that simulate combustion in a pf boiler. The order of slagging propensity shown in Table 4 is broadly in line with power station experience with these coals. Where anomalies do occur these could well be due to a variety of reasons. This method takes no account of boiler parameters or the operating conditions; perhaps, more importantly, it is a measure of the slagging propensity at a particular temperature, which could account for the unexpected values in the EFR index for Silverdale and Bentinck coals. Likewise, the position of Dalquandy gave rise to some concern since this coal would be expected to be of low slagging propensity. The problem may well arise from the analysis of a highly porous microstructure which

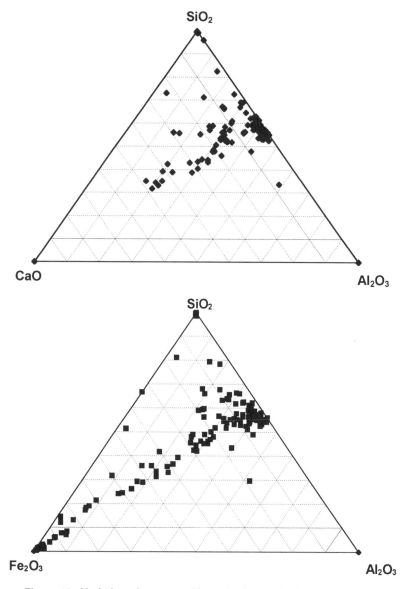

Figure 10. Variations in composition of point analysis obtained from
section through a Butterwell coal-ash deposit on a ceramic
coupon, normalised to CaO-Al$_2$O$_3$-SiO$_2$ and Fe$_2$O$_3$-Al$_2$O$_2$-SiO$_2$

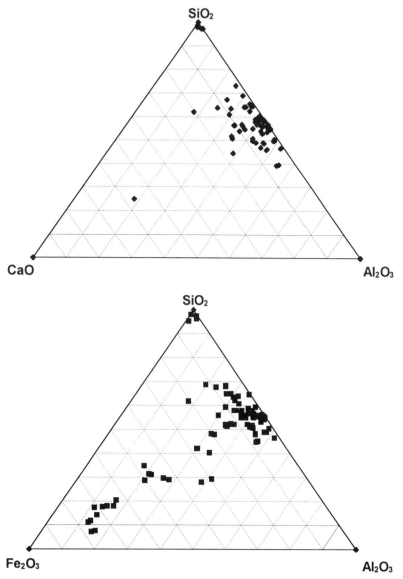

Figure 11. Variations in composition of point analyses obtained from section through a Dalquhandy coal-ash deposit on a ceramic coupon, normalized to $CaO-Al_2O_3-SiO_2$ and $Fe_2O_3-Al_2O_3-SiO_2$

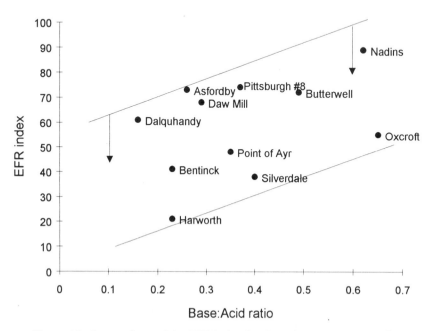

**Figure 12. Comparison of the EFR index for deposits fromed at 1200°C
with base: acid ratio of each coal-ash**

gave rise to a limited number of data points. Such problems may well disappear as the conditions under which deposits are collected are optimised.

In terms of the actual amount of data collected, of the 500 or so analysis points, 70% or more lie on pores within the deposit, thus significantly reducing the accuracy of the assessment. These limitations have been recognised as a likely source of errors, and assessment procedures are currently being developed to provide more data points for each microstructure, improving both the reliability and reproducibility of this technique.

The same coals as shown in Table 3 were used to collect deposits on an air-cooled metal probe inserted at Port 2 in the EFR. The metal surface temperature was maintained at 800-830°C. These deposits showed only very light sintering and had porosities in the range 80-90%. A similar CCSEM analysis was made of the microstructure which allowed the coals to be placed in order of slagging propensity. However, at this temperature the number of particle-particle interactions was much less than had occurred at 1200°C and thus the proportion of data points that lay between the chosen compositional limits was much reduced. While the data was valuable, it was considered to be less accurate than that obtained for the deposits formed at 1200°C.

Concluding remarks

A technique that is based on a microstructural characterisation of a deposit, that takes account of mineral-mineral interactions and mineral distributions in the pf, must hold out the possibilities of improvements on earlier methods based on empirically derived relationships. For example, an assessment based on the base:acid ratio of a coal-ash has to assume that the ash is a homogeneous product of combustion. An EDS analysis of individual particle in a pfa shows how far from homogeneity an ash can be. The fact that few, if any, residues of calcite or dolomite can be seem in a pfa is evidence that these oxides are rapidly assimilated into the aluminosilicate ash particles. On the other hand, particles rich in iron oxides (or silica) are frequently encountered in a pfa and many of these can be detected in a deposit. Thus, the effectiveness of pyrite to form an iron oxide flux must be less than the actual amount of pyrite in the coal. Nevertheless, the base:acid ratio (and the Rs values) have indices that assume that their effectiveness is the same. Fuel technologists frequently use both base:acid and Rs for the prediction of the slagging propensities. Figure 12 shows the EFR index, obtained from CCSEM data from deposits from the EFR, plotted against the base:acid ratio from each of the coal-ashes. Each coal-ash lies within a broad band that shows the slagging propensity increasing with base:acid ratio. The top of the band could be taken as an indication of the maximum slagging propensity, if one assumed that all of the fluxing oxides were homogeneously dispersed throughout the ash. Coal ashes that lay within the band would indicate a reduce effectiveness of the fluxing oxides. Thus, two coals with identical base:acid ratios could exhibit entirely different slagging propensities due to the nature and distribution of mineral matter in the pfa. A CCSEM analysis of a pf may be a valuable precursor in determining ash properties (Wigley et al, 1995).

Table 3. Composition of ashes from the coals used in the Entrained Flow Reactor

Sample	wt% ash	Na_2O	MgO	Al_2O_3	SiO_2	P_2O_5	K_2O	CaO	TiO_2	MnO	Fe_2O_3	B/A ratio	Rs
Bentinck	17.4	1.7	1.4	27.1	53.1	0.1	3.2	1.4	1.1	0.1	11.0	0.23	0.44
DawMill	15.8	1.2	2.0	21.7	55.1	0.2	2.6	5.2	1.2	0.2	11.1	0.28	0.46
Silverdale	15.1	0.6	1.0	25.1	45.5	0.2	2.0	2.5	1.0	0.1	22.5	0.40	1.35
Asfordby	7.5	1.5	1.7	25.4	51.8	0.5	2.1	6.6	1.3	0.1	8.8	0.26	0.21
Butterwell	9.8	0.1	3.3	24.3	41.9	0.2	1.0	7.1	1.0	0.2	20.9	0.49	1.00
Dalquhandy	8.9	0.1	0.8	32.2	52.2	0.6	1.0	1.8	1.3	0.1	9.6	0.16	0.17
Harworth	16.1	1.3	1.3	26.8	53.2	0.2	3.8	1.1	1.0	0.1	10.9	0.23	0.46
Nadins	13.9	0.4	4.5	21.6	39.1	0.2	1.7	9.4	0.8	0.2	21.8	0.62	1.79
Oxcroft	10.8	0.2	1.5	21.9	37.8	0.4	3.0	2.4	0.6	0.1	31.7	0.65	2.40
Pittsburgh #8	6.9	0.8	1.1	24.0	47.6	0.4	1.7	6.2	1.0	0.1	16.9	0.37	0.73
Point of Ayr	12.9	0.3	3.0	24.3	48.4	0.2	2.2	6.7	0.8	0.1	13.6	0.35	0.55

Table 4. CCSEM data on ash deposits collected at Port 2 (1200°C) on a ceramic cupon

Coal	(a) Jones Index	(b) Porosity (%)	(c) % of analyses between 5- 40wt% CaO	(d) % of analyses between 10- 50wt%Fe_2O_3	(e) EFR index
Bentinck	0.5	78	15	26	41
Daw Mill	1.5	70	35	33	68
Silverdale	1.0	73	17	21	38
Asfordby	0.5	71	45	28	73
Butterwell	0.5	67	40	32	72
Dalquhandy	1.0	79	33	28	61
Harworth	0.5	76	5	16	21
Nadins	1.0	74	48	41	89
Oxcroft	1.0	73	28	27	55
Pittsburgh #8	0.5	69	51	23	74
Point of Ayr	1.0	71	27	22	48

(a)   Jones index, a visual slagging index, see Jones and Riley (1987)

(b)   Porosities obtained from point analyses reporting as (deposit)+(total count)

(e)   EFR index, (e) = (c) +(d)

CONCLUSIONS

- An entrained flow reactor has been constructed to simulate the time-temperature conditions that pf particles experience in utility boilers.

- Ash deposits collected on ceramic coupons at 1500°C and 1200°C, and on air cooled metal coupons at 800°C, have provided samples for a CCSEM microstructural characterisation.

- Deposits formed at 1200°C showed a range of slagging propensities based on varying ash compositions.

- CCSEM data on deposits formed from 11 coals has enabled each coal-ash to be ranked in terms of slagging propensity of the coal-ash.

- The new EFR index takes account of the nature and distribution of mineral matter in the pf and subsequent particle-particle interactions between ash particles when forming a deposit.

- The new index promises a significant improvement in the ability to predict the slagging propensity of a coal-ash, without recourse to large-scale combustion rigs and/or full-size utility boiler trials.

ACKNOWLEDGMENTS

This worked formed part of the UK collaborative project entitled "Minimising the Effects of Coal-Ash Deposition in Pulverised Coal-Fired Boilers", supported by the Department of Trade and Industry through the Coal R&D programme. The help, advice and experience of the industrial collaborators, notably from PowerGen plc, National Power plc and the British Coal Corporation (through the Coal Research Establishment, Stoke Orchard) is gratefully acknowledged.

REFERENCES

Couch, G. (1994). "Understanding Slagging and Fouling in pf Combustion", IEA Coal Research, London, IEACR/72.

Gibb, W.H., Jones, A.R. and Wigley, F. (1994). "The UK Collaborative Research Programme on Slagging in Pulverised Coal-Fired Furnaces: Results of Full-Scale Plant Trials", in *Proceedings of the 1993 Engineering Foundation Conference on The Impact of Ash Deposition in Coal-Fired Plants*, Williamson, J. and Wigley, F., (Eds), Taylor and Francis, Washington, pp3-18.

Gibb, W.H. (1995). "The UK Collaborative Research Programme on Slagging in Pulverised Coal-Fired Boilers: Summary of Findings", This Conference.

Jones, A.R. and Riley, G.S. (1987). "Deposition Characterisation of a CEGB 660MWe Coal-Fired Boiler", *EPRI Conference on The Effects of Coal Quality on Power Plant*, Atlanta, Georgia.

Wain, S.E., Livingston, W.R., Sanyal, A. and Williamson, J. (1991) "Thermal and Mechanical Properties of Boiler Slags of Relevance to Sootblowing", in *Proceedings of the 1990 Engineering Foundation Conference on Inorganic Transformations and Ash Deposition during Combustion*, Benson, S.A.(ed), Engineering Foundation, New York, pp459- 470.

West, S.S., Williamson, J. and Laughlin, M.K. (1994). "Mineral Interactions During Fluidised Bed Gasificiation of Coals", in *Proceedings of the 1993 Engineering Foundation Conference on The Impact of Ash Deposition in Coal-Fired Plants* Williamson, J. and Wigley, F., (Eds) ,Taylor and Francis, Washington, pp89-100.

Wigley, F. and Williamson, J. (1994). "The Characterisation of Fly-Ash Samples and their Relationship to the Coals and Deposits from UK Boiler Trials", in *Proceedings of the 1993 Engineering Foundation Conference on The Impact of Ash Deposition in Coal-Fired Plants*, Williamson, J. and Wigley, F., Taylor and Francis, Washington, pp385-398.

Wigley, F. and Williamson, J. (1995). "Modelling Fly-Ash generation for UK Power Station Coals", This Conference.

MECHANISMS OF ASH FOULING DURING LOW-RANK COAL COMBUSTION

Donald P. McCollor, Christopher J. Zygarlicke, and Steven A. Benson
Energy & Environmental Research Center
University of North Dakota
PO Box 9018
Grand Forks, North Dakota 58202-9018

ABSTRACT

Four low-rank coals were investigated for fouling severity using bench and pilot combustion testing and microanalytical examination of fouling deposits. The coals contained varying levels of alkali and alkaline-earth elements that are commonly associated with initiating and accelerating ash fouling, including Na, Mg, K, Ca, and Fe. Combustion testing revealed that fouling deposits generated from these coals shared common chemical and physical properties. Four test coals from western U.S. coal fields were selected, including the Beulah and Gascoyne lignites from western North Dakota and the Colstrip subbituminous coal from Montana, and the Utah Wasatch from Utah. All of these coals contained significant levels of Na, Ca, and Mg, with the Beulah lignite containing the highest levels of sodium. Sodium in the Beulah and Gascoyne lignites was very abundant and was organically bound. The Utah Wasatch coal contained significant levels of sodium, but it was bound in the coal as a zeolite silicate termed analcime. Deposits were ranked from low-fouling to severe-fouling based on deposit build-up rate, deposit strength, and liquid-phase viscosity, which was calculated based on the chemistry and the gas temperature near the deposits at the time of quenching. Deposit build-up rates and crushing strengths were the highest for the Beulah and Gascoyne coals, followed by the Utah Wasatch, during both bench- and pilot-scale fouling deposition simulations. The outer layers of the lignite deposits showed well developed captive liquid surfaces and silicate liquid-phase viscosity distributions that were shifted to much lower values compared with the Utah Wasatch and Colstrip deposits. Microanalysis of the deposits using scanning electron microscopy revealed that the gluing material or phase that was responsible for the cementing of the severe-fouling deposits was a low-melting-point sodium–calcium-rich silicate. The Utah Wasatch coal, which also contained sodium did not form as much of the low-melting-point sodium–calcium silicate, partly because the sodium was locked within an existing coal mineral phase and was not able to interact with the silicate material during combustion.

INTRODUCTION

A significant portion of operational problems encountered during coal combustion are related to the inorganic constituents of the coal. These include deposition on heat-transfer surfaces, formation of difficult-to-collect fine particles, and erosion and corrosion of surfaces. A precise and quantitative knowledge of the chemical and physical transformations of the coal inorganic

Applications of Advanced Technology to Ash-Related Problems in Boilers
Edited by L. Baxter and R. DeSollar, Plenum Press, New York, 1996

223

constituents occurring during combustion is needed to accurately predict ash behavior as a function of coal composition and combustion conditions.

Coal inorganic constituents consist of a complex mixture of discrete mineral grains, coordinated inorganic elements, and cations associated with oxygen functional groups and clay minerals. Low-rank coals have oxygen contents with oxygen in the form of carboxyl groups accounting for approximately 25% of the total oxygen content [Given, 1984]. These groups can act as ion-exchange sites for significant quantities of Na, K, Ca, Mg, and Ba. High concentrations of these alkali and alkaline earth cations are found in many western coals because of percolation of groundwater through the coal seams. The organically bound form of these cations is highly dispersed in the coal matrix, often resulting in the formation of extremely small, reactive vapor and particulate material during combustion.

Study of ash transformations in full-scale utility boilers is challenging because of the cost and difficulty in maintaining close control of experimental parameters. Bench- and pilot-scale testing are relatively rapid, inexpensive, and allow closer control of experimental conditions while replicating key conditions, such as the time–temperature history of particles in a full-scale system. This, combined with microanalysis techniques developed for the coal and ash, provides new insights into ash transformation mechanisms.

In this study, four test coals from western U.S. coal fields were selected, including the Beulah and Gascoyne lignites from western North Dakota, and the Colstrip subbituminous coal from Montana, and the Utah Wasatch from Utah. All of these coals contained significant levels of Na, Ca, and Mg, with the Beulah lignite containing the highest levels of Na. These coals were combusted in bench- and pilot-scale combusters to examine ash deposition propensity and the physical and chemical characteristics of the deposits.

EXPERIMENTAL

Coal Mineralogy

Table 1 details selected physical and chemical characteristics of the test coals. From x-ray diffraction (XRD) analysis of ash produced by a low-temperature process, quartz is common to all four coals. Pyrite is found in the Beulah lignite and the Colstrip subbituminous coal. Clays, calcite, and analcime were found in the Beulah, Colstrip, and Utah Wasatch coals, respectively. At the time these specific samples were analyzed, the computer-controlled scanning electron microscopy (CCSEM) technique had not been developed to quantitatively infer mineral phases. More recent CCSEM analysis of these coals are presented in Table 2, showing an affirmation of the major minerals observed using XRD. These analysis were performed on different samples of these coals from the same mines.

Information on the association of inorganic material in the coals was determined by chemical fractionation. This procedure uses a sequence of extractions with water, ammonium acetate, and hydrochloric acid to categorize elements as water-soluble, ion-exchangeable, acid-soluble, and insoluble.

The bulk of the alkali and alkaline-earth elements present in western U.S. coals are associated with the organic portion of the coal as salts of carboxylic acid groups. This "organically bound" fraction, corresponding to the ion-exchangeable portion of the chemical fractionation process, has a profound effect on ash deposition behavior. Neville and others [1981] have found that up to 60 % of sodium in a lignite can escape during combustion, and

TABLE 1
Composition of Test Coals

X-Ray Diffraction Qualitative Analysis	Beulah	Gascoyne	Colstrip	Utah Wasatch
Quartz	✓	✓	✓	✓
Pyrite	✓	—	✓	—
Kaolinite	—	—	✓	✓
Montmorillonite	—	—	—	✓
Clays	✓	—	—	—
Calcite	—	—	✓	—
Analcime	—	—	—	✓

Chemical Fractionation, percentage of elements removed by water and ammonium acetate extractions

	Beulah	Gascoyne	Colstrip	Utah Wasatch
Sodium	100	90	61	47
Magnesium	86	94	86	72
Aluminum	0	0	0	0
Silicon	1	0	1	0
Potassium	13	0	5	6
Calcium	81	75	97	85
Titanium	0	0	0	0
Iron	0	0	0	0
Strontium	76	78	86	20
Barium	72	0	96	100

X-Ray Fluorescence Analysis of the Coals, normalized oxide percentages

	Beulah	Gascoyne	Colstrip	Utah Wasatch
SiO_2	23.9	20.6	39.2	48.3
Al_2O_3	11.6	8.3	20.9	17.8
Fe_2O_3	8.2	6.6	5.1	4.5
TiO_2	1.1	1.1	1.2	1.1
P_2O_5	1.0	1.0	0.7	0.8
CaO	17.4	23.5	13.2	10.7
MgO	5.5	7.0	4.3	2.5
Na_2O	4.0	5.8	ND[1]	3.9
K_2O	0.4	0.1	0.3	0.3
SO_3	26.8	25.9	15.0	10.0

As Received Proximate Analysis

	Beulah	Gascoyne	Colstrip	Utah Wasatch
Moisture	30.2	33.1	19.6	3.3
Volatile Matter	30.0	29.3	29.8	39.4
Fixed Carbon	31.3	30.3	40.7	46.8
Ash	8.5	7.2	9.9	10.5

Ultimate Analysis

	Beulah	Gascoyne	Colstrip	Utah Wasatch
Hydrogen	6.68	6.70	5.89	5.40
Carbon	43.95	42.31	53.11	68.48
Nitrogen	0.50	0.54	0.80	1.25
Sulfur	1.56	1.09	0.72	0.46
Oxygen	38.80	42.10	29.59	13.87
Ash	8.50	7.23	9.90	10.50

Calorific Value (Btu/lb)

	Beulah	Gascoyne	Colstrip	Utah Wasatch
Btu/lb	7519	7170	9176	12,277

[1] Not determined.

TABLE 2

Computer-Controlled Scanning Electron Microscopy Analysis of Minerals in Coals[1]
(wt% ash)

	Beulah	Gascoyne	Colstrip	Utah Wasatch
Quartz	5	22	14	12
Pyrite	24	8	12	9
Kaolinite	21	15	19	13
Montmorillonite	2	2	3	4
Clays (mixed and illite)	2	8	9	6
Calcite	0	1	5	6
Analcime	0	0	0	4
Total Minerals (% total ash)	66	57	69	79
Organically Bound Elements (% total ash)	34	43	31	21

[1] Analysis from EERC CCSEM database and not the same coals used in the combustion tests.

Quann and Sarofim [1986] have found that up to 20% of the magnesium and a few percent of the calcium in western U.S. coals were vaporized during combustion. Subsequent condensation and reaction particles can coat larger particles with alkali species such as sodium. The more refractory alkaline-earth oxides such as magnesium can agglomerate as a very fine $0.3-0.6$-μm size fraction of the ash [Helble and others, 1986].

Combustion Testing

Deposits were produced by combustion of the test coals in bench-scale, pilot-scale, and full-scale combustion systems.

The bench-scale drop-tube furnace (DTF) is a vertically oriented entrained-flow tube furnace with the ability to combust coal and produce ash under closely controlled conditions. Key combustion parameters such as initial hot-zone temperature, residence time, and gas-cooling rate can be closely controlled and monitored to simulate full-scale boiler conditions.

The furnace assembly consists of five independently controlled, electrically heated zones. Coal fed at a rate of 0.1 g/min and ambient-temperature primary air are introduced from a water-cooled probe assembly at the center of the tube. Secondary air is heated to 1000°C and introduced annularly around the probe. The coal combusts as it travels down the length of the furnace in a laminar flow regime at a gas temperature of 1500°C. A ceramic constrictor is used to accelerate the gas flow to approximately 3–5 m/sec at the furnace exit before it impinges on the substrate plate of the deposition probe. Machined boiler steel substrate plates prepared from 1040 carbon steel are attached to the top of the probe. The temperature of the substrate plates were maintained at a selected temperature between 350° and 540°C to simulate a boiler heat- transfer surface in a zone of 1200°C gas temperature.

The apparatus used to determine the strength of ash deposits formed in the DTF consists of a miniature horizontal translator and a miniature pressure transducer. A rod inserted in the side of the block meets the sensing face of the transducer and transmits the force exerted on the deposit as the translator moves.

Pilot-scale testing was performed in the EERC Combustion Test Facility (CTF) furnace. The furnace capacity is nominally 34 kg/hr (550,000 Btu/hr) in a 0.76-m-diameter by 2.4-m-high refractory-lined combustion chamber. Flue gases pass out of the furnace into a horizontal 0.25-m-square refractory-lined duct. Located in the duct are three 4.2-cm-diameter stainless steel tubes designed to simulate full-scale superheater surfaces. For these tests, the air-cooled tubes were maintained at a temperature of 538°C. Flue gases entered the probe bank at a temperature of 1100°C and a velocity of 7.6 m/sec. At the completion of a nominal 5.5-hour combustion test, deposits were removed from the tubes, weighed, and stored for subsequent analysis.

Characterization Methods

Characterization of the deposits was performed using x-ray diffraction as well as scanning electron microscopy and electron microprobe analysis. Scanning electron microscopy point count (SEMPC) was used to determine the types and quantities of various chemical species present in the samples.

Ash Deposition Experiments

Ash-fouling deposits formed in the DTF from the Beulah Colstrip, Gascoyne, and Utah Wasatch coals were analyzed with the goal of determining key parameters able to explain fouling tendencies. The data on fouling tendencies derived from the DTF work were then compared to those derived from ash-fouling unit testing and full-scale boiler deposits.

Results and Discussion

Four major criteria were selected to examine the DTF deposits: crystalline and amorphous phase composition, viscosity distribution of the amorphous liquid phases, sticking co-efficients, and deposit strength. Compositions of the deposits were derived from SEMPC analysis of the main part of the deposit, which was the portion of the deposit that grew in the direction of the gas–particulate stream. Viscosity distribution profiles were then calculated using the chemistry of the liquid phases obtained from SEMPC, with viscosities calculated using a modified Urbain method [Kalmanovitch and Frank, 1988]. The correlation between viscosity and fouling is based on the assumption that low-viscosity phases in a deposit can facilitate a more highly sintered and strong deposit. Sticking coefficients and deposit strengths give a more physical picture of the fouling propensity of the coals.

In Table 3 the composition of the main deposits is presented for the four test coals. The severe-fouling Beulah and Gascoyne coals had higher concentrations of an iron oxide phase and CaO in the bulk composition, while the lower-fouling Colstrip and Utah Wasatch had greater quantities of quartz and plagioclase. Compositions of the amorphous phases showed enrichment of SiO_2, Al_2O_3, and K_2O for all the deposits except for the Utah Wasatch, which was depleted in SiO_2.

TABLE 3
SEMPC-Determined Phases of the Main Drop-Tube Furnace Deposits

Phase (number %)	Beulah	Colstrip	Gascoyne	Utah Was.
Akermanite	0.0	0.0	0.5	0.0
Gehlenite	5.8	10.2	2.6	0.4
Anorthite	1.8	6.4	0.5	2.1
Albite	0.0	0.0	0.0	3.7
Pyroxene	0.9	0.0	2.1	0.8
Calcium Silicate	0.9	0.0	5.7	0.8
Dicalcium Silicate	0.4	0.4	1.6	0.0
Spurrite	1.3	0.4	0.5	0.0
Calcium Aluminate	0.4	0.4	0.0	0.0
Spinel	0.4	0.0	0.0	0.0
Calcium Titanate	0.0	0.0	0.0	0.4
Quartz	6.6	13.2	1.6	24.0
Iron Oxide	7.1	0.9	5.2	0.4
Calcium Oxide	10.2	5.1	3.1	2.9
Ankerite (Ca,Mg,Fe) Rich	3.1	0.4	3.1	0.8
Aluminum Oxide	0.9	0.9	0.0	0.0
Anhydrite	0.0	0.0	2.1	0.4
Sulfated Ankerite	0.0	0.0	0.5	0.0
Pure Kaolinite (amorphous)	0.4	17.2	0.0	2.1
Kaolinite-Derived	1.8	13.2	0.0	0.0
Illite (amorphous)	0.4	0.9	0.0	0.0
Calcium-Derived	0.0	0.0	2.6	0.4
Unclassified	57.5	30.2	68.4	60.7
Bulk Oxide Composition (wt%)				
SiO_2	35.0	48.9	27.4	63.8
Al_2O_3	10.6	25.4	4.8	10.1
Fe_2O_3	16.4	4.3	15.6	5.5
TiO_2	1.4	0.8	1.4	0.5
P_2O_5	0.5	0.3	0.1	0.1
CaO	29.2	17.0	35.5	11.3
MgO	2.5	1.8	2.9	0.5
Na_2O	0.3	0.0	1.1	6.4
K_2O	1.2	0.5	1.5	0.6
SO_3	2.7	0.7	9.8	1.0
ClO	0.4	0.3	0.1	0.0
Amorphous Oxide Composition (wt%)				
SiO_2	41.8	47.6	32.5	57.5
Al_2O_3	13.2	30.2	5.7	13.8
Fe_2O_3	10.7	3.8	12.2	6.2

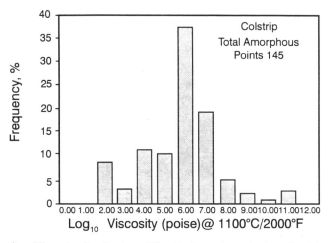

Figure 1a. Viscosity distribution of liquid phases in main deposit of Colstrip.

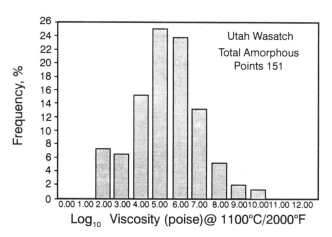

Figure 1b. Viscosity distribution of liquid phases in main deposit of Utah Wasatch.

The viscosity distributions for all four deposits shown in Figs. 1 and 2 show fairly extensive melting, interaction, and sintering of the fly ash grains in the deposit. Liquid-phase viscosities remained fairly low for the Beulah and Gascoyne deposits with median values of 2.0 and 1.5 $\log_{10}$ poise, respectively, but the Utah Wasatch viscosity median vaulted up to 5.5 $\log_{10}$ poise. The Colstrip median viscosity remained high at 6.0 $\log_{10}$ poise.

Sticking coefficient curves and deposit strength profiles for the four test coals are given in Figs. 3 and 4, respectively, where the sticking coeffient is the ash deposit formation rate divided by the ASTM ash firing rate. The Utah Wasatch coal produced a deposit that grew the fastest on the steel substrate. The Beulah sticking curve was only slightly lower than the Utah Wasatch, while the curves for Colstrip and Gascoyne were much lower in slope and magnitude.

The deposit strength profiles indicated that the Gascoyne coal produced the strongest deposit, followed by Beulah, Colstrip, and the Utah Wasatch. These results compare favorably with viscosity and base:acid ratio distribution data, which predicted that the Beulah and Gascoyne coals would have been the most severe-fouling and Colstrip the least severe. Although the sticking coefficient of the Gascoyne ash is low, the deposit strength is high. The high strength suggests that ash, once deposited, would not be readily shed from the probes and would result in the "high" reported CTF fouling tendency. Although the sodium content of the Utah Wasatch coal was significantly high at 5.7%, the overall viscosities of the DTF deposits were high, and the base:acid ratios were fairly homogeneous, which pointed toward a less severe-fouling coal. Normally, high sodium contents imply that a lower-viscosity, highly reactive liquid phase will form in the deposit. It is proposed here that the Utah Wasatch deposit is weak because enough of the total sodium in the coal is uniformly dispersed within individual ash grains, rather than concentrated near the surface or in the liquid phase between grains. This hypothesis is supported by the fact that much of the sodium in the Utah Wasatch coal is present in a mineral form (analcime) and not associated entirely with carboxyl groups, as with the Beulah. Because of the mineral association, the sodium is not liberated during combustion and, therefore, does not subsequently condense on existing ash particles to enrich their surfaces with sodium.

In summary, the results show that coals with high fouling tendencies form strong DTF fouling deposits. Predictions of fouling potentials of the four coals based on strength data collected on the DTF deposits were in agreement with fouling characteristics generated in the pilot-scale CTF. Table 4 summarizes the comparison of DTF and CTF characteristics. Fouling tendency in the CTF was based on the amount and strength of deposits that accumulated on the simulated fouling probes. Figure 5 shows photographs of the Beulah, Colstrip, and Utah Wasatch coals. The Colstrip coal produced very little deposit material, while the Beulah lignite produced the largest quantity of deposits. Although the Beulah and Utah Wasatch coals have similar sodium contents, the mineral-bound Na in the Utah coal was less able to react with the other inorganic species in a volatile phase to produce stickier and more molten deposits. Therefore, less deposit of lower strength collected on the fouling probes for the Utah Wasatch coal.

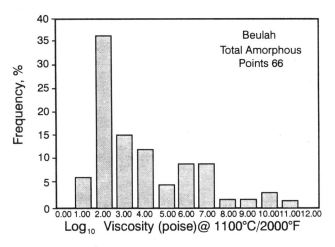

Figure 2a. Viscosity distribution of liquid phases in main deposit of Beulah.

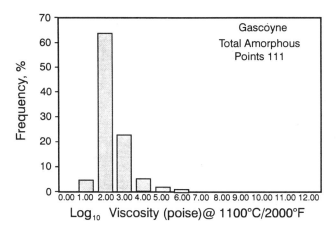

Figure 2b. Viscosity distribution of liquid phases in main deposit of Gascoyne.

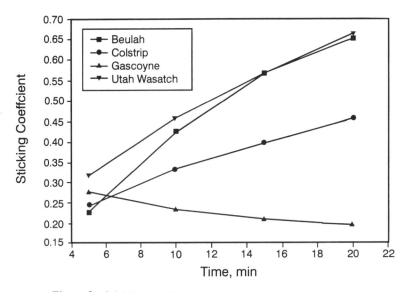

Figure 3. Sticking coefficients versus time for DTF deposits.

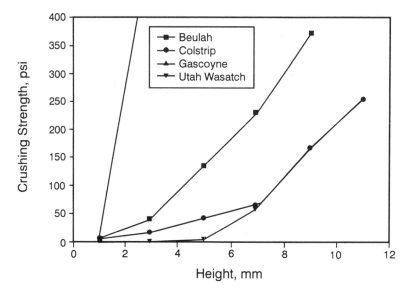

Figure 4. Crushing strength versus height of deposit for DTF deposits.

TABLE 4

Comparisons Between the Drop-Tube Furnace Deposits Produced
from Beulah, Gascoyne, Colstrip, and Utah Wasatch

Coal	CTF Fouling Tendency	Sticking Fraction[1]	Deposit Strength[1]	Median Viscosity	Base:Acid Distribution
Beulah	High	2	2	2.0	Dispersed
Gascoyne	High	4	1	1.5	Dispersed
Colstrip	Low	3	4	6.0	Uniform
Utah	Low–Med.[2]	1	3	5.5	Uniform

[1] 1 for highest, 4 for lowest.
[2] Data from a Utah Wasatch test which had only 3.9% Na_2O in the ash.

Figure 5a. CTF probe bank deposits of Beulah lignite.

Figure 5b. CTF probe bank deposits of Colstrip.

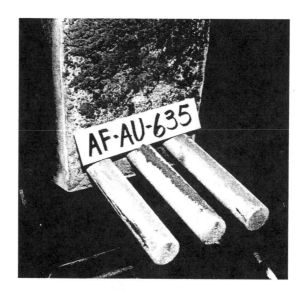

Figure 5c. CTF probe bank deposits of Utah Wasatch

CONCLUSIONS

The four western U.S. test coals all contain significant levels of Na, Ca, and Mg, with the Beulah lignite containing the highest levels of sodium. Sodium in the Beulah and Gascoyne lignites was very abundant and was organically bound. The Utah Wasatch coal also contained significant levels of sodium, but bound in the coal as analcime, a zeolite silicate. The analcime form of the sodium in the Utah Wasatch coal resulted in deposition behavior significantly less severe than would be expected based on the bulk sodium concentration of the ash.

Deposits were ranked from low-fouling to severe-fouling based on deposit buildup rate, deposit strength, and liquid-phase viscosity, which was calculated based on the chemistry and the gas temperature near the deposits at the time of quenching. Deposit buildup rates and crushing strengths were the highest for the Beulah and Gascoyne coals, followed by the Utah Wasatch during both bench- and pilot-scale fouling deposition simulations.

The ranking of the fouling behavior of the coals is in qualitative agreement with operating experience in full-scale utility boilers, with the Beulah and Gascoyne lignites notorious for severe fouling behavior, and the Utah Wasatch being more problematic than the Colstrip, although both of these subbituminous coals exhibit less severe fouling than the lignites.

REFERENCES

Given, P. H. (1984). "An Essay on the Organic Geochemistry of Coal." In M. L. Gorbaty, J. W. Larsen, and I. Wender, (Eds.), *Coal Science*, Vol. 3, p. 137.

Helble, J., Neville, M., and Sarofim, A. F. (1986). "Aggregate Formation from Vaporized Ash During Pulverized Coal Combustion" In *Twenty-First Symposium (International) on Combustion*. The Combustion Institute, pp. 411–417.

Kalmanovitch, D.P., and Frank, M. (1988). "An Effective Model of Viscosity for Ash Deposition Phenomena." In *Mineral Matter and Ash Deposition from Coal*. Engineering Foundation Conferences, Santa Barbara, CA, February 22–26, 1988, pp.89–101.

Neville, M., Quann, R. J., Haynes, B. S., and Sarofim, A. F. (1981). "Vaporization and Condensation of Mineral Matter During Pulverized Coal Combustion." In *Proceedings of the Eighteenth Symposium (International) on Combustion*; The Combustion Institute, pp. 1267–1274.

Quann, R. J., and Sarofim, A. F. (1986). "A Scanning Electron Microscopy Study of the Transformations of Organically Bound Metals During Lignite Combustion." *Fuel*, *65*, 40–46.

CORRELATING THE SLAGGING OF A UTILITY BOILER WITH COAL CHARACTERISTICS

Rod Hatt

Commercial Testing & Engineering Co.
340 South Broadway, #101
Lexington, KY 40508

ABSTRACT

This paper will describe how a utility was able to correlate intermittent slagging problems a boiler was experiencing with coal ash chemistry. The coal ash fusion temperatures alone were not sufficient to be able to separate the good versus poor performing coals. Further investigation revealed that the iron levels in the coal ash, and several indexes such as Base to Acid ratio, slagging factor and iron loading provided good bases for separating the coals that caused problems and those that did not. A total of nine coal supplies were burned in the boiler of which four caused furnace slagging problems. This utility experience further demonstrates the value of the ash chemistry data and the necessity of including more data than the fusion temperatures in coal specifications. The split between successful and unsuccessful coals is specific to this plant and should not be used universally. This work does demonstrate that ash chemistry can provide information for determining weather or not a coal can be successfully used at this plant.

INTRODUCTION

For years the fusion temperature or cone melt down test has been used to evaluate the melting and slagging behavior of coal ash. These tests have been applied to all sorts of situations, some appropriate, others not so. The American Society of Mechanical Engineers (ASME) Research Committee on Corrosion and Deposits from Combustion Gases has long advocated the weakness of the ash fusion test and provided several formats to make advancement and use of other means of evaluating the depositional behavior of coal ash. The ASME ASH FUSION RESEARCH PROJECT (1) provided a good review of the various ash fusion tests used around the world, along with several examples of poor applications of the fusion test. The use of fusion temperatures in contract specifications to eliminate poor performing coals can lead to several detrimental situations. Two common ones are: coals that meet the specifications and cause problems, and the elimination of coals that don't meet the spec., but perform well. The utility experience described in this paper will again demonstrate the inability of the ash fusion temperature test to quantify plant performance. It will further show good correlation of plant experience with ash chemistry.

Applications of Advanced Technology to Ash-Related Problems in Boilers
Edited by L. Baxter and R. DeSollar, Plenum Press, New York, 1996

237

Please note that all coal analyses were conducted at a variety of private and commercial laboratories. All laboratories indicated that they used American Society for Testing and Materials (ASTM) standards (2). No account was taken for any potential laboratory differential. The laboratory results are typical of the type of information available to coal producers and utilities.

THE PLANT

The plant, although anonymous, can be described as a large pulverized coal unit with a high heat release per unit volume. This design constraint limits the coal used to high fusion central Appalachian coal, typically in the seven to ten percent ash range. Regular monitoring of the carbon in the fly ash indicated that the unit was able to achieve good combustion. All boiler slag formation tests were conducted at or near full load conditions during a sustained time frame. Those coals listed as successful had no apparent operational problems, the unsuccessful coals had excessive slag build up rates.

TYPICAL FUEL

The coals utilized by the plant had similar characteristics. Ranges of common analyses are shown below in Table I.

Table I
Typical Coal Analyses

As Received	Typical Value	Range
Moisture	5.5	3.0 - 8.0
Ash	7.4	5.2 - 10.6
Sulfur	0.88	0.67 - 1.12
Btu/lb	13,000	12,400 - 13,300
HGI	43	39 - 45
Fusion Temp. ID F.	2,803	2,742 - 3,000+
Fusion Temp. Fluid F.	2,850+	2,800+ - 3,000+

FUSION TEMPERATURE RESULTS

When the coals are separated into successful and unsuccessful categories and the fusion temperatures are displayed, you cannot set a temperature that separates the categories. Figure 1 shows the initial deformation temperatures for the coals studied. As indicated, it would be hard to distinguish which coals would be successful or not. It should be noted that in many cases the fusion temperature was beyond the maximum furnace temperature, and was reported as 2,800+ or 2,850+ degrees F.

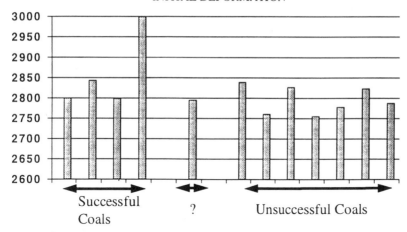

FUSION TEMP DEGREES F.
INITIAL DEFORMATION

Figure 1. Initial deformation temperature for coals by
successful/unsuccessful categories, ? is coal that
had some problems, but not as severe as unsuccessful
coals.

The fluid fusion temperatures are even harder to interpolate as the majority of the results are at the maximum furnace temperature. These results are shown in Fig. 2.

Figures 1 and 2 represent the challenge faced by many utilities today. In this instance the fusion test was asked to differentiate between coal ashes that melted at or over the furnace design limits. In addition, a high fusion temperature did not ensure a successful rating.

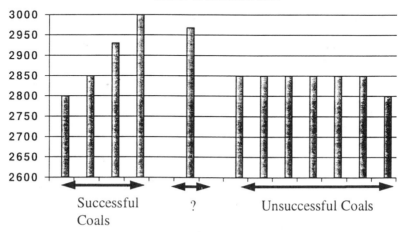

FUSION TEMP DEGREES F.
FLUID TEMPERATURE

Figure 2. Fluid temperature for coals by successful/unsuccessful
categories, ? is coal that had some problems, but not as
severe as unsuccessful coals.

ASH CHEMISTRY

The ash chemistry of all the coals showed high levels of silicon and aluminum. When reported as oxides these to elements make up eighty to ninety percent of the ash. These two elements along with one to two percent titanium dioxide make up the acidic oxides. Iron oxide is the third most abundant element, and the highest percentage of basic oxide. Figure 3 shows the iron oxide levels of the coal ashes in a similar manner as Fig. 1 and 2.

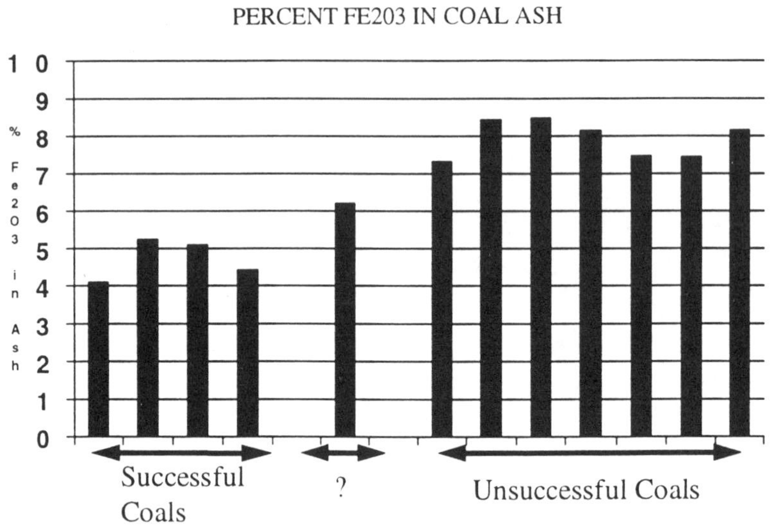

Figure 3. Iron oxide (Fe203) levels in coal ash by successful/unsuccessful categories, ? is coal that had some problems, but not as severe as unsuccessful coals.

As shown in Fig. 3 the iron oxide levels of all the coals that were successful are below six percent. Those coals that were deemed unsuccessful have iron oxide levels greater that seven percent. The coal that had intermediate performance had an iron oxide level between six and seven percent. The Silica Percentage as described by Reid (3) is expressed as

$$\text{Silica } \% \ = \ 100(SiO_2/SiO_2 + Fe_2O_3 + CaO + MgO)$$

Figure 4 shows the Silica % results by slagging performance category. Like the iron oxide levels the Silica % is able to group the coals into successful and unsuccessful categories. The break is at about eighty seven percent, unfortunately the parameter placed the intermediate performing coal with the successful coals.

The ratio of basic to acidic oxides as expressed by the B/A has been described by Winegartner (4). Numerically it is found by using the following formula:

$$B/A = \frac{Fe2O3 + CaO + MgO + K2O + Na2O}{SiO2 + Al2O3 + TiO2}$$

Figure 5 shows the results of the base to acid ratio versus slagging performance. Like the Silica percentage the B/A groups the coals well in terms of separating the successful coals from the

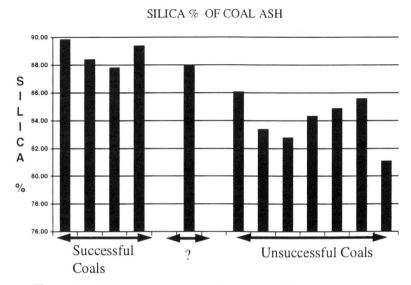

SILICA % OF COAL ASH

Figure 4. Silica percentage for coals by successful/unsuccessful categories, ? is coal that had some problems, but not as severe as unsuccessful coals.

unsuccessful coals. The B/A also placed the intermediate performing coal with the successful coals. The unsuccessful coals all had B/A at or above 0.14, the successful coals were at or below 0.11, with the intermediate coal also at 0.11.

Attig and Duzy (5) have derived an empirical slagging index by multiplying the base to acid ratio by dry sulfur.

BASE TO ACID RATIO OF COAL ASH

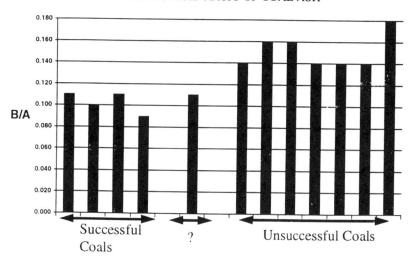

Figure 5.   Base to acid ratio, B/A for coals by successful/unsuccessful
categories, ? is coal that had some problems, but not as
severe as unsuccessful coals.

SLAG INDEX OF COAL ASH
B/A X DRY SULFUR

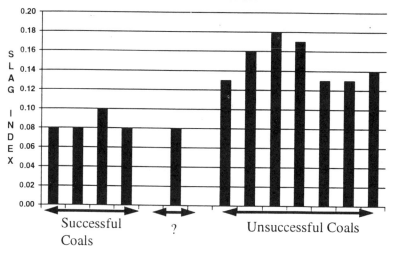

Figure 6.   Slagging Index, Rs for coals by successful/unsuccessful
categories, ? is coal that had some problems, but not as severe as
unsuccessful coals.

Slagging Index, Rs = dry S% x B/A

If a portion of the sulfur in coal exists as pyrite it can be seen that the slagging index is doubling the impact of iron by including both sulfur and iron oxide in the numerator. All of the coals in this study rate low slagging potential by having a slagging index less than 0.6. Figure 6 shows the results of the slagging index for these coals.

Again the slagging index separates the coals into successful and unsuccessful categories. Like the B/A and the Silica percentage it groups the intermediate performing coal with the successful coals.

The author has found on several occasions the usefulness of an iron loading term, typically pound of iron oxide per million Btus. This term has successfully separated the full scale performance of a boiler using high sulfur Illinois basin coals where a low ash-high iron percentage coal had less slagging than a higher ash-lower iron level coal. Figure 7 shows the results of the iron loading calculation for the coals. This term is found using the following equation:

Iron Loading, lb. iron/MBtu = $\dfrac{\%Fe_2O_3 \times (\%Ash/100)}{(Btu/lb /10,000)}$

As shown the iron loading term separates the successful and unsuccessful performing coals. In addition, it also places the intermediate coal with the unsuccessful coals. Most utilities would prefer to use terms that are more conservative and place coals with intermediate performance with the coals that do not perform. It is felt that the iron loading term is useful due to it incorporating the amount of material, rather than just the characteristics of the ash. The fusion temperature, Silica %, base to acid ratio, and slagging index only characterize the ash, not the amounts of material available to form deposits.

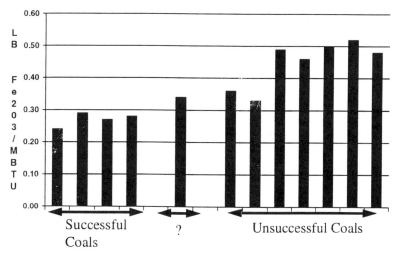

Figure 7. Iron loading expressed as lb. Fe2O3 per million Btu for coals by successful/unsuccessful categories, ? is coal that had some problems, but not as severe as unsuccessful coals.

CONCLUSION

This paper describes one utility's experience with coal slag. The coals studied were similar in nature and all possessed low slagging potential. Due to boiler design with higher than typical furnace heat release rate, the boiler was occasionally subject to slagging deposits. Using only the ash fusion temperatures the slagging potential of a given coal could not be predicted. By incorporating the ash chemistry into the evaluation and reviewing past experience, several techniques were found to better quantify coal characteristics with plant operating experience. Several commonly used expressions were found to be useful in segregating the successful and unsuccessful coals, but could not quantify the performance of an intermediate performing coal. The percent iron oxide and the iron oxide loading were able to better quantify the behavior of the coals.

References

1. ASME ASH FUSION RESEARCH PROJECT, CRTD-18, For availability contact: ASTM/CRTD, 1828 L St. NW, Suite 906, Washington DC 20036.

2. Annual Book of ASTM Standards, Vol. 5.05 Gaseous Fuels; Coal and Coke, ASTM 1916 Race Street, Philadelphia, PA 19103-1187, USA.

3. Reid, W.T. and Cohen, P. "The flow characteristics of coal ash slags in the solidification range" *Jour. Eng. Power, Trans. ASME Series A*, 66, 83, 1944.

4. Winegartner, E. C. (ed.). *Coal Fouling and Slagging Parameters.* ASME research Committee on Corrosion and Deposits from Combustion Gases, ASME pub.,1974.

5. Attig, R. C. and Duzy, A. F., "Coal ash depositional studies and application to boiler design" *Amer. Power Conf.*, Chicago, Ill., Illinois Institute of Technology, 1969.

EXCLUDED / INCLUDED MINERAL MATTER AND ITS SIZE DISTRIBUTION RELATIVE SIGNIFICANCE ON SLAGGING AND FOULING CHARACTERISTICS

R. P. Gupta, X. He, R. Ramaprabhu, T. F. Wall

Department of Chemical Engineering
University of Newcastle, NSW2308 Australia

I. Kajigaya, S Miyamae, Y. Tsumita

Ishijikawa-Harima Heavy Industries Co. Ltd
2-16 Toyosu 3-Chome Koto-ku Tokyo 135, Japan

ABSTRACT

The effect of the nature of mineral matter (extraneous or inherent) on combustion generated ash is investigated by means of thermodynamic calculations and combustion experiments in a drop-tube furnace. Five coals, including one high in silica, one high in calcium and magnesium and one high in iron, were analysed for mineral matter size distribution using the CCSEM technique. The particle size distribution of mineral matter is found to be bi-modal in nature. The coal samples are found to have included mineral matter smaller than 20 µm whereas the excluded mineral is mostly larger than 40 µm in size.

The nature of mineral matter determines its reactions during combustion; the excluded mineral matter is in equilibrium with the bulk flue gases at the gas temperatures whereas the included minerals are in equilibrium with char at the burning char particle temperature. It is predicted from the thermodynamic calculations that almost all the evaporation is either from the included mineral matter or from the atomically dispersed minerals in coal. This is due to the high temperature and reducing atmosphere inside the char particle. The release of the evaporated species is controlled by diffusion through the burning char particle and may be, therefore, estimated theoretically.

The theoretical predictions of the ash character so derived are compared with the experimental data obtained by firing the coals in 10%, 23% and 50% oxygen in a drop tube furnace. The amount of mineral matter that is vaporised may be related to fouling, whereas, the melt phase present on the surface of large particles may be related to slagging. The mineral matter distribution is, therefore, related its effect on these two factors.

1. INTRODUCTION

In previous work (Gupta and Wall, 1993), Newlands, Daido and Miike coals were investigated for their depositional characteristics. The present study investigates the effect of the nature and the size distribution of mineral matter in these coals. It includes two new coals containing high levels of silicon, calcium and magnesium, namely Ulan coal (from Australia), and Medicine Bow coal (from USA). The standard ash analyses are presented in Table 1. The ash from Ulan coal is rich in silica (85%), the ashes from Daido and Miike coals are high in iron (9-10%), and the ashes from Miike and Medicine Bow are high in calcium and magnesium (25-40%).

Applications of Advanced Technology to Ash-Related Problems in Boilers
Edited by L. Baxter and R. DeSollar, Plenum Press, New York, 1996

245

The mineral matter within the coal matrix is called included minerals, whereas the discrete particles of minerals not associated with carbonaceous matter are called excluded minerals. In pulverised fuel, the amount of excluded mineral increases with increased grinding. The physical (eg size and density) and chemical (eg composition of mineral phases) characteristics of mineral matter relate to several factors associated with ash deposition: the evaporation of mineral species, the size distribution of ash, and the composition of the sticky exterior layer of the coarse ash particles formed during combustion. These factors influence the ash transportation to walls and its stickiness.

The coal samples used in the ash formation experiments were sieved to a narrow size fraction of 63-90µm. The experiments were conducted in a drop-tube furnace under different environments (1200°C, 1350°C and 1500°C, and at oxygen concentrations of 10%, 23% and 50% by weight). The ash was separated into fines (< 2µm)which are rich in fumes and coarse ash. The chemical analysis of coarse ash was determined by the XRF technique. The acid soluble oxides in fine ash and coarse ash were determined by the Atomic Absorption technique to obtain the composition and amount of the fines and fumes.

The same size fraction samples were examined for mineral matter size distribution by the Computer Controlled Scanning Electron Microscopic (CCSEM) technique at the Energy and Environmental Research Center (EERC) at University of North Dakota. The CCSEM data include the particle size distribution of both the included and the excluded mineral grains, along with the chemical composition of each mineral grain. Figure 1 shows the heterogeneous composition of mineral grains (detected by the CCSEM only) in four of the coals. Organically associated inorganics are expected to be distributed uniformly within the char matrix.

The mineral matter details from the CCSEM analysis are used to provide the effective ash composition that is in equilibrium with the surrounding gases. The included minerals are in equilibrium with char (expressed as carbon) at the temperature of the char particle for estimating the amount of the minerals evaporated. The equilibrium of the excluded mineral with flue gases at the gas temperatures is used to estimate the characteristics of sticky molten phase on coarse particles. Thermodynamic equilibrium calculations for the excluded and included minerals are thus performed separately.

The experimental results from the ash formation experiments are compared with the theoretical estimates for the amount of fumes generated and the amount of molten phase and its composition.

1.1 Literature Review

Couch (1994) has given an extensive survey of the current status and research related to ash deposition. The following points are the main conclusions from the literature review:

- Mineral matter in coal and the combustion environment are the two key factors that influence the ash formation process.
- The fraction of mineral matter in coal, its form and particle size distribution is also related to the ash formation process. Advanced techniques such as CCSEM (Skorupska and Carpenter, 1993) and Quantitative Evaluation of Materials by Scanning Electron Microscope QEM*SEM (Gottileb et al., 1992) are able to provide detailed information on the mineral matter necessary for development of new indices for fouling and slagging.
- The combustion environment influences the temperature of the burning particle and thus enhances the vaporisation of included minerals within the char matrix and thereby influences the amount of fume formation.
- The particle size distribution of fly ash is known to influence the rate of its transportation. Fine particles are transported by thermophoresis, whereas larger particles are mostly transported by inertial impaction. Models can predict the ash PSD from the mineral matter distribution (Beer et al, 1991; and Wilmenski et al, 1991).

1.2 Mineral Matter in Coal

The three aspects of mineral matter influencing the ash formation are: nature of mineral matter - (included or excluded), particle size distribution and form of mineral matter.

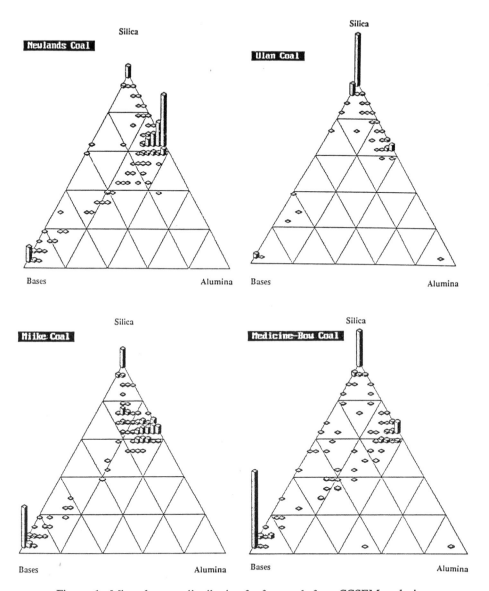

Figure 1. Mineral matter distribution for four coals from CCSEM analysis

Nature of Mineral Matter: The extraneous mineral matter, the mineral matter within the coal matrix and the organically associated inorganics produce ashes of varying character. The association of included mineral matter with char also increases the evaporation of mineral species significantly (Boni et al. 1990). The mineral matter within the coal matrix reacts and evaporates under more severe conditions (a reducing atmosphere and high temperature) as compared to the extraneous mineral matter. The inherent minerals are physically close and therefore can react easily. In addition, a part of these mineral inclusions may melt and coalesce to form dense ash particles.

The mineral particle size distribution (PSD): A large mineral grain is less reactive than a small mineral grain, as a large mineral grain may participate partially in ash reactions. The evaporation of silica increased from 5% to 40% in 50% oxygen and 1650K when the quartz grain size decreased from 10µm to 2µm (Boni etal., 1990). Previous thermodynamic calculations (Gupta and Wall, 1993) also predicted a significant effect of size distribution of mineral matter on the evaporation of alkalis shown in Figure 2. In this figure AC=0.1 for silica and alumina implies that only 10% of these oxides are available for thermodynamic equilibrium.

Mineral Matter Form: The vaporisation of silica and calcium is strongly influenced by their mode of occurrence. Their presence as large discrete extraneous minerals may not enhance slagging. Silica may be present in the form of quartz, illite and kaolinite. However, silica in the form of quartz is relatively least reactive but in the form of illite it will coalesce with other minerals rapidly. Similarly, calcium and magnesium in illite react rapidly, but if it occurs in the form of carbonates (eg calcite and ankerite) large mineral grains fragment increasing the viscosity of the molten phase (Zygarlicke et al. 1991).

This suggests that for high silica and calcium coals, it is not the amount of the species that is critical, but it is the form of that mineral, the fraction present as included minerals and its grain size that would influence slagging or fouling tendencies. The excluded calcium, for example, would contribute to fouling, whereas the included calcium (in presence of silica) would form low melting point complexes enhancing slagging tendencies. Iron present in the form of pyrites is slowly oxidised with some fragmentation in prsence of oxygen, whereas, iron in the form of ankerite or siderite is prone to severe fragmentation to 0.1-1.0µm size particles (Raask, 1985). However, sodium and potassium would vaporise irrespective of how they occur.

1.3 Combustion Environment

The combustion environment influences ash formation in two ways: the temperature of the char particle during combustion and the particle size distribution of ash. Gas temperature and the amount of excess air determine the evaporation of mineral species. The vapour pressure of the mineral species increases exponentially with temperature. Thermodynamic calculations (Gupta and Wall, 1993) predicted sharp increases in fume formation with increase in gas temperature. The vaporisation was also enhanced by a reducing atmosphere (Figure 2). The particle temperature is strongly influenced by the oxygen concentration. A higher particle temperature subsequently increases the evaporation of the included mineral species. Erickson et al. (1992) found the mass median diameter of the ash samples formed from synthetic char at 900°C and 1500°C to be 15µm and 40µm, respectively.

1.4 New Characterisation Techniques

Kalmanovitch (1991) has suggested a technique to present detailed information on mineral matter and ash in form of a distribution curve. The curves show the percent of ash that will be present as a molten phase at a given temperature.

Zygarlicke et al. (1991) have also proposed some performance indices based on CCSEM analysis. The criteria used in developing the indices is based on the following factors: clay content facilitating formation of low melting point compounds (LMPC), fine and included quartz content which is more reactive, calcite content which may increase the viscosity thereby decreasing the deposit strength, and organically bound sodium and calcium content which result in the formation of LMPC.

Gupta and Wall (1993) have suggested the amount of sodium and iron present in gas phase per unit of coal consumed as indicators for fouling and slagging, respectively. These indices account for mineral

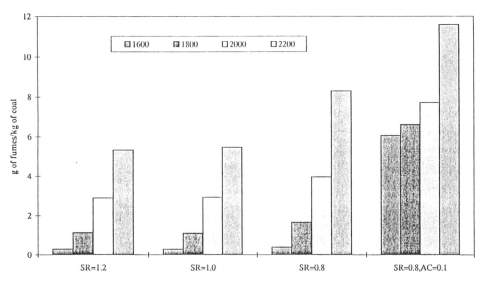

Figure 2. Effect of operating conditions (four combustion temperarures and several stoichiometric ratios) and mineral matter size (AC for silica and alumina being 0.1) on evaporation of alkalis.

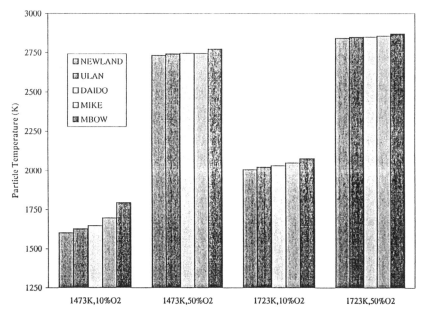

Figure 3. Predicted temperatures of burning char particles of 80μm in different combustion environments for five coals.

matter distribution and the operating conditions of the boiler, which are further investigated and modified in this project.

Stickiness has been considered a function of viscosity of particles at some effective particle temperature so that the most common stickiness models relate to the viscosity of particles (Benson et al., 1993). The stickiness of the ash particles is quantified by the ratio of a critical or reference viscosity to the viscosity of particles at particle temperature. The main limitation in these models is the uncertainty in the value of critical viscosity. The viscosity of the fly ash can be obtained from ash composition by using the correlation of Urbain et al. (1981). The model was developed to estimate the alumio-silicate based ceramics. There have been some modifications for ashes with high iron content.

1.5 High silica and high calcium coals

A study at MIT (Graham, 1991), focussed on the reactions of silica and calcium; the analysis showed the final products formed from calcium consisting of CaO, $CaSiO_3$ and Ca-Si glasses. Calcium and magnesium are present in coal in the form of carbonate clays ie calcite, dolomite and ankerite. In another study on the behaviour of calcium in low rank coal, Beulah lignite, (Shah et al.,1991, and Srinivasachar et al.,1991), the ash samples prepared in a laboratory combustor were size graded by Anderson impactor. Most of the calcium (organically associated) reacts with aluminosilicate to form calcium aluminosilicate of varying composition. Calcium rich particles (mostly unreacted CaO particles from calcination of calcite) were found mainly in the first impactor (>9µm). In the samples from other impactors (<9µm) calcium rich particles decreased. Zygarlicke et al. (1991) also indicated that organically bound calcium quickly reacts with aluminosilicate and quartz within the coal matrix forming lower melting point phases eg $CaSiO_3$. However, large discrete grains of calcite may act as diluent and may increase the viscosity of liquid phases.

In all the five coals investigated in the present study, calcium is mostly in the form of calcite. The above mentioned research indicates that the assessment of high silica or high calcium coal requires detailed knowledge on the form of the minerals, the size distribution of these minerals in the coal matrix and their association with each other. The CCSEM analysis is a critical factor in evaluating their ash deposition characteristics.

2. TEMPERATURE OF BURNING CHAR PARTICLE

The mineral matter associated with carbonaceous matter in pulverised coal (ie included minerals) attains the temperature of the burning particles. It has been shown experimentally that the temperature of a burning particle can be 400°C higher than that of the surrounding flue gases depending on the particle size, oxygen concentration and the combustion kinetics (Timothy et al. 1985). This requires a good knowledge of the particle temperature during combustion. This section deals with the estimation of such temperatures for burning char particles. The excluded minerals, fed to the furnace as discrete particles, are essentially in thermal equilibrium with furnace gases.

2.1 Model for burning char particle

The particle temperature may be determined by equating the heat generated from char combustion to the heat losses due to convection and radiation. The rate of heat generation, Q_{gen}, by a particle is given by the following equation;

$$Q_{gen} = m \, \Delta H_c \qquad (1)$$

where   m   - rate of combustion (gram of char consumed per second)
        ΔH_c - heat of combustion per gram of char consumed by the following reaction:

$$C_{(S)} + \psi O_2 \longrightarrow 2(1-\psi)CO + 2(\psi-0.5) CO_2 \qquad (2)$$

where ψ is the number of moles of oxygen consumed per mole of carbon and determines the ratio of the combustion products CO and CO_2. Values of ψ of 0.5 and 1.0 correspond to the complete combustion to CO and CO_2, respectively. Tognotti et al. (1990) obtained experimentally the value of ψ as 0.52 and 0.59 at 20% and 100% oxygen levels, respectively, for a 130μm particle. Boni et al. (1990) referred to a study where it was found that ΔT_p is proportional to r_p^2 and $(\psi-0.5)^{0.7}$. The value of ψ is, therefore, obtained by the following empirical equation;

$$\psi = 0.5 + [0.502 + 0.0877X_{O2}].(D_p(\mu m)/130)^{1/0.7} \qquad (3)$$

The rate of char combustion is usually expressed by the following expression.

$$m = k_s.P_s^n \qquad (4)$$

$$k_s = A.exp(-E/RT_p) \qquad (5)$$

where m is the combustion rate kg-carbon/(sec-m^2 of external surface area) and n is the global reaction order. The global combustion rate, k_s, includes the effects of pore diffusion, internal surface area and intrinsic surface reactivity, and is given by equation (5).

Hurt and Mitchell (1992) developed a generalised correlation that allows the prediction of char combustion rates for a range of coals as a function of carbon content. These correlations, assuming the reaction order of half with respect to oxygen, reproduced the combustion rates within 30% at gas temperatures of 1250°C to 1750°C. In the absence of kinetic data for the five coals investigated in the present study, the above correlations are used to predict the temperature of the burning char particle.

The rate of heat loss from the particle includes the conduction and radiation losses to the surrounding environment.

$$Q_{loss} = 4\pi\, r_p\, k_{eff}\, (T_p - T_g) + \sigma\, (4\pi r_p^2)\varepsilon_p\, (T_p^4 - T_g^4) \qquad (6)$$

ε_p is the emissivity of the particle. The value of the particle emissivity is assumed to be unity. The effective thermal conductivity, k_{eff}, of the gases in the boundary layer around the particle is evaluated at a mean temperature of T_p and T_g. The dissociation of oxygen is taken into account as it influences the thermal conductivity significantly at high temperatures.

Timothy et al. (1982) measured the particle temperatures of Montana lignite and Illinois #6 coal at 1700 °K in oxygen concentration varying from 10-100%. The two coals have 74% and 70% carbon content on dry ash free basis, respectively. The present model was used to determine the temperatures of 130μm size char particles of Montana lignite and Illinois #6 coals in different oxygen environments during combustion. The predictions for the Montana lignite are slightly (40-60°C) lower than those for Illinois #6 due to its lower reactivity, whereas the measured temperatures for the Montana coals are a few hundred degrees lower than those for the Illinois #6. The results compare very well with those for Illinois #6 coal.

2.2   Particle Temperature of Test Coals

Figure 3 compares the predicted temperature of 80μm for the coals in different combustion environments. The particle temperatures of the five coals vary in the order of reactivity ie Newlands, Ulan, Daido Miike and Medicine Bow; Newlands being the least reactive and Medicine-Bow being the most reactive coal. The particle temperature is clearly influenced to a greater extent by the oxygen concentration than by the surrounding gas temperature.

Low oxygen concentration (10%): In an environment of 10% oxygen and gas temperatures of 1200°C gas temperatures, the particle temperature for Newlands is 125°C, and for Medicine-Bow 320°C higher than the surrounding gas temperature. The difference in coal type is most significant in a kinetics controlled regime at 1200°C; the particle temperature of Medicine-Bow is 195°C higher than that of Newlands coal. At a gas temperature of 1450°C, the particle temperatures of different coals are estimated to differ only by 70°C, (the particle temperature being 2010°C for Newlands and 2080°C for Medicine-Bow coal).

Oxygen Concentration (50%): The combustion regime changes from kinetic control to diffusion control due to an increase in oxygen concentration or gas temperature. The temperature of a burning particle can be as high as 2500°C in the 50% oxygen environment and is not influenced by the gas temperature. The difference between the particle temperatures for the five coals diminishes to 40°C at 1200°C gas temperature and to 30°C at 1450°C gas temperature.

In PF boilers, the oxygen concentration is usually less than 10% during the char combustion. Results from the sensitivity analysis indicated that a finer grind of coal particles would then lead to lower burning particle temperatures resulting in less evaporation of mineral matter. The fine particles (< 10μm) of high rank coals may burn within 50°C of the gas temperatures in a low oxygen (<5%) atmosphere. A finer grind would also result in a smaller proportion of included minerals. The oxygen concentration does not influence the behaviour of the excluded minerals. A further conclusion is: the thermodynamic calculations may be performed at the same particle temperature irrespective of the coal rank while analysing the drop-tube furnace experiments conducted at high temperature and high oxygen concentration.

3.   THERMODYNAMIC CALCULATIONS

The theoretical estimates for the formation of fines and fumes, and for the amount and composition of the molten phases on sticky particles were obtained from thermodynamic equilibrium calculations using a chemical equilibrium program CHEMIX, developed by CSIRO Australia (Gupta and Wall, 1993). The program computes the equilibrium compositions in different phases by minimising Gibbs free energy. The inputs to the CHEMIX program are the products resulting from the combustion of coal including the ash components.

The theoretical estimates of ash vaporisation and the molten phase are modified here by means of three considerations;

- The use of an availability coefficient
- Temperature estimates of the burning char particle
- Presence of char for the included minerals

3.1 Availability Coefficient

The effect of mineral matter grain size on its evaporation has already been demonstrated experimentally (Boni et al., 1990). Wibberley and Wall (1982) estimated the thickness of a reacted sticky layer on silica particles to be 0.1-0.3μm, when silica was fed into the drop-tube furnace with sodium salts. This implied that the core of the particle does not react. The mass fraction of the reacted layer to the total mass of the particle is referred to here as the availability coefficient (AC).

The equilibrium calculations assume the activity of all the ash components as unity, as if all the components are freely available for reaction. The estimates for alkali vaporisation were under-predicted by this approach when compared with results from the drop-tube furnace experiments. With an arbitrary value of availability coefficient of 0.1 for silica and alumina (only 10% of silica and alumina is reactive), these estimates were found to improve (Gupta and Wall, 1993). The availability coefficient (AC), therefore, may also be defined as the fraction of the minerals actively participating in the establishment of the thermodynamic equilibrium. A lower availability of silica and alumina for example increased the vaporisation of alkali species and reduced the evaporation of silica and alumina.

The estimates for evaporation of mineral matter, the amount and composition of the molten phase are made for the five coals (Newlands, Daido, Miike, Ulan and Medicine-Bow). The mineral matter grains in the coal vary in size, and the PSD of the mineral grains from a CCSEM analysis was used to determine the AC of individual species. The CCSEM determines the size distribution of both included and excluded mineral matter.

The CCSEM data are processed into 50 equal size fractions from 0-100 μm. The particle size distribution

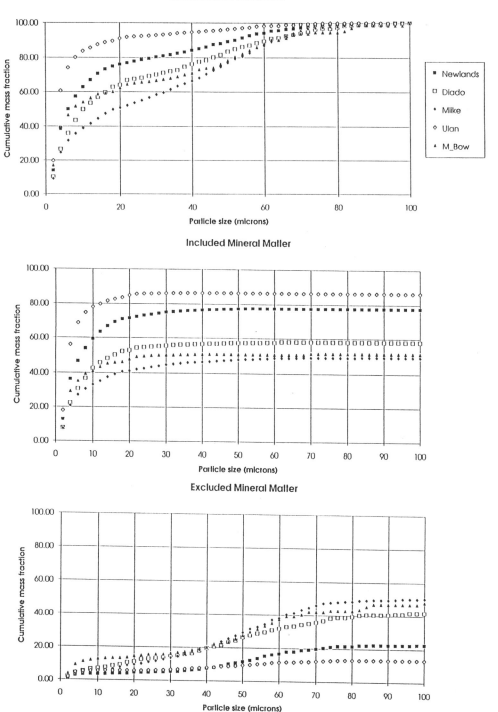

Figure 4. Particle size distribution of total, included and excluded mineral from CCSEM analysis.

of the mineral matter, detected by the CCSEM, shown in Figure 4a, is bimodal in nature due to different mean particle sizes for included and excluded minerals. Figures 4b and 4c show the PSD of the included and excluded mineral for the five coals. These figures suggest that most of the included mineral is in the range of 0-20µm, whereas the excluded mineral matter is mostly above 40µm in size. The particle size distribution of total, included and excluded silica, alumina, sodium oxide, calcium oxide and iron oxide is obtained similarly. The AC is evaluated for the included and excluded minerals independently.

The availability coefficient of excluded or included mineral is then computed from the particle size distribution by using the following correlation, where f_i is the mass fraction of excluded mineral in the size bin of radius r_i with t_i its reactive thickness.

$$AC = \frac{\sum f_i (1.0 - (r_i - t_i)^3 / r_i^3)}{\sum f_i} \tag{7}$$

The availability coefficient for a particular oxide depends on its reactive thickness. The reactive thickness for some synthetic ash particles formed at 1500ºC was determined to be less than 1µm in laboratory experiments (Wibberley and Wall, 1980). The reaction layer thickness for the excluded minerals is assumed here to be 1µm. Since the included mineral matter determined from CCSEM encounters higher temperatures and extremely reducing environment within char, the thickness of the reaction layer for the fine grains of the included minerals is expected to be very high.

The amount of ash components obtained from the CCSEM analysis of minerals in coal is less than the ash determined from the XRF technique. This is mainly due to the fact that CCSEM cannot analyse sub-micron minerals and dissolved salts within the coal matrix. The amount and composition of this finer fraction has been estimated from the difference of those from XRF and CCSEM analyses. In the present calculations the AC for the included mineral together with the soluble salts and fine mineral matter, is assigned a value of unity.

3.2 Thermodynamic Equilibrium for Excluded Minerals

The excluded minerals are expected to be in thermal and chemical equilibrium with the flue gases. As the mineral reactions take place mostly within the flame, the composition of the surrounding flue gases is computed for a stoichiometric ratio of 0.8 (a reducing environment). The equilibrium calculations were performed at temperatures ranging from 1500ºK to 3000ºK.

Gas Phase: Predictions of the mole fraction of silica, iron, sodium and potassium in the gas phase are presented in Figure 5. The constant slope of log(x) vs 1/T in these figures suggests that the mole fraction of SiO and Fe species in the gas phase is directly proportional to the vapour pressure. At temperatures higher than 2500ºK the complete evaporation of silica and iron from the excluded minerals is evident. Among the coals considered, Medicine Bow has the smallest mole fractions of all the ash components in the gas phase, while Miike has the highest mole fractions.

According to the experimental study by Srinivasachar et al.(1990), a higher concentration of alkalis would enhance the stickiness of the particles. The mole fractions of sodium and potassium in the gas phase increase with temperature. However, the mole fraction of these alkali components is not proportional to the vapour pressure of these components, but is controlled by the retention of these alkalis by silica in the molten phase. The mole fraction of sodium and potassium are also highest for Miike coal followed by Ulan, Daido and Newlands coals. Apart from Medicine-Bow coal, the mole fraction of alkalis appears to be one of the factors ranking the coals in the order of slagging.

Molten Phase: Figure 6 presents estimates of the molten phase per kg of ash content. This factor distributes the molten phase over the total ash. If the molten phase per unit of ash is low, it is expected to be less sticky, even if the total amount of molten phase is high. According to this criterion, Newlands and Ulan have the least amount of molten phase per unit of ash content and consequently have less sticky particles.

The calculations suggest that the quantity of the molten phase is not influenced significantly by the

254

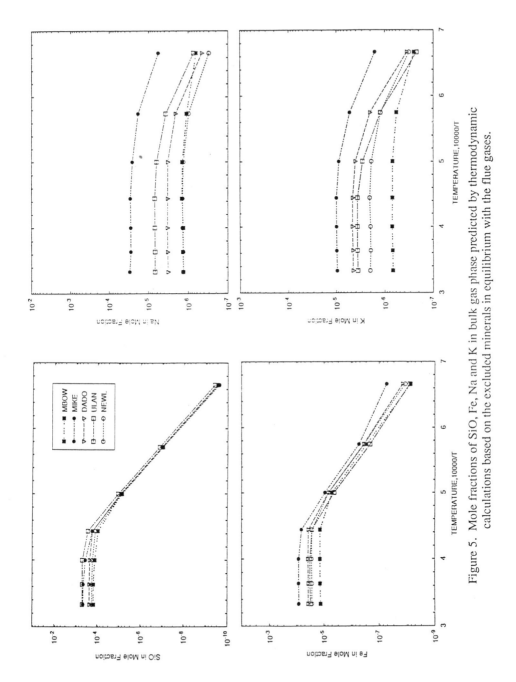

Figure 5. Mole fractions of SiO, Fe, Na and K in bulk gas phase predicted by thermodynamic calculations based on the excluded minerals in equilibrium with the flue gases.

combustion environment. However, the composition of the molten phase is affected. There is successive depletion of sodium, potassium and iron with an increase in the gas temperatures. The mole fractions of sodium, potassium and iron in the molten phase are minimum for Medicine-Bow and are maximum for Miike coal. The minimum values in the case of Medicine-Bow coal are due to high proportion of calcium and magnesium.

The mole fraction of calcium increases in the molten phase with an increase in temperature, which is due to a negligible evaporation of calcium oxide. At lower temperatures the molten phase has low melting point aluminosilicates of sodium and potassium, making the particles sticky. As the temperature rises, the concentration of calcium increases, giving rise to the sticky phase of calcium silicate. Therefore, in high calcium coals, the particles may be sticky even at higher temperature.

3.3 Thermodynamic Equilibrium for Included Minerals

The soluble salts and sub-micron size minerals (not detected by the CCSEM technique) and the included

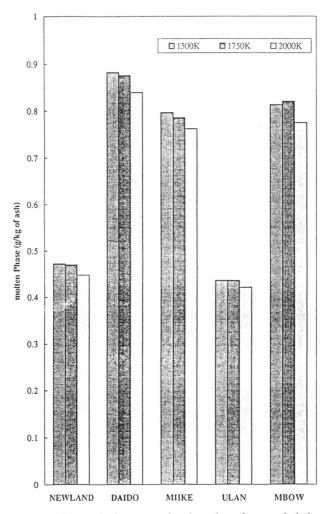

Figure 6. Theoretical amount of molten phase from excluded
minerals per kg of ash.

minerals are in equilibrium with char and with the flue gases within the char particle. Thermodynamic calculations have shown that the mole fraction of SiO in the gas phase is at least two orders of magnitude higher in the presence of carbon. The availability coefficient of the mineral matter within the coal matrix is considered unity. The moles of gases in equilibrium with the included minerals are assumed to be equal to those present in 50% of the char volume at 2000°K.

The predicted mole fractions of silica, iron, calcium and magnesium correspond to the vapour pressure of these species in the presence of carbon. The mole fractions of these species increase exponentially until all is vaporised. A decrease in the amount of alkalis mole fraction with increase in particle temperature from 2000°K to 2500°K is due to relatively higher evaporation of silica, calcium and magnesium.

It is speculated that a part of these species diffuses from the char and condense to form fume. The rest of the species in the gas phase will cool and condense to form fine spherical particles. The concentration of the alkalis in the condensed phase is expected to be proportional to the initial concentration of the alkalis.

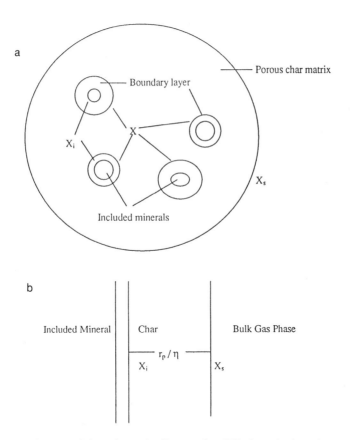

Figure 7. The schematic diagram for diffusion of mineral species from char to bulk gas phase (a) and the simplified model (b).

A high proportion of alkalis in Medicine-Bow and Miike coals results in low melting point silicates, and correspondingly, the included minerals will result in fine spherical ash particles (Figure 8). The size of these spheres observed in the ash from Medicine-Bow is small probably due to very low mineral loading of the coal particles. The ash formed from Ulan coal containing high proportions of silica will not have such low melting point compounds and hence it is expected to result in the formation of fluffy ash as shown in the Figure 8. These speculations on ash character are confirmed from the SEM images of the ash collected in the drop tube experiments.

The predicted mole fractions of the gaseous mixture within the burning char particles of Daido and Newlands coals are not very different. The mole fractions of sodium, potassium and magnesium for Newlands coal are higher than those from Ulan coal, whereas the mole fractions of iron and calcium are higher for Daido coal. A higher proportion of spherical particles present in the ash from Newlands coal is probably due to the formation of low melting point aluminosilicates of sodium and potassium having low melting temperatures.

4.   DEVELOPMENT OF A DIFFUSION MODEL

The concentration of the mineral species in gas phase within burning char particles has been shown to be orders of magnitude higher than that in the bulk phase. The difference in the concentrations results in diffusion of these species into the bulk gas phase. These species will then form fine fumes as they oxidise and condense. At this stage it is expected that a major proportion of fumes would result from this process. The figure 7a shows a schematic diagram for the vaporising mineral inclusions in a char particle. The small hatched circles represent the mineral inclusions within the char particle. A boundary layer of the vaporising minerals is shown by concentric circles. A diffusion model for the transportation of vaporised minerals from the char into the surrounding gases is developed based on the following assumptions:

- The inclusions are of a single grain size and type, and are much smaller than the char particle. These are uniformly distributed in the char matrix.
- Vaporisation is limited by diffusion of mineral species from the surface of the mineral inclusions to the exterior surface of char.
- Vaporisation is driven by the gas phase concentration of the volatile species at the inclusion surface. This is modelled as diffusion from the inclusion surface to a mean concentration in the char, which is a function of the char particle radius.
- Diffusivity is constant through the pore matrix.
- The char combustion is assumed to be in pseudo-steady state in a stagnant gas of infinite extent.

The molar flux of the mineral species (J), determined by this model, is proportional to the driving force: the difference of the mole fractions at the inclusion (X_i) and at the char surface (X_s). The molar flux of the mineral species at the char surface ($r = r_p$) is given by the equation.

$$J = cD_{eff}.4\pi r_p (X_i - X_s)\eta \qquad (8)$$

where $\eta = [\dfrac{\Theta}{\tanh \Theta} - 1]$ and $\Theta = (\dfrac{r_p}{r_i})(3(\rho_p / \rho_i).f)^{0.5}$ $\qquad (9)$

D_{eff} is the effective diffusivity of the mineral species in the char. The term η may be called an enhancement factor, by which the diffusion rate is enhanced. Θ can be considered as a modified Thiele modulus, a parameter characterising the dispersion of mineral matter within the char matrix. It is a function of mineral volume fraction f, the densities and radii of char particle and of mineral inclusions; a higher Thiele modulus inferring a better dispersion. All the complexities related to the mineral matter distribution within the char particle are lumped into a single enhancement factor given by equation (9).

In this simplified treatment the included minerals are on one side and the bulk gas phase on the other side of a char slab of thickness ($\Delta L = r_p / \eta$). A high enhancement factor reduces the path length implying a reduced diffusion resistance. Equation (8) is the equivalent of equation (10) for a simplified one dimensional system as shown in Figure 7b.

$$J = cD_{eff}. (Area=4\pi\, r_p{}^2). (\Delta X = driving\ force)/ (\Delta L) \quad (10)$$

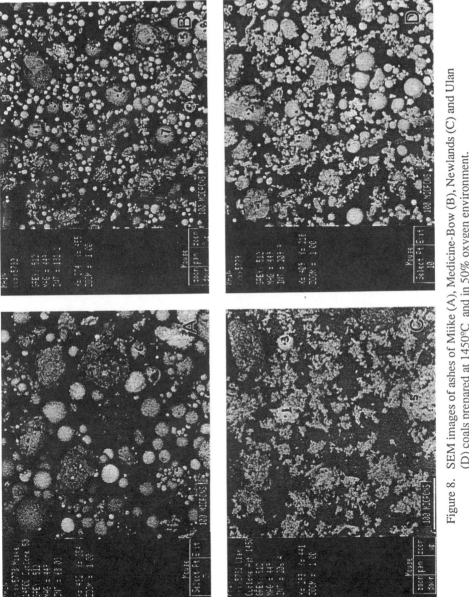

Figure 8. SEM images of ashes of Miike (A), Medicine-Bow (B), Newlands (C) and Ulan (D) coals prepared at 1450°C and in 50% oxygen environment.

Table 1: Ash analyses of the five coals.

Coals	Newlands	Daido	Miike	Ulan	Medicine Bow
Total Ash (%db)	11.95	7.58	17.5	22.0	4.7
SiO_2	51.4	55.9	47.2	88.1	27.0
Al_2O_3	32.0	24.9	20.0	7.9	10.4
TiO_2	1.47	1.29	1.72	0.46	0.71
Fe_2O_3	3.75	7.65	6.69	2.09	6.95
CaO	4.16	3.36	9.26	0.10	26.6
MgO	0.94	0.55	1.04	0.05	5.14
Na_2O	0.3	0.3	2.1	0.3	1.7
K_2O	0.79	1.08	0.76	0.3	0.28
P_2O_5	1.95	0.6	0.23	0.03	0.13
SO_3	3.18	4.23	10.9	0.64	21.26

Table 2. Ranking of coals in terms of ash deposition problems using the thoretical criteria defined in section 5.

Ash Deposition Ranking	Fumes/Fines (Bulk)	Gas Phase Alkalis (Bulk)	Molten Phase Comp (Ex Min)	Molten Phase Comp (In Min)
	Fouling Related	Slagging Related		
	Factor 1	Factor 2	Factor 3	Factor 4
High severity	Miike Ulan	Medicine-Bow Miike	Miike Daido Medicine-Bow	Miike Medicine-Bow
	Daido Newlands	Ulan	-	Daido Newlands
Low severity	Medicine-Bow	Newlands, Daido	Newlands Ulan	Ulan
Comments	Ranking agrees with that from the measured mass of fumes and fines from experiments.	-	-	Physical character observed in SEM images relates to thoretical melt composition

A higher ratio of the radii of char and mineral inclusions implies finely dispersed minerals in the char matrix. A high volume fraction of a finely dispersed phase of included minerals results in a high enhancement factor. For the same mineral loading the enhancement factor increases both with the particle size and the mineral loading. For diffusion of species, the surface area per g of coal, however, is inversely proportional to the radius of the char particles. A decrease in mineral size and increase in volume fraction of mineral matter enhance the diffusion of vapour species.

The molar flux of the mineral species is thus expressed as a function of $(X_i - X_s)$, the difference of mole fraction at the inclusion and char surface. The mole fraction of the species at the char particle surface is also a function of external transport. However, the diffusion of the species through the char particle may be considered as the controlling resistance. One can, therefore, determine the X_i from the equilibrium fraction of the included mineral matter within a char particle and X_s from the equilibrium fraction of the excluded mineral.

The concentration of the species at the char surface can be neglected compared to that at the inclusion surface. The rate of diffusion of these species will, therefore, be determined primarily from the mole fractions within burning char particle resulting from the included minerals. An estimation of the total diffusion of species requires the integration of equation (8) over the burning time of the char particle. The burning time and the particle temperature for 50% oxygen environment do not differ significantly with coal type. Assuming X_i and D_{eff} does not vary during combustion time, the amount of fumes diffusing out of particles is directly proportional to η, the enhancement factor.

The enhancement factor depends on the mineral loading and the ratio of particle size to that of the mineral grain. The total mineral volume fraction in the char particle varies from about 2% for Medicine-Bow coal to about 10% for Ulan coal (based on minerals having twice the density of char). From the CCSEM data, the ratios of particle radius to mineral inclusion radius are taken to be 50 for the fines and dissolved salts and 4 for the other included minerals.

The mole fraction of total inorganics in the gas phase is found to be similar for different coals. This implies that the diffusion rate of total mineral species increases with an increase in the enhancement factor. The enhancement factor thus obtained is 1.2, 1.8 2.6 4.0 and 4.5 for Newlands, Daido, Miike, Ulan and Medicine-Bow coals respectively. The rate of fumes generated is greatest for the Medicine-Bow coal, followed by Ulan coal. Accordingly, the coals are ranked in the following order: Newlands as the least fouling prone coal, Daido, Miike, Ulan and Medicine-Bow coal as the most fouling prone coal. The high levels of alkalis will make the fumes from Medicine-Bow very sticky. The high mole fraction of silica and low levels of alkalis in the fumes from Ulan coal will result in dry ash. This ranking considers the contribution of fumes from the excluded minerals is comparatively negligible.

4.1 Comparison of fumes generated

The ash collected in the drop tube furnace experiments was analysed both for fines and coarse ash compositions. The fine ash rich in fumes, collected on a filter paper, is mostly finer than 2μm in size and results from the diffusion of evaporated mineral species from char and also from fragmentation. The remaining ash collected as coarse ash is mostly greater than 2μm in size. The amount of fines, thus collected, were negligible for Newlands and Daido coals. The fines generated for the Medicine-Bow coal (1g/100g of coal) were a magnitude higher than those for Ulan and Miike coals (0.1-0.12 g/100g of coal) at 1450°C and 50% oxygen.

D_{eff} can vary from 0.05 cm²/s to 5.0 cm²/s for Knudsen diffusion and bulk diffusion limits, respectively. The amount of fumes diffused from 100g of coal, for a D_{eff} of 0.05cm²/s, from the char particles in 0.1s (typical burnout time for 80μm particle in 50% oxygen concentration) is of the order of 0.06g for Miike coal and 0.12g for Medicine-Bow coals, respectively. A high porosity of low rank coals (eg char from Medicine-Bow coal) would probably result in a higher D_{eff} indicating a better match with the experimental value. The experimental values for fines are higher probably due to fines generated by fragmentation of char.

The fumes generated in 1250°C and 50% oxygen environment were estimated to be of similar magnitude from the experiments. Theoretically, there is no significant difference in the temperatures of the burning char particles under the two experimental conditions. The theoretical estimates for the fumes and fines generated would, therefore, be same under the two environments.

4.2 Character of ash particles from Included Minerals

The included minerals transformed at high temperature in a reducing environment are expected to result in spherical particles. The proportion of spherical particles and their size, inferred from the thermodynamic equilibrium composition, may be compared with the SEM images of the ashes collected from the drop tube furnace experiments in Figure 8. This is based on the assumption that a larger proportion (by number) of ash particles are derived from the included minerals.

It has been demonstrated from the thermodynamic calculations that at particle temperatures higher than 2500ºK, (corresponding to 50% oxygen and 1450ºC environment) all the included minerals species except for alumina vaporise. At present it is speculated that a fraction of these species will diffuse from the char and condense to form fume. The rest of the species will cool and condense to form spherical particles - small from the coals with low mineral content and large from the coals with high mineral content. The particles will be perfect spheres for coals having high alkali contents, but may be non-spherical fluffy particles for included minerals of coals low in alkali content.

A high proportion of alkalis in Medicine-Bow and Miike coals would, therefore, result in low melting point silicates and, subsequently, the included minerals will transform into fine spherical ash particles. The size of these spheres observed in the ash from Medicine-Bow coal is rather small probably due to very small mineral loading of the coal particles. The ash formed from Ulan coal containing high proportions of silica should have compounds of high melting point and hence is expected to result in the formation of fluffy ash. The SEM images of the ash collected in the drop tube experiments agree with the predictions. A higher proportion of spherical particles observed in the ash from Newlands is probably due to the formation of low melting point aluminosilcates of sodium and potassium.

The presence of spherical ash particles as those derived from Medicine-Bow and Miike coals indicates sticky particles, whereas, the fluffy ash particles from Ulan coal are not expected to be sticky. Accordingly, Medicine-Bow and Miike coals are ranked as the most difficult regarding ash deposition, Newlands and Daido coals as moderate, and Ulan coal having the least problems in terms of ash deposition.

5. IMPLICATIONS FOR FOULING AND SLAGGING

There are a number of criteria that need to be considered when developing indices for ranking coals for stickiness and subsequently slagging and fouling tendencies. Table 2 gives the ranking of the five coals with respect to each of these criteria.

1. The amount of fumes and fines: The contribution to the fines from the excluded mineral is greatest from Miike and Ulan coals and is minimum for the Medicine-Bow coal. However, the contribution from the included minerals is greatest from the Medicine-Bow coal, followed and Miike, Ulan and is the least for Newlands coal as seen from their respective enhancement factor. Simple calculations show that the contribution from included mineral is the dominant factor and thus ranks the coal in fouling tendencies.
2. The concentration of alkalis in bulk gas phase: The presence of alkalis in the bulk gas phase is expected to cause sticky ash particles. The concentration of alkalis due to excluded minerals would rank the coals in the order of Miike, followed by Ulan, Daido, Newlands and Medicine-Bow coals. The contribution from the included mineral is greatest for Medicine-Bow and Miike coals and would dominate the alkalis in the gas phase.

3. The molten phase associated with excluded minerals: The amount of molten phase on the excluded minerals per unit of total ash is related to the proportion of binding component on the ash particles and this criterion ranks Miike, Medicine-Bow and Daido coals as severely slagging coals compared to Newlands and Ulan coals.

4. The molten phase associated with included minerals: The theoretical estimates of the composition in the gas phase inside the burning char particle, indicate the formation of fine spherical ash particles resulting from combustion of Medicine-Bow coal. The ash particles from Miike coal are also predicted to be spherical but large, and to be comparatively sticky in nature. Ulan coal, due to its composition of gas phase present within the char particle, is expected to be fluffy and non-sticky.

Confirmation with Experiments: The fumes and fines are expected to result mostly from the included minerals and the theoretical ranking of coals based on this criterion agrees with trends from the experiments. The theoretical amount of fumes and the amount of fine ash from from experiments from the Miike coal is of similar magnitude. Also the speculations on the physical character of ash, suggested by the theoretical composition of the molten phase generated by the inherent minerals, agree with the SEM images of the ash collected in the experiments.

6.   BIBLIOGRAPHY

Beer, J. etal. "From Coal Mineral Matter Properties to Fly Ash Deposition Tendencies", in 'Inorganic Transformations and Ash Deposition during Combustion', Benson, S. A. (Ed), Engg. Found. Conf. Palm Coast (1991).

Benson, S. A., M. L. Jones and J. N. Harb, "Ash Formation and Deposition" Chapter 4 in 'Fundamentals of Coal Combustion', Smoot, L. D. (Ed), Elsevier, (1993).

Boni, A. A. et al, "Transformation of Inorganic Coal Constituents in Combustion Systems", DoE Report No. AC22-86PC90751, March (1990).

Couch,G., "Understanding Slagging and Fouling in PF Combustion", IEA Report, (1994).

Erickson, T. A., Fuel, Vol 71, p15-18, (1992)

Gottileb, P. et al, "The Characterization of Mineral Matter in Coal and Fly Ash" in 'Inorganic Transformations and Ash Deposition during Combustion', Benson, S. A. (Ed), Engg. Found. Conf. Palm Coast (1991).

Graham, K. A., "Sub-Micron Ash Formation and Interaction with Sulphur Oxides during Pulverized Coal Combustion", PhD Thesis, MIT, (1992).

Gupta, R. P, and T.F.Wall, " Ash Deposition on the Burner Walls of Coal Fired Furnaces" in 'The Impact of Ash Deposition on Coal Fired Plants', Williamson, J.and F. Wigley (Eds), Engg Found. Conf., Solihull (1993).

Harvey, R. D. and R. R. Ruch, "Mineral Matter in Illinois and other US Coals", in "Mineral Matter and Ash in Coal', Vorres, K. S. (Ed) ACS., (1986).

Hurt, R. H. and R. E. Mitchell, 24th Symp. (Int) on Comb., Comb. Inst., Pittsburgh, p 1243-1250, (1992).

Kalmanovitch, D. P., "Predicting Ash Deposition from Fly Ash Characteristics" in 'Inorganic Transformations and Ash Deposition during Combustion', Benson, S. A. (Ed), Engg. Found. Conf. Palm Coast (1991).

Raask, E., "Mineral Impurities in Coal Combustion", Hemisphere, (1985).

Reid, W.T., Prog. Energy Combust. Sci., Vol 10, p159-175, (1990).

Shah, N. et al, " Graphical Presentation of CCSEM Data for Coal Minerals and Ash Particles" in 'Inorganic Transformations and Ash Deposition during Combustion', Benson, S. A. (Ed), Engg. Found. Conf. Palm Coast (1991).

Skorupska, N. M., and A. M. Carpenter, "Computer Controlled Electron Microscopy of Minerals in Coal", IEA Report, (1993).

Srinivasachar, S. et al, Prog. Energy Comb. Sci., Vol 16, p293, (1991).

Timothy et al, 19th Symp. (Int) on Comb.,Comb. Inst., Pittsburgh, p1123-1130, (1982).

Tognotti et al, 23rd Symp. (Int) on Comb., Comb. Inst., Pittsburgh, p 1207-1215, (1990).

Urbain, G. et al, Trans. J. Br. Ceram. Soc., Vol 80, p139-141, (1981).

Wibberley, L. J. and T. F. Wall, Fuel, Vol 61,p93, (1982).

Wilmenski etal. "Modelling of Mineral Matter Redistribution Ash Formation in Pulverized Coal Combustion" in 'Inorganic Transformations and Ash Deposition during Combustion', Benson, S. A. (Ed), Engg. Found. Conf. Palm Coast (1991).

Zygarlicke et al, "Fly Ash Particle-Size Distribution and Composition: Experimental and Phenomenological Approach" in 'Inorganic Transformations and Ash Deposition during Combustion', Benson, S. A. (Ed), Engg. Found. Conf. Palm Coast (1991).

STUDIES ON ASH SPECIES RELEASE DURING THE PYROLYSIS OF SOLID FUELS WITH A HEATED GRID REACTOR

Tuomas Valmari, Esko I. Kauppinen and Terttaliisa Lind
VTT Aerosol Technology Group, VTT Chemical Technology
P.O. Box 1401, FIN-02044 VTT, Finland

Minna Kurkela and Antero Moilanen
VTT Energy, Gasification
P.O. Box 1601, FIN-02044 VTT, Finland

Riitta Zilliacus
VTT Chemical Technology, Radiation Chemistry
P.O. Box 1404, FIN-02044 VTT, Finland

ABSTRACT

The release of alkalis (sodium and potassium) and five ash matrix elements (silicon, aluminium, iron, calcium and magnesium) during the pyrolysis of Polish coal, Pittsburgh #8 coal, peat and two wood chips was studied with a heated grid reactor using heating conditions relevant to fluidised bed combustion and gasification. Mineral particles or any other super micron fragments were found not to be released from any of the fuels during the pyrolysis. Less than 5 % of any of the elements studied were released from the fuels, with an exception of iron from wood chips. Some iron may have been released from the wood chips (less than 14 % from whole tree chips and less than 8 % from wood chips fuel fraction). Less than 1 % of iron was released from coals and peat.

INTRODUCTION

The behaviour of inorganic ash species, especially alkali metals, released from the solid fuels during the combustion and gasification processes is an important factor affecting deposition and corrosion problems in boiler operation. In coals a large amount of ash forming constituents is found as mineral particles, but also as salts and as organically bound inorganic species [Raask, 1985]. In peat the ash forming constituents are in minerals as well as in ion exchangeable cations and salts [Kurki, 1982]. In biomass the ash is mainly found in soluble form [Kabata-Pendias and Pendias, 1984; Mengel and Kirkby, 1987]. As the behaviour of the ash forming constituents under the fuel conversion is studied, the behaviour has to be known during the different stages of the process. The release of volatiles during the pyrolysis stage that takes place before the oxidation of the char, may account for the majority of the mass loss of a fuel particle

Applications of Advanced Technology to Ash-Related Problems in Boilers
Edited by L. Baxter and R. DeSollar, Plenum Press, New York, 1996

265

and is usually much faster step than the oxidation of the remaining char. The ash species that are released during the pyrolysis escape from the fuel particle as gas phase species or they may react with those ash forming constituents that remained in the char. The knowledge of pyrolysis release is important especially in processes where pyrolysis and char oxidation stages are separated, e.g. in two-stage gasification processes.

The behaviour of volatiles during the pyrolysis of coal has been studied extensively [Solomon et al., 1992]. The research on the behaviour of the ash forming constituents during the pyrolysis is mainly concentrated on their influence on the release of volatiles. However, Manzoori and Agarwal [1992] have studied the behaviour of organically bound inorganic elements and sodium chloride during fluidised bed pyrolysis and combustion of high sulphur, high sodium low rank coal where the inorganic matter were mainly organically bound. During pyrolysis stage the organically bound calcium, magnesium and part of the sodium transformed to acid-insoluble compounds. Sodium chloride was found to dissociate and/or react with other compounds resulting in the formation of non-volatile sodium species and gaseous hydrogen chloride. The release of sodium was small during the pyrolysis, e.g. about 5 % after 40 seconds of pyrolysis at the temperatures of 770 °C and 830 °C. However, over 80 % of the sodium, calcium and magnesium and practically all of the chlorine were lost from the coal particle during 40 seconds of combustion in a temperature of 700 °C.

Our earlier work includes studies of ash species behaviour in a power plant scale circulating fluidised bed combustion of Venezuelan bituminous coal. In the combustion generated fly ash particles sodium and potassium were mostly found as water insoluble compounds in particles larger than 1 μm, most likely bound to aluminosilicates [Lind et al. 1994]. According to aerosol dynamical model calculations, less than 1 % of the total sodium and potassium were found in gas phase at conditions immediately after the process cyclone. Lind et al. [1995] found that less than 1 % of the fly ash particles were in particles smaller than 1 μm. Calcium, aluminium, iron, magnesium and silicon showed no significant vaporisation. However, during the fluidised bed combustion of biomasses, severe ash deposition and corrosion problems occur, indicating that alkali species can vaporise during combustion.

The purpose of this paper is to study the released fractions of alkalis (Na and K) and five matrix elements (Si, Al, Fe, Ca and Mg) during the pyrolysis stage of fluidised bed combustion and gasification of solid fuels with different ash characteristics (bituminous coals, peat, wood chips). The release of the ash forming constituents during the pyrolysis was studied with a heated grid reactor. This type of pyrolysis system was developed by Anthony et al. [1974]. The heating rate and the pyrolysis temperature were chosen to be relevant for the fluidised bed processes. The released fractions of the seven elements studied were evaluated by analysing the particles formed from the released condensable species, majority of which are tars.

EXPERIMENTAL

Fuels

The fuels chosen for this project were two bituminous coals (Polish coal and Pittsburgh

#8), peat and two different kinds of wood chips (whole tree chips and wood chips fuel fraction). Polish coal differs from Pittsburgh #8 in a way that it has much larger overall ash content as well as alkali content than Pittsburgh #8. The peat used in this project has been used as a feedstock in a gasification study [Kurkela et al., 1993]. Whole tree chips were made of whole trees, including bark, knots etc. for fuel purposes. Wood chips fuel fraction was made with Massahake-method, which is a new method under development in VTT for producing raw material for pulp industry. The bark content of Massahake pulp fraction is under 1 %. The Massahake fuel fraction studied here is the remainder that is not used by pulp industry. It is enriched with bark and contains more ash forming constituents than whole tree chips. Both wood chip samples were made of so-called first-thinning trees. The same fuels will be used later on laboratory development unit studies on fluidised bed combustion and gasification. They are stored under controlled conditions in order to minimise changes during storage.

The fuel samples were homogenised and fractionated. The particle size fraction of 44-105 μm was used. Ultimate- and proximate analysis were made to all of the five fuels. The mineral compositions of the two coals were analysed with a Computer Controlled Scanning Electron Microscopy (CCSEM). The CCSEM analysis of the Polish coal was carried out by the Microbeam Technologies Incorporated, ND, USA and the analysis of the Pittsburgh #8 by the University of Kentucky.

The elemental analyses of eight ash matrix species and alkali metals were carried out by ICP-AES method (Inductively Coupled Plasma- Atomic Emission Spectroscopy). Samples were ashed at a temperature of 500 °C. The ashes were fused by lithium metaborate ($LiBO_2$) and the melt was dissolved into the hydrochloric acid. The determinations were made from this solution. In addition, chlorine was analysed from Polish coal, peat and whole tree chips with INAA method (Instrumental Neutron Activation Analysis).

Pyrolysis Experiments

The heated grid reactor used in this study (Fig. 1) is described in detail elsewhere [McKeough et al., 1995]. The experiments were carried out under atmospheric pressure. The samples of about 100 mg were evenly distributed between three times folded 400 mesh stainless steel screens. The screens were heated resistively with a heating rate of ≈ 800 °C/sec to the final temperature of 850 - 950 °C. The holding time at the final temperature was 20 seconds. A through-flow of nitrogen (flow rate 5 l/min, NTP, temperature 20 °C prior to contact with the heated grid) was employed through the reactor during the experiments. The nitrogen carrier gas was introduced from the heated grid reactor through a funnel. Some of the nitrogen was introduced outside the pyrolysed sample to act as a sheat gas in a laminar flow field region between the funnel walls and the sample carrying nitrogen in order to prevent particle losses on the furnace walls. Outside the reactor the sample carrier gas was diluted with a secondary nitrogen flow (Fig. 1).

The particles were sampled from the carrier gas with a Tapered Element Oscillating Microbalance (TEOM Series 1200 Ambient Particulate Mass Monitor, Rupprecht & Patashnick Co., Inc., Albany, NY, USA) and with a Berner-type low pressure impactor (BLPI) [Berner 1984; Hillamo and Kauppinen, 1991]. BLPI and TEOM were operated

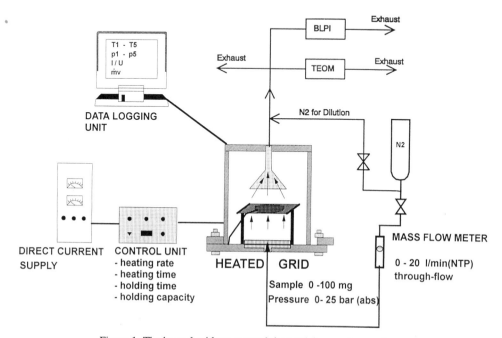

Figure 1. The heated grid reactor and the particle sampling system.

at a temperature of 20 °C. The particles observed with BLPI and TEOM were thus mainly tar, but the particles collected included also any mineral particles and other condensed ash species released during the pyrolysis. TEOM mass monitor was used to measure the mass release rate of the condensable pyrolysis products. TEOM collects particles on a quartz fiber filter, whose mass is evaluated frequently. The filter is located on the top of a vibrating hollow tapered tube, whose vibration frequency (about 200 Hz) depends on the mass collected on the filter [Patashnick and Rupprecht, 1991]. BLPI was used to measure the mass size distribution of the pyrolysis product particles in the size range of 0.03 - 50 μm. BLPI classifies particles into 11 size fractions. Each fraction is collected on a collection substrate that is weighed before and after the experiment. In this study particles were collected on either aluminium or polycarbonate (Nuclepore) collection substrates.

At least two and no more than five samples of each fuel were pyrolysed. TEOM was used with all the samples. BLPI was used with 2 - 4 samples of each fuels. In addition, two blank samples were made with BLPI. One blank sample was made with Polish coal without heating the screen. Another blank sample was made by heating empty grid without fuel with similar heating rate and final temperature than during pyrolysis experiments.

The elemental released fractions of silicon, aluminium, iron, calcium, magnesium, sodium and potassium were analysed from one BLPI measurement for each fuel. Samples collected on Nuclepore collection substrates were used for elemental analyses. Only the eight smallest size fractions (particle diameter smaller than 4 μm) were analysed since the particle masses collected on the coarser fractions were negligible in all cases. A quarter of each substrate was dissolved in HNO_3 until the spots formed of collected particles had completely disappeared from the substrate. Dissolutions were analysed with ICP-MS- method. The elemental concentrations of the blank BLPI sample when the grid was heated without fuel was also measured to set the detection limit for the evaluation of released fractions.

The morphology of the particles produced with the pyrolysis of the Polish coal was studied by analysing BLPI- collected particles in size fraction 0.5 - 1.0 μm with a Scanning Electron Microscope (SEM, Hitachi S-800) at the University of New Mexico. The elemental composition of those particles were analysed with an energy dispersive X-ray analyser (EDX).

RESULTS

Fuel Analyses

The results from the proximate and ultimate analyses of all the five fuels are presented in Table 1. The results from the CCSEM analysis of the coals are presented in Table 2. The CCSEM results for Pittsburgh #8 have been reported by Helble et al. [1994]. Most of the minerals (68 %) in the Polish coal were quartz and aluminosilicates, mainly kaolinite and illite. Pittsburgh #8 contained 44 % of quartz and aluminosilicates and 32 % Fe-rich minerals, mainly pyrite. About one third (36 %) of the minerals in the Polish

Table 1. The composition of the fuels, particle size fraction 44-105 μm.

	Polish coal	Pittsburgh #8	Peat	Whole tree chips	Wood chips, fuel fraction.
Moisture, %	2.3	1.6	7.4	3.6	4.0
Volatiles (dry, %)	30.7	35.4	69.1	78.5	75.9
Fixed carbon (dry, %)	55.0	55.6	19.4	16.6	17.7
Ash (dry, %)	12.0	7.4	4.1	1.3	2.4
C (dry, %)	73.2	77.3	55.0	52.1	52.9
H (dry, %)	4.6	5.2	6.0	6.3	6.3
N (dry, %)	1.2	1.4	1.6	0.3	0.5
S (dry, %)	0.9	2.1	0.17	0.05	0.05
O (dry, %) (difference)	8.1	6.6	33.13	39.95	37.85

coal were smaller than 10 μm. The minerals in Pittsburgh #8 were slightly smaller than those in the Polish coal, 50 % of them being smaller than 10 μm. The Mössbauer analysis of Pittsburgh #8 showed that 90 % of the iron in coal minerals was in pyrite [Helble et al. 1994].

The results of the matrix element and alkali concentration analyses are presented in Table 3. Polish coal has more alkalis than the other fuels studied here. The sodium and potassium concentrations in Pittsburgh #8 are approximately one third of the ones in the Polish coal. Wood chips contain a large amount of potassium (0.2 %), but their sodium content is low.

Pyrolysis Experiments

Blank samples

The released mass of a Polish coal sample without heating the screen observed with BLPI was -0.06 % (-53 μg) of the total fuel sample mass in particles with diameter under 4 μm and +0.09 % (80 μg) in size range over 4 μm. The negative released fraction is associated with the accuracy of the gravimetric analysis, whereas a small amount of large particles was released, even if the sample was not heated. The mass released from the empty grid without a fuel sample heated with similar heating rate and final temperature than during pyrolysis experiments was 0.09 mg with BLPI. That would be 0.09 % of a 100 mg sample.

Total Mass Release

The total masses released during the pyrolyses are presented on Table 4. The char yields presented are the average of all the 2 - 5 samples for each fuel. The mass of the particles formed from released species is presented as an average from the BLPI-measurements. The mass of the particles as measured with TEOM was 160±30 % for coals and 105±15 % for the other fuels as compared to the values measured concurrently with BLPI. This may indicate that some gaseous species released during coal pyrolysis react with the quartz fiber TEOM-filter thus increasing the mass

Table 2. The mineral analysis of the Polish coal and Pittsburgh #8 made with a computer controlled scanning electron microscopy (CCSEM). The concentration of each mineral class is presented as the total weight per cent of the mineral matter on a coal.

Mineral	Polish coal (%, wt)	Pittsburgh #8 (%, wt)
Quartz	9	9
Kaolinite	20	4
Illite	20	9
Other silicates	19	22
Pyrite	10	27
Other Fe-rich	-	5
Dolomite	8	-
Other carbonates	-	9
Other minerals	14	15

observed by TEOM. The released mass that is present as gaseous species at the ambient temperature was calculated from the mass balance based on BLPI measurements.

Figure 2 shows the particle mass (tar) as the function of time for the pyrolysis of the Polish coal as measured with TEOM. The time-resolution of the system is limited by the gas residence time distribution in tubing between the heated grid and the TEOM-sensor, since the gas velocity in the middle of the tube is larger than near the walls of the tube. Thus the particles formed from the species released simultaneously from the grid arrive

Table 3. The matrix element and alkali concentrations of the fuels (particle size 44-105 μm). The elements were analysed with the ICP-AES method, except chlorine with the INAA method. The concentrations are presented in mg/kg.

	Polish Coal	Pittsburgh #8	Peat	Whole tree chips	Wood chips, fuel fraction.
Matrix Elements					
Si	24000	14800	9460	1950	5040
Al	15600	8340	3480	510	1220
Fe	8160	9180	6620	330	770
Ca	4800	3330	4810	1820	2520
S	2760	1300	900	100	220
Mg	2400	470	770	470	480
Cl	1500	–	380	< 170	–
Ti	850	460	120	18	46
P	440	140	600	340	500
Alkalis					
Na	1130	410	470	170	360
K	2040	740	560	1530	1820

to the TEOM-sensor at different times. The average gas residence time was 4.5 seconds in case of all the measurements with the Polish coal (Fig. 2). Most of the particles were observed within 10 seconds from the beginning of the pyrolysis. The time resolution of the system was obviously too limited for evaluating the mass release rate during the pyrolysis, since under a high heating rate, most of the mass is released in less than 1 second [Solomon et al., 1993]. However, it can be observed that no slow secondary reactions with significant formation of condensable species took place. The other fuels studied showed similar behaviour.

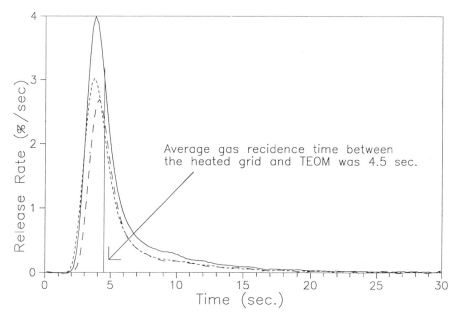

Figure 2. The release rate of the Polish coal pyrolysis product particles (condensed tar) observed by TEOM mass monitor. The coal heating began at t = 0. The average penetration time of pyrolysis gases between the heated grid and TEOM was 4.5 seconds.

The mass size distributions measured with BLPI as well as the samples chosen for ICP-MS -analysis are shown on Fig. 3. The size distributions are presented as a function of the aerodynamic diameter of the particles, which means that a particle density of 1 g/cm^3 is assumed. The size distributions observed are unimodal with the majority of particles in the size range of 0.3 - 3 μm for all the fuels. The reproducibility of the size distributions are reasonable with an exception of peat. In case of peat the mass of the particles formed from released species varied from 4.7 % to 13.6 % of the original fuel sample mass in three different runs. The peat sample chosen for ICP-MS -analysis was from the experiment with most particles formed (13.6 %).

Table 4. The fractions remained in char and released from the fuel samples during the pyrolysis experiments with a heated grid reactor.

Fuel	Char, %	Released, %		
		Particles formed from the released species (BLPI).	As gas in the temperature of 20 °C (difference).	Total released
Polish coal	62.9	4.9	32.2	37.1
Pittsburgh #8	57.9	8.0	34.2	42.1
Peat	27.6	8.2	64.2	72.4
Whole tree chips	13.2	14.2	72.6	86.8
Wood chips, fuel fraction.	16.3	11.6	72.1	83.7

No major release of unfragmented mineral particles from the coals took place, since the amount of released particles larger than 3 μm was small. More than 70 % of the minerals in both coals were larger than 4.6 μm according to the CCSEM analyses of the coals.

A SEM-micrograph of the BLPI- collected product particles from the pyrolysis of Polish coal in the size fraction of 0.5 - 1.0 μm is shown in Fig. 4. Particles are spherical. They have partially fused together on the collection substrate. The particles are composed of tar species that are volatilised and then condensed as nitrogen was cooled. No irregular shaped particles that could consist of released minerals or other fragments that would have been released from the fuel particle without volatilisation were observed. The EDX- analysis showed that the pyrolysis product particles contained some sulphur and chlorine. No other species were observed, with the exception of carbon and platinum. Carbon comes not only from the particles, but also from the polycarbonate collection substrate. Platinum is present since the sample was coated with platinum prior to analysis.

The Release of Si, Al, Fe, Ca, Mg, Na and K

The masses of silicon, aluminium, iron, calcium, magnesium, sodium and potassium released from the fuel samples during pyrolysis as well as the masses analysed from the blank samples during the heating of the empty grid are shown on table 5. The masses from the empty grid measurement include the background from the polycarbonate collection substrates as well as the mass released from the empty grid during heating. The numbers given are evaluated by adding together the content of the element analysed with ICP-MS method in all the eight size fractions of the BLPI samples. The content of silicon, iron, calcium, sodium and potassium were below detection limit in some fractions. In these cases the lower and upper limit of the released mass of the element in question is given assuming the mass of the element in those fractions to be zero and equal to the detection limit, respectively.

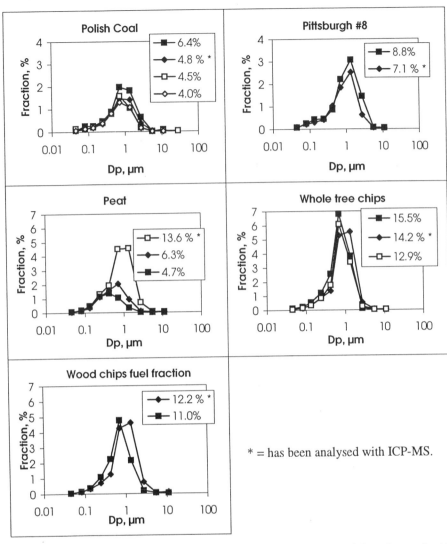

Figure 3. The mass size distributions of the pyrolysis product particles observed with BLPI. The numbers given are the total fractions of particles formed from the released species.

The elemental released fractions of all the seven elements studied were evaluated by dividing the mass of the element in question in the BLPI-collected samples with the mass of that element in the original fuel sample (Table 6). The released mass of all the elements studied, except silicon, was on the same level or even lower with a fuel sample than with the empty grid. For this reason, only upper limits for the released fractions for

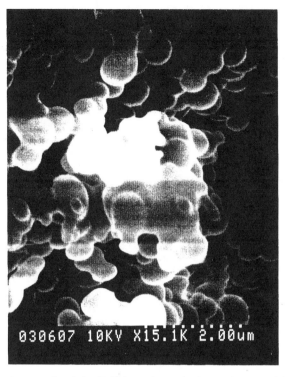

Figure 4. SEM-micrograph of the Polish coal pyrolysis product particles collected on BLPI.

these elements can be given. Upper limits are given based on the masses analysed from the BLPI samples without subtracting the background values from the empty grid sample and assuming the mass of the elements in fractions below detection limit to be equal to the detection limit. The released fractions of silicon evaluated by subtracting the background values observed during the empty grid measurement from the values observed during fuel sample pyrolysis are given in paranthesis.

Less than 3 % of all the elements studied were released from coals and peat. In case of wood chips the detection limits were higher for some elements since their content were small in these fuels. Less than 5 % of all the elements studied except iron were released from the wood chips samples. Under 14 % of iron was released from whole tree chips and under 8 % from wood chips fuel fraction.

The observed elemental mass size distributions of released silicon are shown on Fig. 5. The data shown is scattered due to variation in background values of each individual collection substrate, variation in the amount of silicon released from the empty grid and also by the limited accuracy of the analysis method when the fraction of analysed element is near the detection limit. However, most of the silicon observed is found in large particles with all the fuels except peat. In case of peat the data is too scattered for observing a trend in size distribution curve.

Table 5. The mass of Si, Al, Fe, Ca, Mg, Na and K released from the fuel samples during the pyrolysis experiments with a heated grid reactor. The released masses were evaluated as a sum of the mass from the eight size fractions of the BLPI-samples analysed with the ICP-MS- method. Also the detection limit for one size fraction is shown for each element.

	Released during pyrolysis, µg						
	Si	**Al**	**Fe**	**Ca**	**Mg**	**Na**	**K**
Detection limit for one size fraction.	0.2	< 0.03	0.1	0.2	< 0.01	0.04	0.1
Empty grid	1.5 - 2.5	1.7	5.4	8.1 - 8.6	0.5	1.0	< 0.9
Polish coal	7.1 - 7.2	1.2	4.4 - 4.6	5.2 - 5.8	0.3	2.4 - 2.5	1.3 - 1.8
Pittsburgh #8	8.7	1.3	2.4 - 2.7	4.9 - 5.5	0.4	0.7 - 0.9	0.8 - 1.2
Peat	8.2	1.2	4.2 - 4.5	4.9 - 5.8	0.4	1.2 - 1.4	0.6 - 1.4
Whole tree chips	9.1	1.2	4.2 - 4.4	6.8 - 7.5	0.3	< 0.4	< 0.8
Wood chips, fuel fraction.	9.3	2.2	6.0 - 6.1	8.4 - 8.6	0.5	1.5 - 1.6	0.6 - 1.3

Discussion

The released fractions were evaluated by studying the products released from the fuel samples. The alternative way would have been to analyse the char remained in the grid during pyrolysis. However, it was not possible to carry out char analysis with a good accuracy. The char remained in the steel screen during the pyrolysis is heterogenously distributed. It can not be completely separated from the screen, but it has to be analysed together with the screen. This decreases the accuracy of the results, since the presence of large amount of steel makes dissolving and analysis of the samples difficult. Studying the released species directly gave also a possibility to study elements that the screen itself contains, such as iron.

The heated grid method is reported to have a disadvantage that the temperature history of the pyrolysed particles is not well known with high heating rates, as it does not necessary follow the temperature of the thermocouple attached to the grid [Suuberg et al. 1992]. Small fuel particle sizes were used to minimise this problem. The steel screens were heated with similar power output with each individual measurement to ensure comparativity between different measurements. The holding temperatures monitored by thermocouples showed variation between 850 and 960 °C in different measurements. The heating rates measured were 750 - 1000 °C/sec (with one exception of 550 °C/sec). The two thermocouples attached to different locations showed typically 30 °C difference. The measured values during the heating of the empty screen (holding temperature was 990 °C and the heating rate 1000 °C/sec) were only slightly higher than' typical values during the pyrolysis experiments, suggesting that the pyrolysis temperatures were similar with different fuels. No dependence of the char yield or of the released particles on the measured temperatures was observed. It seems that the main reason for the differences in measured temperature values was the variation in

Table 6. The upper limits of the fractions of Si, Al, Fe, Ca, Mg, Na and K released from the fuel samples during the pyrolysis experiments with a heated grid reactor. The released fractions of Si evaluated by subtracting the background values observed during the empty grid measurement from the values observed during pyrolysis of the fuel samples are given in paranthesis.

Fuel	Released during pyrolysis, % (wt)						
	Si	Al	Fe	Ca	Mg	Na	K
Polish coal	< 0.3 (0.2)	< 0.07	< 0.6	< 1.2	< 0.13	< 2.2	< 0.9
Pittsburgh #8	< 0.5 (0.4)	< 0.16	< 0.3	< 1.6	< 0.8	< 2.2	< 1.7
Peat	< 0.9 (0.6)	< 0.4	< 0.7	< 1.2	< 0.5	< 3	< 2.4
Whole tree chips	< 5 (3)	< 2.2	< 14	< 4	< 0.6	< 2.3	< 0.6
Wood chips, fuel fraction.	< 1.9 (1.3)	< 1.8	< 8	< 3.4	< 1.1	< 5	< 0.7

contact between the thermocouple, grid material and the fuel particles, as well as local temperature differences within the grid. Even if it is not possible to set the heating conditions completely similar to those in a real scale process, this experimental set up still gives a possibility to compare a pyrolysis behaviour of different kinds of fuels during rapid ($\approx 10^3$ °C/sec) heating conditions.

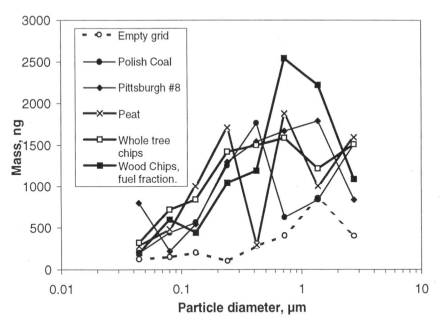

Figure 5. The mass of silicon observed in different size fractions of the particles formed from the released species during the pyrolysis experiments with a heated grid reactor. Data points under detection limits are drawn assuming them to be equal with the detection limit.

Heat transfer from the particle to the environment is different in a heated grid reactor than in a real scale process. In heated grid, the fuel particles are directly in contact to each other and flushed with nitrogen that is much cooler than the particles, where as in real scale processes the fuel particles are pyrolysed individually and they are surrounded by hot gas. The bed material, which may form 95 % of the solid material in a fluidised bed, was not present in our experiments. However, as the pyrolysis stage is dominated by the volatiles escaping from the fuel particle and thus isolating the particle from the environment, the bed material is not likely to directly affect the pyrolysis release, even if it is important for the whole picture of the behaviour of ash during combustion and gasification.

CONCLUSIONS

The release of alkalis (Na, K) and five matrix elements (Si, Al, Fe, Ca, Mg) during the pyrolysis stage of the fluidised bed conversion processes was studied with a laboratory scale heated grid reactor. The unfragmented mineral particles, or any other super micron fragments, were not carried away from the pyrolysed fuel particles with the volatiles escaping from the fuel. Significant fractions of quartz and aluminosilicates were not carried away from coals even as a sub-micron fragments, since less than 0.5 % of silicon and less than 0.2 % of aluminium was found to be released from the two coals studied here. Pyrite is a mineral that may decompose on the temperature of under 1000 °C. However, even if the pyrite on coals may have decomposed, no significant amount of iron bound to pyrite was released from the coals studied.

Less than 5 % of any of the elements studied were released from any of the fuels during pyrolysis stage with the exception of iron in case of wood chips samples. Some iron may have released from the wood chips (under 14 % from whole tree chips and under 8 % from wood chips fuel fraction). The detection limit for the released fraction of iron was high for these fuels, since their iron content was low.

ACKNOWLEDGMENTS

We would like to thank the Ministry of Trade and Industry of Finland, IVO, Enviropower, Tampella Power and Ahlström for funding this study via research program LIEKKI 2. We would like to thank Geological Survey of Finland for the ICP-MS analyses, Mr. Sulo Piepponen from VTT Chemical Technology for carrying out the ICP-AES analyses and Dr. Jorma Jokiniemi, Mr. Jussi Lyyränen, Mr. Esa Kurkela and Mr. Vesa Arpiainen from VTT Energy for their valuable contribution during the project. We thank Mr. A. Gurav and Prof. T. Kodas from the University of New Mexico, Albuquerque, NM, USA, for their contribution when carrying out SEM-analyses.

REFERENCES

Anthony, D.B., Howard, J.B., Meissuer, H.P. and Hottel, H.C. (1974). Rev. Sci. Instrum., 45, 992-996.

Berner, A. (1984). "Design Principles of AERAS Low Pressure Impactor." In Liu, B.Y.H., Pui, D.Y.H., Fissan H.J. (Eds.), Aerosols: Science, Technology and Industrial Applications of Airborne Particles. New York: Elsevier.

Helble, J.J., Bool, L.E., Sarofim, A.F., Zeng, T., Peterson T.W., Gallien, D., Huffman, G.P., Huggins, F.E. and Shah, N. (1994). Fundamental Study of Ash Formation and Deposition: Effect of Reducing Stoichiometry. Quarterly Report No. 3. U.S. Department of Energy, DOE Contract No. DE-AC22-93PC92190.

Hillamo, R.E. and Kauppinen, E.I. (1991). "On the Performance of the Berner Low Pressure Impactor." Aerosol Sci. Technol. 14, 33-47.

Kabata-Pendias, A. and Pendias, H. (1984). Trace elements in soils and plants. CRC Press, Inc., Boca Raton, Florida.

Kurkela, E., Ståhlberg, P., Laatikainen, J. and Simell, P. (1993). "Development of simplified IGCC processes for biofuels - supporting gasification research at VTT." Bioresource Technology 46, 37-48.

Kurki, M. (1982). "Main chemical characteristics of peat soils. Peatlands and their utilization in Finland." Finnish Peatland Society - Finnish National Committee of the International Peat Society, Helsinki, 37-41.

Lind, T., Kauppinen, E., Jokiniemi, J.K., Maenhaut, W. and Pakkanen T. (1994). "Alkali metal behaviour in atmospheric circulating fluidised bed coal combustion." In J. Williamson and F. Wigley (Eds.) "The Impact of Ash Deposition on Coal Fired Plants" Proceedings of the Engineering Foundation Conference, Taylor & Francis, 77-88.

Lind, T., Kauppinen, E., Maenhaut, W., Shah, A., and Huggins, F. (1995). "Ash vaporization in circulating fluidized bed coal combustion." Accepted for publication in Aerosol Science and Technology.

Manzoori, A. R. and Agarwal, P. K. (1992). "The fate of organically bound inorganic elements and sodium chloride during fluidized bed combustion of high sodium, high sulphur low rank coals." Fuel 71, 513-522.

McKeough, P., Pyykkönen, M. and Arpiainen, V. (1995). "Rapid Pyrolysis of Kraft Black Liquor. Part 2. Release of Sodium." Paperi ja Puu-Paper and Timber 77, 39:44.

Mengel, K. and Kirkby, E. A. (1987). Principles of plant nutrition. 4th Ed., International Potash Institute, Bern.

Patashnick, H. and Rupprecht, G. (1991). "Continuous PM-10 Measurements Using the Tapered Element Oscillating Microbalance." Journal of Air Waste Management Association 41, 1079-1083.

Raask, E. (1985). Mineral Impurities in Coal Combustion. Washington: Hemisphere Publishing Corporation.

Solomon, P.R., Fletcher, T.H. and Pugmire, R.J. (1993). "Progress in coal pyrolysis." Fuel 72, 587-597.

Solomon, P.R., Serio, M.A. and Suuberg, E.M. (1992). "Coal Pyrolysis: Experiments, Kinetic Rates and Mechanisms." Prog. Energy Combust. 18, 133-220.

IRON OXIDATION STATE AND ITS EFFECT ON ASH PARTICLE STICKINESS

L.E. Bool III and J.J. Helble
PSI Technologies, a division of Physical Sciences Inc.
20 New England Business Center
Andover, MA 01810 USA
email: bool@psicorp.com

ABSTRACT

The oxidation state of iron in glassy particles dramatically affects the deposition of those particles in pulverized coal fired utility boilers. In this work, two synthetic ashes were fabricated and used to explore the effect of iron oxidation state on ash stickiness in a bench scale experimental study. The iron in these synthetic ashes, approximately 20 wt%, was initially 100% in the Fe(II) state. Experiments conducted with these ashes in an electrically heated entrained flow reactor were used to determine the stickiness of the synthetic ashes under different oxidizing environments and temperatures. Particulate samples were collected to measure the conversion of Fe(II) to Fe(III) under various conditions, and the results used to determine the rate of iron oxidation in glassy ash particles. This information was used to identify the rate limiting step for iron oxidation and to develop a simple preliminary model to predict the fraction of Fe(II) oxidized to Fe(III) as a function of time.

INTRODUCTION

One of the major difficulties associated with the combustion of pulverized coal for power generation is the formation of slag deposits on heat exchange surfaces in the boiler. Slag deposits form when ash particles impact on and adhere to tube surfaces. These deposits inhibit heat transfer (decreasing unit efficiency), cause or accelerate tube corrosion, and decrease boiler capacity and availability.

Although boiler ash deposits generally have a chemical composition that is similar to the ash composition of the coal being fired, iron has long been identified as a major initiator of slagging problems in utility boilers. For example, slag deposits near the burner can often be traced to iron in the form of molten pyrite or pyrrhotite [ten Brink et al., 1992]. Iron may also dramatically affect slagging farther downstream in the boiler. When iron-containing minerals coalesce with aluminosilicate minerals during char burnout, a molten iron containing glass is formed [Bool et al, 1995]. Previous research has shown that when iron is incorporated into an aluminosilicate melt it dramatically affects the melting point and

Applications of Advanced Technology to Ash-Related Problems in Boilers
Edited by L. Baxter and R. DeSollar, Plenum Press, New York, 1996

281

viscosity of the melt. For example, Kalmanovitch et al. [1986], demonstrated that increasing the iron content of a glass decreases its melting point. Austin et al. [1982] in laboratory measurements demonstrated that the addition of Fe(II) as FeO to a coal ash reduced the surface temperature at which the ash particles began to stick by 100°C, from nearly 500°C to under 400°C. More recently, analysis of detailed field tests in the U.K. have implicated iron containing aluminosilicate ash particles in slag formation during the combustion of certain bituminous coals [Wigley and Williamson, 1994].

Iron in aluminosilicate-derived ash particles contributes to slag deposition under typical boiler conditions through Fe(II) acting as a weak base and reducing ash particle and slag viscosity [Raask, 1985]. Using the ash viscosity model of Senior and Srinivasachar [1995] the effect of iron oxidation state can be seen. Figure 1 also illustrates this effect as a function of temperature for the range 600 to 820°C. When the percentage of iron in the reduced Fe(II) state is increased from 20 to 76%, the melt viscosity decreases dramatically for a given temperature. For example,

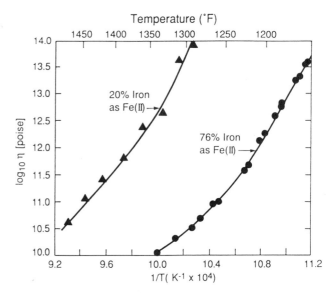

Figure 1. Iron oxidation state effects on ash particle viscosity [Cukierman and Uhlmann, 1974].

for a melt temperature of 700°C (1300°F), the viscosity decreases by approximately 3.5 orders of magnitude. This decrease in viscosity leads directly to an increase in particle stickiness, i.e., tendency of impacting particles to adhere to a surface at a given temperature [Srinivasachar et al., 1990a]. Srinivasachar et al. [1990a,b, 1992] demonstrated that the viscosity of an ash particle is a critical parameter in determining deposit initiation on a clean surface. By using a synthetic ash and several coal fly ashes to measure the adhesion efficiency of ash as a function of composition and temperature, it was shown that particles with viscosities above a critical range, 10^6 to 10^9 poise, rebounded from an impaction tube surface without sticking. Particles with viscosities lower than the critical range typically adhered to the tube. Therefore, any increase in particle viscosity due to oxidation of Fe(II) to Fe(III) in glassy particles should cause a decrease in particle stickiness at a given temperature.

It is clear that the oxidation state of iron in glassy particles plays a critical role in determining the deposition of these particles. Gaining a better understanding of the effect of the iron oxidation state on the deposition of ash under pc boiler conditions, and understanding the mechanisms governing the oxidation of Fe(II) to Fe(III) in glassy ash particles are critical to better predictions of slag location, coverage, and thickness in pc boilers. In this paper, we address this problem by presenting experimental data on the effect of changing iron oxidation state on ash particle deposition, discussing the mechanism governing oxidation of the iron in these particles, and developing a simple model to predict the oxidation of Fe(II) in ash particles.

EXPERIMENTAL

To gain a better understanding of the role of iron in controlling ash deposition under reducing conditions, a number of experiments were performed to measure the adhesion efficiency of iron containing glassy ash particles. To avoid the particle-to-particle variations in ash composition and iron oxidation state typical of coal ash, two synthetic ashes, with identical compositions, were used in these experiments. Each synthetic ash was produced by a commercial laboratory from a homogeneous melt containing iron (II) oxide, silica, alumina, and potassium oxide (see Table I). By producing the synthetic ashes from homogenous melts, particle to particle variations in composition were avoided. The resulting glasses were then pulverized to a particle size distribution approximating that of coal fly ash. As shown in Fig 2. the size distributions of the two samples measured by the Malvern technique were similar, with mass mean diameters of 28 and 30 μm for glass 1 and glass 2, respectively.

Table I. Composition Specifications for Iron Containing Glass

Oxide	Weight %
SiO_2	51
Al_2O_3	21
FeO	20
K_2O	8

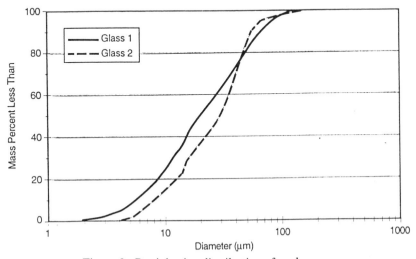

Figure 2. Particle size distributions for glasses.

Samples of each glass were analyzed by Mössbauer spectroscopy and by Fe(II) titration to measure the percentages of Fe(II) and Fe(III) present. Although the iron was expected to be present exclusively as FeO, the initial Mössbauer analysis indicated that both glasses also contained metallic iron. The FeO titration, however, indicated the glasses each contained approximately 19.5% FeO by weight, the target amount. To resolve this discrepancy, samples of the glasses were examined for total iron concentration by atomic absorption spectroscopy. The total iron analyses (18.64% and 19.95%) agreed with both the manufacturing specifications and the FeO titration. It was therefore concluded that the Mössbauer indication of metallic iron was incorrect; all of the iron in the each glass was initially in the Fe(II) state.

The adhesion efficiency experiments were conducted in the entrained flow reactor (EFR) shown schematically in Fig. 3. Each glass was fed into the top of the reactor at 0.06 g/min using a calibrated syringe pump-type dry materials feeder. The glass was entrained at the exit of the syringe with a transport gas with a typical flow rate of 4.2 slpm. At the top of the converging section shown in the diagram, glass and transport gas mixed with the remaining gases, yielding a total gas flowrate of 23.3 standard liters per minute (slpm). The gas and entrained glass particles were then concentrated in a venturi and rapidly expanded to disperse the glass across the entire tube cross section. Upon exiting the converging section, the feed stream entered the laminar flow reactor. This reactor consists of an externally heated 8.57 cm (3.375 in.) inner diameter mullite tube approximately 1.57 m (62 in.) long. The reactor is divided into three independently controlled temperature zones. For these experiments the wall temperature in the center zone was varied between 900 and 1400°C. Maximum gas temperature was typically 50°C lower than the furnace setpoint.

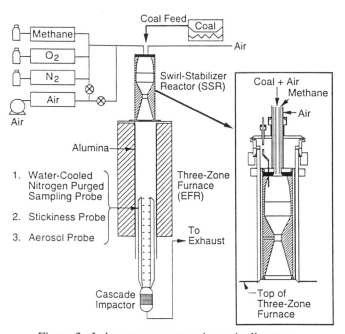

Figure 3. Laboratory reactor schematic diagram.

For the deposition experiments the particle laden gas stream exited the reactor and entered a conical converging section where the cross-sectional area was reduced to approximately 5 cm^2. Two deposition probes, consisting of uncooled mullite tubes, were placed perpendicular to the direction of the flow. The deposition probes were placed in the conical section such that the velocity of the approaching particles was either 1 m/s (top probe) or 5 m/s (bottom probe). These particle velocities yield particle kinetic energies consistent with those of particles approaching the tube walls of coal fired boilers. Particles that did not adhere to one of the tubes were subsequently collected on a quartz filter for analysis. The mass collection efficiency, defined as the mass fraction of particles approaching the probe that are collected by the probe, was determined from the mass change of each probe, the mass collected on the after filter, the projected area of the probe, and the cone cross sectional area at the tube location.

Deposition experiments were performed over a range of temperatures and carrier gas compositions as shown in Table II. In the first set of experiments one of the glasses was passed through the EFR under pure nitrogen at gas temperatures between 900 and 1400°C. For the second set of experiments the oxygen concentration was set at 20% and the furnace setpoint again varied from 900 to 1400°C. Similar experiments were also performed under 100% oxygen. The experiments were repeated at selected conditions with the second glass sample. At each condition ash collection efficiency at gas velocities of 1 and 5 m/s was determined using the deposition apparatus described above. In addition, particle samples were collected at the exit of the EFR using a nitrogen quenched probe for three different residence times (for use in the modeling effort). The residence time was controlled by either changing the gas flow rate or by injecting particles into the lower section of the EFR. Selected samples were then analyzed by FeO titration to determine the conversion of Fe(II) to Fe(III).

Table II. Experimental Conditions

	Data Set		
	1	2	3
Furnace Setpoint Temperatures (°C)	900 1000 1150 1300 1400	1000 1150 1300 1400	1000 1150 1300 1400
O$_2$ Concentration (Mole %)	0	20	100
Glass Used	Glass 1* Glass 2	Glass 2	Glass 1* Glass 2

*Measurements only made at 1400°C for this glass

The results of the deposition measurements are shown in Fig. 4 and Fig. 5. Both the temperature and the iron oxidation state (as affected by the gas composition) play a major role in determining the collection efficiency of the ash. In almost all cases the collection efficiency increases with increasing temperature. This increase however, is much more pronounced for the pristine glasses (all iron present as Fe(II)). When the iron in the glass is oxidized to Fe(III), as occurs at the higher oxygen concentrations, the collection efficiency decreases by a factor of 2 to 4. This decrease in collection efficiency is likely due to the increase in particle viscosity associated with the conversion of Fe(II) to Fe(III). As shown in Fig. 6, measurement of the post-experiment iron oxidation state in selected samples suggests that very little conversion of Fe(II) to Fe (III) is required to cause the observed decrease in ash stickiness. These measured values are based upon the bulk concentration of Fe(III); concentrations at the particle surface are likely higher.

The conversion of Fe(II) to Fe(III) for the two glasses is shown in Fig. 7 as a function of the percentage of oxygen in the bulk gas. The data for glass 2, measured at 1400°C and oxygen concentrations of 0, 20, and 100%, suggest that the conversion is independent of oxygen concentration.

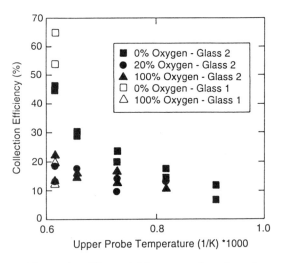

Figure 4. Glass stickiness as a function of particle temperature (1 m/s).

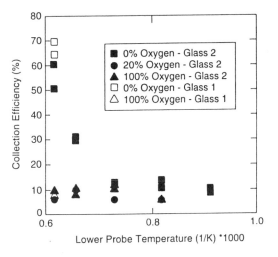

Figure 5. Glass stickiness as a function of particle temperature (5 m/s).

For glass 1 measurements were only made at oxygen concentrations of 0 and 100%. These data are included in Fig. 7. Although both glasses were of the same compostition, the Fe(II) conversion was slightly higher for glass 1. This difference may be attributed to the fact that glass 1 was slightly smaller than glass 2 - increasing the reaction rate as discussed later. At approximately 2.9 s, 100% oxygen, and a furnace setpoint of 1400°C the conversion ranged between 9 and 28%. Additional experiments (data not shown) indicated that the conversion was 0% at lower gas temperatures (1150°C) under 100% oxygen. The independence of conversion from oxygen concentration suggests that the oxidation of the iron in glass is melt diffusion limited.

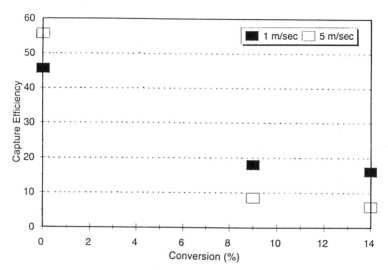

Figure 6. Effect of Fe(II) oxidation on particle stickiness.

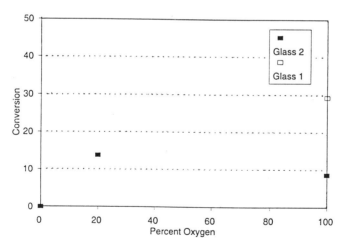

Figure 7. Oxidation of iron containing glassy particles.

MODELING

To test the hypothesis that melt diffusion controls the oxidation of iron in glassy ash particles, a simple model was developed. It was assumed that the oxygen concentration in the melt (C_0) at the gas-liquid boundary was constant, and equaled some fraction of the oxygen solubility in the melt. The model further assumed that the glass droplets were static as they oxidized. Oxygen transport was therefore assumed to occur by diffusion alone - no convective transport by liquid recirculation takes place inside the particle. By making this assumption it was possible to use a shrinking core model to describe the oxidation of iron in the glass particles. This mechanism is described schematically in Fig. 8. Oxygen diffuses through the molten glass containing Fe(III) (defined as Fe_2O_3) to the reaction interface where it reacts as follows:

$$2FeO + 1/2O_2 \rightarrow Fe_2O_3 \tag{1}$$

$$-\frac{dV}{dt} = \frac{b \; 4 \; \pi \; D_e \; C_0}{\left(\dfrac{1}{r_c} - \dfrac{1}{R}\right) \rho_b} \tag{2}$$

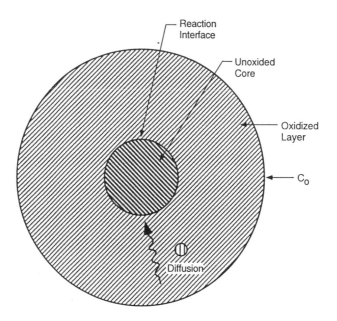

Figure 8. Schematic diagram of glass oxidation model.

The reaction rate is then defined by Eq. (2), where V is the volume of unreacted glass, and r_c and R are the radii of the unreacted core and the particle, respectively. D_e is the effective diffusivity (the diffusivity of oxygen through the melt) and is a function of temperature as described later. C_0 is the concentration of oxygen in the melt at the surface of the particle, b is the stoichiometric conversion, and ρ_b is the concentration of Fe(II) (FeO) in the unreacted glass. When we substitute the definition of the conversion (X_b), Eq. (3), we obtain an equation that describes the conversion of Fe(II) in glass as a function of time (Eq. (4)).

$$X_b = 1 - \frac{V}{V_0} = 1 - \left(\frac{r_c}{R}\right)^3 \tag{3}$$

$$\frac{dX_b}{dt} = \frac{3\, b\, D_e\, C_0}{\rho_b\, R^2}\left(\frac{(1-X_b)^{1/3}}{1-(1-X_b)^{1/3}}\right) \tag{4}$$

The necessary parameters were either calculated or obtained from the literature. For example, the FeO concentration (approximately 0.00738 mol/cc) was determined from the Fe(II) mass fraction in the glass. The oxygen concentration (C_o) was defined as a fraction of the oxygen solubility in molten glass, 7.2×10^{-4} g/cc, or 2.2×10^{-5} mol/cc, as presented by Cable [1961] for the 1400°C particle temperature considered here. The oxygen solubility was assumed to be independent of temperature. The oxygen diffusivity (D_e) was obtained from the data of several investigators compiled by Turkdogan [1983] for oxygen diffusion in a CaO-Al_2O_3-SiO_2 melt. The temperature dependence of the oxygen diffusivity can be modeled with an Arrhenius type equation [Turkdogan, 1983]:

$$D_e = D_{e_0} \exp\left(-\frac{A}{T}\right) \tag{5}$$

The constants are approximately:

$$
\begin{aligned}
D_e &= 1.3283 \times 10^3 \quad [\text{cm}^2/\text{sec}] \\
A &= 3.50 \times 10^4 \quad [\text{K}]
\end{aligned}
$$

Because there is very little heat generated by the iron oxidation, the temperature of the glassy particles can be assumed to equal the gas temperature. For the entrained flow reactor described above the gas temperature profile was fitted in three segments as a function of residence time in the furnace. The three segments consisted of a linear heating zone (from room temperature to peak gas temperature), a flat constant temperature zone (maintained at the peak gas temperature), and a linear cooling zone (from peak gas temperature to exit temperature).

The model was integrated numerically to determine the Fe(II) conversion as a function of the particle residence time for two model extremes; 0.1 and 0.9 times the oxygen saturation. The results from these simulations are shown in Fig. 9. The data included on this plot were collected at various residence times under conditions of a furnace setpoint equal to 1400°C and a gas composition of 100% O_2. Qualitative agreement between experimental data and the model is indicated, suggesting that melt diffusion is the limiting mechanism. The trend shown by the experimental data suggests that the oxygen solubility may increase with temperature. At the shortest residence time the measured conversion was very similar to that predicted for 10% of the maximum oxygen saturation at the surface of the particle. At the longest residence time the measured value was similar to that predicted for 90% of the maximum saturation. In our experiments the gas temperature, and therefore the particle temperature, is still fairly low (less than the peak value) at the residence time of the first measurement. Therefore, the oxygen solubility may be lower at these lower temperatures. At longer residence times the particle temperatures are higher - therefore the mean oxygen solubility may be higher. However, more experimental data and a better value for the temperature dependence of the oxygen solubility in aluminosilicate melts are required to test this hypothesis and to assess the quantitative agreement between the model and the data.

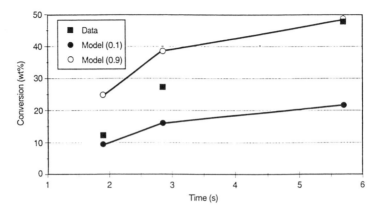

Figure 9. Results of iron in glass oxidation model.

SUMMARY AND CONCLUSIONS

An experimental investigation of the adhesion of iron-containing glass particles inertially impacting on tube surfaces demonstrated that particle deposition under post-combustion conditions is highly dependent upon the oxidation state of the iron in the glass. At temperatures of 1050°C and higher, the conversion of Fe(II) to Fe(III) in glass particles decreased the adhesion efficiency of the particles by 2 to 10 times. This result is attributed to the increase in glass viscosity that occurs upon oxidation of Fe(II) to Fe(III). Only a small level of Fe(II) conversion to Fe(III) - from 9 to 28% - was required to cause this decrease in ash stickiness.

290

Examination of experimental data collected as a function of temperature, oxygen partial pressure, and residence time revealed that the oxidation rate of Fe(II) in glass is slow. The data further suggest that the oxidation rate is relatively independent of the gas oxygen concentration. This finding suggests that the oxidation of Fe(II) is limited by the diffusion of oxygen through the glassy molten particles. A simple melt-diffusion limited model of Fe(II) oxidation yielded Fe(II) conversions as a function of time that were in qualitative agreement with experimental data.

These results suggest that under reducing conditions, iron present in glassy ash particles as Fe(II) may persist for long periods. Given the relatively slow rate of oxidation of Fe(II), systems operating under low-NO_x or staged combustion conditions may find increased levels of iron-glass deposition (relative to standard pc combustion systems). The finding that only small amounts of oxidation are required to significantly reduce particle stickiness suggests that deposition of iron-rich particles in the upper furnace pendants in conventional pc boilers is less likely than in the lower furnace, where iron may be present as Fe(II).

ACKNOWLEDGMENTS

This work was generously supported by the U.S. Department of Energy, Pittsburgh Energy Technology Center, under contract DE-AC22-93PC92190 (P. Goldberg, Contract Manager).

REFERENCES

Austin et al. (1980); "Study of the Mineral Matter Distribution in Pulverized Fuel Coals with Respect to Slag Deposit Formation on Boiler Tubes", DOE Report Number DOE/ET/10560-T1

Bool, L.E., Peterson, T.W., Wendt, J.O.L (1995), "The Partitioning of Iron During the Combustion of Pulverized Coal.", *Comb. and Flame*, 100, 262-270

Cable, M., (1961), "Study of Refining: II", *Glass Technol.*, 2, 60-70

Cukierman, M., Uhlmann, D.R., (1974), "Effects of Iron Oxidation on Viscosity, Lunar Composition 1555," *J. Geo. Research*, 79, 1594-1598

Kalamanovitch, D.P., Sanyal, A., Williamson, J. (1986), "Slagging in Boiler Furnaces-a Prediction Technique Based on High Temperature Phase Equilibria"; *J. Inst. Energy*, 20, 20-23

Raask, E. (1985), *Mineral Impurities in Coal Combustion*; Washington: Hemisphere Publishing Corporation.

Senior, C.L., and Srinivasachar, S. (1995), "Viscosity of Ash Particles in Combustion Systems for Prediction of Particle Sticking", *Energy and Fuels* (in press).

Srinivasachar, S., Helble, J.J., Boni, A.A., (1990a), "An Experimental Study of the Inertial Deposition of Ash Under Coal Combusion Conditions", *Proc. of Twenty Third Symposium on Combustion*, The Combustion Institute, Pittsburgh, 1305-1312

Srinivasachar, S., Helble, J.J., Katz, C.B., and Boni, A.A. (1990b), "Transformations and Stickiness of Minerals During Pulverized Coal Combustion," in *Mineral Matter and Ash Deposition from Coal*, R.W. Bryers and K.S. Vorres, ed., United Engineering Trustees, New York, 201-213.

Srinivasachar, S., Senior, C.L., Helble, J.J., and Moore, J.W. (1992), "A Fundamental Approach to the Prediction of Coal Ash Deposit Formation in Combustion Systems," *Proc. of Twenty Fourth Symposium on Combustion*, The Combustion Institute, Pittsburgh, 1179-1187.

ten Brink, H.M., Eenkhoorn, S., and Hamburg, G. (1992), "Mineral Matter Behaviour in Low-NO$_x$ Combustion," Proc. EPRI Conference on Coal Quality, the Electric Power Research Institute, Pal Alto CA.

Turkdogan, E.T. (1983), *Physicochemical Properties of Molten Slags and Glasses*; London: The Metals Society

Wigley, F., and Williamson, J. (1994), "The Characterisation of Fly Ash Samples and Their Relationship to the Coals and Deposits from UK Boiler Trials," in *The Impact of Ash Deposition in Coal Fired Plants*, J. Williamson and F. Wigley, eds., Taylor and Francis Publishers, London, 385-398.

INVESTIGATION OF MECHANISMS FOR THE FORMATION OF ASH DEPOSITS FOR TWO POWDER RIVER BASIN COALS

Galen H. Richards* and John N. Harb[†]
Department of Chemical Engineering and
Advanced Combustion Engineering Research Center
Brigham Young University
Provo, UT 84602

Larry L. Baxter
Combustion Research Facility
Sandia National Laboratories
Livermore, CA 94550

ABSTRACT

Two subbituminous coals from the Powder River Basin were fired in a pilot-scale combustor to study ash deposit formation under low-temperature fouling conditions. The two coals were chosen as they are similar in composition, but behaved differently when fired in a utility boiler. This study focuses on the mechanisms governing the formation of the initial deposit layer and is limited to deposition times of approximately 3 hours.

Both the elemental composition and phase composition of the fly ash and deposits were examined to gain insight into the deposition process. The dominant phase found in the fly ash from both coals was a calcium-rich phase formed from organically bound Ca, Fe, Al, and Mg in the coals. SEM analyses of both the surface and cross sections of fly ash samples showed little evidence for sulfation during the approximately 1.5s residence time in the furnace. However, ASTM analyses of the deposits showed significant sulfur enrichment. X-ray maps of the deposit cross sections revealed that the Ca-rich fly ash particles contained rings of sulfur, evidence of in situ sulfation. None of the silicate or aluminosilicate ash particles incorporated in the deposit showed evidence of sulfation, suggesting that the dominant mechanism for sulfation was the reaction of gas-phase sulfur species with the Ca-rich fly ash particles, and not condensation of sulfate species. However, a thin layer of sodium sulfate was present on the surface of the probe, probably due to condensation of sodium hydroxide followed by reaction with gas phase sulfur containing species to form the sulfate.

The particle capture efficiencies were similar for both coals, but varied significantly with temperature. However, the deposits formed from Coal A contained a larger fraction of the Ca-rich ash particles and had a higher concentration of sulfur than the Coal B deposits. The increased propensity of the Coal A deposits to undergo sulfation is consistent with the observation that Coal A caused more severe deposition problems when fired in a utility boiler.

---

* Currently at ABB Power Plant Laboratories, Windsor, CT 06095
[†] Author to whom correspondence should be addressed

Applications of Advanced Technology to Ash-Related Problems in Boilers
Edited by L. Baxter and R. DeSollar, Plenum Press, New York, 1996

INTRODUCTION

The use of subbituminous coals from the Powder River Basin (PRB) to generate electricity has increased due to their low sulfur content and relatively low cost. Ash deposition problems have sometimes been associated with the use of these low-rank coals, especially in the horizontal tube banks at the back end of the convective pass in utility boilers. Deposition in this region of the boiler has been associated with PRB coals which contain high levels of organically associated calcium and low levels of sodium, and is commonly referred to as low-temperature fouling. Most utility boilers burning bituminous coals do not experience severe deposition problems in this region of the boiler and are not equipped with soot blowers to manage the problem. Hence, utilities that switch to or blend with low-rank western coals to reduce sulfur emissions may encounter deposition problems that they are not prepared to handle.

The purpose of this paper is to report the findings of a recent study on the formation of fly ash and low-temperature fouling deposits from two Powder River Basin coals. The inorganic constituents in the coals were characterized using standard ASTM tests and computer-controlled scanning electron microscopy. A series of combustion tests, approximately 3 hours in length, was performed on the two coals in a pilot-scale combustor. Both fly ash and deposit samples were collected and analyzed. A comparison of the elemental and species composition of the coal inorganics, the fly ash samples and the deposits was used to increase our understanding of the mechanisms which control ash formation and deposition.

COAL ANALYSES

Two US subbituminous coals from the Wyodak-Anderson seam of the Powder River Basin were used in this investigation. The coals will be referred to as Coal A and Coal B. These two coals were chosen because they are relatively similar, but showed different fouling tendencies when fired in a full-scale utility boiler. The proximate and ultimate analyses of the coals are listed in Table 1.

TABLE 1

Proximate and ultimate analyses of the coals used in this investigation.

	Coal A	Coal B
Moisture (%)	22.9	21.3
Ash (% dry)	5.5	6.5
Proximate (% daf)		
Fixed Carbon	54.2	42.0
Volatile Matter	45.8	58.0
Ultimate (% daf)		
C	72.4	74.8
H	5.2	5.4
O	20.9	18.3
N	1.1	1.0
S	0.4	0.5
Heating Value (daf) (Btu/lb)	13562	13119

The bulk ash chemistry of the two coals is shown in Table 2. In both cases, the ash is comprised mainly of the oxides of Si, Ca, Al, Fe, and Mg. Coal A contains a higher percentage of calcium, magnesium, and iron than Coal B. Calcium, magnesium, and iron are all elements that can be associated with the organic matrix [Benson et al., 1993a]. Coal B contains more silicon and aluminum than Coal A, indicative of more aluminosilicate minerals.

TABLE 2

Ash chemistry of the coals used in this study on a SO_3-free basis.

% of Ash	Coal A	Coal B
SiO_2	30.4	36.7
Al_2O_3	13.8	19.7
Fe_2O_3	9.9	6.1
TiO_2	0.8	1.5
CaO	35.1	25.5
MgO	6.8	5.7
Na_2O	1.9	1.7
K_2O	0.3	0.4
P_2O_5	1.0	1.1
SO_3	17.7	17.5
Undetermined	0.0	1.4

Table 3 presents the distribution of discrete mineral phases found in both coals on a weight percent basis as determined by computer-controlled scanning electron microscopy. These phases are actually composition groups which are based solely on X-ray analysis of the elemental composition of each particle. As shown in the table, Coal A contains primarily quartz (43.1%), aluminosilicate (16.2%), calcium aluminosilicate (5.0%), a calcium-aluminum-phosphorous compound (8.1%), and iron oxide/iron carbonate (7.3%). A significant fraction of the mineral content (11.8%) was not classified as a particular mineral species and is referred to as unknown. The unknown category includes minerals whose composition was not included in the classification scheme. Also, since the excitation volume of the electron beam is larger than the small (~1μm) mineral grains, the unknown minerals may reflect the presence of more than one mineral grain in the diagnostic volume. Additionally, organically associated calcium, magnesium, aluminum, and iron in the coal may also be detected in the CCSEM analysis of the small particles and incorrectly interpreted as part of the discrete mineral composition. This incorrect composition may lead to classification of the particle as "unknown" or may cause the particle to be incorrectly placed into another category. For example, some of the mineral grains identified as calcium aluminosilicate may be the result of a combination of small aluminosilicate grains and organically bound calcium from the surrounding coal matrix. The above problems with small mineral grains are common to this type of analysis and not expected to have a significant impact on the results presented in this paper.

A similar analysis indicated that Coal B contained more aluminosilicate minerals (31.3%) than quartz (26.9%). This is consistent with the bulk ash analysis presented in Table 2 which showed more aluminum in Coal B, as well as a higher Al/Si ratio. Coal B also contains more pyritic iron with 7.9% pyrite and 5.3% FeS/FeSO4. Most of the minerals in the FeS/FeSO4 category are probably pyrite grains that have been partially oxidized. Coal B also contains a significant fraction of minerals that were not classified (11.7%). Note that neither coal contained a significant amount of calcite.

TABLE 3

Major mineral phases (wt%) as determined by CCSEM

Major Mineral Phases	Coal A	Coal B
Quartz	43.1	26.9
Aluminosilicate	16.2	31.3
Ca-Al-Silicate	5.0	7.8
Fe-Al-Silicate	1.5	0.4
$Fe_2O_3/FeCO_3$	7.3	0.4
Pyrite	0.0	7.9
$FeS/FeSO_4$	2.3	5.3
Calcite/CaO	0.5	0.3
Ca-Al-P	8.1	3.0
Unknown	11.8	11.7

EXPERIMENTAL APPARATUS AND PROCEDURE

Multifuel Combustor

A series of combustion tests were performed on the two coals in Sandia's Multifuel Combustor (MFC), a pilot-scale reactor. Both fly ash and deposit samples were collected during each test for later analysis using ASTM tests and scanning electron microscopy.

The MFC is a 0.1 MBtu/hr, down-fired combustor consisting of 7 individual sections, each 2 feet high with a ceramic liner approximately 15cm in diameter. The top 6 sections of the combustor are independently heated by resistance heaters (SiC) to provide the desired temperature profile in the combustor. A natural gas-air flame may be used to generate a vitiated gas stream for the combustor. All sections of the combustor contain multiple access ports for both fuel lances and thermocouples. Also, the MFC has an open section at the exit of the furnace to provide access for the fly ash collection and deposition probes. Additional details on the MFC can be found in [Baxter et al., 1990].

By injecting coal near the top of the MFC, approximately 14 ft from the furnace outlet, the particle and gas temperature histories, and coal particle burnout approximate the conditions found near the backpass of a utility boiler where low-temperature fouling may occur. The MFC was operated with the furnace wall temperature set at either 900°C or 1300°C which resulted in gas temperatures of approximately 750°C and 900°C and residence times of about 1.5 seconds at the deposition probe. Coal was fed at approximately 1.8 kg/hr with 3% excess oxygen.

Sample Collection

A sampling system was used to collect fly ash samples from the open test section of the MFC. The water-cooled collection probe used helium gas to quench the fly ash before it was collected on a 1 μm Nucleopore filter (6" diameter). The ash sampling probe was typically operated 30 minutes during each combustion test to collect sufficient amount of sample for analysis (approximately 0.5 g). Two deposition probes were also inserted into the open test section at the outlet of the MFC. The probes were constructed of stainless steel pipe (5/8" OD and 1/2" ID) and were approximately 3 feet long. The air-cooled deposition probes were instrumented with thermocouples to monitor the surface temperature which was controlled by adjusting the mass flow rate of cooling air. The probes were polished with an emery cloth, washed with water, and dried before each combustion test.

After approximately 3 hours of deposition, the probes were removed from the furnace and allowed to cool. The deposit was removed from one of the probes using a stainless steel brush. The ash that deposited on the leading edge (top) of the probe due to inertial impaction was removed separately from the ash that deposited on the sides of the probe. These deposit samples were weighed and used for elemental analysis. The second probe, with its deposit, was cast in epoxy, cross-sectioned and polished for SEM analysis.

Sample Analysis

Elemental analysis of the coal inorganics, the fly ash and the deposits was performed at CONSOL, Inc. with use of standard ASTM techniques. Scanning electron microscopy analysis of the ash and deposit samples was performed at Brigham Young University using a JSM-840A scanning electron microscope (JEOL, USA, Inc.) equipped with an eXL-FQAI microanalysis system, a *Pentafet* LZ5 light element detector, and a LEMAS stage automation system (Link Analytical Ldt). The SEM was also equipped with an ultrathin window (MOXTEX). Detailed descriptions of the SEM techniques used in this study are provided in [Richards, 1994].

FLY ASH RESULTS

Tables 4 and 5 contain the bulk elemental oxide compositions of fly ash generated from the two coals under both sets of combustion conditions. Each entry in the tables represents the average composition of from 3 to 6 different fly ash samples. The standard deviation of the measurements was typically less than 10% of the average value, indicating the reproducibility of the results. Note that the fly ash compositions for both coals are similar to the bulk ash composition of the coals as noted in the tables. Also listed in the tables is the percent of carbon in the fly ash. Both coals experienced approximately 99.5% burnout.

TABLE 4

Average elemental oxide compositions (SO_3-free) of fly ash generated from
Coal A in the MFC at furnace wall temperatures of 900 and 1300 °C.

wt% of Ash	T = 900°C	st dev	T = 1300°C	st dev	Coal
SiO_2	29.6	3.8	25.1	1.0	30.4
Al_2O_3	17.1	2.1	15.4	0.4	13.8
Fe_2O_3	10.2	0.6	10.7	0.3	9.9
TiO_2	0.9	0.1	0.9	0.1	0.8
CaO	31.3	4.5	36.5	1.3	35.1
MgO	6.1	0.7	6.8	0.3	6.8
Na_2O	2.5	0.3	2.3	0.1	1.9
K_2O	0.9	0.4	0.6	0.1	0.3
P_2O_5	1.4	0.2	1.8	0.2	1.0
SO_3	4.6	0.7	3.5	0.6	17.7
% Carbon in Ash	2.3	0.3	1.8	1.4	

TABLE 5

Average elemental oxide compositions (SO_3-free) of fly ash generated
from Coal B in the MFC at furnace wall temperatures of 900 and 1300 °C

wt% of Ash	T = 900°C	st dev	T = 1300°C	st dev	Coal
SiO_2	34.8	0.5	34.0	0.6	36.7
Al_2O_3	20.7	0.3	20.4	0.2	19.7
Fe_2O_3	5.8	0.2	5.8	0.5	6.1
TiO_2	1.4	0.0	1.4	0.1	1.5
CaO	27.6	0.7	28.6	0.5	25.5
MgO	5.9	0.1	5.8	0.1	5.7
Na_2O	1.9	0.2	1.9	0.0	1.7
K_2O	0.6	0.2	0.6	0.1	0.4
P_2O_5	1.3	0.1	1.4	0.1	1.1
SO_3	2.8	0.6	1.9	0.2	17.5
% Carbon in Ash	1.0	0.3	0.4	0.1	

Fly ash samples were mounted in epoxy, cross-sectioned, and polished for analysis with the SEM. The major phases present in the fly ash were determined using the CCSEM technique for both coals at the two furnace wall temperatures (Table 6). As shown in the table, the fly ash from both coals contained approximately 15% quartz. The ashes from Coal B contained significantly more Al-silicate particles (11.5-19.5%) than the ashes from Coal A (3.5-4.6%). The Coal B ash also contained approximately 3 times more Ca-Al-silicate particles than the Coal A ash, while the Coal A ash contained more Ca-Silicate particles. The higher fraction of Ca-Al-Silicate ash particles is consistent with the increased quantities of Al-Silicate mineral grains found in Coal B. Coal A contained more quartz particles than Al-Silicate minerals and tended to form more Ca-Silicate ash particles. The included quartz and Al-Silicate minerals reacted with the organically associated calcium to form the Ca-Silicate and Ca-Al-Silicate phases. It should be noted, however, that the calcium silicate and calcium aluminosilicate ash particles are a relatively small fraction of the total ash, approximately 10% by weight. This result indicates that relatively little mixing occurred between the discrete quartz and aluminosilicate particles and organically associated calcium. A significant fraction of the quartz ($\approx$ 50%) and aluminosilicate ($\approx$ 30%) particles are excluded from the coal matrix and may pass through the furnace relatively unchanged [Richards, 1994]. Note that the "quartz" category referred to above includes amorphous silica since classification was made based on composition above.

The dominant phases in ashes from both coals, however, were particles that were rich in calcium, magnesium, aluminum, and iron. These Ca-rich particles were formed from the coalescence of the organically associated elements in the coal [Richards, 1994]. The fraction of these Ca-rich particles was significantly higher in Coal A for fly ash collected at both furnace wall temperatures.

The Ca-rich classification was subdivided into three separate groups as shown in Table 6. Points classified as Ca-Mg-Al-Fe contain at least 40 mole % Ca (on an oxygen free basis) and 5% each of Mg, Al, and Fe. The sum of the 4 elements must be larger than 90% and have less than 10% Si. The Ca-Mg-Al-Fe-P classification is distinguished by points which also contain at least 5% P. The Ca-Mg-Al-Fe-P-Si points must contain at least 70% Ca-Mg-Al-Fe and more than 10% Si.

TABLE 6

Fly ash phase distributions from Coal A and Coal B as determined by CCSEM

Major Phases	Coal A Fly Ash		Coal B Fly Ash	
	T = 900°C	T = 1300°C	T = 900°C	T = 1300°C
Quartz	10.9	19.2	13.3	15.4
Al-Silicate	4.6	3.5	11.5	19.5
Ca-Silicate	3.8	6.3	3.4	2.0
Ca-Al-Silicate	1.4	2.6	6.6	7.4
Mixed Silicate	0.7	0.8	1.5	3.6
Fe_2O_3	0.5	0.7	1.7	1.3
Ca-Rich	72.7	63.2	51.2	45.5
Other	5.4	3.7	10.8	5.3
Ca-Rich Points				
Ca-Mg-Al-Fe	52.0	41.5	29.9	27.6
Ca-Mg-Al-Fe-P	2.6	11.0	8.0	7.3
Ca-Mg-Al-Fe-Si	18.1	10.7	13.3	10.6

A comparison of the results presented in Table 6 with the elemental composition of the two ashes from Coal A (Table 4) illustrates a problem associated with the SEM analyses. The two ashes generated from Coal A showed similar elemental compositions, although the ash generated at 900 °C had a slightly higher percentage of SiO_2. The SEM results indicate, however, that the ash generated at 1300 °C contains a higher percentage of quartz and of many of the silicon containing species, including calcium silicate and calcium aluminosilicate. The relatively high quartz concentration in the high temperature ash (19.2%) was due primarily to the presence of some large quartz particles (diameter > 30 μm). In fact, eliminating just the largest quartz particle (diameter ≈ 55 μm) from the analysis decreased the amount of quartz in the sample to 13.9%. Large particles are problematic because of the relatively small numbers of such particles in a typical sample. Problems with poor large particle statistics can be minimized by analyzing a large sample area at low magnification. However, care should always be used when interpreting CCSEM results.

DEPOSIT CHARACTERIZATION

Deposition Rates

Ash deposits were formed on a deposition probe at conditions simulating low temperature fouling in a utility boiler. Deposits were generated at two furnace wall temperatures 900°C and 1300°C, resulting in gas temperatures at the deposition probe of approximately 750°C and 900°C, respectively. The deposits formed with a furnace wall temperature of 1300°C were partially sintered, but would crumble easily when handled. The deposits formed at the lower temperature showed no evidence of sintering. Accumulation was greatest on the top of the probe, with a relatively thin layer of a tan-colored deposit extending around the sides of the probe. Most of the ash particles that deposited on the top of the probe were 10-50 μm in diameter, while the thin deposit on the tube sides contained mostly small ash particles (less than 10 μm in diameter). Because the two regions of the deposit were so different in appearance, they were collected and analyzed separately. Typically, a 250-1000 mg sample was collected from the top deposit, and only 25-150 mg from the sides of the probe. The quantity of ash deposited on the probe was a function of both temperature and deposition time.

The average rate of deposit growth was estimated for each experiment from the mass of deposit collected and the deposition time. Typical ash deposition rates were on the order of 0.002 g/min. The capture efficiency (η) of the probe (expressed as a percentage) is calculated by

$$\eta = \frac{D \cdot A_f}{F \cdot A_p} \cdot 100$$

where D is the measured deposition rate (g/min), F is the rate at which ash was fed into the reactor (g/min), A_f is the cross sectional area of the furnace, and A_p is the projected area of the probe. The ash feed rate is calculated from the measured coal feed rate by assuming the ash fraction in the coal remains constant at the ASTM value.

A comparison of the capture efficiencies for each type of deposit as a function of coal type and furnace wall temperature is shown in Figure 1. The capture efficiency was very similar for the deposits formed from the two coals. It should be noted, however, that the actual deposition rates for the Coal B deposits were greater than the rates for the Coal A deposits by approximately 20% because Coal A contained 5.5% ash while Coal B contained 6.5% ash. As illustrated in the figure, the capture efficiency of the top deposit increased by a factor of approximately 2.5 as the combustor wall temperature increased from 900 °C to 1300 °C. The probe surface temperature also increased as the furnace wall temperature increased, from approximately 475 °C to about 600 °C. At the higher temperature, more of the fly ash particles were deposited due to the increased stickiness of both the particles and the deposit surface. The rate of growth of the deposits that formed on the sides of the probe also increased with temperature as illustrated in Fig. 1. These deposits were comprised mainly of small particles (diameter < 10 μm) that deposited due to thermophoresis and eddy impaction.

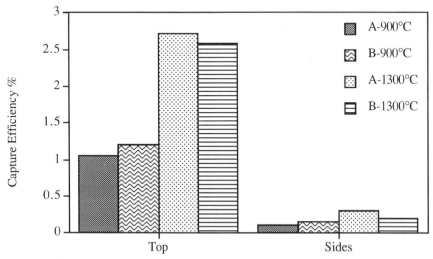

Figure 1   Comparison of the capture efficiency as a function of coal type and furnace wall temperature

Bulk Deposit Compositions

Table 7 lists the average compositions of deposits formed from both coals. As previously noted, the deposits that formed on the top of the probe were analyzed separately from the deposits that formed on the sides of the probe. The number of samples that were analyzed at each condition is also noted in the table. Although the elemental composition of the deposits was different for the two coals, the same trends were observed for both coals. The deposits formed on the top of the probe had a significantly higher SiO_2 content than deposits formed on the sides of the probe. The top deposits were also enriched in calcium and magnesium relative to the side deposits. The CaO and MgO enrichment, however, appears to be the result of higher SO_3 in the side deposits. Note that the Ca/Si ratio was significantly lower in the top deposits relative to the sides, in spite of the apparent calcium enrichment in the top. Sodium and sulfur were greatly enriched in the deposits formed on the sides of the probe. No significant difference in composition due to temperature was observed.

TABLE 7

Summary of average ASTM deposit compositions for Coal A, SO_3 included

	Coal A				Coal B			
	T = 900 °C		T = 1300 °C		T = 900 °C		T = 1300 °C	
Elem. Oxide	Top	Sides	Top	Sides	Top	Sides	Top	Sides
SiO_2	34.3	16.6	33.1	23.7	41.5	27.5	40.9	29.2
Al_2O_3	12.1	11.3	11.5	10.5	16.2	18.3	16.5	17.8
Fe_2O_3	9.5	11.0	9.2	9.3	6.2	7.4	6.0	6.1
TiO_2	0.8	0.7	0.8	0.7	1.5	1.3	1.6	1.2
CaO	32.2	28.6	31.1	28.6	25.8	21.9	25.1	18.6
MgO	6.3	5.6	6.0	5.5	5.5	4.7	5.3	4.0
Na_2O	0.6	5.1	0.9	3.9	0.4	3.2	0.4	3.3
K_2O	0.2	0.6	0.2	0.9	0.2	0.7	0.2	1.3
P_2O_5	1.2	1.7	1.5	2.0	0.9	1.3	1.2	1.9
SO_3	3.1	18.8	5.7	15.2	1.8	13.6	2.9	16.8
# of analyses	5	4	5	3	7	6	2	2

Figure 2 compares the average deposit compositions from the top of the probe to the compositions of the fly ash generated from Coal A. The fly ash compositions were similar to the bulk ash compositions. SiO_2 was enriched in the deposits by approximately 20%, while the Al_2O_3 and Na_2O concentrations in the deposit were less than those of the fly ash. However, the composition of deposits that formed on the sides of the probe was significantly different than the fly ash composition. The biggest difference was the increased amount of sulfur in the side deposits. The ash contained approximately 5% SO_3 while the deposits contained approximately 18%. The side deposits were also enriched in sodium (~4.5% in side deposit vs. 2.4% in the ash). Similar trends were observed for Coal B.

The difference between the behavior of the sodium and calcium is clearly evident in these results. It is therefore expected that different mechanisms dominate the behavior of these two elements. The fact that organically bound sodium is much more likely to vaporize during combustion than calcium is undoubtedly important. In any case, the enrichment of sulfur over that found in the ash indicates that sulfation takes place in situ after deposition has occurred.

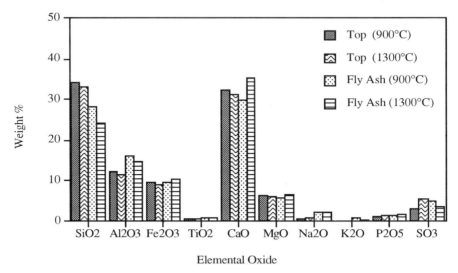

Figure 2   Comparison of the elemental oxide compositions of the top deposits with the fly ash from Coal A

Phase Composition of the Deposit

Ash deposits were also formed on sacrificial probes that were cast in epoxy and cross-sectioned to permit analysis using scanning electron microscopy point count (SEMPC). The SEMPC technique involves taking X-ray spectra at uniformly spaced grid points on a deposit sample and assigning a particular phase to each point based on its elemental composition [Jones et al., 1992]. The frequency of occurrence of each phase is converted to a weight percent with use of the density corresponding to each phase. A typical analysis may contain 500-1000 (or more) points.

The distribution of major phases identified in the deposits from Coal A is shown in Table 8. Note that only the deposits that formed on the top of the probe were analyzed at the low temperature. The dominant phases in each type of deposit are the Ca-rich phases (Ca-Mg-Al-Fe, Ca-Mg-Al-Fe-P, Ca-Mg-Al-Fe-Si) that form from the organically associated elements in the coal. Approximately 60% (by weight) of each deposit is comprised of these Ca-rich particles, comparable to their fraction in the fly ash (60-70%). Although the sides of the deposit (T = 1300 °C) contained only approximately 45% Ca-rich particles, a significant fraction of some of the other phases identified were derived from the Ca-rich particles. For example, the deposits on the sides of the probe contained 6.8% Ca/Mg sulfate. Also, some of the analysis points that were not classified were Ca-rich points which contained significant quantities of sulfur, but not enough to be classified as a sulfate.

The increased quantity of unclassified points in the deposit on the sides of the probe may also have been due to the smaller particle size, which increases the probability of including more than one particle in a single analysis. Note that no points were seen in any of the deposits that contained over 80% Ca (oxygen free), the amount needed to be classified as CaO.

The distribution of the major phases in the Coal B deposits are also listed in Table 8. Similar to the Coal A deposits, the dominant phases are derived from the Ca-rich ash particles. The Ca-rich phases account for approximately 50% of the mass of the Coal B deposits, less than the 60% seen in the Coal A deposits but similar to the Coal B fly ash (45-50%).

TABLE 8

Major phases in deposits from Coal A as determined by SEMPC

	Coal A			Coal B		
	900 °C	1300 °C		900 °C	1300 °C	
Elemental Oxide	Top	Top	Sides	Top	Top	Sides
Quartz	21.4	15.3	14.3	18.5	15.0	4.0
Al-Silicate	3.8	3.9	0.5	9.5	9.5	4.4
Ca-Silicate	4.4	7.1	2.4	3.2	3.5	0.7
Ca-Al-Silicate	1.9	1.7	2.2	5.8	7.1	0.6
Mixed Silicate	5.9	13.8	14.9	9.2	10.4	42.4
Fe_2O_3	0.6	0.1	5.1	1.5	0.9	1.2
Ca-Rich	60.5	57.6	44.1	50.8	42.1	27.5
Ca/Mg Sulfate	0.2	0.7	6.8	0.1	3.5	2.9
Other	0.6	1.1	9.5	1.5	8.0	16.1

Each of the deposits generated from both coals contained significant quantities of quartz, silicates, and aluminosilicates. Approximately 15-20% of each deposit was quartz, while approximately 1-3% of the deposits from Coal A and 4-10% of the deposits from Coal B were Al-Silicate. Calcium silicate was a major component of most of the Coal A deposits, while the Coal B deposits contained a larger fraction of Ca-Al-Silicates ($\approx$ 6%). The increased fraction of Al-Silicate and Ca-Al-Silicate in the Coal B deposits is consistent with the fact that Coal B contained more Al-Silicate minerals than Coal A. A large fraction of the silicate species in the deposits were classified as mixed silicates. Mixed silicates include particles that contain significant quantities of silicon, no sulfur, and were not classified as other types of silicates or aluminosilicates. The increased fraction of mixed silicates in the deposits on the sides of the probe may be due, in part, to the smaller ash particle size, which increases the probability of including more than one particle in a single analysis.

Sulfate Formation in Deposits

Low-temperature fouling deposits are characterized by large quantities of sulfates that can lead to the formation of strong deposits. The sulfation of particles in the deposits was investigated with elemental composition maps on the scanning electron microscope. Figure 3 shows a backscattered electron image of a portion of the top deposit from Coal A (T = 1300°C) with the corresponding X-ray composition maps for calcium, silicon, and sulfur.

Examination of the Ca and S maps reveals sulfur rings (sulfated edges) around the outside of many of the calcium-containing particles. Further comparison with the Si map shows that the calcium containing particles which also contained significant amounts of silicon did not form a sulfur ring. Observations such as these show that only the Ca-rich fly ash particles contain rings of sulfur. Neither the quartz particles nor the calcium silicate particles contained detectable quantities of sulfur. Therefore, the X-ray maps suggest that the gas-phase sulfur will react with calcium that is not part of a silicate matrix.

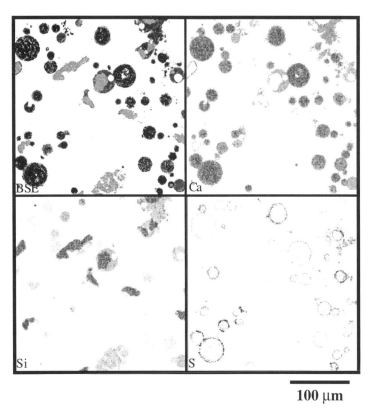

100 μm

Figure 3   A backscattered electron image of part of the deposit on top of the probe from Coal A (T = 1300 °C) and the corresponding Ca, Si, and S X-ray maps

In order to support observations made from the X-ray maps, the SEMPC data for all of the deposits formed were used to determine the average composition of points in the deposits that contained at least 5% sulfur (by weight). An average composition similar to that of the Ca-rich fly ash particles would provide evidence that these particles are more likely to undergo sulfation. Figures 4 and 5 show the average composition of the sulfur-containing points in deposits formed from each of the coals. Ca, Fe, Mg, and Al were the dominant elements, although approximately 5-10% SiO_2 was also present. These average compositions are in good agreement with the composition of the Ca-rich ash particles formed from the organically associated elements in the coals, supporting the suggestion that these particles are preferentially sulfated. Based on these observations, a greater extent of sulfation is expected for deposits which have a larger fraction of ash particles formed from organically bound elements

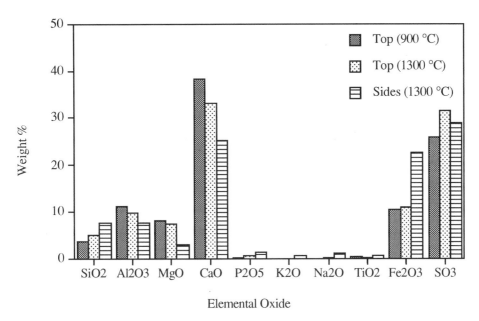

Figure 4   Average composition of the sulfur-containing phases in the Coal A deposits

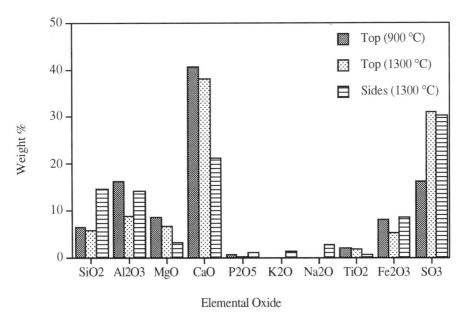

Figure 5   Average composition of the sulfur-containing phases in the Coal B deposits

The fraction of analysis points containing sulfur varied with temperature and coal type as shown in Fig. 6. The extent of sulfation was significantly greater for Coal A in both cases. The increase in the fraction of points containing significant quantities of sulfur with temperature was expected because the rate of the sulfation reaction increases with temperature. According to Benson *et al.* [1993b], the sulfate species become unstable and are not found in deposits where temperatures exceed 1030 °C. The maximum gas temperature in these experiments was approximately 850-900 °C for the furnace wall temperature of 1300 °C. The maximum deposit temperatures were somewhat lower because the probe surface was air-cooled (surface temperature ≈ 600 °C) and the deposits were also cooled by radiative exchange with the laboratory surroundings. However, the experimental conditions (i.e., gas temperature, particle residence time) used in this study approximate the conditions in the back end of a utility boiler where low-temperature fouling may occur.

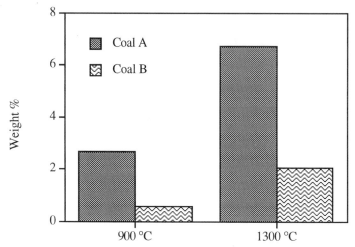

Figure 6   A comparison of the weight percent of sulfur-containing points in the top deposits from both coals

The next logical question is whether the sulfation occurs before or after deposition. X-ray analyses of hundreds of fly ash particles showed little evidence of sulfation. Apparently, the sulfation reaction is sufficiently slow that the fly ash particles are not significantly sulfated before they deposit on the heat transfer surfaces. In addition, most of the particle residence time in the furnace is spent at temperatures above 1030 °C where the sulfate species are not stable. Hence, the sulfation of the Ca-rich fly ash particles occurs after they have been deposited upon the heat transfer surfaces.

As the Ca-rich fly ash particles in the deposits are preferentially sulfated, the extent of deposit sulfation is a function of the fraction of Ca-rich particles in the fly ash. Of the two coals that were fired in this study, Coal A had a larger fraction of Ca-rich fly ash particles. The analyses of the deposits generated from these two coals show increased sulfation of the Coal A deposits. A higher sulfation rate in the Coal A deposits is also consistent with the observation that Coal A caused more severe deposition problems when test fired in a utility boiler. As this study focused on the formation of the initial deposit layer, additional experiments should be performed to corroborate the utility observations.

X-ray composition maps of the deposits also provided insight into the different behavior of sodium and calcium. Na and S maps showed increased quantities of both of these elements on the tube surface, indicating the presence of a condensed sodium sulfate layer on the tube surface. This observation is consistent with in situ, real-time emission spectroscopy analyses of the deposits which detected the presence of sodium sulfate on the probe surface just minutes into the deposition process [Richards et al., 1994]. In contrast, calcium was deposited as discrete particles.

CONCLUSIONS

Two-subbituminous coals from the Powder River Basin were fired in a pilot-scale combustor to study ash deposition under simulated low-temperature fouling conditions. Both the elemental composition and phase composition of the fly ash and deposits were examined in order to gain additional insight into the deposition process. The dominant phase found in the fly ash from both coals was a calcium-rich phase formed from organically bound Ca, Fe, Al, and Mg in the coals. SEM analyses of both the surface and cross sections of fly ash samples showed little evidence for sulfation during the approximately 1.5s residence time in the furnace. However, elemental analyses of the deposits showed significant sulfur enrichment. X-ray maps of the deposit cross sections revealed that the Ca-rich fly ash particles contained rings of sulfur, evidence of in situ sulfation. None of the silicate or aluminosilicate ash particles incorporated in the deposit showed evidence of sulfation, suggesting that the dominant mechanisms for sulfation was the reaction of gas-phase sulfur species with the Ca-rich fly ash particles and not condensation of sulfate species. However, a thin layer of sodium sulfate was present on the surface of the probe, probably due to condensation of sodium hydroxide followed by reaction with gas phase sulfur containing species to form the sulfate.

The capture efficiencies were similar for both coals, but varied significantly with temperature. However, the deposits formed from Coal A contained a larger fraction of the Ca-rich ash particles and had a higher concentration of sulfur than the Coal B deposits. The increased propensity of the Coal A deposits to undergo sulfation is consistent with the observation that Coal A caused more severe deposition problems when fired in a utility boiler.

ACKNOWLEDGMENTS

The combustion tests in this study were performed in the Multifuel Combustor Laboratory, part of the Combustion Research Facility at Sandia National Laboratories in Livermore, CA. This work performed at Sandia was sponsored by the US Department of Energy through the Pittsburgh Energy Technology Center's (PETC) Direct Utilization Advanced Research and Technology Development Program. Philip Goldberg and James Hickerson are the PETC project and program managers, respectively. The sample analysis performed on the scanning electron microscope was performed at Brigham Young University with support from the Advanced Combustion Engineering Research Center. Funds for this Center are received from the National Science Foundation's Division of Cross Disciplinary Research, the State of Utah, 28 industrial participants, and the US Department of Energy. The contributions of Peter Slater to the development of the analysis routines at BYU is gratefully acknowledged. The ASTM analyses of the coal, ash, and deposit samples were performed by CONSOL Inc.

REFERENCES

Baxter, L.L., Mitchell, R.E., and Fletcher, T.H. (1990). "Experimental Determination of Mineral Matter Release During Coal Devolatilization." 7th Annual International Pittsburgh Coal Conference, Pittsburgh, PA, Sept. 10-14.

Benson, S.A., Jones, M.L., and Harb, J.N. (1993a), "Chapter 4: Ash Formation and Deposition," in Fundamentals of Coal Combustion for Clean and Efficient Use, L.D. Smoot, ed., Elsevier, Amsterdam, 1993.

Benson, S.A., Hurley, J.P., Zygarlicke, C.J., Steadman, E.N., Erickson, T.A. (1993b) "Predicting Ash Behavior in Utility Boilers." *Energy & Fuels*, **7**; 746-754.

Jones, M.L., Kalmanovitch, D.P., Steadman, E.N., and Benson, S.A. (1992), "Application of SEM Techniques to the Characterization of Coal and Coal Ash Products," Advances in Coal Spectroscopy , H.L.C. Meuzelaar, ed., Plenum Press, NY, 1992.

Richards, G.H. (1994). *Investigation of Mechanisms for the Formation of Fly Ash and Ash Deposits for Two Powder River Basin Coals.* Ph.D. dissertation, Brigham Young University, Provo UT.

Richards, G.H., Harb, J.N., Baxter, L.L., Bhattacharya, S., Gupta, R.P., and Wall, T.F. (1994). "Radiative Heat Transfer in Pulverized-Coal-Fired Boiler — Development of the Absorptive/Reflective Character of Initial Ash Deposits." *25th Symposium (International) on Combustion/The Combustion Institute*, 511-518.

CHARACTERIZATION OF ASH DEPOSIT STRUCTURE FROM PLANAR SECTIONS: RESULTS AND CHALLENGES

Everett R. Ramer and Donald V. Martello

Pittsburgh Energy Technology Center
U. S. Department of Energy
P. O. Box 10940
Pittsburgh, PA 15236

ABSTRACT

Three-dimensional structural parameters were measured for fouling ash deposits from two Powder River Basin subbituminous coals using direct microscopical methods on planar sections through the deposits. These parameters included solid and pore volume fractions; specific surface area; particle contiguity; and mean solid, particle, and pore chord lengths. Spatial trends in the results for the two deposits indicated that the solid volume fraction remained relatively constant from the tube side to the flame side, but the solid phase coarsened in this direction. An increase in the contiguity between ash particles indicated that the coarsening mechanism was sintering, both via increased particle agglomeration and encapsulation of particles by a glassy phase. The bulk averages of the results were identical for both deposits, with a solid volume fraction of 0.23, a specific surface area of 0.6×10^6 m^{-1}, a contiguity of 0.23, a mean solid chord length of 16 µm, a mean particle chord length of 12 µm, and a mean pore chord length of 53 µm. In addition, a two-dimensional structural parameter, the density-density correlation function, was measured for one of the deposits. This result indicated that the cross-sectional profiles of the solid regions with diameters less than 20 µm were isotropically oriented, and that the larger solid region profiles were preferentially oriented in the direction of the incoming particle trajectories.

INTRODUCTION

The physical properties of the ash deposits that form in pulverized-coal-fired boilers are critically important to the design and operation of these boilers. Since deposit structure has an enormous effect on properties such as strength and heat transfer, its quantitative characterization is the key to understanding these properties. Quantitative structural measurements also yield information about the mechanisms of deposit formation and growth, and provide the geometric description necessary for the computational modeling of deposit properties.

Applications of Advanced Technology to Ash-Related Problems in Boilers
Edited by L. Baxter and R. DeSollar, Plenum Press, New York, 1996

This paper describes a study of the physical structures of ash deposits produced by two subbituminus coals from the Powder River Basin, located in the western United States. The structural parameters measured include volume fraction, specific surface area, particle contiguity, mean chord lengths, and the density-density correlation function.

The methods used in this work were direct. They are based on the analysis of photomicrographs of planar sections through the deposit specimens. Direct methods have several important advantages over physical methods, such as porosimetry, gas adsorption, and compressive strength. These include elimination of the need for an idealized model of the material and the capability of obtaining localized results, allowing the study of physical structure gradients within the deposits.

EXPERIMENTAL PROCEDURE

Specimen Preparation

Two ash deposit specimens were studied in this work, one from coal A and the other from coal B. Both specimens were generated under identical conditions in the Multi-Fuel Combustor in the Combustion Research Facility at Sandia National Laboratories. In particular, the combustor wall temperature was maintained at 1300°C, and the specimens were collected on a 17-mm diameter, stainless steel collection probe, maintained at a constant surface temperature of 500°C. During the three-hour collection time, the thicknesses of the deposits reached 5.0 mm for coal A and 4.2 mm for coal B.

At the conclusion of the collection time, the probe was withdrawn from the combustor, its end was placed in a mold. The probe and deposit were then impregnated with epoxy under ambient pressure. After the epoxy cured, the end of the probe was cut off, yielding an undisturbed deposit specimen still in contact with the probe surface. The probe and specimen were cross sectioned and mounted in cylindrical molds for grinding and polishing. A series of silicon carbide papers (to a finest grit size of 600) was used for grinding, with water as the lubricant. Further grinding was done using a series of water-based diamond suspensions (9 μm, 3 μm, and 1 μm) mixed with colloidal silica on a napless cloth. For polishing, colloidal silica was used on a napless cloth. After polishing, the specimen was cleaned by flushing with water.

Image Analysis

The polished cross sections of the deposit specimens were plasma coated with gold and imaged on a scanning electron microscope (SEM) using back-scattered electrons. The nominal magnification of the microscope was 500X, which resulted in image dimensions of 180 μm by 240 μm and a pixel size of 0.31 μm square. The total numbers of digital SEM images analyzed were 67 for the specimen from coal A and 66 for the specimen from coal B.

A representative SEM image is shown in Fig. 1. In this image the bright regions are cross-sectional profiles of the solid phase of the specimen, and the dark regions correspond to the void phase. It is important to remember that even though the solid regions appear to be isolated, the solid phase forms a continuous structure in three dimensions.

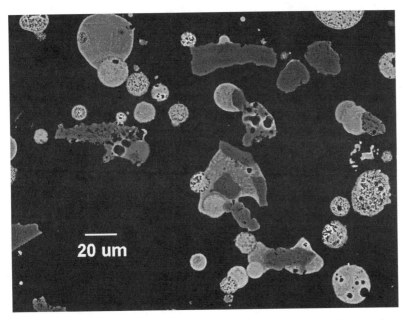

Fig. 1. Back-scattered electron micrograph of the ash deposit from coal B.

In Fig. 1 the direction of the bulk gas flow, and the incoming-particle trajectories, is from top to bottom. Deformations from collisions between particles are evident. In most cases the deformed profiles are on top, indicating that the incoming particles were softer than those already incorporated into the deposit.

Before the analysis of an SEM image, three processing steps were employed. First, the luminance scale of the image was converted from gray to binary. This step segmented the image into solid and void phases. The threshold value used to distinguish the solid phase was the lowest gray-scale level above the random background noise of the image. In the second step, solid regions with a boundary length of less than 40 pixels (12 μm) were removed. The purpose of this step was to get rid of features that were smaller than the resolution of the boundary length measurement, and to eliminate background noise. The value of the minimum boundary length was chosen to maintain the relative uncertainty in the boundary length measurement at less than 10%. Finally, the closed pores within the solid regions were filled. This step was taken to make the ash particle the fundamental unit of the solid phase. The internal microstructures of particles (e.g., mineralogy, crystallinity, and density) were outside the scope of this work.

Structural Parameters

The paragraphs that follow describe the structural parameters used in this work. These parameters are divided into two categories, two-dimensional and three-dimensional, depending on whether they are defined for the planar section of the specimen or for its three-dimensional structure.

A list of the symbols used here is provided at the end of this document. These symbols employ a notation that is commonly used in stereology [Weibel, 1979]: the first capital letter identifies the measured quantity and the second capital letter specifies the measurement space. For example, a volume fraction is written as V_V, a specific surface area as S_V, and a boundary length per unit area as B_A. To distinguish between phases subscripts are added. In this work α was used for the solid phase and β for the pore phase. The subscripts 2 and 3 are added to the symbols if there is a need to distinguish between two- and three-dimensional parameters.

Volume Fraction

The solid volume fraction, $V_{V\alpha}$, is the volume of solid phase per unit volume of deposit. This is the single most important structural parameter in correlating the effective thermal conductivity of two-phase materials, such as the fouling and slagging deposits in boilers [Wain, et al., 1992]. It has also been shown to have a marked effect on the mechanical properties of ash deposits that affect their removability, namely the compressive strength, the elastic modulus, and the thermal shock resistance [Wain, et al., 1992].

In this work, $V_{V\alpha}$ was determined from the area fraction of the solid phase in the planar section, $A_{A\alpha}$, using the principle of Delesse [Weibel, 1980]

$$V_{V\alpha} = A_{A\alpha} \equiv \frac{A_\alpha}{A_T} . \tag{1}$$

The solid area fraction was taken as the fraction of pixels in the α phase of the processed binary image.

Since there are only two phases present in the binary image, the porosity, $V_{V\beta}$, was determined by difference

$$V_{V\beta} = 1 - V_{V\alpha}. \tag{2}$$

Specific Surface Area

The specific surface area, $S_{V\alpha\beta}$, is the surface area per unit volume of deposit. This parameter describes the fineness of the deposit, and has a dominant effect on its optical properties, such as reflectivity [Wall, et al., 1993].

In this work $S_{V\alpha\beta}$ was determined from the length of the solid boundary per unit test area, $B_{A\alpha\beta}$, using the relation developed by Saltykov, and by Smith and Guttmann [Weibel, 1980]

$$S_{V\alpha\beta} = \frac{4}{\pi} B_{A\alpha\beta} \equiv \frac{4}{\pi} \frac{B_{\alpha\beta}}{A_T} . \tag{3}$$

The α-β boundaries were automatically created in the processed binary images using an image processing algorithm. The lengths of the artificial boundaries created at the edges of the image were not included in $B_{A\alpha\beta}$.

Contiguity

The contiguity, $C_{\alpha\alpha}$, describes the amount of contact between particles. This is a measure of particle sintering, which is an important mechanism of strength development in ash deposits [Nowok, et al., 1990; Hurley, et al., 1994].

The contiguity is defined as the average fraction of particle surface area that is shared with neighboring particles [Gurland, 1958]. This can be expressed in terms the specific particle contact area, $S_{V\alpha\alpha}$, and the specific surface area of the specimen, $S_{V\alpha\beta}$, as

$$C_{\alpha\alpha} \equiv \frac{S_{V\alpha\alpha}}{S_{V\alpha\alpha} + S_{V\alpha\beta}} = \frac{2B_{\alpha\alpha}}{2B_{\alpha\alpha} + B_{\alpha\beta}}. \tag{4}$$

In this work the second relationship, which was derived from the first using Eq. 3, was used to measure $C_{\alpha\alpha}$. Note that because the contact area is counted for each particle, the length of the contact boundary between particles, $B_{\alpha\alpha}$, is counted twice.

The contact lines between the particles were manually drawn on the gray-scale SEM images. Contact lines around particles that appeared to be completely embedded within the solid phase were included in $B_{\alpha\alpha}$, because these particles might not be completely embedded when viewed from the third dimension.

Mean Chord Lengths

The mean chord lengths are measures of the solid, particle, and pore sizes in the deposit. These quantities have significant effects on the transport of heat across the deposits by thermal conduction and radiation. The relationships that express the mean chord lengths in terms of deposit volume fraction and specific surface area were first developed by Fullman [Weibel, 1980].

The mean chord length in the solid phase is

$$l_{3\alpha} = 4 \frac{V_{V\alpha}}{S_{V\alpha\beta}}. \tag{5}$$

By using the definition of contiguity given in Eq. 4, the mean particle chord can be derived as

$$L_{3\alpha} = 4 \frac{V_{V\alpha}}{S_{V\alpha\alpha} + S_{V\alpha\beta}} = l_{3\alpha}(1 - C_{\alpha\alpha}). \tag{6}$$

Finally, the mean chord length of the pore phase is

$$l_{3\beta} = 4\frac{V_{V\beta}}{S_{V\alpha\beta}} = 4\frac{1-V_{V\alpha}}{S_{V\alpha\beta}}. \tag{7}$$

Density-Density Correlation

The density-density correlation function, $\gamma(r)$, describes the average variation in the density of a material. It can identify, for example, if the material is periodic or random, homogeneous or heterogeneous, isotropic or anisotropic. The density-density correlation function has been used to describe the fabric, or texture, of rocks [Lin, 1982; Berryman and Blair, 1986; Pfleiderer et al., 1993] and cement-based materials [Lange, et al., 1994]. Fabric has an important effect on the strength and permeability of these porous materials, which are structurally similar to fouling and slagging deposits in boilers.

The significant quantity in the density-density correlation function is $S_2(r)$, the two-point probability function. This function is defined as [Torquato, 1992]

$$S_2(r) \equiv \langle \rho(x)\,\rho(x+r) \rangle, \tag{8}$$

where $\rho(x)$ is the density of the material at point x, $\rho(x+r)\rangle$ is the density at the point $x + r$, and the angle brackets indicate that the product of these two densities is averaged while x is allowed to be every point in the material.

For a binary image $\rho(x)$ has two possible values

$$\rho(x) = \begin{cases} 1, & \text{for } x \text{ in phase } \alpha \text{ (solid)} \\ 0, & \text{for } x \text{ in phase } \beta \text{ (pore)} \end{cases}. \tag{9}$$

This means that for a line segment r drawn on a binary image, the product $\rho(x)\,\rho(x+r)$ will be 1 if both endpoints are in the solid phase, and 0 otherwise. Thus for a binary image, the function $S_2(r)$ is the probability that a line segment r randomly placed on the image will have both end points in the solid. This physical interpretation of $S_2(r)$ is very useful for understanding its properties.

For $|r| = 0$, the segment r is a point. The probability of a point landing in a solid region of a binary image is equal to the solid area fraction of the image, $A_{A\alpha}$. Thus,

$$S_2(0) = A_{A\alpha}. \tag{10}$$

As the length of r increases, the probability of finding both endpoints in the same solid region decreases, causing $S_2(r)$ to decrease. Furthermore, this probability decreases faster for the small solid regions than for large ones. Thus, the rate at which $S_2(r)$ decreases is governed by the size distribution of the solid regions. If the solid regions are not circular, and are preferentially oriented, the rate of decrease in $S_2(r)$ will depend on the direction of r.

When r gets large with respect to the solid regions, its endpoints may fall in different solid regions. If these regions are randomly distributed, the probability that one endpoint lands in the solid is independent of the other. In this limit the probability that both endpoints are in the solid is the probability that the first lands in the solid, $A_{A\alpha}$, times the probability that the second lands in the solid, also $A_{A\alpha}$. Thus,

$$S_2(|r| \to \infty) = A_{A\alpha}{}^2. \tag{11}$$

For convenience in comparing $S_2(r)$ from different images, it is useful to normalize $S_2(r)$ so its range is from 1 to 0 instead of from $A_{A\alpha}$ to $A_{A\alpha}{}^2$. The normalized $S_2(r)$ is the density-density correlation function

$$\gamma(r) = \frac{S_2(r) - A_{A\alpha}^2}{A_{A\alpha} - A_{A\alpha}^2}. \tag{12}$$

Because of the tremendous computational effort required to determine $S_2(r)$ using the direct approach described in Eq. 8, in this work $S_2(r)$ was evaluated using the two-dimensional fast Fourier transform [Press et al., 1988]

$$S_2(r) = \text{FFT}^{-1}(|\text{FFT(ROI)}|^2). \tag{13}$$

The ROI, or region of interest, was a 512 x 512 pixel subregion (160 μm square) located in the center of the processed binary image. The dimensions of the ROI are about three times the dimension of the larger solid regions in the images.

RESULTS AND DISCUSSION

The first issue that must be addressed in interpreting the image analysis results is the sampling uncertainty associated with the measurements. This uncertainty exists because a single image represents only a portion of the specimen. For example, if an image contains only one solid region, the confidence that the quantities obtained from it will be representative of the specimen is low. However, if an image contains 100 solid regions, this confidence is greatly improved. Thus, the sampling uncertainty is inversely proportional to the number of solid regions in the image. For a given specimen, increasing the number of solid regions means increasing the size of the image. However, to maintain the required resolution this cannot be accomplished by simply decreasing the magnification of the microscope. As an alternative, it is common to pool data from adjoining images, effectively creating larger, composite images.

An estimate of the standard deviation of the solid volume fraction for a single image is given by [Russ, 1990]

$$\sigma(V_{V\alpha}) = \frac{A_{A\alpha}}{\sqrt{N}}, \tag{14}$$

315

where N is the number of solid regions in the image. For the deposits from coals A and B, the relative value of the standard deviation for a single image was 22%. To reduce this uncertainty, images were grouped into blocks of 12, three horizontal by four vertical, each 720 μm square. The solid area measurements from the 12 images were pooled as follows

$$V_{V\alpha} = \frac{\sum_{i=1}^{12} A_{\alpha i}}{\sum_{i=1}^{12} A_{Ti}} . \tag{15}$$

The relative standard deviation of $V_{V\alpha}$ for a block of 12 images was ±6%.

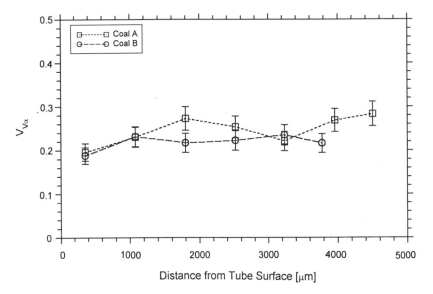

Fig. 2. Spatial profiles of $V_{V\alpha}$ in deposits from coals A and B.

The variation of $V_{V\alpha}$ with distance from the tube surface is shown in Fig. 2 for the deposits from coals A and B. The points plotted on the graph are the results obtained from the 12-image blocks described above. The surprising feature of these curves is the lack of a strong increase in $V_{V\alpha}$ from the tube surface to the flame surface: $V_{V\alpha}$ increases only slightly for coal A and remains constant for coal B. Larger increases would have been expected because of sintering and the formation of a molten capture phase at the flame surface of the deposit. Both curves show large-scale fluctuations in the solid volume fraction. Perhaps these are due to oblique sectioning of vertical (i.e., parallel to the gas flow) channels in the deposits.

The error bars in Fig. 2 represent the estimated uncertainty of ±10% in the precision of the measurement. This estimate is based on a ±8% uncertainty in the area of a solid region due to a ±1 pixel uncertainty in the locations of its boundary pixels, determined experimentally by eroding and dilating solid regions, and by the ±6% sampling uncertainty discussed above. In addition to these random errors, a systematic error was introduced by the choice of the threshold used to convert the SEM images from gray-scale to binary. This threshold was chosen as the lowest possible value above the background noise, because this value could be consistently identified for all images. Since the area of a solid region grows as the threshold decreases, the measured $V_{V\alpha}$ is the maximum possible value from an estimated range of 18%. However, because all images were treated in the same manner, this systematic error does not affect trends in the results or comparison of the two specimens.

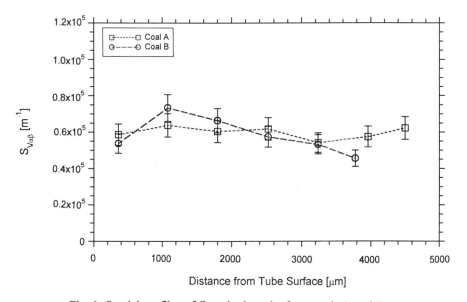

Fig. 3. Spatial profiles of $S_{V\alpha\beta}$ in deposits from coals A and B.

The spatial profiles of $S_{V\alpha\beta}$ are presented in Fig. 3. It is evident that $S_{V\alpha\beta}$ remained constant for coal A and decreased slightly for coal B. Combining the results for $S_{V\alpha\beta}$ and $V_{V\alpha}$, it is apparent that the surface to volume ratio of the solid phase decreased for both coals: for coal A, $S_{V\alpha\beta}$ remained constant while $V_{V\alpha}$ increased and for coal B, $S_{V\alpha\beta}$ decreased while $V_{V\alpha}$ remained constant. This indicates a coarsening of the solid phase.

The estimated uncertainty in $S_{V\alpha\beta}$ of ±10% is reflected in the error bars in Fig. 3. This uncertainty is based on a ±8% uncertainty in the boundary length, determined experimentally by smoothing the boundaries, and by the ±6% sampling uncertainty.

Compared to the previous two parameters, Fig. 4 shows that $C_{\alpha\alpha}$ increased dramatically across the deposit from tube surface to flame surface. This increase is evidence that sintering, not larger particles, was responsible for the coarsening of the solid phase. The visually observed sintering was via both increased particle agglomeration and encapsulation of particles by a glassy phase.

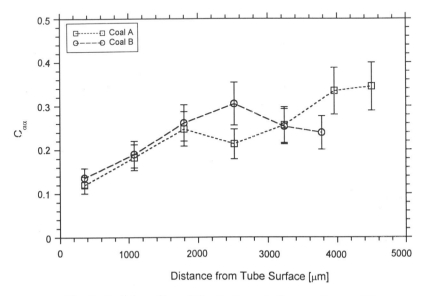

Fig. 4. Spatial profiles of $C_{\alpha\alpha}$ in deposits from coals A and B.

The uncertainty in $C_{\alpha\alpha}$ was estimated at ±16%. This uncertainty is primarily due to an ambiguity in identifying dark regions within solid blobs as encapsulated particles or differences in solid density.

The spatial profiles of the mean chord lengths $l_{3\alpha}$, $L_{3\alpha}$, and $l_{3\beta}$ are presented in Figs. 5 and 6 for coals A and B, respectively. The most interesting feature in these figures is the divergence of $l_{3\alpha}$ and $L_{3\alpha}$, the mean chord lengths in the solid phase and in the particles. This is evidence of increased particle agglomeration from tube surface of the deposit to the flame surface. The small increase in $L_{3\alpha}$ for coal B may indicate increased encapsulation of particles by a glassy phase.

The uncertainties in $l_{3\alpha}$ and $l_{3\beta}$ are estimated at ±14% by combination of the ±10% uncertainties in $V_{V\alpha}$ and $S_{V\alpha\beta}$. The uncertainty in $L_{3\alpha}$ is estimated to be ±21%, through the combined uncertainties in $l_{3\alpha}$ and $C_{\alpha\alpha}$.

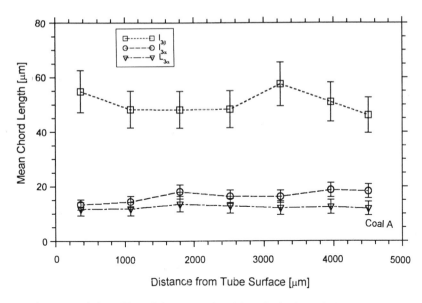

Fig. 5. Spatial profiles of the mean chord lengths in deposit from coal A.

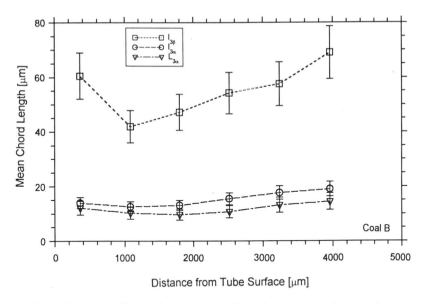

Fig. 6. Spatial profiles of the mean chord lengths in deposit from coal B.

319

The bulk averages of the measured parameters are presented in Table 1. These results are very similar for both coals. This outcome was expected because the compositions of the coals are similar, and because the combustion conditions were identical. Coal A has a greater low-temperature fouling tendency than coal B [Hurley, et al., 1991]; however, the combustion conditions were in the high-temperature fouling regime.

Table 1. Bulk values of structural parameters.

	Coal A	Coal B
$V_{V\alpha}$	0.25	0.22
$S_{V\alpha\beta}$	0.60×10^6 m^{-1}	0.58×10^6 m^{-1}
$C_{\alpha\alpha}$	0.24	0.23
$l_{3\alpha}$	16 μm	15 μm
$L_{3\alpha}$	12 μm	12 μm
$l_{3\beta}$	51 μm	55 μm

For mono-dispersed spherical particles, $L_{3\alpha}$ is 2/3 the diameter. Since most of the particles in the deposits from coals A and B are spherical, the $L_{3\alpha}$ values in Table 2 indicate a mean particle diameter of 18 μm.

The density-density correlation function was computed for the deposit from coal B. A contour plot of the result is presented in Fig. 7. This is the average of $\gamma(r)$ over a block of 12 images located close to the flame side of the deposit. The central dot represents the peak value of 1.0, and the contour lines decrease in steps of 0.1 to the outer contour line at 0. The inner contours are circular, indicating that solid regions 20 μm and smaller are isotropic. The noncircular outer contour indicates that larger solid regions are preferentially oriented in the vertical direction. Thus, the structure of the deposit preserves the direction the incoming particles were traveling when they arrived at the deposit surface.

CONCLUSIONS AND CHALLENGES

The results obtained here demonstrate the usefulness of direct, microscopical methods in characterizing ash deposit structure; however, challenges remain in both the measurement of structure and in its use to solve ash-related problems in boilers.

One difficult aspect of the structural characterization is the complex morphologies of some deposit particles, particularly low-density nebular particles. These appear to be space-filling units when viewed in the transparent epoxy mounting medium through an optical microscope. However, because these structures are so open, they loose their external boundaries on planar sections and disintegrate into isolated features. These may be so small that they are overlooked in the SEM images.

Another challenge is the need for serial sectioning and three-dimensional imaging. Given the low solid volume fractions of deposits, models of their properties will depend strongly on the connectivity parameters of the solid-phase network, such as coordination number and particle contact area. This information cannot be obtained from the random planar sections used in this work, but requires the more difficult methods of serial sectioning and three-dimensional imaging.

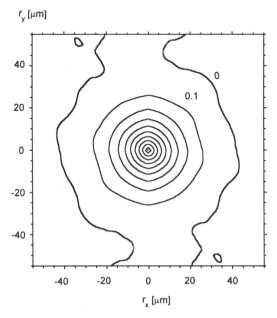

Fig. 7. Density-density correlation function for the deposit from coal B.

The biggest challenges, however, lie in using the measurements of ash deposit structure. According to DeHoff [1986], there are three principal uses for the structural information obtained by stereology: (1) monitoring the evolution of microstructure during processing, (2) comparing structures that are presumed similar, but which behave differently, and (3) developing structure-property correlations. An additional application is, (4) providing the geometric description necessary for computational modeling of material formation processes and material properties.

In the literature, there are numerous examples of uses (1) and (2). Stereology is ideally suited to these applications because it yields quantitative results, with confidence limits, and provides a rigorous sampling methodology. The remaining applications of structural information have not been widely used. Structure-property correlations are difficult because properties may depend on extreme values of structural parameters, not the average values obtained from stereology, and physical properties are often defined by accepted standardized measurements that have little structural basis [DeHoff, 1986]. The literature describes many structural studies, physical property measurements, and physical property models; however, examples of *practical* structure-property and structure-model relations are almost nonexistent [DeHoff, 1986; Liu, 1993].

ACKNOWLEDGMENTS

The authors wish to thank Larry L. Baxter of Sandia National Laboratories and Galen H. Richards of Brigham Young University for the ash deposit specimens. Special thanks are due George Kinter, student at Geneva College, for his work in analyzing images.

REFERENCES

Berryman, J. G. and Blair, S. C. (1986). "Use of Digital Image Analysis to Estimate Fluid Permeability of Porous Materials: Application of Two-Point Correlation Functions." *J. Appl. Phys., 60*(6), 1930-1938.

DeHoff, R. T. (1986). "Quantitative Microstructural Characterization and Description of Multiphase Ceramics." in *Tailoring Multiphase and Composite Ceramics*, New York: Plenum Press, 207-222.

Gurland, J. (1958). "Spatial Distribution of Discrete Particles." in R. T. DeHoff and F. N. Rhines (eds.) *Quantitative Microscopy*, New York: McGraw-Hill, 278-290.

Hurley, J. P., Benson, S. A., and Mehta, A. K. (1994). "Ash Deposition at Low Temperatures in Boilers Burning High-Calcium Coals." in Williamson, J. and Wigley, F. (eds.) *The Impact of Ash Deposition on Coal Fired Plants*, Washington: Taylor and Francis, 19-30.

Hurley, J. P., Erickson, T. A., Benson, S. A., and Brobjorg, J. N. (1991). "Ash Deposition at Low Temperatures in Boilers Firing Western U. S. Coals." Presented at the International Joint Power Generation Conference, San Diego, California, October 7-10, 1991.

Lange, D. A., Jennings, H. M., Shah, S. P. (1994). "Image Analysis techniques for Characterization of Pore Structure of Cement-Based Materials." *Cement and Concrete Research 24*(5), 841-853.

Lin, C. (1982). "Microgeometry I: Autocorrelation and Rock Microstructure." *Mathematical Geology, 14*(4), 343-360.

Liu, G. (1993). "Applied Stereology in Materials Science and Engineering." *J. Microsc., 171*(1), 57-68.

Nowok, J. W., Benson, S. A., Jones, M. L., and Kalmanovitch, D. P. (1990). "Sintering Behaviour and Strength Development in Various Coal Ashes." *Fuel, 69*, 1020-1028.

Pfleiderer, S., Ball, D. G. A., Bailey, R. C. (1993). "AUTO: A Computer Program for the Determination of the Two-Dimensional Autocorrelation Function of Digital Images." *Computers & Geosciences 19*(6), 825-829.

Press, W. H., Flannery, B. P., Teukolsky, S. A., and Vetterling, W. T. (1988). *Numerical Recipes in C The Art of Scientific Computing*, Cambridge: Cambridge University Press.

Russ, J. C. (1990). *Computer-Assisted Microscopy The Measurement and Analysis of Images*, New York: Plenum Press.

Torquato, S. (1992). "Connection between Morphology and Effective Properties of Heterogeneous Materials." in S. Torquato and D. Krajcinovic (Eds.) *Macroscopic Behavior of Heterogeneous Materials from the Microstructure AMD-Vol. 147*, New York: ASME.

Wain, S. E., Livingston, W. R., Sanyal, A., and Williamson, J. "Thermal and Mechanical Properties of Boiler Slags of Relevance to Sootblowing." in Benson, S. A. (Ed.) Inorganic Transformations and Ash Deposition During Combustion, New York: Engineering Foundation, 459-470.

Wall, T. F., Bhattacharya, S. P., Zhang, D. K., Gupta, R. P., and He, X. (1993). "The Properties and Thermal Effects of Ash Deposits in Coal-Fired Furnaces." *Prog. Energy Combust. Sci. 19*, 487-504.

Weibel, E. R. (1979). *Stereological Methods Vol. 1 Practical Methods for Biological Morphometry*. London: Academic Press.

Weibel, E. R. (1980). *Stereological Methods Vol. 2 Theoretical Foundations*. London: Academic Press.

LIST OF SYMBOLS

Symbol   Units   Definition

1. Phases

  α Solid

  β Pore

2. 2-D parameters

Symbol	Units	Definition
A_T	m^2	Test area
A_α	m^2	Solid area
$A_{A\alpha}$		Solid area per unit test area
$B_{\alpha\beta}$	m	Solid boundary length (perimeter)
$B_{A\alpha\beta}$	m^{-1}	Solid boundary length per unit test area
$B_{\alpha\alpha}$	m	Particle contact boundary length (counted twice, once for each particle)
$\rho(x)$	m^{-2}	Local density at point x
$\gamma(r)$		Density-density correlation function: variation in local density around a point
$S_2(r)$		Two-point probability function: probability that a randomly placed line segment r will have both end points in the solid

3. 3-D parameters

Symbol	Units	Definition
$V_{V\alpha}$		Solid volume fraction
$V_{V\beta}$		Pore volume fraction (porosity)
$S_{V\alpha\beta}$	m^{-1}	Surface area per unit test volume (specific surface)
$S_{V\alpha\alpha}$	m^{-1}	Particle contact area per unit test volume (counted twice, once for each particle)
$C_{\alpha\alpha}$		Contiguity: average fraction of particle surface area shared with neighboring particles
$l_{3\alpha}$	m	Mean solid chord length
$L_{3\alpha}$	m	Mean particle chord length
$l_{3\beta}$	m	Mean pore chord length (mean free path)

DANISH COLLABORATIVE PROJECT
ON MINERAL TRANSFORMATIONS AND ASH DEPOSITION
IN PF-FIRED BOILERS
AND RELATED RESEARCH PROJECTS

Ole Hede Larsen

Faelleskemikerne, ELSAM, I/S Fynsværket
DK-5000 Odense C, Denmark

Karin Laursen

Geological Survey of Denmark and Greenland
DK-2400 Copenhagen NV, Denmark

Flemming Frandsen

Department of Chemical Engineering
Technical University of Denmark
DK-2800 Lyngby, Denmark

ABSTRACT

A 3-year Danish collaborative research project on ash deposition in PF-fired boilers was initiated in 1993. The project is targetted towards the power plants in Denmark, including the new USC-boilers under construction, with steam parameters 290 bar/580 °C and using coals of worldwide origins. The purpose of the project and the activities of each of the participants, i.e. the Jutland-Funen Electricity Consortium, ELSAM (project coordinator, full-scale tests), the Geological Survey of Denmark and Greenland (coal, ash and deposit characterization) and the Combustion and Harmfull Emission Control (CHEC) research programme at the Technical University of Denmark (modelling of ash deposition propensities) are presented. Projects and activities on related topics will also be presented. These include biomass aspects, fuel organics, high temperature corrosion, trace elements, ash clean-up and residues.

Applications of Advanced Technology to Ash-Related Problems in Boilers
Edited by L. Baxter and R. DeSollar, Plenum Press, New York, 1996

325

COLLABORATIVE PROJECT ON ASH DEPOSITION IN PF-FIRED BOILERS

The power plants in the ELSAM area of Denmark are mainly coal PF-fired boilers using coals of worldwide origins, see Figure 1.

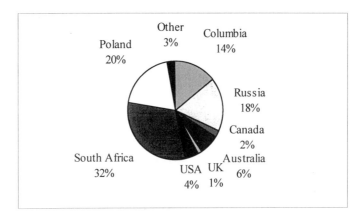

Figure 1. 1993 Danish coal import

At present, ash deposition problems in boilers in the ELSAM area are handled by blending coals (dilution effect). Although, it is recognized, that the trend of increasing steam parameters and use of low NO_x burners may change this procedure and that the slagging and fouling indices used today, will be insufficient for forecasting the coal ash impact on especially the new generation high-efficiency boilers. Thus, more basic understanding of the physical and chemical reactions involved in mineral transformations and ash deposition is necessary.

A three-year research and development (R&D) project on ash deposition in PF-fired boilers was initiated in 1993. The R&D project is targetted towards new USC-boilers under construction, with steam parameters 290bar/580°C and is carried out in collaboration between ELSAM, the Geological Survey of Denmark and Greenland (DGGU) and the Combustion and Harmfull Emission Control (CHEC) research programme, the Technical University of Denmark. The purposes of the R&D project are outlined in Table 1.

Table 1. The main purposes of the 3-year Danish collaborative R&D project on mineral transformations and ash deposition in PF-fired boilers.

➢ Etablishment of a better knowledge of the mechanisms involved in ash deposition
➢ Development of analysis necessary for evaluation of ash deposition propensity
➢ Development of a computational tool for prediction of ash deposition propensity

The main experimental work in this project is full-scale tests at different power plants in the ELSAM area and using different coal types and operational variables. Ash deposits are collected using temperature regulated probes, and furthermore pulverized coal, bottom ash and fly ash is sampled for different analysis. For the next tests, also an in-furnace video probe will be used for direct visual observation of ash build-up. The tools for characterization of coal inorganics, ash and deposits are developed primarily at the Geological Survey of Denmark and Greenland. The main analytical tools are CCSEM, other SEM-based analysis and XRD. The modeling work is perfomed as part of the CHEC research programme. The first step implies empirical prediction of ash deposition propensities based on coal and boiler characteristics and the target is development of a mechanistic model for prediction of fly ash composition and ash deposition based on CCSEM analysis of the coals.

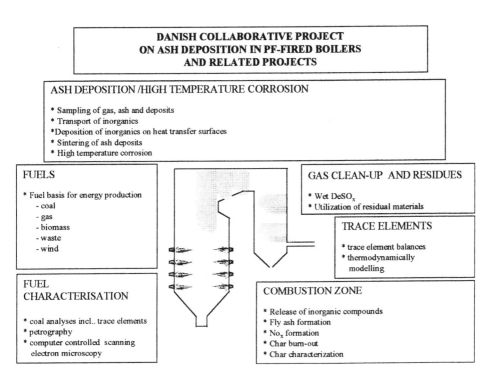

Figure 2. Sketch of poster.

The content and lay-out of the poster is shown in Figure 2. Each box represents a major zone or item related to fuel ashforming constituents and within each box, headlines for specific topics are given. More information on these topics is given below.

ELSAM

ELSAM is a power utility pool, owned by the six power plants of western Denmark. The main activities of ELSAM are listed in Table 2.

Table 2. The main activities of ELSAM (the power plant organization covering the western part of Denmark).

➢ Manage and coordinate the production and combined heat and power
➢ Manage the fuel supply for power stations
➢ Buy and resell the production of the power stations
➢ Plan, establish and finance new plants and transmission network
➢ Participate in research and development of new energy technologies

Many of the R&D activities are initiated due to foreseen and practized governmental environmental regulations. All newer units are equipped with $DeNO_x$ (low NO_x burners, SCR) and $DeSO_x$ (Spray Dry Absorption, Wet Gypsum and SNOX) equipment. The Danish parliament has decided on a 20 % reduction in CO_2 emission in year 2005 as compared to year 1988. ELSAM follows two main roads to reach that goal: high efficient coal-fired PF-boilers and high-efficient biomass combustion. The 400 MW_{el} coal-fired unit at Nordjyllandsvaerket to be commisioned in 1997 will produce main steam at 290 bar/580 °C, double reheat at 580°C and net efficiency of 47 % is expected. Further improvements are expected to enable plants in operation in year 2005 with net efficiency at 50 % - see Kjaer (1994).

The major biomass source in Denmark is and probably remains to be straw. Straw is presently used in several steam producing boilers in Denmark, but at rather low steam parameters and hence low efficiencies. The very high concentrations of chlorine and potassium in straw can cause superheater corrosion, being one of the major limiting factors for the efficient use of this fuel (Henriksen et al. (1995)). Several concepts are under investigation including strawfiring on grate and co-combustion of coal and straw in PF boilers and CFB boilers.

Ash related projects at ELSAM

Many of the R&D projects in ELSAM are related to ash, but only one will be mentioned here: high-temperature corrosion. In the study of high temperature corrosion is has been realized, that the mechanisms of liberation of fuel inorganics during combustion and the following reactions in ash deposition are very important and therefore build-up of basic knowledge on high temperature corrosion and ash deposition are preferably done together.
The fuel/boiler combinations considered are listed in Table 3.

Table 3. Fuel/boiler combinations considered in ELSAM area.

> Coal in PC-USC boiler with steam temperatures up to 620°C
> Coal + straw in PC boiler with steam temperatures up to 580°C
> Coal + straw in CFB boiler with steam temperatures up to 580°C
> Straw on grate with steam temperatures up to 580°C

For high temperature corrosion at coal combustion in PC-USC boilers, a Brite-EuRam project has been started, involving a test-loop at the Vestkraft Unit 3 power plant for quantification of corrosion at different temperatures and superheater materials. Coal + straw in PC boilers has been investigated at full scale tests at Vestkraft Unit 1 and will be continued later this year with a comprehensive full scale testing at Midtkraft Unit 1. This involves corrosion studies with temperature regulated probes and test tubes, studies on deposit collected on probes, measurements of in-furnace flue gas chemistry and aerosoles and also effects on fly ash composition and $DeNO_x$, $DeSO_x$ are considered. Similar investigations are being performed at coal + straw in CFB boilers, with testing at Grenå CFB and also on strawfiring on grate at Rudkoebing. Some results will be presented in Henriksen et al. (1995).

THE GEOLOGICAL SURVEY OF DENMARK AND GREENLAND (DGGU)

The Geological Survey of Denmark and Greenland (DGGU) is an advisory and research institution under the Danish Ministry of Energy and Environment. The primary functions of the survey is to provide essential geological service for the utilization and protection of the natural resources of Denmark and Greenland. DGGU has significant knowledge of coal as a mineral, and has extensive experience with petrographical characterization of coal. In the recent years DGGU has studied coal characteristics in relation to the combustion of coals through a number of research projects.

Coal and biomass analytical facilities at DGGU

DGGU has excellent facilities for organically and inorganic characterization of coal and biomass. The facilities on the organic side includes both chemical and microscopic analysis. Based on different microscopes the following analysis are being carried out at DGGU: maceral analyses, combined maceral and microlithotype characterization, rank determination and automatic image analysis (total reflectance analysis).Chemical analysis of the organic parts of coal includes: various pyrolysis techniques together with GC and GCMS analysis on extracts. On the inorganic parts a new scanning electron microscope (Phillips XL 40) equipped with a Noran Instrument (Voyager) analytical system has been set up for computer controlled scanning electron microscope analysis (CCSEM). Other inorganic analysis includes XRD and XRF.

Coal and biomass projects at DGGU

The Geological Survey of Denmark and Greenland has significant knowledge of coal as a mineral, and has extensive experience with petrographical characterization of coal. In the recent years DGGU has studied coal characteristics in relation to the combustion rate of coals through a number of research projects. These projects has mainly been supported by the Danish power station consortiums (ELSAM and ELKRAFT) and the Danish Ministry of Energy.

In a recently finished research project on "Determination and correlation of coal data", 10 coals were extensively characterized by chemical and petrographical methods - see Thomsen et al. (1992). These data were compared to results of high temperature coal and char burn out measurements taken in a pilot-scale test reactor. Using multivariate techniques a statistical model has been established from which char burn out parameters can be predicted from relatively simple petrographical and chemical data. A project on "Char characterization" aims at correlating the raw coal composition to the morphology and macroporosity of chars from various reactors including full-scale. Significant variations in the char morphology are found as a function of the reactor type and temperature. The project "Coal Characterization and Chars Reactivity" is closely related to the earlier projects. The existing database is extended with more coals covering a wider range of geographical and geological origins, more sophisticated analytical methods are utilized and the experience obtained in char classification is applied in the prediction of the combustion reactivity of the coals.

A project on "Characterization of ash from thermal conversion of biomass fuels" has recently been initiated. The scope of the project is to study the fate of inorganic constituents during combustion in various full-scale boiler systems. The results from the project will form the basis of development of a new laboratory routine for characterization of biomass ashes.

A running research project on "Characterization of Slagging and Fouling Properties of Coal Minerals in PC- fired Boilers" includes characterization of the inorganic constituents in coals consumed in Danish power stations, characterization of ash deposits from boilers and characterization of the physical-chemical reactions controlling ash deposition in the boilers -see Laursen et al. (1995).

THE CHEC RESEARCH PROGRAMME

CHEC is the acronym for a fundamental and applied research programme on Combustion and Harmful Emission Control at the Department of Chemical Engineering, Technical University of Denmark. The CHEC research programme is cofunded by ELSAM (the Jutland-Funen electricity consortium), ELKRAFT (the Zealand electricity consortium), the Danish Technical Research Council, the Danish Energy Research Programme and the Nordic Energy Research Programme. In Table 4, the objectives of CHEC are listed.

The research activities engage about 30 people (permanent staff, technicians, post docs, Ph.D.- and M.Sc.-students). In Table 5, the major research areas within CHEC are listed.

Table 4. Objectives of the Combustion and Harmful Emission Control (CHEC) research programme.

> ➤ To provide new fundamental information on combustion and gasification processes and on formation and control of harmful emissions.
> ➤ To train scientists and engineers.
> ➤ To assist industry and public authorities.
> ➤ To catalyze cooperation between Danish and foreign research organizations and business enterprises.

Table 5. Major research activities within the Combustion and Harmful Emission Control (CHEC) research programme.

> ➤ Kinetics and Mechanisms of High-Temperature Gas-Solid and Gas Phase Reactions
> ➤ Kinetics and Mechanisms of Low-Temperature Gas Cleaning Processes
> ➤ Chemical Kinetic Modeling
> ➤ Reactant Mixing in Gas Phase Reactions
> ➤ Empirical Modeling of Reacting Systems
> ➤ Fluid Dynamics in Fluidized Beds
> ➤ Measuring Techniques
> ➤ Inorganic Metal Compounds in Combustion and Gasification

For general introduction to the CHEC research activities, including an update of recent research activities, see Dam-Johansen (1994). Below, an introduction to activities within the fate of inorganic metal compounds in combustion and gasification is given.

Inorganic Metal Compounds in Combustion and Gasification

Within this research area, laboratory-, pilot- and full-scale measurements and modeling (thermodynamic, empirical and mechanistic) activities are carried out in close collaboration with power plant industry. Subtitles on CHEC-activities concerning inorganic metal compounds in combustion and gasification are listed in Table 6.

Table 6. CHEC projects concerning inorganic metal compounds in thermal fuel conversion systems.

> ➢ Ash deposition in coal, biomass and coal-biomass co-fired thermal fuel conversion systems.
> ➢ Sampling and analysis of alkali metals.
> ➢ Agglomeration in straw-fired FBC.
> ➢ Phase transformations and sintring in ash deposits.
> ➢ High temperature corrosion.
> ➢ Trace element partitioning in thermal fuel conversion systems.

Hansen et al. (1995) investigated the fate of alkali metals in a 20 MW_{th} multi-circulating coal-straw co-fired fluidized bed combustor. As part of the project, a probe for sampling of gaseous alkali metals from flue gases with a high particulate loading was developed. A systematic comparison of fuel ash, fly ash, deposits and gas phase alkali metal concentrations was performed and a semi-empirical model was set up for the system. The model results were found to correlate with the performed experiments, except for the riser section where the model predicted higher concentrations of potassium than measured.

Laursen et al. (1995) performed 'cold-finger' deposition probe experiments at two Danish power stations firing coal of different rank and origin. Deposits, fuel-, bottom- and fly ash were analyzed by CCSEM and SEM-EDX. A computational tool for empirical prediction of ash deposition propensities in PC-fired utilities were developed (Frandsen et al. (1995)). The programme is based on the concept of comparing a bulk ash chemistry index (characterizing the chemical potential of a coal as, eg. a content of S or Na) with a plant parameter characterizing the combustion system (operation, furnace size and geometry, burner belt geometry and/or boiler configuration). Experimental as well as theoretical predictions showed a very low rate and extent of ash deposition in the two boilers firing their respective coals, see Laursen et al. (1995) and Frandsen et al. (1995).

Frandsen et al. (1994) performed an extensive theoretical study of the fate of trace elements in PC-fired boilers. The equilibrium composition of a well defined system at constant total composition, pressure and temperature is calculated by minimizing the total Gibbs free energy of the system. This type of calculation provide the user with valuable information about the distribution of a particular element among and within fluid phases in a chemically reacting system. The study included equilibrium distributions of 18 trace elements as a function of the temperature and the redox conditions. The model for trace element partitioning was tested against full-scale mass balance measurements and it was shown that the model predictions correlated reasonably with the measurements (Frandsen et al. (1993), Frandsen (1995)).

For further information, please do not hesitate to contact:

CHEC
Prof. Kim Dam-Johansen
Dept. of Chemical Engineering
Technical University of Denmark
Building 229
DK-2800 Lyngby, Denmark.
Fax: + 45 45 88 22 58.

DGGU
Per Rosenberg
Geochemical Dept.
Geological Survey
Thoravej 8
DK-2400 Copenhagen NV
Fax: + 45 31 19 68 68

ELSAM
Ole Hede Larsen
Faelleskemikerne
I/S Fynsvaerket
Havnegade 120
DK-5000 Odense C
Fax: + 45 65 90 38 12

REFERENCES

Dam-Johansen, K. "CHEC Annual Report 1994", Dept. Chem. Eng., Techn. Univ. of Denmark, 1995.

Frandsen, F. J. *"Trace Elements from Coal Combustion";* Ph.D.-Thesis, Dept. Chem. Eng., Techn. Univ. of Denmark, 1995. ISBN-87-90142-03-9.

Frandsen, F. J.; Dam-Johansen, K.; Laursen, K.; Larsen, O. H. "Prediction of Ash Deposition Propensities at two Coal-Fired Power Stations in Denmark"; Paper presented at the 12th Int. Ann. Pittsburgh Coal Conference, Pittsburgh, PA, September 9-11, 1995.

Frandsen, F. J.; Dam-Johansen, K.; Rasmussen, P.; "Trace Elements from Combustion and Gasification of Coal - An Equilibrium Approach"; Prog. Energy Combust. Sci.; 1994; 20; 115 - 138.

Frandsen, F. J.; Dam-Johansen, K.; Rasmussen, P.; Sander, B.; "Trace Elements from Combustion: Thermodynamic Equilibrium Approach and Full-Scale Measurements -A Comparative Study"; Proc. Swedish-Finnish Flame Days 1993, Gothenburg, September 7-8, 1993.

Hansen, L. A.; Michelsen, H. P.; Dam-Johansen, K.; "Alkali metals in a coal- and biomass-fired CFBC - Measurements and Thermodynamic Modeling"; Proc. 13th Int. Conf. on FBC, ASME, Orlando, FA, May 1995.

Henriksen, N.; Larsen, O.H.; Blum, R.; "High-Temperature Corrosion at Co-combustion of Coal and Straw in PC Boilers and CFB Boilers"; Paper to be presented at the VGB Conference 'Corrosion and Corrosion Protection in Power Plants', Essen, Germany, November 29-30, 1995.

Kjaer, S; "The Advanced Pulverized Coal-Fired Power Plant - Status and Future"; ASME Joint Power Generating Conference. October 1994, Phoenix, Arizona.

Laursen, K.; Frandsen, F. J.; Larsen, O. H.; "Slagging and Fouling Propensity: Full-scale Tests at Two Power Stations in Western Denmark"; Paper presented at the Eng. Found. Conf. on 'Application of Advanced Technology to Ash-Related Problems inBoilers', Waterville Valley, NH, July 16-21, 1995.

Thomsen, E.; Rosenberg, P.; Boisen-Koefoed, J.; Guvad, C.; "Determination and Correlation of Coal Data - Petrographical and Chemical Coal Characterization"; Danish Ministry of Energy, Energy Research Project No. 1323/87-16, 1992.

EMITTANCE OF LAYERS OF UNIFORM ASH-LIKE SLAG PARTICLES: SPECTRAL NATURE AND EFFECTS OF PARTICLE SIZE, COMPOSITION, AND HEATING

S.P. Bhattacharya[1], T.F. Wall, and R.P. Gupta - Department of Chemical Engineering
The University of Newcastle, NSW 2308, Australia
A.M. Vassallo - CSIRO Division of Coal and Energy Technology
North Ryde, Sydney, Australia

Abstract

This paper presents results of a theoretical and experimental investigation on the effects of particle size, iron content and heating on the emittance of particulate layers of uniform ash-like slag particles. Theoretical predictions have been made on the emittance of particulate layers to illustrate the effect of particle size and composition. Controlled spectral emission measurements are carried out between wavelengths of 2 and 12 μm on layers of ground synthetic slag particles having known composition and particle size. Emission spectra are acquired between 600 and 1200°C at every 200°C during both heating and cooling cycles. An increase of emittance with particle size and iron content is observed. Emittance measured during the cooling cycle is found to be higher than the emittance measured during the heating cycle. All the effects are found to be limited to wavelengths below 6 μm which is also the region of primary interest to radiative heat transfer in pulverised fuel fired furnaces. The trends from the predictions matched well with the measured values, except for some differences in magnitude primarily in the wavelength region below 5 μm. Possible reasons for this difference are discussed.

1   Introduction

Knowledge of the spectral emittance of the particulate deposits is necessary for monitoring of deposit temperature and condition, and prediction of heat transfer through deposits in a pulverised fuel (pf) fired furnace (Wall et al., 1993). As the need increases for higher efficiency and better pollution control in pf fired furnaces, there is a need of increased understanding of the factors that affect the emittance of deposits. Studies on emittance of particulate deposits are few and far between. Studies by Boow and Goard (1969) and Brajuskovic et al. (1991) are on *total* emittance measurements of deposits at surface temperatures ranging from 200 to 1000°C showing the effect of temperature on total emittance. The study by Wall and Becker (1984) are on spectral *band* measurements at temperatures ranging from 700 to 1000°C and shows the effect of iron content and

Keywords: Ash Deposits, Emmitance, and Radiative Properties

[1]Present Address: Cooperative Research Centre for Power Generation from Low ran Coal, 8/677 Springvale Road Mulgrave, Victoria 3170, Australia.

Applications of Advanced Technology to Ash-Related Problems in Boilers
Edited by L. Baxter and R. DeSollar, Plenum Press, New York, 1996

335

temperature. The study by Markham et al. (1992) shows the effect of physical state of a particulate deposit on its spectral emittance. This article presents theoretical predictions on the effects of particle size, physical state, and iron content in ash on spectral emittance of particulate deposits, and also presents measurements using well characterised particles to illustrate these effects as well as the effect of heating. The results of this study are expected to provide a theoretical basis for on-line monitoring of ash deposits.

2. Emittance : Definition and Factors Affecting It

Emittance (ε) of a surface is commonly defined as the ratio of its emission rate (E) to that from a blackbody (E_B) at the same temperature ($\varepsilon = E/E_B$). A grey surface is one for which the emittance is independent of wavelength and therefore of temperature, but the 'effective' emissivity of a deposit depends on the spectral distribution of emittance and the surface temperature. Emittance can be defined as *hemispherical* (ε_h) or *normal* (ε_n) depending on the direction of measurement, and as *spectral* (ε_λ) or *total* (ε_t) in which case it is referred to part or all of the wavelength spectrum with reference to the direction and spectral range. It can be defined as a combination of any two of the above. Total emittance can be calculated from spectral emittance (Wall and Becker, 1984).

Emittance of deposits is known to be affected by its physical character (slagged, sintered or particulate), composition of the material, and size and concentration of the particles in the deposits. Several trends are noted in the data of Boow and Goard (1969) and Brajuskovic et al. (1991) on layers of particles prepared from synthetic slag, which are given in figures 1 and 2 respectively. These are similar to some other studies reported by Wall et al. (1993) on powdery ash deposits and slag layers. The trends observed are :

(a)   A reduction of ε_t of particulates with temperature until sintering and fusion occurs. On sintering, ε_t increases. Chemical composition determines whether sintering and fusion will take place. Samples with low silica and alumina and high iron oxide and condensible salts are prone to sintering and fusion.

(b)   A systematic increase of ε_t with increase in particle size prior to sintering.

(c)   An increase of ε_t with iron content (Fe_2O_3) prior to sintering. Measurements on slag layers by Markham et al. (figure 3) suggest that the emittance attained on complete fusion (noted in (a) above) is likely to exceed 0.9. As may be seen in figure 2, sample no. 5 (12.1% Fe_2O_3, 6.6% SiO_2 2.3% Al_2O_3, 71.4% $CaSO_4$) reported by Brajuskovic et al. (1991) recorded normal emittance in excess of 0.9 during the cooling cycle.

Emittance of deposits may also be affected by *dependent effects*, studies on which are scarce. The scattering and absorption are called *dependent* if the scattering and absorption characteristics of a particle in a medium are influenced by neighbouring particles. Two mechanisms are known to be responsible for dependent effects in radiation characteristics of a particulate system. The *first mechanism* is the inter particle effect in the *near field*, where the net incidence on a particle as well as its internal field is changed. This modifies the amount of radiation absorbed as well as the radiation scattered by each particle due to the proximity of other particles. The *second mechanism* is the interference among the scattered waves in the *far field* which affects only the scattering characteristics of the medium (Tien, 1988). Dependent effects have been analysed in the literature for either very small particles (Rayleigh size) or very large particles (Geometric size, 0.2 mm and above). No studies

336

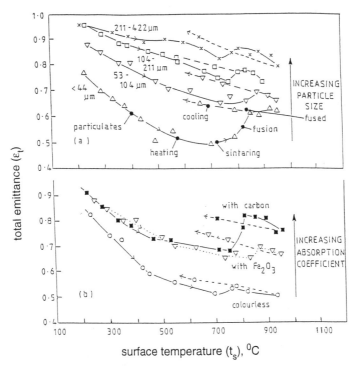

FIGURE 1. Influence of Particle Size and Heating on the Total Emittance of Particles Prepared by Crushing Synthetic Slag of 5% Fe₂O₃ (Boow and Goard, 1969).

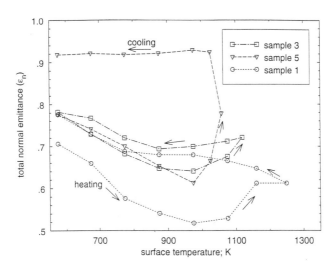

FIGURE 2. Influence of Heating on Total Normal Emittance of Ash Particles, (Adapted from Brajuskovic et al., 1991). Fe₂O₃ contents are: Sample 5: 12.1%
Sample 3: 4.1%
Sample 1: 9.7%
Particle Size: 44 - 350µm. Same Sample Designation as Uued by the Authors.

appear to have been reported for the intermediate size range of particles typical of the wall deposits in pf fired furnaces, although an analysis can in principle be made from the Maxwell equations of electromagnetic theory (Bohren and Huffman, 1983). However, the net effect is believed to be an increase in the absorption of incident radiation and hence an increase in emittance to an unknown extent.

3. Predictions of Emittance of Smooth Slag and Particulate Layers

3.1 *Properties of Samples Used in Model Prediction*

Complex refractive index data of the following samples (Goodwin , 1986) were used in model predictions. These samples, prepared from melting oxide mixtures and grinding thereafter, have composition similar to flyash generated in pf fired furnaces, and shown in table 1.

Table 1

Composition of Sample Materials Used in Model Predictions

(Goodwin, 1986)

Sample	SiO_2	Al_2O_3	CaO	Fe_2O_3
SA05	53.53	26.32	15.50	4.65
SA10	50.64	25.04	14.82	9.51

3.2 *Emittance of Opaque slag with Smooth Surface*

For a melted and solidified slag layer, an approximation to an optically smooth surface may be postulated. A surface is smooth if $h < \lambda/(8\cos\theta)$ where h is the roughness height, θ is the angle of incidence and λ is the wavelength of incident radiation (Bohren and Huffman, 1983). If the surface is not smooth, the incident beam may be diffusely reflected over a distribution of angles, rather than specularly reflected at a single angle. For normal emittance the following expression from Siegel and Howell (1992) may be used :

$$\varepsilon_\lambda = 1 - [(n_\lambda - 1)^2 + k_\lambda{}^2] / [(n_\lambda + 1)^2 + k_\lambda{}^2] \quad \text{where,}$$

n and k are the spectral real and absorption index of the sample material.. The hemispherical emittance was calculated by integrating the directional spectral values corresponding to an unpolarised beam over a hemispherical envelope covering the surface (Siegel and Howell, 1992), details of which are available in Bhattacharya (1995). Results of the calculation are presented in figure 4. Since such slabs do not scatter and the slag is opaque, the predictions are insensitive to the absorption index values ie., composition of the sample. The emittance values are mostly grey as may be seen, and the ratio ($\varepsilon_h / \varepsilon_n$) is relatively constant at a value of about 0.92. The predicted normal emittance values are closely similar to the values reported by Markham et al. (1992) from their measurements, and reproduced in figure 3, with slight underprediction may be attributed to the surface roughness which may still exist.

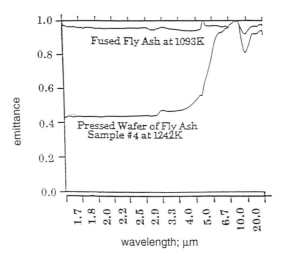

FIGURE 3. Difference in Character of Spectral Normal Emittance of Slagged and Particulate Deposits, Experimental Data by Markham et al., (1992).

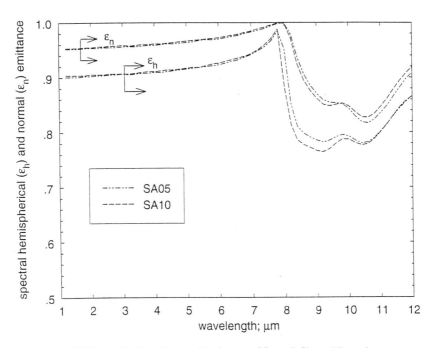

FIGURE 4 Predicted Spectral Emittance of Smooth Slagged Deposits.

3.3   Prediction for Opaque Particulate Layers

A layer is modelled to be one dimensional, and the discrete ordinate approximation for radiative transfer is obtained by discretising the directions of transfer into 16 directions, with weighting fractions corresponding to the solid angle represented (Chandrasekhar, 1958). Both spectral normal and hemispherical emittances are calculated, details of which are available in Bhattacharya (1995). The following assumptions are made in the model :

(i)   the particles are spherical in shape
(ii)   all particles have same chemical composition in the sense they have same complex refractive index
(iii)   only independent and multiple scattering is considered; no possible dependent effects due to the proximity of the particles have been taken into account.

The predictions have been presented in figure 5 for spectral normal and hemispherical emittances and for three particle sizes of 2 , 10 and 100 μm.

Effect of Size

The emittance of particle layers have a distinct spectral character. With particle size, the emittance increases. For small particles and samples of low absorption index (such as SA05), the emittance could be as small as 0.2 at low wavelengths. This is because of the fact that small particles reflect a significant portion of the incident radiation while the large particles absorb most of it. Consequently, layers of small particles which are typical of those formed during the initiation of deposit formation, are more reflective *ie* less emissive. For the weakly absorbing sample SA05, even 100 μm particulate layer shows an emittance of around 0.4 at low wavelengths. The spectral character of emittance associated with particulate deposits is supported by the measurement reported by Markham et al. (1992) and shown in figure 3.

Effect of Composition

Figure 5 also shows the effect of composition on spectral emittance of opaque deposits. Samples of higher iron content recorded higher emittance mainly because of the higher absorption index associated with these samples. Both the effects are limited to the same wavelength region below 6 μm, which is also important for high temperature combustion systems so that for predictions of emittance of particulate deposits, knowledge of both size and optical properties are essential.

3.4   Emittance - Dependence on Physical Character of Deposits

As is evident from figures 4 and 5, there is a distinct difference between the nature of the spectral emittance of particulate and slagged deposits. Particulate deposits are characterised by a low emittance at low wavelengths up to 4 μm. A sharp rise occurs thereafter up to 8 μm before decreasing again. Slagged deposits on the other hand show little variation with wavelength and remain high throughout. Thus the nature of the measured emittance of a deposit is a distinctive pointer to the physical character of the deposit. If the deposit is neither completely slagged nor completely powdery, the emittance is expected to have an

340

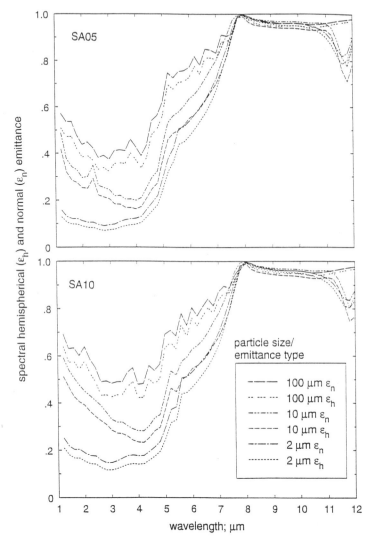

FIGURE 5. Predicted Effect of Particle Size on Spectral Emittance of Opaque Particlulate Deposit. (Sample Designation as Per Table 1).

intermediate value. This is because the rough surfaces of such deposits could result in higher reflection and hence lower emission compared to a smooth surface but not quite to the extent of a completely powdery surface. For the range of iron content investigated, the effect on emittance of a melted slag and particulate layer is relatively insignificant compared to the particle size. The physical state of the deposit therefore has the greatest influence on the emittance, with deposit chemistry being to determine the state.

4. Experimental Work

The experimental equipment is located at the Division of Coal and Energy Technology, CSIRO at North Ryde in Sydney. It has primarily been used to study the structural changes occurring in coal and other minerals on heating (Vassallo et al., 1992; Vassallo and Finnie, 1992).

The infrared emission cell consists of a modified atomic absorption graphite rod furnace and is illustrated in Figure 6. The furnace was driven by a thyristor controlled ac power supply capable of delivering up to 150 amps at 12 volts. A platinum disk (6 mm diameter), which acts as a hotplate to heat the sample, is placed on the graphite rod. An insulated 125 μm type R thermocouple was embedded inside the platinum in such a way that the thermocouple junction was < 0.3 mm below the surface of the platinum (fig. 6 inset). Temperature control of $\pm$ 3°C at the maximum operating temperature of 1500°C was achieved by using a Eurotherm Model 808 proportional temperature controller, coupled to the thermocouple.

An off-axis paraboloidal mirror with a focal length 25 mm mounted above the heater captured the infrared radiation and directed it into a Digilab FTS-60A Fourier transform infrared spectrometer. The heater assembly was located so that the surface of the platinum was slightly above the focal point of the off-axis paraboloidal mirror. With this geometry, a spot of approximately 3 mm diameter is sampled by the spectrometer. The modifications to the spectrometer involved the removal of the aperture assembly and replacing it with a mirror. Reflection of modulated radiation back from the interferometer to the sample with subsequent additional modulation has been identified as a problem with this type of measurement. Very little distortion of the spectra due to this effect is observed however, possibly because of the low reflectivity of the powdery samples used in the experiments. A room temperature DTGS detector was used and as a result is not responsive to emission from surrounding objects. The entire spectrometer and sample furnace was covered with a perspex box which was purged with nitrogen to remove any IR absorption by water vapour and carbon dioxide in the path of the sample emission.

4.1   Samples Used and Their Preparation

Two different particulate samples were used in the measurement. These are slag samples ground after melting oxide mixtures in proportion similar to those found in power station ash. The details of the procedure is available in Bhattacharya (1995). Particles of such samples are expected to be of uniform composition which is presented in table 2. Slag sample S2 (2% Fe_2O_3) was separated into two size groups, one having a size range of 5.6 - 35.4 μm (Sauter mean diameter of 12.9 μm), the other being sieve sized 45 - 53 μm. One

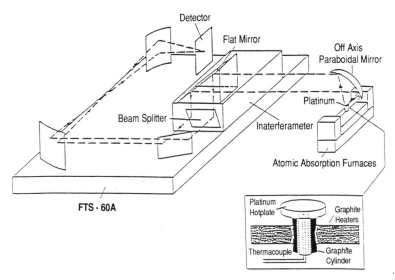

FIGURE 6. Schematic Diagram of the Spectral Emission Cell and the Modified FT-IR Spectrometer. The Details of the Platinum Hotplate and the Thermocouple are Shown in the Insert. (Reproduced with Permission from Applied Spectroscopy).

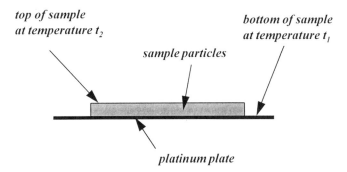

FIGURE 7. Sample Layer on the Platinum Plate and Associated Temperatures.

343

size group of the slag sample S10 was used in the experiment, with a size range of 6.8 - 33.5 μm and Sauter mean diameter of 14.9 μm.

Table 2
Composition of the Samples Used in the Experiment

Sample	SiO_2	Al_2O_3	CaO	Fe_2O_3
S2	54.95	27.25	15.80	2.00
S10	50.46	24.87	14.67	10.00

4.2   Experiment Procedure

The sample to be analysed is carefully spread over the platinum surface. The sample heated to the desired temperature , was held at this temperature while accumulation of single beam spectrum took place. The temperature range during the measurements varied between 600°C and 1200°C at an interval of 200°C. The spectra were recorded with the use of 64 scans at a nominal resolution of 4 cm^{-1}. After accumulation (normally less than one minute) of the single beam spectrum, the sample was heated to the next desired temperature for acquiring spectra again.

A graphite plate with the same geometry as that of the platinum plate was used to approximate a blackbody source. Spectra from the graphite plate were acquired at the same temperatures as those of the samples and stored for later use. For the platinum plate alone, spectra were acquired at the same temperatures as those of the samples and stored for later use.

The samples were first subjected to transmission measurements before undertaking emission experiments. Thus, for the same sample, two sets of measurements ie. transmission and emission were carried out. An infrared microscope was used for the purpose and the sample particles were spread horizontally over a 6 mm diameter area of a KBr window.

4.3   Calculation Procedure

For a sample particulate layer distributed on the platinum plate during measurement (as shown in figure 7) emittance was determined as follows :

$$\varepsilon_{s,\lambda} = \left(\frac{I_{s, t2} - \tau \times I_{bg, t1}}{I_{bb, t2}} \right) \times \varepsilon_{g,\lambda} \qquad (1)$$

where,

I_s = the intensity from the sample, as measured
I_{bg} = the intensity from the platinum plate, as measured
I_{bb} = the intensity from the graphite plate, as measured
τ = transmission across the sample, average of three measurements from the IR microscope

344

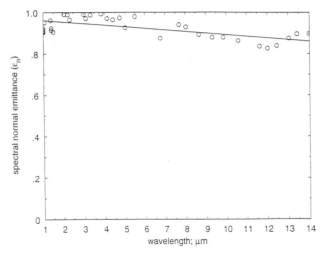

FIGURE 8. Spectral Normal Emittance of Graphite. (Toloukian and Ho, 1989).

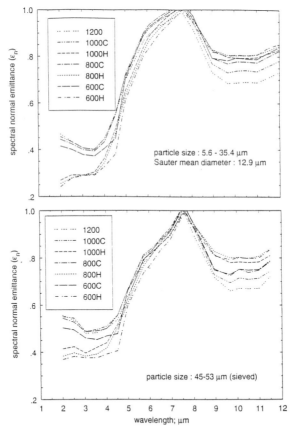

FIGURE 9. Measured Effect of Particle Size and Heating on Spectral Emittance of Semi-Transparant
Particulate Deposit Sample S2: Sample Designation as Per Table 2. Code Indicated
Temperature (°C) and0 Whether Measurements were Made During Heating (H) or Cooling (C)
Cycle.

ε_g = spectral emittance of graphite, from Toloukian and Ho (1989), reproduced in figure 8.

The temperature t_1 is measured during the experiment, whereas t_2 depends on the thermal conductivity (k) and thickness (t) of the sample layer, and radiative (and/or convective) loss from the surface. The transmittance (τ) depends primarily on the layer thickness. A procedure to estimate the top temperature t_2 and its effect on sample emittance is outlined in Bhattacharya (1995). It was observed that for the sample mass used in the experiment, the temperature t_2 was not significantly below the temperature t_1, and therefore, $I_{bb,t1}$ (which was measured) instead of $I_{bb,t2}$ was used in equation (1) to estimate the sample emittance. Relevant calculations have been presented and the error in sample emittance resulting from this assumption is discussed in Bhattacharya (1995).

The fraction in the parenthesis (in equation 1) represents the sample emittance relative to the emittance of the graphite plate. Therefore, the right side of the equation is multiplied by the spectral emittance of graphite plate.

For opaque sample, transmittance across the layer is zero so that the equation (1) reduces to

$$\varepsilon_{s,\lambda} = \left(\frac{I_{s,t2}}{I_{bb,\,t2}} \right) \times \varepsilon_{g,\lambda} \qquad (2)$$

For all the samples, concentration was such that the transmittance across the layers were between 5 and 10%. The contribution from the intensity emitted from the platinum plate was taken into account using equation (1).

4.4   Estimation of Error

The error estimation procedure has been formulated follwing Coleman and Steele (1989) The sources of uncertainties involve uncertainties in transmission measurement, uncertainties in the spectral emittance of the graphite plate, uncertainties in measurement of the intensity form the graphite plate, platinum plate and the samples. The details are presented in Bhattacharya (1995). The overall error in the measured emittance is estimated to be ±0.06.

5.   Experimental Results and Discussion

5.1   Effect of Size

Slag sample S2 (2% Fe_2O_3) was separated into two size groups, one having a size range of 5.6 - 35.4 μm (Sauter mean diameter of 12.9 μm), the other being sieve sized 45 - 53 μm. Measurements were performed between 600 and 1200°C. Spectra were acquired during heating and cooling cycle. Results are presented in figure 9.

In both figures, particulate character, as predicted in section 3.1, of the emissive layer is evident as also the region of *Christiansen effect* between 7 and 8 μm wavelength. The slag particles are expected to be of homogeneous composition because of melting during preparation, and therefore, figure 8 shows the effect of size on the measured emittance. For reasons explained in section 3, the larger particles recorded higher emittance. As indicated earlier in section 4.3, transmittance across the layers varied between 5 and 10%. These measurements indicate that spectral emittance of particulate layers could be as low as 0.3 at low wavelengths, supporting predictions made in section 3 and industrial measurements reported by Carter et al. (1992). The effect of particle size is primarily limited to the wavelength region below 6 μm, supporting the predictions made in section 3.

In a pf fired furnace, low wall emittance such as 0.3 may result in a decrease in heat absorption and therefore in an increase in furnace exit temperature. The approximate changes in furnace exit temperature may be estimated by a well-mixed furnace model (Hottel and Sarofim, 1967) which has also been used by Richter et al. (1986). The model assumes an uniform build-up of ash deposits over the complete furnace walls, and requires, among others, total absorbance of the wall deposits as input. Spectral absorbance of a deposit is equal to the spectral emittance and the total absorbance of a deposit may be estimated once the spectral emittance and the flame temperatures are known (Wall et al., 1993). It is estimated that for a 660 MWe pf fired furnace, a change in wall emittance from 0.7 (clean tube) to a low value of 0.3 affects the furnace heat absorption in a way that the exit gas temperature increases by around 150^0C (Bhattacharya, 1995).

A brief comparison of these measured values with those by Boow and Goard (1969) may be made. Total normal emittance calculated from the spectral emittance of the 45-53 μm sized S2 particles at 600°C during heating cycles is 0.46 and 0.57 during the cooling cycle. The sample was heated to 1400°C during the heating cycle before cooling. Therefore, it had sintered and resulted in higher emittance during the cooling cycle. In figure 1, total emittance values reported by Boow and Goard (1969) for particle sizes < 44 μm are 0.5 and 0.65 respectively during the heating and cooling cycles. The sample of Boow and Goard (1969) had 5% Fe_2O_3 as against 2% in the sample of the present study. A detailed analysis of the results from the present study and comparison with the total emittance reported in literature is in progress and will be the part of a future publication.

More experiments involving larger particles are recommended in order to ascertain the particle size which results in emittance being independent of size. Slagged particles may preferably be used in such experiments to eliminate any effect of variation in composition among particles.

5.2   *Effect of Iron Content*

Figure 10 shows the effect of iron content on the emittance of particulate deposits. The ground particles of samples S2 and S10 had Sauter mean diameters of 12.9 μm and 14.9 μm respectively. Measurements carried out during heating cycle between 600°C and 1200°C are reported. Both these samples had comparable amounts of silica, alumina and calcium oxide, and also particle sizes of similar magnitude, so that the effect noticed on emittance is predominantly due to the variation in iron content. As is evident, sample S10 with the higher iron content gave a higher emittance. The effect of iron is limited to wavelengths below 5 μm, supporting predictions made earlier in section 3. The increase in emittance with iron content is due to the higher absorption index associated with high iron bearing

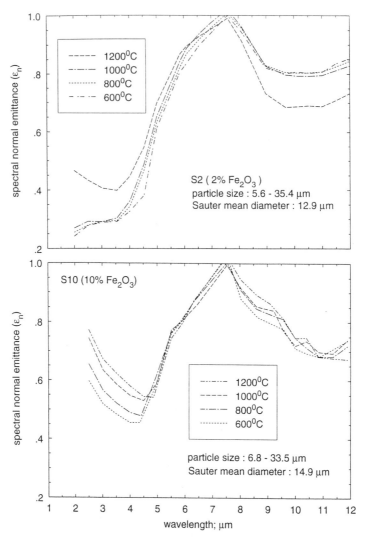

FIGURE 10. Measusred Effect of Fe₂O₃ Content on Spectral Emittance of Particulate Deposits; Sample Designation as per Table 2. Measurements Made During Heating Cycle.

samples (Goodwin, 1986) at low wavelengths. As may be seen from figure 10, sample S10 showed a sharp drop in emittance at all temperatures going from 2 to 4 μm wavelength, a trend similar to the prediction made in figure 5 for sample SA10 of particle sizes 10 μm and above. During the measurements, the sample S10 (figure 10) recorded an increase in emittance with temperature starting at 800°C. This was not noticeable for the sample S2 (figure 10) which started recording higher emittance above 1000°C and had slightly higher silica and one-fifth of the iron content in S10. This is expected since samples with higher iron content are expected to sinter earlier. Thus effect of iron content on emittance is possibly due to both the reasons, *ie.*, higher absorption index associated with higher iron content, and sintering/fusion at lower temperatures.

5.3 *Effect of Heating*

Figures 9 and 10 also show the effect of heating on the spectral emittance of the particulate layers. Emission spectra were acquired during heating as well as during the cooling cycle. For both sizes, at 1000°C and above, a rise in emittance is evident. Also evident is the increased emittance at a particular temperature during cooling cycle as compared to its value during heating. Clearly these increases are due to thermal transformations (sintering) of the sample during heating. The increase in emittance during cooling cycle are more significant at lower temperatures as may be seen from figure 9. Furnace deposits exposed to higher temperatures for a long time are thus expected to produce higher emittance due to thermal transformations that they undergo.

These results support the observations made by Boow and Goard (1969) and Brajuskovic (1991) that at surface temperatures in excess of 700^0C, total emittance increases irreversibly as particles are heated and then cooled. In their experiments, they used opaque deposits having thickness of 5 mm or more. Thus a surface temperature of 700^0C might correspond to a bottom temperature of 1100^0C or more as temperature gradients of 100^0C/mm are known to exist in particulate ash deposits (Anderson et al., 1987). The increase in emittance during cooling cycle in the present experiment may be attributed to irreversible structural change associated with higher temperature (Baxter et al., 1993), and /or the onset of sintering and fusing which is known to result in increase of emittance (Boow and Goard, 1969; Markham et al, 1992).

6   **Conclusions**

The emittance of ash deposits in pf fired furnaces depends on the physical and chemical character of deposits. Emittance of slagged layers is grey with values expected to exceed 0.9. Emittance of particulate layers on the contrary is highly spectral. The measurements and theoretical predictions show that fine particulate layers could be reflective, with emittances of less than 0.3 specially at wavelengths below 4 μm, as has been reported by industrial on-line measurements. Such layers are found during the initiation of deposits. Layers of coarse particles are predicted to have higher emittance, a trend matched by measurements.

The effect of iron content in ash on emittance of deposits is found to be significant, with samples having higher iron content recording higher emittance. Emittance of the particulate layers was observed to increase with heating presumably because of the thermal transformations they undergo. Measurements during the cooling cycle of the same layer indicated higher emittance compared to the measurements during the heating cycle. This is due to the sintering of the layer at high temperature.

All effects (particle size, physical state, heating, composition) on emittance of the deposits are limited primarily to the wavelength region below 6 μm. This region is also of primary interest to radiative heat transfer in pf fired furnaces. This clearly dictates that care must be exercised in either measurements or predictions of spectral emittance of deposits. Emittance of a deposit depends on strongly coupled ways with its thermal conductivity and also on the composition, porosity and physical structure. Further measurements using a wider range of samples and particle sizes are recommended to examine the effects in detail. Slagged particles may preferably be used in such experiments to eliminate any effect of the variation in composition among particles.

References

Anderson, D. W., Viskanta, R., and Incropera, F. P. (1987), *Effective Thermal Conductivity of Coal Ash Deposits at Moderate to High Temperatures,* J. of Eng. Gas Turbines Power, 109, pp. 215-221.

Baxter, L. L. (1993), *In Situ Real Time Emission FTIR Spectroscopy as a Diagnostic for Ash Deposition During Coal Combustion,* Engineering Foundation Conference, Solihull, UK, June 20-25, 1993.

Bhattacharya, S. P. (1995), *The Radiative Properties and Thermal Effects of Ash Clouds and Deposits In Pulverised Fuel Fired Furnaces,* PhD Thesis, The University of Newcastle, Australia.

Bohren, C. F. and Huffman, D. R., (1983), *Absorption and Scattering of Light by Small Particles*, John Wiley and Sons, USA.

Boow, J. and Goard P. R. C. (1969), *Fireside Deposits and their effect on heat transfer in a pulverised fuel fired boiler*, Fuel, pp 412.

Brajuskovic, B., Uchiyama, M., and Makino, T. (1991), *Experimental Investigation of Total Emittance of Power Plant Boiler Ash Deposits,* Experimental Heat Transfer, Fluid Mechanics, and Thermodynamics (Keffer J. F., Shah R. K., and Ganic E. N. , eds.,), Elsevier.

Carter, H. R., Kokdsal, C. G., Garabrant, M. A. (1992*), Furnace Cleaning in Utility Boilers Burning Powder River Basin Coals*, International Power Generation Conference, Oct. 18-22, 1992, Atlanta, Ga.

Chandrasekhar, S. (1960), Radiative Transfer, Dover Publications, NY.

Coleman, W., and Steele, G. (1989), Experimentation and uncertainty Analysis for Engineers, John Wile, NY.

Goodwin, D. G. (1986), *Infrared Optical Constants of Coal Slags* , PhD thesis, Stanford University.

Hottel, H. C. and Sarofim, A. F. (1967), Radiative Transfer, McGraw Hill

Markham J. R., Solomon P. R., Best P. E., Yu Z. Z. (1992), *Measurement of Radiative Properties of Ash and Slag by FT-IR Emission and Reflection Spectroscopy*, ASME J. of Heat Transfer, vol 114, pp 458-464.

Richter W., Payne P., Heap M. P. (1986), Mineral Matter and Ash in Coal (ed. Karl Vorres), American Chemical Society, Washington DC, pp 375-383.

Siegel R., and Howell J. R. (1992), Thermal Radiation Heat Transfer, Hemisphere Publishing, Washington.

Tien C. L. (1988), *Radiative Heat Transfer in Packed and Fluidised Beds*, J. of Heat Transfer, 110, pp 1230-1241.

Touloukian Y.S. and Ho C. J. (1989), *Thermal Radiative Properties of Nonmetallic Solids*, vol. 8, Plenum Press, New York.

Vassallo, A. M., Cole-Clarke, P. A., Pang, L.S. K., and Palmisano, A. J. (1992), *Infrared Emission Spectroscopy of Coal Minerals and Their Thermal Transformations,* Applied Spectroscopy, 46(1), 73-78.

Vassallo, A. M. and Finnie, K.S. (1992). *Infrared Emission Spectroscopy of Some Sulfate Minerals*, Applied Spectroscopy, 46, 10, 1477-1482.

Wall, T. F., and Becker, H. (1984), *Total Absorptivities and Emissivities of Particulate Coal Ash from Spectral Band Emissivity Measurements*, J. of Engg. for Gas Turbines and Power, 106, pp 771.

Wall, T. F., Bhattacharya, S. P., Zhang, D. K., Gupta, R. P., He, X. (1993), *The Properties and Thermal Effects of Ash Deposits in Coal Fired Furnaces* , Progress in Energy and Combustion Science, *19*, pp 487-504.

AGGLOMERATION AND DEFLUIDIZATION IN FBC OF BIOMASS FUELS - MECHANISMS AND MEASURES FOR PREVENTION

Anders Nordin and Marcus Öhman

Energy Technology Centre in Piteä
Department of Inorganic Chemistry
University of Umeä, Sweden

Bengt-Johan Skrifvars and Mikko Hupa

Department of Chemical Engineering
Åbo Akademi University, Turku, Finland

ABSTRACT

The use of biomass fuels in fluidized bed combustion (FBC) and gasification (FBG) is becoming more important because of the environmental benefits associated with these fuels and processes. However, severe bed agglomeration and defluidization have been reported due to the special ash forming constituents of some biomass fuels. Previous results have indicated that this could possibly be prevented by intelligent fuel mixing. In the present work the mechanisms of bed agglomeration using two different biomass fuels as well as the mechanism of the prevention of agglomeration by co-combustion with coal (50/50 %$_w$) were studied. Several repeated combustion tests with the two biomass fuels, alone (Lucerne and olive flesh), all resulted in agglomeration and defluidization of the bed within less than 30 minutes. By controlled defluidization experiments the initial cohesion temperatures for the two fuels were determined to be as low as 670 °C and 940 °C, respectively. However, by fuel mixing the initial agglomeration temperature increased to 950°C and more than 1050° C, respectively. When co-combusted with coal during ten hour extended runs, no agglomeration was observed for either of the two fuel mixtures. The agglomeration temperatures were compared with results from a laboratory method, based on compression strength measurements of ash pellets, and results from chemical equilibrium calculations. Samples of bed materials, collected throughout the experimental runs, as well as the produced agglomerated beds, were analysed using SEM EDS and X-ray diffraction. The results showed that loss of fluidization resulted from formation of molten phases coating the bed materials; a salt melt in the case of Lucerne and a silicate melt in the case of the olive fuel. By fuel mixing, the in-bed ash composition is altered, conferring higher melting temperatures, and thereby agglomeration and defluidization can be prevented.

Applications of Advanced Technology to Ash-Related Problems in Boilers
Edited by L. Baxter and R. DeSollar, Plenum Press, New York, 1996

353

INTRODUCTION

The use of traditional and "new" biomass fuels for both heat and electricity production is becoming more important because of the environmental benefits associated with these fuels. The most promising energy conversion technologies are based on fluidized bed combustion (FBC) or gasification (FBG). These processes enable higher electrical and total efficiency as well as greater flexibility concerning the use of various fuels. In addition, the relatively low combustion temperatures reduces the risk of operating problems due to the special ash forming constituents of the biomass fuels. A large number of FBC boilers in Sweden and Finland are presently using biomass fuels with very little or no problem. However, the ash characteristics vary considerably between different biomass fuels [Nordin 1994, Skrifvars and Hupa 1995] and severe bed agglomeration has been reported from FBC and FBG of several types of biomass fuels [Nordin 1995, Salour et al. 1993, Viktorén 1991, Bitowft and Bjerle 1986, Gulyurtlu et al. 1991].

In a recent work the importance of fuel mixing to prevent both harmful emissions and agglomeration was indicated [Nordin 1995]. When coal was co-combusted with an agricultural fuel (Lucerne) in a small pilot scale FBC, an optimal inherent sulphur retention of more than 90 % was obtained. No signs of agglomeration were observed whereas combustion tests of Lucerne alone, resulted in agglomeration and defluidization of the bed within less than one hour of normal operation. Control of in-bed agglomeration by fuel blending therefore seems to be an interesting approach and has also previously been demonstrated by Salour et al [1993]. However, despite some research work made in recent years, of which most has concerned coal combustion, the mechanisms of agglomeration and defluidization are not fully understood. The purpose of the present work was to present a closer study of the mechanisms of both bed agglomeration using two different biomass fuels and prevention of bed agglomeration by co-combustion of these fuels with coal.

EXPERIMENTAL

Fuels and bed material used

The fuels chosen for the present study were the previously studied agricultural fuel (Lucerne), an imported biomass fuel (olive flesh) and coal. Large scale test runs using the olive fuel in a pulverized fuel boiler showed extreme slagging [Höglund et al. 1995] and interest was shown in using the fuel in a FBC instead. The biomass fuels were pelletised to a diameter of 8-10 mm and the coal was crushed and sieved to sizes of 5-15 mm. The characteristics of the fuels and fuel ashes are presented in Table 1.

Ashing of the fuels preceding elemental analysis was performed at 550 °C to reduce the loss of volatile ash elements. Molar compositions of the ashes, in forms as predicted by chemical equilibrium calculations for ashing conditions, are illustrated in Fig 1. All fuels were sieved before the experiments so that the fraction of fines could be kept to less than 2 % after the feeding. The bed material (quartz sand, 99.9%) was initially sieved to sizes between 200 and 250 um.

Table 1 Fuel characteristics (wt%)

	Lucerne	Olive flesh	Coal	Lucerne/Coal	Olive/Coal
Dry substance	92.1	85.6	92.9	92.5	89.3
Ash[a]	8.5	9.9	6.6	7.5	8.2
HV (MJ/kg$_{Fuel}$)	16.1	16.0	27.2	21.6	21.6
C[a]	46.7	50.2	77.3	62.0	63.8
H[a]	5.9	6.3	4.9	5.4	5.6
N[a]	3.1	1.4	1.3	2.2	1.4
S[a]	0.16	0.14	0.74	0.45	0.44
Cl[a]	0.35	0.15	0.08	0.22	0.12
O[a]	35.6	32.1	6.6	21.1	19.4
SiO_2[b]	3.9	36.2	43.6	23.8	39.9
Al_2O_3[b]	0.4	3.6	32.9	16.7	18.3
Fe_2O_3[b]	0.3	4.25	11.0	5.7	7.6
MgO[b]	3.2	12.4	1.8	2.5	7.1
CaO[b]	28.2	18.2	4.3	16.3	11.3
K_2O[b]	26.5	18.2	1.8	14.2	10.0
Na_2O[b]	0.8	1.7	0.8	0.8	1.3
P_2O_5[b]	8.2	4.0	1.9	5.1	3.0

[a]wt% dry matter
[b]wt% of ash

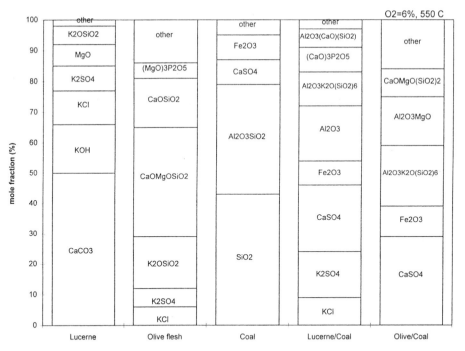

Figure 1 Illustration of the ash composition (550°C), as predicted by chemical equilibrium calculations

Pilot scale FBC tests

A pilot scale FBC reactor is used to study bed agglomeration and defluidization in a highly controlled way and under homogeneous conditions. The objects of the tests are; i) to quantify agglomeration tendencies of biomass fuels, as well as for blends of different fuels; and ii) to study the mechanisms of both agglomerate formation and prevention of agglomeration by co-combustion. The reactor is utilised in either of two operational modes. By normal combustion operation, a relatively accurate simulation of a full-scale case is accomplished. However, these tests are not suited for comparisons as the conditions are generally inhomogeneous and the actual temperature of the agglomeration cannot be determined. Our other mode of operation is therefore initiated by loading of the bed with a certain ash to bed material ratio, during normal combustion at a sufficiently low temperature. Then the fuel feeding is stopped and the bed is externally heated in a homogeneous and controlled way. The agglomeration/-defluidization temperature is taken as a measure of the agglomeration/defluidization tendency for that specific fuel or fuel blend. Samples of bed and ash material for evaluation of mechanisms are collected periodically during the operation.

The pilot FBC reactor

A small pilot scale FBC reactor (5kW), previously constructed in order to facilitate well designed experiments under highly controlled conditions, is used (Fig. 2). The reactor was constructed from stainless steel, being 2 m high, 100 mm and 200 mm in bed and freeboard diameters, respectively.

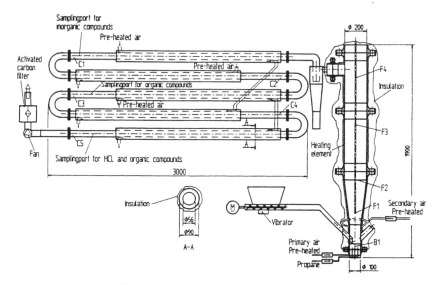

Figure 2 Schematic view of the pilot scale fluidized bed reactor

Bed temperatures at four locations in the bed are measured by shielded thermocouples (type S) and four differential bed pressures are determined by differential pressure transmitters. To minimise the significant influence of cold walls in such a small scale unit, the reactor is equipped with electrical wall heating elements. To allow for a constant increase of the bed temperature with a homogeneous temperature profile, much effort was spent on improving the bed section of the reactor as well as the controlling system (Fig. 3). A preheater, allowing primary air temperatures of more than 1050 $^\circ$C, was constructed. Forced convection is utilised in a cyclone-like stainless steel cylinder equipped with Kanthal electric heating elements. All temperatures are controlled by Eurotherm temperature controllers and the maximum temperature deviations within the bed were found to be less than $\pm$ 3 $^\circ$C. The gases O_2, CO_2, CO, SO_2, NO and NO_2 are continuously monitored, using conventional instruments.

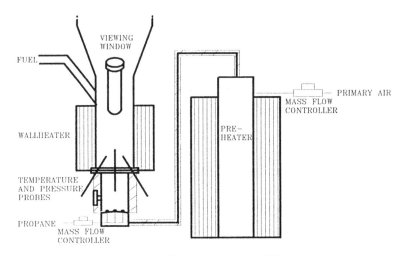

Figure 3 Detailed illustration of the modified bed section

The previous construction work, the controlling possibilities, as well as the analytical details of the reactor have been reported elsewhere [Nordin et al. 1995].

Normal combustion tests
To initially determine the tendencies for bed agglomeration, several attempts were made to combust the two biomass fuels in the pilot FBC, using normal operating conditions. Totally three repeated experiments were performed for each fuel alone. A bed temperature of 850 $^\circ$C and an oxygen concentration of 6 %$_{dry}$ were used. Samples of the resulting bed agglomerates were analysed with SEM EDS.

To verify the previous findings of the possibilities to control bed agglomeration and defluidization by co-combustion with coal, four ten hour extended test runs were performed. Mixtures of 50%$_{wt}$ coal and 50%$_{wt}$ of either of the two biomass fuels were combusted under the same operating conditions as for the biomass fuels alone. Samples of the final bed materials were analysed with SEM EDS.

Controlled agglomeration tests
The most important individual parameter for bed agglomeration is the actual process temperature. However, the particle temperature in FBC is an unknown variable and

may exceed the bed temperature by 20-600 °C, the largest difference being found for the smallest particles and highest particulate phase oxygen concentrations [c.f. Yates 1983, Roscoe 1980, Halder and Basu 1991]. To avoid this uncertainty in the surface temperature of burning particles in FBC, controlled bed agglomeration tests were performed in the pilot FBC without actual combustion. Each bed experiment was started by loading the bed with ash (6%wt). Initial experimental runs showed that less then 1.5% ash is sufficient for agglomeration and defluidization to occur. In the present work, the loading temperature was chosen to be 600 °C for the biomass fuels and 750 °C for the coal blends. In addition, one experimental run was performed using 750 °C also for the olive fuel. Subsequent to this "ashing procedure" the fuel feeding was stopped and the bed temperature was externally raised by 3 °C per minute. The onset of bed agglomeration is indicated by measuring differential pressures and temperatures of the bed as well as by video filming of the process. The detection of initial bed particle cohesion by considering all bed related variables simultaneously is performed by on-line multivariate statistical process control (MSPC), i.e. on-line principal component analysis (PCA).

Throughout the experimental runs, samples of bed material were collected by an air cooled suction probe, equipped with a cyclone separator. These samples, as well as the produced agglomerated beds, were analysed by ICP, SEM EDS and X-ray diffraction analysis. The samples for SEM EDS were mounted in epoxy, cut by a diamond saw, polished and the resulting cross-sections were carefully examined. Prior to the X-ray analysis, the samples were sieved, to reduce the quartz content (increase ash reflections), and then ground.

Laboratory compression strength method

The agglomeration temperatures obtained in the pilot FBC were compared with the results from a laboratory method based on compression strength measurements of ash pellets, previously described by Skrifvars et al. [1992] and Skrifvars [1994]. The fuels were ashed at 550 °C and well homogenised samples of ash were crushed and screened. A narrow, controlled particle size fraction then was pressed to a pellet with a diameter of 10 mm and a height of 10 mm. These pellets were heat treated in a laboratory furnace under controlled conditions (gas atmosphere, different temperatures). After the heat treatment, the pellet samples were crushed in a standard crushing device. The compression strength was taken as a measure of the degree of sintering and the transition temperature as a measure of predicting the bed agglomeration temperature.

CHEMICAL EQUILIBRIUM CALCULATIONS

Agglomeration of ash and bed particles in a fluidized bed will take place if a liquid phase is present to form a bond between particles. Thus, the temperature at which a liquid phase would first form, i.e. the solidus temperature, and the effect on the solidus temperature of changing the ash composition, is of importance. Further, the smaller the difference between the solidus and liquidus temperatures, the greater the increase in the amount of liquid present for a small temperature rise. Therefore, the melt content of the ash, assuming chemical equilibrium, was determined as a function of temperature for the fuels and fuel mixtures used. The equilibrium calculations were performed using the computer program CHEMSAGE 3.0 and stoichiometric data were taken mainly from the Scientific Group Thermodata Europe (SGTE). Only stoichiometric condensed species were considered in these first calculations. Phase diagrams were then consulted for a rough estimate of the melting behaviour of the major ash constituents.

In a second set of calculations non-ideal solution models were also included. In addition to 110 stoichiometric species from SGTE, ten solid solutions and two liquid solutions (melts) were considered. Of the two liquid phases assumed, one comprises alkali salts, and the other oxides and silicates. As described previously by Nordin et al. [1994b] this is not fully correct but probably an applicable approach, considering the limited data on liquid models available at present. The two phases rarely co-exist, as the liquid salt solution dominates at lower temperatures and the liquid silicate solution at higher temperatures.

RESULTS AND DISCUSSION

Normal combustion tests

The repeated experiments of normal combustion (850 °C, 6% O_2) of the two biomass fuels alone, all resulted in severe bed agglomeration within less then 30 minutes of operation. The progress of initial agglomeration to total defluidization was found to be very rapid. The time from the first signs of any disturbance of the bed (as indicated by MSPC) to total defluidization was estimated to only a few minutes. In a full scale process this would probably quickly lead to severe defluidization and catastrophic shut-down of the plant.

The results from SEM EDS analysis of the agglomerated beds of both fuels showed a clear assosiation of the elements Si, K, Al and Ca, indicating a potassium rich silicate layer, fusing the quartz particles together (Fig. 4). However, it should be mentioned that due to the agglomeration, loss of fluidization and heat transfer the final fusing temperature of the agglomerates are not known.

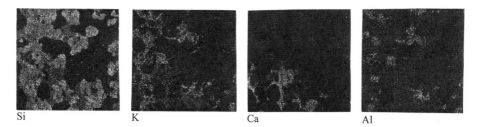

Si   K   Ca   Al

Figure 4 **Typical elemental composition of a bed agglomerate (Lucerne) after severe defluidization**

359

The experiments also confirmed that the surface temperature of a burning particle is significantly higher than the bed temperature. When the process was rapidly quenched after initial agglomeration, silicate aggregates with the same appearance as the original fuel pellets were found in the bed material, indicating that the agglomeration starts at the surface of the fuel particles due to the higher temperature there.

When co-combusted with coal, both biomass fuels could be successfully processed during more than ten hours (one working day) of normal combustion (850 $^\circ$C, 6% O_2). No tendency for agglomeration was discovered and, in addition, high sulphur retentions (> 80%) were obtained throughout the tests.

Controlled agglomeration tests of the biomass fuels

The principles of a typical controlled pilot FBC agglomeration test for the Lucerne fuel, as well as the corresponding X-ray maps of some of the periodically removed samples, are shown in Fig. 5. Examination of the maps shows continuous build up of a salt matrix (Ca, K, S, P, CO_3 and some Cl) in the bed. By the external increase of the bed temperature, the critical agglomeration temperature was determined to be 680 $^\circ$C. At this and higher temperatures, the salt ash matrix will partly melt and "glue" the bed material together. Agglomeration and defluidization in FBC of Lucerne under normal operating conditions would therefore probably be initiated by formation of a liquid (sticky) potassium-rich salt cohering the quarts particles.

Results from the controlled agglomeration tests of the olive fuel are presented in Fig. 6. For this fuel, a continious enrichment of mainly K and Ca (and Si) on the quarts surfaces was detected. A partial melt of a K-rich silicate phase, already covering the bed particles, is therefore the most likely active "glue" for this fuel. The initial agglomeration temperature was determined to 940 $^\circ$C. This temperature is in good agreement with the solidus temperatures of the corresponding area in the SiO_2-CaO-K_2O-phase diagram.

The results from repeated experiments with different ash to bed material ratios for the Lucerne fuel are presented in Fig. 7. Different ratios did not seem to influence the actual agglomeration temperature significantly. The spread in some of the data can be explained by the state of the experimental set-up during these first series of tests in our work. However, the process controlling possibilities were significantly improved during the course of the work and the six recent reproducibility tests, utilising constant operating conditions, indicated that repeated agglomeration temperatures within ±10 $^\circ$C can now be obtained.

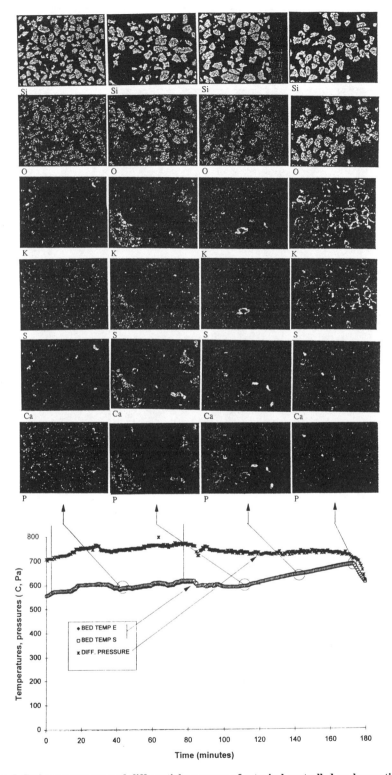

Figure 5 Bed temperatures and differential pressures of a typical controlled agglomeration run for the Lucerne fuel, as well as SEM EDS elemental maps of some of the collected samples

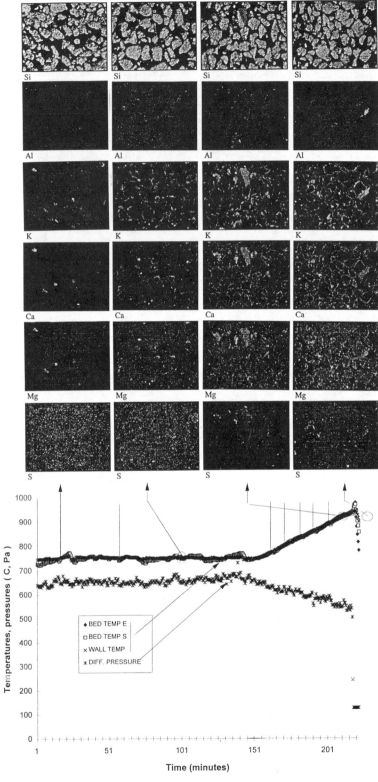

Figure 6 Bed temperatures and differential pressures of a typical controlled agglomeration run for the olive fuel, as well as SEM EDX elemental maps of some of the collected samples.

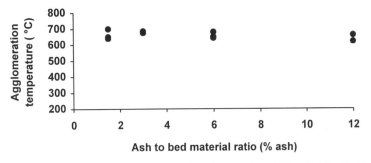

Figure 7 FBC agglomeration temperature as a function of ash to bed material ratio (Lucerne)

Controlled agglomeration tests of the fuel blends

The controlled agglomeration tests of the Lucerne/coal blend resulted in an agglomeration temperature of 950 °C, whereas the olive/coal blend did not agglomerate at any of the studied temperatures (< 1050 °C).

Laboratory compression strength method

The results from the laboratory compression strength method for the two biomass ashes are shown in Fig. 8. Sintering temperatures of about 650 and 900 °C were obtained for Lucerne and olive flesh, respectively and different ash to bed ratios did not influence the actual sintering temperature. A higher ash fraction in the bed results in a stronger agglomeration but does not significantly influence the sintering temperature. All these results are in good agreement with the results obtained from the pilot FBC.

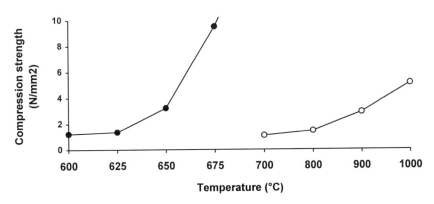

Figure 8 Compressive strength of heat treated ash pellets from Lucerne (closed symbols) and olive flesh (open symbols)

Chemical equilibrium calculations

The general results from the chemical equilibrium calculations verified the results obtained by SEM EDS analysis. The Lucerne fuel will predominantly form salts with low melting temperatures whereas the Olive fuel rather forms silicates with somewhat higher melting temperatures. Although a complete non-ideal data-set was not available for consideration in the calculations, the general results, as well as the trends obtained agree well with the experimental results obtained.

DISCUSSION

The results from the controlled agglomeration tests with the Lucerne fuel, where a molten salt matrix was identified as the "glue" in the agglomerate, seem to be inconsistent with corresponding results obtained during actual combustion, where a silicate "glue" was found. This is probably an effect of the actual agglomeration temperature and the following explanation is proposed. In a normal process with actual combustion, the salt matrix will melt and agglomerate. When these oversized agglomerates of particles exceed a critical dimension, the degree of mixing is reduced and higher temperatures will be reached locally, resulting in further agglomeration and finally complete bed defluidization. If not controlled, this self accelerated process will finally lead to very high temperatures, increased relative stability of silicate melt and thereby severe viscous sintering. Such transformations from salts to silicates at the interface between the quarts and the salt were identified using SEM EDX and high resolution line mapping over some of the agglomerates. A sufficiently high surface temperature of the fuel particles would also result in a silicate melt and viscous sintering.

The conclusion from the equilibrium calculations was that the salt and silicate liquid solutions rarely co-exists as salts are more stable at "low" temperatures whereas silicates dominate at "higher" temperatures. The transformation temperature is further highly influenced by the fuel compositions, or more correctly the amounts of available ash forming elements of the fuels. The dominant fraction of inorganic constituents of biomass fuels are organic or ionic bound and will be released in a highly reactive form during the combustion of the fuel [Nordin, 1993 and Baxter 1993]. The potential for chemical equilibrium calculations for prediction of the formation of inorganic ash products is therefore higher for biomass fuels than for coal. At the relatively low temperatures found in FBC, the organically bound ash forming elements are more likely to undergo major transformations than the mineral inclusions in the coal. It was therefore difficult to judge which elements should be regarded as available and which should be regarded as inert in the equilibrium calculations of the fuel blends.

The experimental results indicated that the explanation for prevention of agglomeration by co-combustion could be that the in-bed ash composition is altered towards higher melting temperatures. By the introduction of additional sulfur from the coal, somewhat higher melting temperatures (within the salt system) could be expected. However, supplementary combustion experiments with the biomass fuels and addition of SO_2 to the primary air rejected this hypothesis. A more likely explanation could be that the fluxing elements from the biomass fuel are chemically and/or physically adsorbed on the large total surface of the coal ash particles (clay minerals). This was also confirmed by adding corresponding amounts of kaolninite to the biomass fuels. Hereby, no agglomeration was experienced.

CONCLUSIONS

- The two biomass fuels studied, Lucerne and Olive flesh, will rapidly form agglomerates and cause severe defluidization when combusted in a fluidized bed. The chemical explanation is a coating of the bed material by sticky molten or partly molten ash layer, consisting of salts in the case of Lucerne and silicates in the case of the Olive fuel.

- Different ash to quartz ratios did not influence the actual agglomeration temperature significantly, but rather the physical effects of the agglomeration process. This is also in agreement with the compression strength method; a higher ash fraction in the bed results in a stronger agglomeration but does not influence the sintering temperature.

- When co-combusted with coal, agglomeration can be effectively prevented. The in-bed ash compositions in both cases are altered towards higher melting temperatures and the reason may be that the coal ash will act as a physical and/or chemical adsorbent for the "fluxing" elements in the biomass ashes. Other measures to reduce the agglomeration tendencies therefore also includes using coal minerals (e.g. kaolinite) as additives.

- A relatively good agreement was obtained between chemical equilibrium results, results from the compression strength method and actual bed agglomeration results from the pilot FBC. However, future validation work, involving improved non-ideal solution models and several different biomass fuels, will give more accurate and quantitative results.

ACKNOWLEDGMENTS

Financial support from the Swedish National Board for Industrial and Technical Development (NUTEK) and LIEKKI Combustion Research Program in Finland are gratefully acknowledged.

REFERENCES

Baxter, L. L. (1993). "Ash deposition during biomass and coal combustion: a mechanistic approach." *Biomass and Bioenergy 4,* 2, 85-102

Bitowft, B. K. and Bjerle, I. (1986). "A generic study of the sintering aspects of biomass in a fluid-bed gasifier." *Energy from Biomass and Waste XI*, 511-529

Dawson, M. R. and Brown, R. C. (1992). "Bed material cohesion and loss of fluidization during fluidized bed combustion of midwestern coal." *FUEL, 71,* 585-592

Gulyurtlu, I., Reforco, A. and Cabrita, I. (1991). "Fluidised bed combustion of corkwaste." *Fluidized Bed Combustion ASME,* 1421-1424

Halder, P. K. and Basu, P. (1991). "The temperature of burning carbon particles in a fast fluidized bed of fine particles." *Chem. Eng. Comm. 104*, 245-255

Nordin, A. (1993). "On the chemistry of combustion and gasification of biomass fuels, peat and waste - environmental aspects." Thesis, Dept. of Inorganic Chemistry, Umeå University

Nordin, A. (1994). "Chemical elemental characteristics of biomass fuels." *Biomass and Bioenergy, 6*(5), 339-347.

Nordin, A. (1995). "Optimization of sulfur retention in ash when cocombusting high sulfur fuels and biomass fuels in a small pilot scale fluidized bed." *FUEL, 74*, 615-622.

Nordin, A., Marklund, S., Wikström, E. (1995). "Construction of a pilot-scale fluidized bed combustor for combustion chemistry applications." Submitted for publication.

Höglund, K., Wikström, A. Hägerstedt, L-E., Medin, K., Hagström, U., Nordin, A. (1995). "Illustration of slagging and fouling problems during combustion of biomass fuels", Poster presentation at the Int. Conf. on Application of Advanced Technology to Ash-related Problems in Boilers, Waterville Valley, July 16-21.

Roscoe, J. C., Witowski, A.R. and Harrison, D. (1980). *Trans. I. Chem. Engrs., 58*, 69.

Salour, D., Jenkins, B. M., Vafaei, M. and Kayhanian, M. (1993). "Control of in-bed agglomeration by fuel blending in a pilot scale straw and wood fueled AFBC." *Biomass and Bioenergy, 4,* (2), 117-133

Skrifvars, B-J., Hupa, M. and Hiltunen, M. (1992). "Sintering of ash during fluidized bed combustion.", *I&EC Research, 31,* 1026-1030

Skrifvars, B-J. (1994). "Sintering tendency of different fuel ashes in combustion and gasification conditions." Thesis, Dept. of Chemical Engineering, Åbo Akademi University.

Skrifvars, B-J. and Hupa, M. (1995) *"Characterization of biomass ashes."* Proc. of the Int. Conf. on Application of Advanced Technology to Ash-related Problems in Boilers, Waterville Valley, July 16-21.

Viktorén, A. (1991). "Combustion of salix in a CFB-boiler." Report no. 416, Thermal Engineering Research Foundation, Stockholm, 36 p.

Yates, J. G. (1983). *Fundamentals of fluidized-bed chemical* processes. Butterworths, London

UTILIZATION OF ESTONIAN OIL SHALE AT POWER PLANTS

Arvo Ots - Thermal Engineering Department
Tallin Technical University, Tallinn, Estonia

ABSTRACT

Estonian oil shale belongs to the carbonate class and is characterized as a solid fuel with very high mineral matter content (60-70% in dry mass), moderate moisture content (9-12%) and low heating value (LHV 8-10 MJ/kg).

Estonian oil shale deposits lie in layers interlacing mineral stratas. The main constituent in mineral stratas is limestone. Organic matter is joined with sandy-clay minerals in shale layers.

Estonian oil shale at power plants with total capacity of 3060 MW_e is utilized in pulverized form. Oil shale utilization as fuel, with high calcium oxide and alkali metal content, at power plants is connected with intensive fouling, high temperature corrosion and wear of steam boiler's heat transfer surfaces.

Utilization of Estonian oil shale is also associated with ash residue use in national economy and as absorbent for flue gas desulphurization system.

1.   INTRODUCTION

There are some large thermal power plants burning oil shale in the Estonian Republic. Oil shale used at power plants belongs to the low-grade fuel group, and is characterized as fuel with high mineral matter content. Oil shale is peculiar to high calcium oxide, alkali metals' (mainly potassium) and chlorine content in the ash and moderate nitrogen quantity in organic matter.

Wide use of Estonian oil shale in power plants dates back to 1924 when Tallinn Thermal Power Plant was shifted to oil shale with 22 MW_e capacity. Steam boilers with grate furnaces were the most typical of that time.

The first PF oil shale boilers were the modernized coal boilers with 14-18 kg/s steam output and steam parameters 3.5-4 MPa and 420-450°C.

A new period in the development of oil shale power plants began in the years 1959-1960, when the first power units were applied at the Baltic Power Plant. The plant's designed capacity was 1624 MW_e. There are eight condensing turbines, each with 100 MW_e capacity and two back pressure turbines both with 12 MW_e capacity, receiving steam from 18 boilers with 61 kg/s steam output and steam parameters 9.8 MPa and 525°C. Each of the four 200 MW_e power units have two boilers with 89 kg/s steam output and parameters of superheated/reheated steam 13.8/2.2 MPa and 520/520°C.

In 1973 the Estonian Power Plant was run with 1610 MW_e. Seven turbines with 200

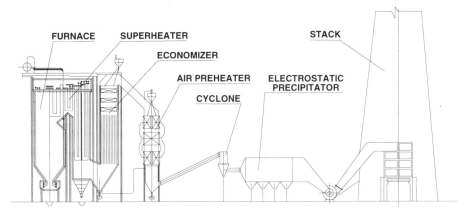

Figure 1. Scheme of the Estonian Oil Shale PF Boiler Flue Gas Ducts.

MW_e capacity and one 210 MW_e as well as boilers with 89 kg/s steam output, pressure 13.8/2.2 MPa and temperature 520/520°C have been installed.

Layout of gas ducts with ash separators in oil shale boiler has been presented in Fig. 1. The conventional PF technique for burning oil shale is characterized by very intensive boiler heat transfer surfaces' fouling with ash deposits and comparatively high sulphur dioxide emission in spite of high binding capacity of sulphur dioxide by calcium oxide in boiler's gas passes. Approximately 15-20% of oil shale total sulphur goes in the stack as SO_2 [Ots, 1992]. In spite of great calcium oxide amounts and very high Ca/S molar ratio, the PF technique does not provide complete SO_2 absorption in calcium oxide of ash.

Oil shale ash does not cause only intensive deposits' formation on boiler heat transfer surfaces, but it is also characterized by very high corrosive activity [Ots, 1994]. KCl presence in ash is the main reason for such high corrosion activity.

The boilers are equipped with cleaning systems to stabilize heat transfer between flue gases and heated medium. The cleaning cycles not only remove ash deposits from heat transfer surfaces, and also incur damage the protective oxide film on tubes' surfaces and accelerate high temperature corrosion of tubes.

2. CHARACTERISTICS OF ESTONIAN OIL SHALE AND MINERAL MATTER BEHAVIOR

2.1. Characteristics of Oil Shale

Estonian oil shale belongs to the carbonate class. Oil shale deposits are separated in North-East Estonia. Estonian oil shale is characterized as a solid fuel with high mineral matter content (60-70% in dry mass), moderate moisture content (9-12%) and low heating value (LHV 8-10 MJ/kg).

Estonian oil shale deposits lie in layers interlacing mineral stratas. Limestone is the main constituent in mineral stratas. Organic matter is joined with sandy-clay minerals in shale layers.

About 50% of oil shale is mined underground and the rest in open-pit mines.

Oil shale dry mass consists of the following three components: organic, sandy-clay and carbonate. Chemical composition of oil shale separate parts has been given in Table 1.

Carbonate and sandy-clay parts' mineralogical composition on the bases of their chemical composition (Table 1) and XRD method analyses has been given in Table 2.

Table 1. Chemical Composition of the Estonian Oil Shale Components

Organic matter		Sandy-clay matter		Carbonate matter	
Component	Amount %	Component	Amount %	Component	Amount %
C	77.45	SiO_2	59.2	CaO	53.5
H	9.70	CaO	0.7	MgO	2.0
S	1.76	Al_2O_3	16.3	FeO	0.2
N	0.33	Fe_2O_3	2.8	CO_2	44.3
Cl	0.75	TiO_2	0.7		
O	10.01	MgO	0.4		
		Na_2O	0.8		
		K_2O	6.3		
		FeS_2	12.3		
		SO_3	0.5		
Total 100.00		Total 100.00		Total 100.00	

Table 2. Mineralogical Composition of Carbonate and Sandy-Clay Parts of the Estonian Oil Shale

Groups of minerals	Minerals	Formula	Amount %
Carbonates	Calcite	$CaCO_3$	90.5
	Dolomite	$CaMg(CO_3)_2$	9.2
	Siderite	$FeCO_3$	0.3
			Total: 100.0
Sandy-clay	Quartz	SiO_2	23.2
	Rutile	TiO_2	0.7
	Orthoclase	$K_2O \cdot Al_2O_3 \cdot 6SiO_2$	28.1
	Albite	$Na_2O \cdot Al_2O_3 \cdot 6SiO_2$	5.8
	Anortite	$CaO \cdot Al_2O_3 \cdot 2SiO_2$	1.4
	Hydromuscovite	$[K_{1-x}(H_2O) \cdot Al_3Si_3O_{10}(OH)_2]_2$	23.0
	Amphibole	$[NaCa_2Mg_4(Fe,Al)Si_8O_{22}(OH)_2]_2$	2.0
	Marcasite	FeS_2	12.0
	Limonite	$Fe_2O_3 \cdot H_2O$	2.8
	Gypsum	$CaSO_4 \cdot H_2O$	1.0
			Total: 100.0

Hydrogen and oxygen high content and nitrogen low quantity are the most characteristic to the Estonian oil shale. C/H mass ratio is approximately 8 and it is close to the liquid fuels. The volatile matter content in organic part of oil shale is 85-90%. Chlorine combined with the organic matter is one of the peculiar features of the Estonian oil shale.

Calcium oxide is the main component of the carbonate part.

Quartz, aluminium oxide, marcasite and potassium oxide are the essential components of the sandy-clay matter. Potassium oxide quantity proves to be about 12 times the amount of sodium oxide.

The inorganic matter of oil shale does not contain chlorine, but chlorine is a constituent of the organic matter (chlorine content in organic matter is about 0.75%, Table 1).

Calcite is the main mineral in the carbonate constituent of oil shale. Quartz, feldspars (mainly orthoclase) and hydromicas (mainly hydromuscovite) are the sandy-clay part's constituents.

The approximate ratio between organic, sandy-clay and carbonate parts in dry mass of oil shale is following: 33%, 24% and 43%.

The total sulphur content in dry mass of the Estonian oil shale is in the range of 1.5-1.6%, and Ca/S molar ratio is approximately 8. The chlorine content in dry mass of oil shale is in the range 0.25-0.30%.

The mineral matter's inherent/extraneous ash ratio and also its chemical and mineralogical composition determine the dependence of CaO and SiO_2 in ash on ash quantity. This is illustrated by Figure 2, which shows the dependence of CaO and SiO_2 pertcentage in laboratory ash of the Estonian oil shale on ash dry mass quantity. If the amount of oil shale ash rises, SiO_2 percentage in ash will decrease, while CaO quantity will increase. Oil shale origin causes this kind of SiO_2 and CaO dependence in ash on fuel's ash quantity because the calcium oxide in fuel connected with extraneous mineral matter.

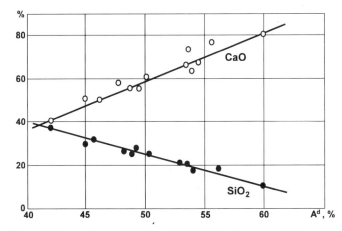

Figure 2. Dependence of CaO and SiO_2 in Laboratory Ash on Ash Content in the Estonian Oil Shale Dry Mass

If the calcium oxide in fuel is bound with extraneous mineral matter, as in oil shale, then the CaO content in the particles will increase and amount of quartz will decrease with particle sizes' increase. This is illustrated by Figure 3, which shows fly ash

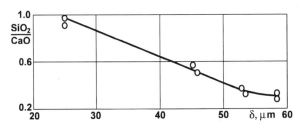

Figure 3. Dependence of the Estonian Oil Shale Fly Ash Particles' SiO_2/CaO Ratio on Their Sizes

particles' SiO_2/CaO ratio dependence on their sizes. This leads to CaO content decrease in fly ash formed during oil shale combustion. Quartz is distributed between fly ash and bottom ash contrary to calcium oxide.

2.2. Mineral Matter Behavior

During combustion, the fuel's mineral matter is subjected to the influence of high temperature and flue gas. The ash formed in the combustor is different from the initial composition of the fuel's mineral matter. The process involved in the conversion of the fuel's mineral matter during combustion may be divided into the following groups:

- decomposition of initial minerals into simple compounds;
- formation of new minerals under the influence of high temperature and flue gas atmosphere;
- contacts between separate minerals with the particles themselves or between separate particles and the gaseous medium;
- volatilization of some components of mineral matter and formation of fine dispersed minerals;
- fuel mineral matter' separate particles' or total mass conversion into plastic or liquid state.

Calcium oxide is one of the most active components of the ash formed from the carbonates. Reacting with quartz, calcium oxide will form bi-calcium silicate $2CaO \cdot SiO_2$ (belite). Belite is an important mineral in ashes with high calcium oxide content. Tri-calcium bi-silicate $3CaO \cdot 2SiO_2$ and tri-calcium silicate $3CaO \cdot SiO_2$ formation is also possible at the same time with belite formation. These reactions will occur if the temperature in combustor is above 1300-1350°C, and they will be important in clinker mineral forming processes. Also some part of calcium oxide will react with aluminium and iron oxides.

2.2.1. Calcium Oxide Behavior

Free CaO is the main ash component resulting from firing fuels rich in calcium, which binds sulphur dioxide in boiler gas passes and on the heat transfer surfaces (fouling). Free CaO amount in ash depends on the fuel combustion technique. Only a part of calcium oxide remains free in the PF conditions, because of the high temperature level in the combustor.

Dissociation of calcite and dolomite is the main source of calcium oxide at firing oil shale. Part of calcium oxide released from the carbonates, as it was mentioned above, will transform into clinker minerals. Calcium oxide balance at carbonate fuels' firing may be divided into three parts as follows: a) CaO combined with carbonates, CaO_c; b) CaO combined with clinker minerals, CaO_b; c) CaO in free form, CaO_f. Marking $K_c = CaO_c/CaO$, $K_b = CaO_b/CaO$, $K_f = CaO_f/CaO$, where $CaO = CaO_c + CaO_b + CaO_f$, and $K_c + K_b + K_f = 1$.

The balance of calcium oxide versus time, in the flame of the Estonian pulverized oil shale, has been given in Figure 4. Combined CaO amount in carbonates is reducing continuously, but binding with clinker minerals is increasing. Free calcium oxide content in ash grows first and then decreases slowly. Free CaO and total CaO ratio in fly ash after combustor will remain in the range of 25-30%.

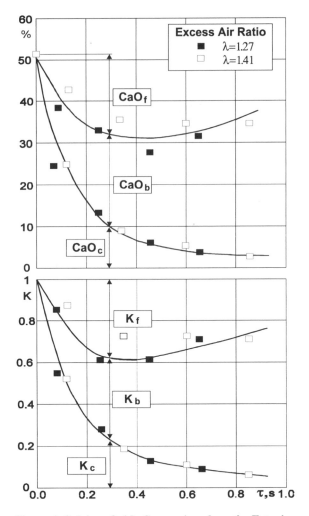

Figure 4. Calcium Oxide Conversion along the Estonian Pulverized Oil Shale Flame

2.2.2. Alkali Metals Behavior

Volatilization intensity of alkali metals from fuel particles depends significantly on temperature, composition of surrounding medium and form of alkali metals in fuel. When alkali metals occur as constituents of feldspars and micas, for instance in the Estonian oil shale (Table 2), they can volatilize only after minerals' decomposition or after their chemical reaction with other ash components.

Alkali compounds solubility in boiling water is one indicator, by which we can characterize their behavior during combustion. Alkali metals' compounds subjected to deeper changes during fuel combustion, are also more soluble in water. Proceeding from that, alkali metals' compounds can be divided into the following three groups: a) compounds which dissolve in boiling water within two hours (easily soluble) - M_e; b) compounds which dissolve in boiling water after two hours solution to constant level (hard soluble) - M_h; c) compounds insoluble in boiling water - M_i. Marking $p_e=M_e/M$, $p_h=M_h/M$, $p_i=M_i/M$ ($M=M_h+M_e+M_i$ - total amount of alkali metals in fuel), then $p_e+p_h+p_i=1$.

The balance of alkali metals versus time, in the flame of the Estonian pulverized oil shale, has been given in Figure 5 by different excess air ratio λ.

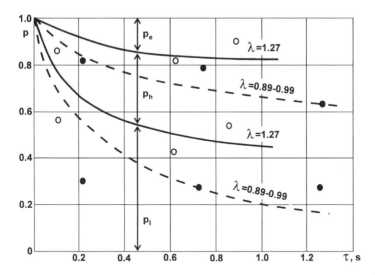

Figure 5. Alkali Metal Conversion along the Estonian Pulverized Oil Shale Flame

Oxygen concentration increase in flue gas is concurrent with increase of easily and hardly dissolving alkali metals parts' amount and simultaneous decrease of insoluble alkali metals share in ash. Such characteristic dependence of alkali metals' dissolution in boiling water on excess air ratio shows a very great influence of the air regime in combustor on alkali metal's behavior.

3.   FOULING OF STEAM BOILER CONVECTIVE HEAT TRANSFER SURFACES

3.1.   Ash Deposits Characterization

One of the main issues concerning utilization of the Estonian oil shale in power plants is intensive ash deposits' formation on boiler's convective heat transfer surfaces [Öpik, 1961; Ots, 1977; Ots, Arro, 1980].

In the Table 3 the chemical composition of fly ash and ash deposits on the convective heating surfaces' tubes are given.

The character of ash deposits on heat transfer surfaces' tubes depends on the flue gas temperature level and also on the aerodynamic conditions.

All the deposits compare with fly ash were enriched with sulphur (SO_3) and potassium. Peculiar of back side deposits are very high potassium and sulphur content. The chemical composition of ash deposits' samples from parallel flow air preheater tubes is very similar to chemical composition of fly ash.

Table 3. Chemical Composition of Fly Ash and Ash Deposits on Convective Heat Transfer Surfaces

Compo-nents, %	Fly ash	Superheater / reheater						Econo-mizer	Air heater
		Flue gas temperature, °C							
		1100-1000°C			950-850°C			650-550°C	350-250°C
		Front side		Back side	Parallel flow			Front side	Parallel flow
		Outer layer	Inner layer		Outer layer	Inner layer			
SiO_2	26.6	14.2	15.5	4.9	12.0	12.7		21.3	29.4
Fe_2O_3	4.2	5.0	5.1	5.1	2.4	3.9		3.7	4.3
Al_2O_3	7.2	4.4	4.3	3.5	3.8	4.6		5.7	6.4
CaO	48.3	31.4	31.5	6.4	35.8	28.6		31.8	42.2
MgO	5.3	2.3	2.5	1.1	2.2	2.9		2.1	2.4
K_2O	3.4	8.4	7.8	38.6	5.7	13.0		7.3	5.2
Na_2O	0.1	0.2	0.3	0.5	0.1	0.2		0.1	0.2
SO_3	6.9	33.1	33.0	40.5	36.8	33.8		26.4	11.2
Cl	0.3	-	-	0.17	0.11	0.26		0.3	0.4

3.2. Flue Gas Velocity Influence

Formation of bound ash deposits on convective heat transfer surfaces may be considered a result of two processes - setting of neutral (e.g., SiO_2, Al_2O_3) and binding of chemically active ash particles (e.g., CaO, MgO).

Flue gas velocity (kinetic energy of particles) has a strong influence on the characteristics of deposit formation. At a low flue gas velocity, the deposition rate of small ash particles and the destructive impact of larger particles are negligible. Due to the negligible erosive effect of large particles, the deposits formed under such conditions consist many neutral ash particles. Since the latter prevent the contact and binding of active ash particles' existing in the deposits, such deposits are loosely bound.

If flue gas velocity rises, ash particles will settle on the tube surface more densely, but the erosive effect of large particles will increase at an even higher rate. Therefore, ash deposit formation rate on tubes' heat transfer surfaces, as neutral ash particles' contribution, tends to decrease gradually after obtaining the maximum value at a certain flue gas velocity. At the same time, the content of active components in the deposit increases, leading to their consolidation with a further flue gas velocity increase. Above this flue gas velocity, the destructive effect of ash particles will increase so that only active ash particles will remain in the deposits, and, as a result, very hard deposits will be formed. Consequently, the kinetic energy of ash particles as well as the factor determining the chemical composition of ash deposits in heat transfer surfaces will influence the mechanism of ash deposit formation. It is the reason why the chemical compositions of ash deposits differ from that of fly ash in flue gas (Table 3).

3.3. Calcium Oxide Influence

Formation of the above-mentioned type of bound ash deposits, mainly based on calcium oxide content in fly ash, is a result of numerous calcium oxide compounds' sulphation under the influence of sulphur oxides and oxygen in flue gas with formation of calcium sulphate. Calcium sulphate is the main binding component in ash deposits forming on the tubes of steam boilers' heat transfer surfaces at burning oil shale with high calcium content.

Free CaO is the essential ash component, forming calcium sulphate in ash deposits. The process of binding ash particles with calcium sulphate into deposits can proceed only if free calcium oxide sulphation occurs directly in the deposits' layers. The calcium sulphate formed from free CaO in flue gas prior to surface has bad sintering impact. It should be mentioned that a very close contact between ash particles has great importance in deposits' consolidation. This explains well enough why the back side deposits containing a great deal of substances inactive to sulphation and not being impacted by large ash particles, even with a similar degree of sulphation, are less bound.

However, such compounds as calcium sulphide, calcium chloride and other numerous calcium silicates, aluminates and ferrites have got sulphation ability as well. This trend of calcium oxide bound with quartz, aluminium and iron oxides towards sulphation is demonstrated in the results obtained by the studies of some fly ash sulphation. It was established that the reduction of vitreous (glassy) substance content took place in ash with the simultaneous increase in SO_3 (in ash). Sulphation of ash minerals combined with calcium oxide proceeds many times slower compared to free CaO sulphation. The sulphation process of bound calcium oxide does not have a remarkable effect on ash deposits' first stage formation, but influence general consolidation of deposits.

3.4. Alkali Metals' Influence

Alkali metals' compounds also play a significant role as binding substances in ash deposits formation processes at oil shale burning. Alkali metals' compounds are usually accumulated in the lower and back side layers of deposits. Penetration of alkali metals' sulphates into deposits during the fouling of boiler heat transfer surfaces' tubes may proceed in two ways: immediately during deposit's formation initial stages and gradually after such formation.

Temperature limits and alkali metals' compounds' condensation/desublimation intensity on heat transfer surfaces' tubes depend mainly on the partial pressure of their vapours in flue gas. In this case, determing factor is the dew point (the initial point of condensation/desublimation) of the particular alkali metal compounds. For instance, the alkali metals' compounds' dew points (partial pressure) corresponds to alkali metals' compounds' solubility characteristics in boiling water (Figure 5). If temperature of heat transfer surfaces' tubes or deposit layer outside temperature is lower than dew point of alkali metal compound, the condensation/desublimation of vapour occurs simultaneously with deposit formation. Figure 6 illustrates deposition intensity and dew points of the condensing/desublimating components in flue gas on heat transfer surfaces depending on surface temperature during the Estonian oil shale burning. Alkali metal compounds at the Estonian oil shale burning are connected to potassium behaviour. Potassium sulphate direct condensation on surface occurs mostly in flue gas temperature range higher than 900-950°C. Potassium chloride vapour intensive condensation will occur starting from temperatures 550-650°C. This is the reason why deposits' lower layers (at more favourable temperature conditions) contain potassium chloride higher quantity than the upper ones.

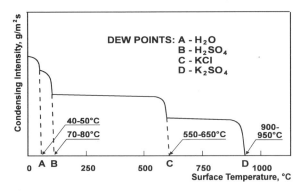

Figure 6. Condensing Compounds' Condensation Intensity and Dew Points
Depending on Surfice Temperature at the Estonian Oil Shale Burning

Alkali metals' chlorides, condensed on the surfaces will transform in time, under the influence of flue gas sulphur oxides, into alkali metals' sulphates. That is why the ratio between chlorine and sulphur trioxide content in oil shale ash deposits will change in time. Chlorine and sulphur trioxide content in the Estonian oil shale ash deposits on heating surfaces' tubes depending on time have been given in Figure 7 (surface temperature is lower than potassium chlorine dew point). Due to potassium chloride sulphation amount of sulphur (SO_3) in deposits will increase in time, but chlorine content will decrease simultaneously. Chlorine from deposits will volatilize as HCl under the influence of water vapour.

Potassium sulphate as well as calcium sulphate is a binding substance of ash deposits.

In Figure 6, dew points of sulphur acid (70-80°C) and water vapour (45-60°C) have also been given.

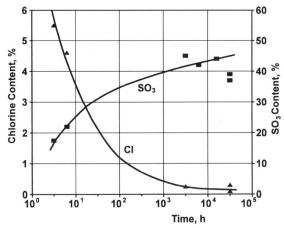

Figure 7. Chlorine and Sulphur Trioxide Content Change in Ash Deposits
Depending on Time at the Estonian Oil Shale Burning

4. HIGH TEMPERATURE CORROSION AND WEAR OF HEAT TRANSFER SURFACES

4.1. High Temperature Corrosion

High temperature corrosion of boilers' heat transfer surfaces' metal depends largely on ash deposits composition on tubes. Potassium, chlorine and sulphur are the most important components affecting metal corrosion intensity at the Estonian oil shale burning.

High temperature corrosion intensity of metal depends mainly on the alloy type and temperature level at the same composition of ash deposits on tubes [Ots, 1977].

The comparative diagram of high temperature corrosion depth of two metal types, under the influence of two separate fuel ashes, has been given in Figure 8. First of them is the Estonian oil shale fly ash with remarkable potassium oxide ($K_2O=6\%$) and chlorine (Cl=0.5%) content, and the second is brown coal fly ash with low alkali metal content ($Na_2O=0.4\%$, $K_2O=0.6\%$, chlorine is absent) in air atmosphere depending on temperature and time. Experiments were also performed in pure air medium (without ash). It becomes clear that both metals corrode with greater intensity under the influence of oil shale fly ash than in pure air medium or due to presence of chlorine free brown coal fly ash. We can see that a relative accelerating impact of oil shale ash on Cr-Ni austenitic steel is more vivid than on Cr perlitic ones, in spite of the fact that the absolute value of corrosion depth in the first case is lower. Such behavior of Cr-Ni austenitic steel, compared to perlitic one, is induced by an extremely strong action of alkali metal chlorides on chromium.

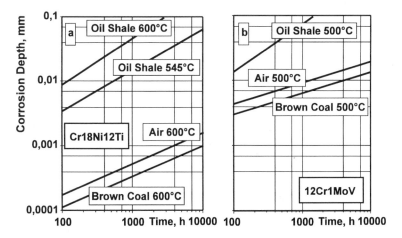

Figure 8. Comparative Diagram of High Temperature Corrosion Depth of Austenitic Cr18Ni12Ti (a) and Perlitic 12Cr1MoV (b) Alloys Depending on Time and Temperature

4.2. Wear

Modern boilers are equipped with different types of cleaning systems to stabilize heat transfer in steam boiler heat transfer surfaces. The problem of efficient cleaning of heat transfer surfaces from ash deposits due to their fast growth, is of paramount importance for boilers burning oil shale as fuel with high calcium oxide, alkali metal and chlorine content. Cleaning cycles do not only remove ash deposits from boiler's tubes, but also incur damage of protective oxide film on metal, diminishing it's diffusion resistance and inevitably accompaning with corrosion intensity growth.

Decrease in corrosive-erosive wear of heat transfer surfaces, due to cleaning effect, can be achieved either by increasing the period between cleaning cycles or by decreasing cleaning force effect on oxide film. Hence, it follows that heat transfer intensity is also intensified by intensifying cleaning of boiler's tubes, however, heat transfer surfaces' tubes corrosive-erosive wear is accelerated as well. Problem of the choice of a heat transfer surfaces' optimal cleaning regime from ash deposits, particularly the interaction between cleaning intensity and its implementation conditions, will arise in this connection. The right solution of the problem will be conclusive for design, operating conditions and economical parameters of the boiler.

5. AIR POLLUTING COMPOUNDS IN OIL SHALE FLUE GAS

5.1. Nitrogen Oxide Concentration

Nitrogen content in the fuel and oxygen concentration are the main factors affecting nitrogen oxide's concentration in oil shale flue gas [Loosaar, Jegorov, 1985]. NO_x formation depends on temperature level in combustor to a certain extent. Influence of excess air ratio on NO_x concentration in oil shale flue gas has been shown in Figure 9. The amount of nitrogen bound with oxygen in flue gas in normal conditions at excess air ratio $\lambda=1$ is in the vertical axis. These results have been obtained on a laboratory device by burning oil shale samples with different nitrogen content (0.21-0.41% in organic mass). NO_x concentration has been reduced to nitrogen amount in the fuel's organic matter - 0.3% taking into account the linear dependence of nitrogen oxide's concentration in flue gas on nitrogen content in fuel.

Depending on excess air ratio (in the range 1.1-1.4) 12-24% of fuel's nitrogen is converted into oxides. The effect of flue gas temperature on fuel's NO_x concentration is negligible.

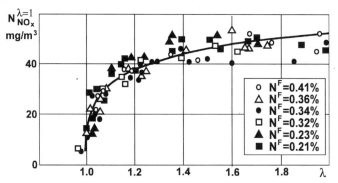

Figure 9. Influence of Excess Air Ratio on Concentration of Nitrogen Bound with Oxygen in the Estonian Oil Shale Flue Gas

Sulphur is one essential factor affecting NO_x concentration in flue gas. Presence of sulphur dioxide in flue gas reduces NO_x concentration [Ots, 1992].

Thermal NO_x concentration in flue gas, leaving boiler as NO_2, is not more than 50-70 mg per normal cubic meter at oxygen 6% concentration at the Estonian oil shale burning in thermal power plants, is due to relatively low combustion temperature (maximum flame temperature in the range of 1400-1450°C). Total NO_x concentration as NO_2, depending on excess air ratio and nitrogen content in oil shale, lies in the range of 150-200 mg per normal cubic meter at oxygen concentration 6%. Relatively low NO_x concentration in oil shale flue gas is caused by low nitrogen content in oil shale organic part (Table 1) and also by some influence of sulphur dioxide on NO_x formation processes.

5.2. Sulphur Dioxide Concentration

Sulphur oxides concentration (SO_x) in flue gas at oil shale burning does not depend only on the total sulphur content in fuel, but also on calcium oxide presence in fuel ash. Due to sulphation reactions between sulphur oxides and calcium oxide in flue gas stream SO_x concentration will reduce remarkably compared to conditions when all sulphur goes into flue gas as oxides.

Ash's sulphur binding factor (relative amount of total sulphur bound with fuel's ash) depends essentially on calcium oxide behavior conditions (Figure 4) in combustor and total sulphur content in fuel. Free calcium oxide is the most active sulphur binding component in oil shale ash. Calcium oxide in clinker minerals may also react with sulphur oxides, but this reaction intensity is very low compared to sulphating on free calcium oxide base.

Sulphur binding reaction with ash depends on temperature level and oxygen concentration (excess air ratio) in flue gas.

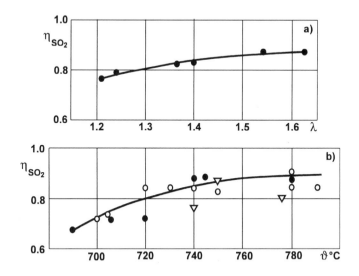

Figure 10. Dependence of Sulphur Binding Factor on Excess Air Ratio (a) and Gas Temperature After Economizer (b) at the Estonian Oil Shale Burning

Figure 10 shows effect of excess air ratio and flue gas temperature after economizer on sulphur oxides' binding factor. Sulphur binding factor with ash will increase with oxygen concentration increase in flue gas. Gas temperature plays some role in sulphur's binding reactions with ash. Sulphur's binding factor with ash depends also on ash particles' sizes. The smallest particles of oil shale ash contain more sulphur in a sulphate form than the biggest ones.

In spite of high sulphur binding effect of ash, it does not guarantee sufficiently low sulphur dioxide concentration in oil shale flue gas. SO_2 concentration in oil shale flue gas, leaving boiler, depending on excess air ratio, boiler load, etc., is in the range 1.1-1.8 g per normal cubic meter at oxygen 6% concentration.

6.   ASH UTILIZATION

Presence of calcium oxide and other appropriate minerals' components in ash enables it's utilization in national economy [Kikas, 1988; Öpik, 1990]. A very good example of that is utilization of ash, formed in Estonian oil shale power plants, in cement industry, building materials industry, agriculture, road construction, etc. Complex utilization possibilities of ash in national economy are determined, on the one hand, by fuel's adequate combustion arrangement, and on the other hand, by ash separation system. The maximum temperature in combustor should be at least 1350-1400°C for recovering cement clinker minerals in oil shale ash.

PF boilers in oil shale power plants are equipped with two stage ash separation systems: cyclone and four-fielded electrostatical precipitators (Figure 1). Combustor and flue gas ducts work as ash separators too. The most valuable ash is separated in electrostatical precipitators. Ash particles with sizes less than 15 μm are used for manufacturing high-grade Portland cement. Ash particles below 30 μm, separated from electrostatical precipitators and cyclone, are utilized to produce middle-grade shale-ash cement, autoclave concrete and as lime fertilizer for agriculture. Generally, ash used in agriculture has the lowest requirements to fractional composition. Bottom ash and ash from gas ducts of oil shale boilers are handled hydraulically and are not utilized in national economy.

It is possible to use fuel ash for desulphurization of flue gas at burning oil shale as a fuel with high calcium and alkali metal's content. Construction of the scrubber using clarified alkaline water from hydraulic ash removal system with pH=11-13, is one alternative of cleaning flue gas from sulphur dioxide. It is also possible to use an ash-soaking basin to obtain alkaline water.

REFERENCES

Ots, A (1992). "Formation of Air-Polluting Compounds while Burning Oil Shale": *Oil Shale*, Vol. 9, N°1, 63-75.

Ots, A (1994). *"The Influence of Cleaning on Corrosive-Erosive Wear of Steam Boiler Heating Surfaces Tubes, The Impact of Ash Deposition on Coal Fired Plants"*: Proceedings of the **Engineering Foundation Conference**, Taylor & Francis, 759-766.

Öpik, I (1961). *The Influence of Oil Shale Mineral Matter on the Conditions of Boiler Operation*: Estonian State Publishing House, Tallinn, p. 249 (in Russian).

Ots, A (1977). *The Processes in Steam Boilers in Burning Oil Shale and Kansk-Achinsk Basin Coals*: Energia, Moscow, p. 312 (in Russian).

Ots, A, Arro, H (1980). *The Mechanism of Fouling Steam Generators Heating Surfaces under the Combustion Products of Solid Fuels. Fouling and Corrosion in Steam Generators*: Beograd, 29-41.

Loosaar, J; Jegorov, D (1985). *"Nitrogen Oxides Formation while Burning Pulverized Estonian Oil Shale in the Experimental Device"*: Trans. of Tallinn Technical University, N°600, 33-39 (in Russian).

Kikas, V (1988). *"Mineral Matter of Kukersite Oil-Shale and Its Utilization"*: *Oil Shale*, Vol. 5, N°1, 15-28 (in Russian).

Öpik, I (1990). *"Ash Utilization After Combustion and Thermal Processing of Low-Grade Fuels"*: VTT Symposium *Low Grade Fuels*, Vol. 2, 117-128.

CHARACTERIZATION OF BIOMASS ASHES

Bengt-Johan Skrifvars, Mikko Hupa
Åbo Akademi University, Turku, Finland

Antero Moilanen
VTT Energy, Espoo, Finland

Ragnar Lundqvist
Foster Wheeler Energia OY, Helsinki, Finland

ABSTRACT

Energy production from combustion or gasification of biomass has recently attracted increased interest. These fuels form a valuable indigenous energy resource for some countries. They represent also an attractive way to decrease CO_2 emissions from the energy production. The development of new combustion and gasification techniques, such as atmospheric or pressurized fluidized bed combustion and gasification has also made it possible to utilize biomass in a more feasible way than before. The availability of these new energy conversion systems is, however, still unknown. Among other questions the behavior of ash can be critical.

In this paper we present some initial results from an ash characterization work performed on 10 different types of biomass ashes with the focus on the new energy conversion systems. The characterization methods were the following:

1) Ash thermal behavior analyzed with a combined differential thermal, thermo-gravimetric analyzer (DTA/TGA).
2) Ash sintering tendency with the compression strength sintering testing method. Ashes tested in both oxidizing and reducing conditions, in temperatures ranging from 500 - 1000°C. Selected ashes tested further in 100 % CO_2(g).
3) Chemical analyses of the ashes and sintering tested samples.
4) Standard fuel characterization analyses.
5) Reactivity analyses for selected biomasses.

The results showed clear differences in the thermal behavior of the ashes. The sintering tendencies varied significantly. The chemical analyses showed that ashes rich in silicon started to sinter at 800 - 900°C both in oxidizing and reducing conditions, while ashes with low silicon content did not. These ashes showed instead an increase in sintering at approximately 700°C and a decrease above 700°C when CO_2(g) was

Applications of Advanced Technology to Ash-Related Problems in Boilers
Edited by L. Baxter and R. DeSollar, Plenum Press, New York, 1996

383

present in the gas atmosphere. In some cases the sintering tendency of the ash also correlated with the gasification reactivity of the corresponding biomass.

The results and their relevance to full scale conversion systems are discussed in the paper.

INTRODUCTION

Biomasses are for some countries indigenous fuels that form a valuable energy resource. Combustion or gasification of biomass is also an attractive way of decreasing CO_2 emissions from energy production. New combustion and gasifications techniques, such as atmospheric and pressurized fluidized bed processes, have broadened the possibility to fire biomass in a more feasible way than before.

Little is, however, known yet about the long term use of biomasses in new energy conversion systems such as fluidized bed combustion or gasification. Especially the mineral matter behavior in these new energy production techniques are today still to some extent unknown. There are indications from fluidized bed combustion tests that the mineral matter from biomasses, the ash, may sometimes cause ash related problems, such as bed agglomeration problems, deposit formation etc. /Dawson et. al. 1991, Baxter 1993, Nordin 1995, Nordin et. al. 1995/.

We wanted to collect some initial data on the ash behavior of biomasses. We focused on the ash sintering behavior, since it has earlier shown to correlate fairly well with bed agglomeration problems of circulating fluidized bed boilers /Skrifvars et. al. 1992, 1994/.

EXPERIMENTAL

Ten different types of biomasses were tested on their sintering tendency with an ash sintering test. Details of the 10 biomasses are presented in Tables 1 and 2. The sintering test method is based on compression strength measurements of heat treated cylindrical ash pellets and it has been used before at several occasion /Barnhart et. al. 1956, Hupa et.al. 1989, Skrifvars et. al. 1992 and 1994/.

Usually the ash to be studied is recieved through a standardized ashing procedure /DIN, ASTM, BS/. For biomasses, however, the temperature used in the standardized procedure is considered too high. A significant loss of volatile inorganic matter may jeopardize the applicability of the prediction made on this ash. Consequently a revised ashing procedure is used where the ashing temperature is kept at 550°C.

This ashing procedure was also used here as a firts approach. Three biomasses were used for this test, the biomasses #3, #9 and #10. These three ashes were then sintering tested with the compression strength test. This ashing procedure lead to problems with the sintering test for the ash #9. The pellets decomposed during the sintering test when the sintering temperature exceeded 800°C. This lead to porous ash pellets with cracks. From these pellets no compression strength could be measured.

The explanation to this behavior was assumed to be that the calcium rich ash #9 formed calcium carbonate during the ashing procedure which decomposed during the following sintering test. According to chemical analyses presented in Table 2 the ash #9.1 (being the one tested in this case) contained 27.6 wt-% CO_3.

Table 1. The studied biomasses.

#	Biomass type	Ash	Volat.	C	H	N	S	Cl	O_{diff}
					wt-%, t.s.				
1	grass	8.9	73.5	45.0	5.7	1.4	0.14	n.a.	38.9
2	straw	4.9	77.4	47.3	5.8	0.5	0.10	0.17	41.2
3	other	15.2	71.1	42.4	5.2	0.3	0.03	0.02	36.8
4	other	5.5	80.0	47.0	5.9	0.7	0.10	0.00	40.8
5	straw	4.5	76.9	47.1	5.9	0.6	0.10	n.a.	41.8
6	other	8.7	72.3	50.0	6.0	1.1	0.10	0.08	34.0
7	wood	1.9	79.4	50.6	5.8	0.4	0.03	0.00	41.3
8	grass	8.4	n.a.	46.7	5.9	3.1	0.25	0.60	35.6
9	wood	0.7	80.5	49.9	5.8	0.2	0.02	0.11	43.3
10	wood	1.2	79.9	49.7	6.1	0.4	0.03	n.a.	42.6

Table 2. The ash analyses of the studied biomasses, expressed as weight-% of their corresponding oxides.

#	SiO_2	Al_2O_3	Fe_2O_3	P_2O_5	CaO	MgO	Na_2O	K_2O	SO_3	CO_2	Cl	SUM
					wt-%							
1*	65.7	1.7	1.4	5.0	2.7	1.8	0.3	4.6	1.4	0.0	0.23	84.7
2*	56.3	0.5	0.6	1.9	6.3	1.4	0.2	12.6	2.9	1.3	2.10	84.0
3*	48.8	6.4	1.9	3.0	3.9	5.5	0.8	18.9	3.5	3.1	n.a.	95.7
4*	44.1	5.1	2.3	2.1	6.7	3.6	0.3	16.0	5.2	1.9	n.a.	87.3
5*	34.2	0.4	0.5	2.1	9.1	2.0	0.1	24.7	2.3	6.1	3.15	81.5
6*	34.0	4.7	4.4	3.0	17.3	11.4	0.4	17.1	1.2	0.5	n.a.	94.2
7	7.3	1.5	1.3	4.8	44.8	5.8	0.8	11.3	1.5	5.2	n.a.	84.3
8*	3.3	0.3	0.3	7.6	27.7	3.1	0.7	28.4	1.9	22.3	4.39	95.7
9*	0.6	0.2	0.3	6.0	35.1	10.4	2.3	13.6	2.0	27.6	0.24	98.2
9	1.0	0.5	0.7	7.1	39.3	9.1	3.2	15.3	1.3	10.1	n.a.	87.6
9	1.0	0.5	0.7	7.6	38.8	8.7	3.8	15.2	1.3	12.4	0.18	89.9
10*	0.6	0.1	0.7	10.7	28.8	4.4	0.2	22.5	5.9	23.5	0.18	97.4

n.a.: not analysed   * ashed in 550°C only.

A surprising result was, however, that the ash #10, which also contained CO_3 (23.5 wt-%), did not show any above described behavior. From these pellets the compression strength could be measured even if they had been sintered in temperatures as high as 900°C.

The ashing procedure was in any case changed so that it included a short temperature increase to 850°C for 10 minutes directly after the ashing. During this extended ashing stage most of any calcium carbonate formed during the actual ashing stage was assumed to decompose to calcium oxide.

With this procedure the problems with calcium carbonate present in the ash #9 was partly avoided. The extended ashing procedure introduced, however, a new problem, the melting of the ash. Some of the ashes contained such elements which formed a first melt already at temperatures below 850°C. This was the case for the ashes #2, #5 and #8.

The ashing tests finally resulted in a dual ashing procedure where biomasses rich in calcium and poor in silicon where ashed at 550°C with a following temperature increase to 850°C and those biomasses rich in silicon ashed at 550°C with no further temperature increase. The biomass #8 made an exception. Even if its ash was rich in calcium (27.7 wt-% in ash) and poor in silicon (3.3 wt-% in ash) the biomass was ashed in 550 °C only since it contained also large amounts of chlorine (4.39 wt-% in ash).

The laboratory ash was then pressed to cylindrical pellets of the size of 11 mm in height and 10 mm in diameter. The ash pellets were after this heat treated in a tube furnace for four hours in five different temperatures. Two different gas atmospheres were used; oxidizing gas (dried air) and reducing gas (5% CO, 95% N_2). Some tests were also performed in a 100% CO_2(g).

The sintering tendency measurements were then compared to full scale operating experiences for corresponding biomasses when available. The goal was to see if the sintering, detected with lab test, correlated with the ash behavior experiences achieved in the full scale tests.

All of the ashes as well as selected heat treated pellets were also subject to quantitative wet chemical analyses. DTA/TGA runs were further made with the ashes #1, 2, 5, 8, 9, and 10. The combined DT/TG analysis reveals any process in a sample, in this case the biomass ash, that consumes or releases heat and detects the sample weight simultaneously. Typical, well detectable processes are melting processes, decomposition processes and chemical reactions. In this case the DT/TG analyses were performed from room temperature up to 1100°C with a heating rate of 10°C/min. in an atmosphere that contained 100 % N_2(g).

Two of the biomasses, #2 and #5, were also subject to gasification reactivity measurements performed in a separate thermogravimetric analyzer. The gasification agent in these tests was H_2O(g), the temperature 800°C and the pressure was 1 bar.

Table 3. The analyses of the heat treated pellets, expressed as weight-% of their corresponding oxides.

#	Samp	SiO_2	Al_2O_3	Fe_2O_3	P_2O_5	CaO	MgO	Na_2O	K_2O	SO_3	CO_2	Cl	SUM
						wt-%							
1	Ash*	65.7	1.7	1.4	5.0	2.7	1.8	0.3	4.6	1.4	0.0	0.2	84.7
1	600°C*	63.1	1.9	1.7	4.9	2.7	2.0	0.3	4.8	1.3	0.2	0.0	83.0
1	800°C*	67.6	1.9	2.3	5.2	2.6	1.9	0.3	4.8	0.9	0.0	0.0	87.6
1	900°C*	67.8	1.9	1.9	5.6	2.7	2.0	0.3	4.9	0.2	0.0	0.0	87.5
1	1000°C*	70.0	1.9	1.9	5.5	2.7	2.1	0.2	4.8	6.0	0.0	2.1	95.1
2	Ash*	56.3	0.5	0.6	1.9	6.3	1.4	0.2	12.6	2.9	1.3	2.1	84.0
2	600°C*	58.4	0.6	0.9	2.6	6.2	1.5	0.2	12.5	3.8	0.6	1.9	87.2
2	700°C*	55.2	0.6	0.4	2.2	7.1	1.5	0.1	12.9	3.5	0.1	0.7	83.7
2	800°C*	62.0	0.7	0.6	2.3	7.3	1.7	0.1	12.3	2.6	0.0	0.4	89.6
3	Ash*	48.8	6.4	1.9	3.0	3.9	5.5	0.8	18.9	3.5	3.1	n.a.	95.7
3	600°C	49.2	2.5	3.6	4.1	4.6	7.6	0.3	14.5	5.5	0.3	n.a.	92.1
3	700°C	32.3	2.5	4.4	4.4	4.6	7.5	0.3	14.5	4.7	0.2	0.9	75.3
3	800°C	38.9	2.3	8.1	3.9	3.6	6.3	0.3	11.4	3.2	0.0	0.4	78.2
4	Ash*	44.1	5.1	2.3	2.1	6.7	3.6	0.3	16.0	5.2	1.9	n.a.	87.3
4	650°C*	46.4	5.3	2.1	2.5	7.1	4.0	0.5	16.4	5.7	0.4	1.8	90.5
4	750°C*	47.7	5.3	2.3	2.5	7.0	4.0	0.1	16.1	5.7	0.1	1.0	90.9
4	1050°C*	49.6	5.7	2.4	2.7	7.4	4.3	0.1	15.2	2.7	0.0	0.1	90.3
5	Ash*	34.2	0.4	0.5	2.1	9.1	2.0	0.1	24.7	2.3	6.1	3.15	81.5
5	600°C*	39.1	0.3	0.5	2.2	9.3	2.1	0.2	25.5	2.3	3.0	3.3	84.5
5	700°C*	42.6	0.3	0.5	2.5	10.5	2.3	0.2	27.9	2.4	0.6	2.3	89.9
5	800°C*	42.4	0.4	0.5	2.5	11.1	2.7	0.2	27.1	1.9	0.4	2.1	89.3
5	800°C*	39.6	0.3	0.5	2.3	9.6	2.2	0.2	25.8	2.3	1.9	3.1	84.6
5	800°C*	38.7	0.3	0.5	2.1	9.5	2.1	0.2	25.9	2.3	2.7	3.1	84.4
6	Ash*	34.0	4.7	4.4	3.0	17.3	11.4	0.4	17.1	1.2	0.5		94.2
6	650°C*	37.0	4.5	4.1	2.7	15.5	11.1	0.5	17.3	1.0	0.7	0.5	94.7
6	750°C*	37.7	4.7	4.1	2.5	15.7	11.4	0.7	16.3	1.2	0.6	0.2	94.9
6	1050°C*	36.4	4.5	4.6	2.7	16.2	11.8	0.7	16.4	1.2	0.1	0.1	94.6
6	1050°C*	37.4	5.1	4.3	2.9	15.4	11.8	0.8	16.4	1.2	0.2	0.0	95.6
7	Ash	7.3	1.5	1.3	4.8	44.8	5.8	0.8	11.3	1.5	5.2	n.a.	84.3
7	650°C	5.3	1.5	1.0	5.5	48.3	6.1	0.9	11.6	1.2	9.8	0.0	91.3
7	750°C	5.1	1.5	1.0	5.5	47.6	6.3	0.9	11.6	1.2	4.7	0.2	85.5
7	850°C	4.9	1.7	1.1	5.5	49.8	6.3	1.1	9.5	1.5	2.0	0.0	83.4
8	Ash*	3.3	0.3	0.3	7.6	27.7	3.1	0.7	28.4	1.9	22.3	4.4	95.7
8	600°C*	3.2	0.3	0.3	7.6	27.8	3.1	0.7	29.0	2.0	19.2	3.4	93.2
8	700°C*	3.1	0.3	0.3	7.9	28.5	3.2	0.8	29.5	2.0	23.6	3.7	99.1
9.1	Ash*	0.6	0.2	0.3	6.0	35.1	10.4	2.3	13.6	2.0	27.6	0.2	98.2
9.2	Ash	1.0	0.5	0.7	7.1	39.3	9.1	3.2	15.3	1.3	10.1	2.4	87.6
9.2	Ash	1.0	0.5	0.7	7.6	38.8	8.7	3.8	15.2	1.3	12.4	0.2	89.9
9.2	650°C	1.1	0.4	0.7	7.8	41.1	10.8	3.4	16.0	1.2	12.7	0.3	95.2
9.2	750°C	2.4	0.6	0.9	8.5	41.7	10.3	3.8	16.4	1.2	7.8	0.3	93.4
9.2	1050°C	1.5	0.6	0.9	9.2	45.9	11.1	4.0	14.1	1.5	2.3	0.0	91.1
10	Ash*	0.6	0.1	0.7	10.7	28.8	4.4	0.2	22.5	5.9	23.5	0.2	97.4
10	600°C*	0.7	0.2	0.6	10.7	31.2	4.9	0.3	23.5	5.6	21.6	0.2	99.3
10	700°C*	0.7	0.2	0.7	10.8	32.0	4.8	0.3	24.2	6.1	16.9	0.2	96.6
10	800°C*	0.8	0.2	0.4	11.8	34.3	5.4	0.3	25.4	5.9	11.7	0.3	96.2

n.a.: not analysed   * ashed in 550°C only.

The goal of these analyses, i.e., the quantitative wet chemical analyses, the DT/TG analyses and the reactivity measurements, were to find explanations to the ash sintering behaviors.

RESULTS

The quantitative analyses

The wet chemical analyses of the ashes and heat treated pellets are shown in Table 3. The table contains the analyzed elements recalculated to their corresponding oxides and the sum of the oxides. The asterisk in the table indicates that the ash was received by ashing the biomass at 550°C only. Two reproducibility tests were performed. The ash #9.2 was analyzed two times while the pellets of ash #6, heat treated at 1050°C and the pellets from ash #1, heat treated at 800°C were analyzed twice. All these tests show good reproducibility.

The elemental composition of the ashes varied widely. The silicon content in the ashes, calculated as oxides, varied between 0.6 wt-% in ash #10 and 65.7 wt-% in ash #1. The calcium content seemed to correlate with the silicon content in the ash in an inverse way, i.e. ashes high in silicon were usually low in calcium. The calcium content in the ashes varied between 44.8 wt-% in the ash #7 and 2.7 wt-% in the ash #1.

All the ashes were rich in potassium. The amounts varied between 4.6 wt-% in the ash #1 and 28.4 wt-% in the ash #8.

Sulfur was detected in small amounts in all the ashes. The highest amount of sulfur, 5.9 wt-%, was analyzed in the ash #10 and the lowest, 1.2 wt-%, was analyzed in the ash #6. The amount of chlorine was analyzed from selected samples. The highest amount of chlorine, 4.4 wt-%, was analyzed in the ash #8 and the lowest, 0.18 wt-%, was found in the ashes #9 and #10.

No major differences between the ashes and the sintering tested ash pellets could be detected. The chlorine content decreased somewhat with increasing heat treatment temperature in the ashes #2 and #8 as well as the sulfur and carbonate content in the ash #9.

The effect of the extended the ashing procedure could be seen for the ash #9. The ash #9.1 which was ashed in 550°C only, contained 27.3 wt-% carbonate while the ash #9.2 which was ashed according to the extended procedure, contained 10.1 wt-% carbonate.

DTA/TGA analyses

The DTA analyses are summarized in Table 4.

For the calcium rich ashes #9 and #10 a two stepped weight decrease was seen in combination with a broad endothermic peak in 100% $N_2(g)$. For the ash #9 a first, fast weight decrease took place at approximately 590-770°C and a second, clearly slower

388

Table 4. DTA peak temperatures and sintering temperatures for the tested ash samples.

#	DTA measurements in $N_2(g)$					Sintering measurements		
	$T_{trans\,1}$ °C	$T_{trans\,2}$ °C	$T_{trans\,3}$ °C	Δm_1 @ temp mg @ °C	Δm_2 @ temp mg @ °C	$T_{sint,\,red}$ °C	$T_{sint,\,ox}$ °C	$T_{sint,\,CO2}$ °C
1*			853			800	800	800
2*		565-639	843	1.50 @ 639-1100		700	700	n.a.
3*	n.a.	n.a.	n.a.	n.a.	n.a.	800	800	n.a.
4*	n.a.	n.a.	n.a.	n.a.	n.a.	750	n.a.	n.a.
5*	450-489	530-590	848	4.0 @ 25-1100		650	650	n.a.
6*	n.a.	n.a.	n.a.	n.a.		950	n.a.	n.a.
7	n.a.	n.a.	n.a.	n.a.		>1050	>1050	n.a.
8*		629-676	848	3.2 @ 629-1100		625	600	n.a.
9		592-742	883	4.0 @ 592-770	1.0 @ 770-900	650	n.a.	n.a.
9*	n.a.	n.a.	n.a.	n.a.	n.a.	n.a.	n.a.	n.a.
10	n.a.	n.a.	n.a.	n.a.	n.a.	n.a.	>900	700
10*		665-748	843	2.8 @ 665-785	0.5 @ 785-915	700	700	700

n.a.: not analysed   * ashed in 550°C only

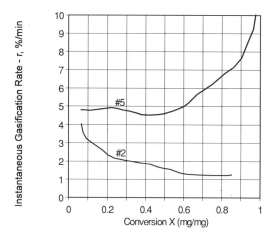

Figure 1. The gasification reactivity of the biomasses #2 and #5, measured with a thermogravimetric analyzer as the instantaneous reaction rate vs the char burn-off

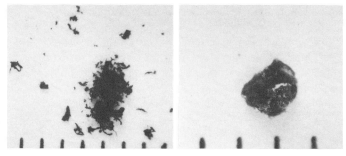

Figure 2. The ash residue of biomass #2 (left) and biomass #5 (right) after the gasification   test. Both ashes have started out from same sized biomass particles. Magnification x10.

one, at 770-900°C. For the ash #10 the corresponding temperature range for the first weight decrease was approximately 670-790°C and for the second one 790-900°C. The second weight decrease detected in both samples may have been due to decomposition of calcium carbonate in the ash. The first weight decrease at 592°C indicated some other decomposition reaction. This temperature is clearly too low to account for calcium carbonate decomposition. Both ashes had also a second, sharp endothermic peak, at 883°C for the ash #9 and at 843°C for the ash #10. The cause for this peak may have been some melting in the sample but this is very uncertain since the peak occurs in the middle of a weight decrease, i.e. a decomposition process.

Also for the calcium rich ash #8 a weight decrease was detected. It started at 629°C and continued to the end of the heating period, i.e., up to 1100°C. The decomposition process was slower than in the case with the ashes #9 and #10. It is not clear if the broad endothermic peak at 629 - 676°C is only due to the decomposition taking place in the sample or if the peak also gets contribution from a melting process in the sample. A second, sharp endothermic peak was detected at 848°C. Again it may indicate some melting but the same uncertainty as in the case with the ashes #9 and #10 is valid.

For the ash #1 a clear, sharp endothermic peak was detected at 853°C with no corresponding weight change in the sample during the $N_2(g)$ run. This peak may indicate a first melting point of the sample at this temperature.

Both ashes #2 and #5 showed a weight decrease throughout the run. At 490°C the ash #5 showed a broad exothermic peak, followed by another at 710°C. One sharp endothermic peak could be seen at 848°C and another broad peak could be reconstructed from the DTA curve at 530 - 590°C. The ash #2 showed two endothermic peaks, one broad peak at 565 - 639°C and another sharp at 843°C. The reason for these peaks are unclear.

Gasification reactivity measurements

For the biomasses #2 and #5 the gasification reactivity was measured in a thermogravimetric analyzer. A clear difference was found for these two biomasses. The reactivity was clearly higher for the biomass #5 than for the biomass #2 even if both biomasses were fairly similar on a standard fuel analysis base (Tables 1 and 2).

The biomass #2 was more difficult to gasify completely which can be seen from Figure 1 as the decreasing trend of the gasification rate with increasing conversion, the gasification rate given as the reaction rate divided by residual ash free mass. For the biomass #5 the gasification rate increased with increasing conversion.

Micrographs presented in Figure 2 show a clearly more compacted structure for the ash from the biomass #5 than for that of biomass #2 after the gasification measurements. In both cases the start-out particle size for the gasification measurements were the same. The residue ash of the biomass #2 contained also rest carbon, while that of biomass #5 did not.

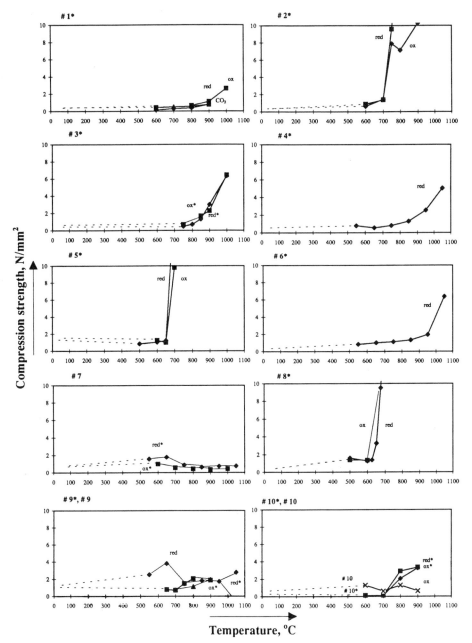

Figure 3. Sintering tendency for ten biomass ashes, tested with the compression strength sintering test in oxidizing and reducing gas atmosphere. O-test indicated in the left corner of each figure. An asterik indicated that the ash has been received by ashing in 550°C only.

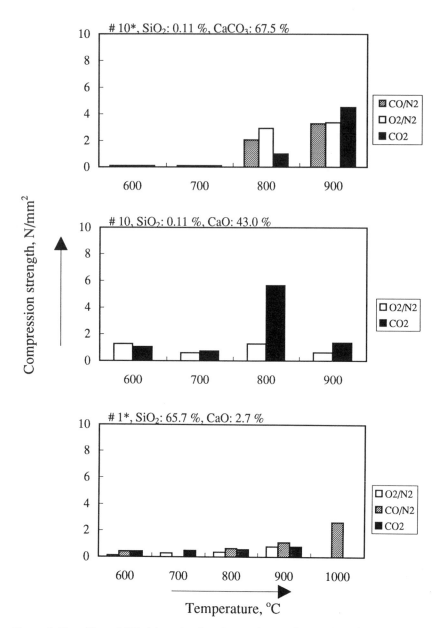

Figure 4. The effect of $CO_2(g)$ on the sintering tendency of two ashes. Gas atmospheres: O_2/N_2, CO/N_2, CO_2. Sintering time: 4h.

Above, #10*, ashed in 550°C only.

Middle, # 10, ashed in 550°C and calcined in 850°C.

Below, #1, ashed in 550°C only.

Sintering tendency of the ashes tested in reducing atmosphere

The results from the sintering measurements are summarized in Table 4 and presented in Figures 3 and 4. All ashes were at first sintering tested in a reducing gas atmosphere (5% CO, 95% N_2).

Clear differences could be seen in the sintering tendencies between the ashes. The temperature for the on-setting of sintering was at its lowest at 625°C for the potassium and calcium rich ash #8 followed by the potassium and silicon rich ashes #2 (700°C) and #5 (650°C).

The calcium rich ashes #7 and #10, which were ashed according to the extended procedure, did not sinter at all in the tested temperature range of 600 - 1000°C. The ash #10 which was ashed in 550°C only, sintered at 800°C.

Ash #9 showed a somewhat increased sintering at 650°C (3.79 N/mm^2) but at heat treatment temperatures above 650°C, the sintering decreased back to a 0-level (1.69 N/mm^2).

The silicon rich ashes #1, 3, 4 and 6 started to sinter at temperatures around 800°C.

The effect of the gas atmosphere on the sintering tendency

The ashes were also sintering tested in oxidizing gas atmosphere (100% air). No major effect of the oxidizing gas on the sintering tendencies could be detected when compared to reducing atmosphere.

Two ashes, the ashes #1 and #10, were further tested in $CO_2(g)$ atmosphere. Here a clear difference could be seen in sintering tendency for that calcium rich ash #10 which had received the extended ashing. For that ash a clear increase in sintering could be detected locally at 800°C when the gas atmosphere contained $CO_2(g)$. At higher temperatures no increase could be seen. A corresponding behavior could not be seen for the ash #10* that had been ashed in 550°C only. Neither could it be detected for the silicon rich ash #1*.

DISCUSSION

Sintering mechanisms

The results reveal significant differences in the thermal behavior of the tested ashes. In both reducing and oxidizing conditions the ashes #2, #5 and #8 show a sudden increase in sintering at 700, 650 and 625°C respectively. The ashes #3, #4, #6 and #10 show a more moderate sintering increase at 800, 850 and 950°C respectively and the ashes #1 , #7 and #9 low or no sintering. Some of these differences may be explained by the chemical composition in the ashes. Figure 5 shows an estimation of different compounds that may be present in the ashes. The estimation is based on stoichiometric calculations of the elements, analyzed in the ashes (Tables 2 and 3).

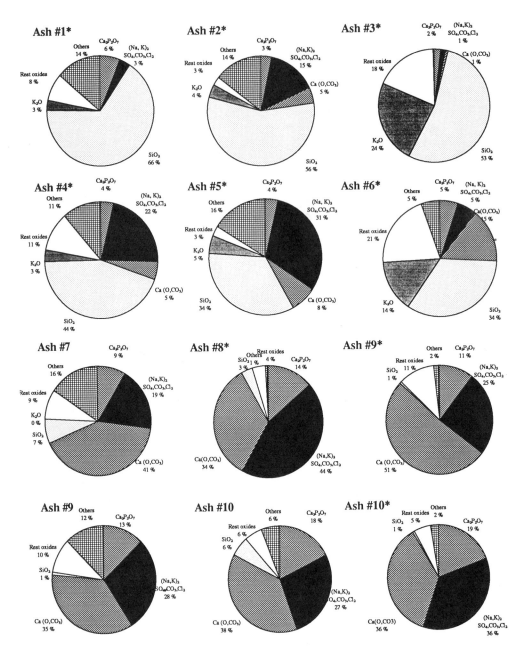

Figure 5. Stoichiometrically estimated amounts of different components (weight-%) present in the 10 biomass ashes, presented in Table 2. The calculations assume that all phosphour is present as $Ca_2P_2O_7$, carbonate as both K_2CO_3 and $CaCO_3$, rest calcium as CaO, all chlorine as KCl, sulfur as both Na_2SO_4 and K_2SO_4, rest potassium as K_2O, all silicate as SiO_2, rest analysed elements as oxides and unanalysed elements as others.

A possible explanation to the sudden and extensive sintering, detected with the compression strength test for the biomass ashes #2, #5 and #8 is a significant melt formation in the ash. All these three ashes show a large portion of possibly low melting alkali salts. Assuming a 15 % share of melt in the ash as being enough to cause significant strength increase in the ash pellets one can estimate the temperature at which this occurs. This temperature is called the T_{15} temperature. Making such an estimation one finds that the T_{15} for ash #2 is approximately 740°C, for ash #5 approximately 620°C and for the ash #8 605°C. Comparing these T_{15} temperatures with the sintering temperatures one finds a fairly good correlation. For the ash #5 the fact that the ash particles further were fused together after the steam gasification tests, supports also a possible melt formation.

The ashes #4, #7, #9 and #10 are interesting to compare here since they also contain large portions of alkali salts (Fig. 5). A rough estimate of the T_{15} for the ash #4 shows that it is in the same range as for the ash #2. For the ashes #7, #9 and #10 it is somewhat higher, roughly 800°C. These T_{15} -estimates, however, do not correlate well with the measured sintering tendencies.

Neither do DTA measurements clearly support the melt formation. No clear endothermic peak in combination with a stable weight could be detected during the DTA tests with the ashes #2, #5, #9 or #10. Differences can be seen for ex. between the ashes #2 and #5 but whether those are due to melting or something else, is not clear.

Another sintering mechanism may be dominating for the ashes from biomasses #7, #9 and #10. All these ashes contain fairly large amounts of calcium, presumably in oxide or carbonate form. If the calcium in the ash is found in an oxide form, it may be assumed that the ash will sinter mainly by reacting with the gas phase. If the calcium on the other hand is found readily as calcium carbonate in the ash, no effect of gas atmosphere on the sintering should be detected. This was found in the case with the ash #10. Comparing the sintering tendencies in 100% CO_2(g) for the ash received through the extended ashing procedure (Fig. 4, middle) with that received without the extended ashing procedure (Fig. 4, up) one finds a clear difference. The ash from the extended ashing sintered clearly more at 800°C than the ash that hadn't gone through the extended ashing procedure.

For the ashes #3 and #6 the moderate sintering tendency detected at 800, and 950°C respectively, may be due to an early stage of "viscous flow" sintering. Both these ashes have fairly high silicon content. Both ashes have also significant amounts of potassium and low amount of salts (sulfate, carbonate, chlorine). Potassium in silicate glass is known to enhance "viscous flow" sintering.

Making an estimate of the T_{15} temperature for these ashes, one finds that the ash #3 has a T_{15} of 770°C and the ash #6 1000°C. For the ash #6 it correlates well with the measured sintering temperature i.e., at 950°C the strength of the ash pellet is still low, at 1050°C it has clearly increased. For the ash #3 the estimated T_{15} is lower than the measured sintering temperature.

Neither of the ashes #3 or #6 were tested with DTA.

For the ash #1 which consists mainly of silicon the chemistry suggests that no significant melting (>15 % melt formation) occurs at temperature below 900°C. The endothermic peak at 850°C detected with the DTA may be due to the complete melting of the alkali salt part of the ash. The amount of this melt, i.e. the amount of the alkali salt part in the ash, is, however, so low that it doesn't seem to affect the compression strength of the pellets.

Ash behavior predictions for full scale FBC:s

If the sintering temperature, measured with the lab test is taken as an initial sign of bed agglomeration in a full scale unit and if we further assume a bed temperature of 800 - 900°C to be typical for a fluidized bed boiler, then the biomasses #2, #5, and #8 are predicted to cause extensive bed agglomeration, the biomasses #3, #4, and #6 moderate bed agglomeration and the biomasses #1, #7, #9 and #10 none or only minor bed agglomeration problems in a full scale atmospheric fluidized bed combustor.

The results indicate further that a biomass ash rich in calcium may form agglomerates at such conditions when the calcium forms $CaCO_3$ from CaO. The other case, i.e., when $CaCO_3$ is present in the ash, should not be relevant to atmospheric combustion, since the calcium always forms CaO there.

Since no major difference could be seen in sintering tendencies between oxidizing and reducing conditions a same kind of bed agglomeration behavior was predicted for both atmospheric combustion and gasification conditions.

The experiences from full scale FB combustion of the biomasses #7, #9 and #10 indicate no or only minor bed agglomeration problems. Also biomass #6 has been burned with no bed agglomeration problems. This is based, however, on a fairly short experience time. The combustion of biomass #3 has caused bed agglomeration problems in full scale FB combustion.

The biomasses #1, #2, #5 and #8 have been combusted in lab scale FB furnaces only. The experiences from these tests indicate that the ashes from the biomasses #2, #5 and #8 form extensive bed agglomerates while the ash from biomass #1 does not form bed agglomerates at all.

In pressurized systems a difference in bed agglomeration was predicted compared to atmospheric systems. Since the partial pressure of $CO_2(g)$ in a pressurized system is higher than that of an atmospheric it was concluded that any sintering mechanism dependent of the gas atmosphere would be intensified in a pressurized system. By interpreting the results achieved with the sintering measurements in 100% $CO_2(g)$ it was predicted that on top of those ashes that sintered due to a melt formation or viscous flow also those ashes rich in calcium would have a higher tendency to sinter. In this specific case an additional requirement was that the pressurized system worked under such conditions that the calcium in the ash could at first form calcium oxide. The bed agglomeration was then predicted to be caused by the carbonation of the calcium oxide in the ash. The ashes from the biomasses #7, 9 and 10 were those which were predicted to have the highest sintering tendency through this mechanism.

It is important to recognize that these predictions take into account mainly the effect of chemistry on the bed agglomeration process when the ash has formed. In full scale FB units also other phenomena, such as mineral matter release, particle size distribution and ash reactions with other material streams in the furnace, will affect the bed agglomeration. The prediction made here are to be taken as approximate ones.

CONCLUSIONS

Ten different types of biomass ashes were characterized with respect to their ash behavior in fluidized bed processes. The ash characterization methods were i) wet chemical analyses, ii) compression strength measurement based sintering tests, and iii) combined DT/TG analyses and iv) gasification reactivity measurements.

Significant differences could be found in the ash compositions. The silicon content varied between 0.6 - 65.7 wt-% oxides in the ash. Silicon rich ashes were usually poor in calcium except for the ash #6 which contained moderately high amounts of both elements. The calcium content varied between 2.7 - 44.8 wt-% oxides. The potassium content varied 4.6 - 28.4 wt-% oxides. The ashes #2, 5, and 8 had moderately high chlorine contents.

No major changes could be seen in the elemental compositions of the ashes as a function of heat treatment. The clearest trend could be found for the carbonate content in the ashes, e.g. it decreased as the heat treatment temperature exceeded some 800°C.

Significant differences could be found in the sintering tendencies for the ashes heat treated in reducing conditions. The highest sintering tendency was found for the ash #8, T_{sint} = 625°C in red. gas, followed by the ashes #5, T_{sint} = 650°C, and #2, T_{sint} = 700°C. Moderate sintering tendencies were found for the ashes #3, T_{sint} = 800°C, and #4, T_{sint} = 850°C. The lowest sintering tendencies were found for the ashes #1, T_{sint} = 900°C, and #6, T_{sint} = 950°C. The ashes #7 and 9 did not sinter at all in the temperature range of 500 - 1100°C. A change from reducing gas to oxidizing gas atmosphere did not change the sintering behavior of the ashes. A change to 100% CO_2(g) in the atmosphere lead to significant sintering for the calcium rich ash #10 at 800°C, but not at 900°C anymore. The sintering of the silicon rich ash #1 did not show any dependence of the CO_2(g).

For atmospheric FB conditions the sintering results indicated bed agglomeration for the biomasses #2, 5 or 8. A molten phase present in the ashes was suggested as the explanation even if the DTA measurements didn't clearly support this fact.

Some bed agglomeration was predicted for the biomasses #3, 4, and 6. For the ashes #3 and #6 a silicate melt could form from these ashes. However, for the ash #6 the melt was assumed not to cause any major bed agglomeration problems under normal running conditions since the amount formed at these conditions would not be high enough. For the ash #4 both a silicate and a salt melt was possibly present in the ash. It was not clarified which of these caused the sintering.

Minor or no bed agglomeration was predicted from the sintering results for the biomasses #1, 7 , 9, and 10. The ash #1 had a possibility to form silicate melts but the amount was assumed to be low. The ashes #7, 9, and 10 were predicted to sinter through a reaction with the gas phase.

For those ashes where full scale FB combustion experiences were available, the bed agglomeration problems experienced there correlated well with the predictions made, based on the sintering tests.

ACKNOWLEDGMENTS

This work is part of the Finnish National Combustion Research Program LIEKKI 2. Financial support from the Finnish Ministry of Trade and Industry and A. Ahlström Corporation through the LIEKKI 2 program is gratefully acknowledged. The ashing of the fuels used in this work was done partly by Mr. Matti Nieminen, VTT Energy, partly by Ms Päivi Mänttäri, A. Ahlström Corp., R&D Laboratory (now Foster Wheeler Energia OY, R&D Center). This as well as the melting characteristic calculations by Dr Rainer Backman and the careful laboratory work by Mr. Ove Holm, Mr. Sören Karlsson, Ms Hanna Malm and Ms Katrine Alhonen at Åbo Akademi University is also acknowledged.

REFERENCES

Barnhart, D. H., Williams, P. C.: Trans. ASME 78, 1229 (1956).

Baxter, L.: Biomass and Bioenergy 4 (2), 89 (1993)

Dawson, M., Brown, R., C.: Fuel 71, 585 (1991).

Hupa, M., Skrifvars, B-J., Moilanen, A.: J. Inst. Energy 62, 131 (1989)

Moilanen, A., Kurkela, E.: Gasifications reactivity of solid biomass fuels, Proc. of the 210th ACS Div. of Fuel Chemistry Conference, Vol. 40, No 3, pp. 688 - 693, 1995.

Nordin, A.: Fuel 74, 615 (1995)

Nordin, A., Skrifvars, B-J, Öhman, M., Hupa, M.: "Agglomeration and defluidization in FBC of biomass fuels - Mechanisms and measures for prevention", presented at the Eng. Found. Conf., July 16-21, 1995, Waterville Valley, NH, USA

Skrifvars, B-J., Hupa, Hyöty, P.: J. Inst. Energy 64, 196 (1991)

Skrifvars, B-J., Hupa, Hiltunen, M.: Ind. & Eng. Chem. Res. 31, 1026 (1992)

Skrifvars, B-J., Hupa, Backman, R., Hiltunen, M.: Fuel 73, 171 (1994)

THE SAFETY AND ECONOMICS OF HIGH ASH ANTHRACITE FIRED MIXING WITH PETROLEUM-COKE IN PULVERIZED COAL-FIRED FURNACE

Zhiguo Zhang, Xuexin Sun, and Fujin Li

National Laboratory of Coal Combustion
Power Engineering Department
HuaZhong University of Science and Technology
WuHan HuBei, 430074
People's Republic of China

ABSTRACT

Petroleum-coke was fired only in CFB because of its content of high S and low volatile. It will bring environment and flame stability problem if petroleum-coke fired in pulverized coal-fired furnace. As the research of coal fire theory has been developed in last decade, low rank anthracite is fired in many pulverized coals fired furnaces without flame stability problem. Here we blend high ash anthracite with petroleum-coke as the fuel for pulverized coal-fired furnace to decrease the ash content in fuel. Experiment result had shown that mixing with petroleum-coke the combustion behavior of blending fuel were improved and ash deposition characteristic would not change compared with high ash anthracite. Using coal/petroleum-coke as the fuel for furnace can bring great benefits for environment and furnace. But S content in blend fuel must be controlled under the regulation of S content in coal and the volatile content should not decrease too low to coal fired the furnace design to avoid the environment and flame stability problems occur.

INTRODUCTION

In south of China many power plants fired high anthracite coal (ash content in coal $A^f > 45\%$) because of economics pressure. This fuel has brought a lot of problems for furnace operation and maintenance. High expenses were using for fuel transportation, for furnace operation and ash storage or transportation. Especially there are many power plants fire coal which contents more ash than its designing. Ash content increased in coal play great role in furnace economics and safety of furnace operation. Ash content increasing in coal also reduces the volatile content and heat value of fuel. It is necessary to change the fuel but unable to do like this because of economics. So we consider to blend some other fuel or coal to improve the combustion behavior of high ash anthracite.

In other way thousands' tons of petroleum-coke are produced in chemical petroleum

Applications of Advanced Technology to Ash-Related Problems in Boilers
Edited by L. Baxter and R. DeSollar, Plenum Press, New York, 1996
399

works every day. Some petroleum-coke were fired to produce the carbon pipe Others are sailed or storage as fuel for fired . Comparing with high ash anthracite petroleum-coke has excellent burning characteristics and low ash content. Property of some petroleum-coke is shown in table 1.

Table 1. Property of Petroleum-Coke.

	Moi	A^f	V^f	C^f	C	H	O	N	S	Q(kj/kg)
PC1	6.4	0.5	12.7	86.8	90.3	4.16	0.6	1.24	3.67	36267
PC2	7.6	0.96	11.0	88.0	89.16	4.11	0.82	1.56	4.35	35937
PC3	7.4	1.64	9.93	88.4	90.24	3.9	1.93	1.11	2.82	33499

Here we consider to mixing the petroleum-coke with high ash anthracite to improve the fuel combustion . behavior of high ash anthracite in furnace from many aspects. As the S content in petroleum-coke is higher than many coals, we should control the blending rate of petroleum-coke to avoid SOx emission increasing when the blend fuel is fired in furnace.

Blending with petroleum-coke the blend fuel combustion characteristics and ash deposition will change neither similar to coal nor to petroleum-coke. We had conducted many experiments of fuel combustion and ash deposition to understand the changing of fuel's characteristic. Theory analyzing was performed to economics of furnace fired blending fuel (anthracite/petroleum-coke). All results show that if blending petroleum-coke with high ash anthracite the ash exportation to environment will reduce and SOx emission will not increase if the blending rates of petroleum-coke are controlled. The apparatus erosion will reduce because the ash concentration in flue gas reducing.

EXPERIMENT

All experiments were performed to investigate the combustion characteristics and ash deposition characteristic in our research work.

A. Combustion Characteristics of Blending Fuel

Comparing with other coal the ignition and flame stability of high ash anthracite is very difficulty. In our research subject the property of high ash anthracite. Petroleum-coke and blending fuel are shown in table 2.

Table 2. Property of Coal, Petroleum-Coke, and Fuel Blends.

	Moi	V^f	A^f	C^f	N	S	Q(kj/kg)
coal	1.85	9.22	41.44	48.7	0.6	1.5	16500
pc	0.57	7.08	0.42	91.93	1.24	2.82	36257
10% cpc	1.73	9.01	36.8	52.46	0.56	1.59	18950
30% cpc	1.42	8.4	29.5	60.68	0.46.	1.81	22531
50% cpc	1.19	8.0	22.1	68.71	0.91	2.01	25031

1.1 Relation between ash content and ignition, flame stability
Many experiments have shown that the fuel's heat value will increase and the ignition temperature will decrease if the ash content decrease. The ignition heats of fuel also decrease if ash content decrease of fuel. Fig.1 shows the relation between ash content and ignition temperature, heat value. Ash content also influences the ignition velocity when the volatile of fuel are same. If the ash content of fuel decreases theory

combustion temperature of coal particle will increase 142-168 C°. So if we decrease the ash content in high ash anthracite, it will improve the behavior of fuel combustion in pulverized coal-fired furnace.

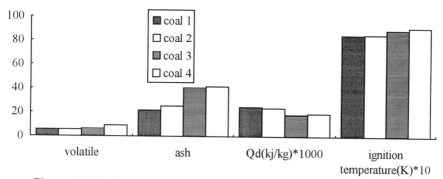

Figure 1. Relation between Ash-Content to Qd and Ignition Temperature.

1.2 Experiment

A. Combustion characteristic of CPC

Coal combustion process in pulverized coal-fired furnace is a complex physical and chemical reaction. Many factors have play role in coal combustion. The influence factors may act in aspects of coal ignition, flame stability coal burn-out etc. As the developments of the coal combustion theory the proximate and utility analyzed are no longer satisfactory to the demand of understand the coal combustion characteristic. Here we introduce two new technologies to investigate coal combustion process and prove two methods to compared the different coal combustion to different coal: thermogravimeter behaviors and pore structure of fuel.

a: Thermogravimeter behavior of CPC

According to thermogravity behavior of fuel we can get DTGA curves from coal process. As the DTGA curves can demonstrate the coal combustion velocity, we can define two indexes of combustion : devolatilization index D and combustion characteristic index S.

From DTGA curves we can get following parameters about coal combustion:

T_s: Temperature at which the volatile matter just start to release;

$(\dfrac{dw}{d\tau})^V_{max}$: maximum rate of devolatilization;

T_{max}: temperature at which $(\dfrac{dw}{d\tau})^V_{max}$ occurs;

$\Delta T_{1/3}$: temperature region corresponding to $(\dfrac{dw}{d\tau})^V/(\dfrac{dw}{d\tau})^V_{max}=1/3$;

Obviously the larger the value of $(\dfrac{dw}{d\tau})^V_{max}$, the more intensive the devolitilization process of coal; the lower the T_{max} and $\Delta T_{1/3}$, the earlier and the more active the devolitilization. Therefore it is more favorable for the ignition process of coal. So we define D as devolatilization index:

$$D= (\frac{dw}{d\tau})^V{}_{max} /Th \times \Delta T_{1/3}$$

Table 3 shows D value change for different coal.

Table 3. D Value for Different Fuel.

	bituminous	sub-bituminous	anthracite
D	$>5*10^{-6}$	$<2*10^{-6}<5*10^{-6}$	$2*10^{-6}<$

Table 4. D Value of Experiment Results.

	anthracite	PC	90% coal 10% cp	70% coal 30% cpc	50% coal /cpc
D	$0.51*10^{-6}$	$0.49*10^{-6}$	$0.498*10^{-6}$	$0.502*10^{-6}$	$0.505*10^{-6}$

According to the devolatilization index of different fuels we know that blending with petroleum-coke will not change the devolatilization property of fuels.

In the study of coal combustion process two main aspects: ignition and burn-out are closed related to the following parameters of coal: 1). ignition temperature; 2). flame stability; 3). activation energy; 4). burn-out rate. As these parameters could not represent the comprehensive combustion behavior of coal by using any single parameter, we conduct following equation to connect all parameters to coal combustion characteristic.

The combustion velocity may be expressed in Arrhenius form:

$$\frac{dw}{d\tau}=A \times \exp(\frac{-E}{RT}) \qquad (1)$$

Differentiating equation with respect to T yield:

$$\frac{R}{E} \times \frac{d}{dT}(\frac{dw}{d\tau})=\frac{dw}{d\tau}*(1/T^2) \qquad (2)$$

At ignition point we have $T=T_i$, so the equation can change to:

$$\frac{R}{E} \times \frac{d}{dT}(\frac{dw}{d\tau})T=Ti \times \frac{(dw/d\tau)^{max}}{(dw/d\tau)^{T=Ti}} \times \frac{(dw/d\tau)^{mean}}{Th}=\{(\frac{dw}{d\tau})_{max} \times (\frac{dw}{d\tau})_{mean}\}/Ti^2 \qquad (3)$$

Th : burn-out temperature; In above equation we can know:

$(\frac{dw}{d\tau})_{max}$: maximum combustion rate; $(\frac{dw}{d\tau})mean$: mean combustion rate;

$(\frac{dw}{d\tau})T=Ti$: combustion rate at ignition point;

so we can define combustion characteristic index as S:

$$S=\{(\frac{dw}{d\tau})_{max} \times (\frac{dw}{d\tau})mean\}/T^2 \times Th \qquad (4)$$

as E represent the activity energy of coal,$\left\{ \dfrac{d\left(\frac{dw}{d\tau}\right)}{dT_{T\,=\,T_i}} \right\}$ represent the differential of combustion rate at ignition point, the large numerical value of this term, the more

intensive ignition process and $\left. \dfrac{\left(\frac{dw}{d\tau}\right)_{max}}{\left(\frac{dw}{d\tau}\right)_{mean}} \right/ T_h$ is the ratio of mean combustion rate to the burn-out temperature, the larger the ratio the faster the burn-out process. So index S represents the comprehensive characteristic's of coal. Obviously the larger the numerical value of S the better the combustion behavior of fuel. Fig. 2 shows the combustion behavior of fuel in experiment.

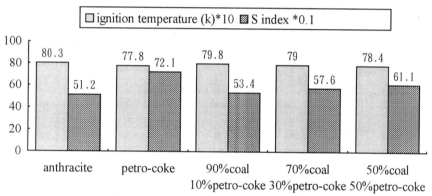

Figure 2. Combustion Characteristic of Fuel.

b. Pore structure of CPC

In the ignition and combustion process of pulverized coal, many physical and chemical reaction are occurring on the surface and pore of coal particle. Especially in the process of coal ignition, the combustion rate is at passive ratio to the coal particle surface area. So pore structure play an important role in coal combustion process. For same diameter of pulverized coal particle the large the pore diameter, the intensive the reaction of coal combustion because it is easy for gas to enter particle inner. The larger the surface area, the more enough connecting of coal and gas to reaction.

In our research work the surface area and pore structures are measured by using ASAP (Accelerated Surface Area and Porosimetry System) instrument that made by Micrometer Co. U.S.A. This instrument that made isotherm adsorption at the temperature of saturated liquid nitrogen to determine the volume of nitrogen adsorbed by the sample of the related pressure P/Po from 0.01 to 0.95 to obtain the surface area and pore distribution. Five samples were measured to study the pore structure of CPC. Fig. 3 shows the surface area of fuel of experiment. Pore volume and pore diameter of fuel show that there is a little change for anthracite and blend .The pores diameter of research fuel is less than 2 nm of blend fuel.

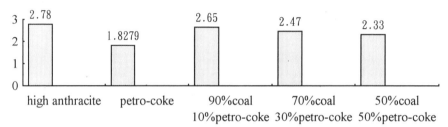

Figure 3. Surface are of fuel (m²/g).

B. Ash Deposition of High Ash Anthracite/Petroleum-coke

Ash deposition which form on heat transfer surface in a pulverized coal-fired furnace can cause significant operation problem and loss efficiency. These are the leading of both derating and outage. Without considering the fluid influence on ash deposition, ash slagging in furnace may concern with ash's composition. Eight major constituents make up most coal ash SiO_2, Al_2O_3, Fe_2O_3, Na_2O, K_2O, SO_3. Different ash composition will have different feasibility. Table 5. and Table 6. show the melting temperature of common coal ash oxides and eutectics.

Table 5. Melting Temperature (F) of Coal Ash Oxides.

SiO_2	Al_2O_3	MgO	CaO	Fe_2O_3	Na_2O	K_2O
3128	3722	4532	4660	2849	2330(sublimes)	660(decomposes)

Table 6. Melting Temperature of Coal Ash Eutectics.

FeO-FeS	1724 F
FeO-Al_2O_3-SiO_2	1963 F
CaO-FeO	1967 F
Na_2O-Al_2O_3-$6SiO_2$	2012 F
Na_2SO_4-$CaSO_4$	1675 F

It can be predicted the fusibility and slagging for ash by using following four parameters:

1. T_2 : ash melting temperature;

2. Silica percentage G : $(\dfrac{SiO_2*100}{SiO_2+Fe_2O_3+CaO+MgO})$;

3. Based-acid ratio B/A: $(\dfrac{Fe_2O_3+CaO+MgO+Na_2O+K_2O}{SiO_2+Al_2O_3+TiO_2})$;

4. SiO_2/Al_2O_3 ratio;

Table 7 is the ash composition of fuel in our experiment. The value of four parameters to predicts ash fusibility and slagging are shown in table 8.

Table 7. Ash Composition of Coal, Petroleum-Coke, and CPC.

	Fe_2O_3	CaO	MgO	TiO_2	K_2O	Al_2O_3	SiO_2	SO_3	P_2O_5
coal	5.86	1.22	1.20	1.72	2.4	28.32	57.44	0.53	0.07
pc	5.79	7.51	1.26	0.67	1.52	27.31	47.69	7.94	0.0
cpc10%	5.84	1.83	1.21	1.64	2.31	28.21	56.67	1.31	0.06
cpc30%	5.83	3.17	1.22	1.35	2.14	28.03	53.93	2.73	0.05
cpc50%	5.82	4.36	1.23	1.14	1.90	27.89	52.20	4.13	0.03

Table 8. Principal of Prediction for Ash Slagging Propensity.

slagging index	range of index	slagging propensity	anth- racite	petro- coke	blend fuel 1	blend fuel 2	blend fuel 3
$t_2(°C)$	>1390	low	>1500	1450	1500	1485	1480
	1260-1390	medium					
	<1260	sever					
SiO_2/Al_2O_3	<1.87	low	2.028	2.6	2.06	1.923	1.872
	2.65-1.87	medium					
	>2.65	serve					
B/A	<0.206	low	0.122	0.24	0.129	0.148	0.163
	0.206-0.4	medium					
	>0.4	sever					
G	>78.8	low	87.4	76.6	86.45	84.15	82.08
	66.1-78.8	medium					
	< 66.1	sever					

According to all experiment result we can get some conclusion about the combustion characteristic and ash deposition of high ash anthracite / petroleum-coke:

1). The ignition of blend fuel had not changed because of its volatile level.

2). The burn-out characteristics of coal are improved when petroleum-coke were added in.

3). The fuel combustion characteristic of blend fuel would not cause the changing of furnace apparatus to fit for CPC.

4). The fusibility of blend fuel would not change when in our experiment even though petroleum-coke is a slagging fuel. So there would be no ash deposition problem for pulverized coal-fired furnace use CPC as fuel. If there are no other factor causing slagging problem, The mixing with petroleum-coke for high ash anthracite is applicable for pulverized-coal fired furnace.

C Economics of Pulverized coal-fired furnace Fired Blend Fuel

Ash content in coal plays an important role in coal ignition, burnout and ash deposition characteristic. If we can decrease the ash in coal, the furnace operation and combustion efficiency will be improved in many aspects. In industry experiment , when ash content decreased from 40% to 17% , the combustion temperature in center of flame has increased 145 C° and the efficiency of furnace increase 1.98%. The fuel consumption has decreased from 39.1t/h to 24.8t/h and the ash concentration in flue gas has decreased. In other way the abrasion of tube in furnace is equal to following equation:

$$I_{max} = \alpha \times M \times \eta \times U_Y^3 \times K\mu^3 \times K\mu \times \mu \times \tau \text{ (mm)} \qquad (5)$$

I_{max} : maximum abrasion amount of tube;

α: abrasion index of ash in fuel;

M: fray property of metal of tube;

η: efficiency index of ash attack the tube;

U_Y : velocity of flue gas near the tube;

$Ku, K\mu$: index of gas velocity and ash concentration distribution before tube;

μ: ash concentration in flue gas

τ : abrasion time of tube

Table 9. Relation between Ash Content, Ash Concentration, and Volume of Flue Gas.

	ash	fuel consumebtion(t/h)	volume of flue gas (Nm^3/h)	velocity before super heater (m/s)	velocity before air heater (m/s)	ash concentration in flue gas(g/m^3)
fuel a	40	37.54	223345	7.17	11.65.	63.13
fuel b	17	23.8	207274	6.72	10.6	17.4

From table 9 and equation （5） we can know that if ash content decreased in fuel the ash concentration in flue gas and velocity of flue gas on heater will decrease. This will be very beneficial for furnace operation and apparatus maintenance.

CONCLUSION

All experiments shown that blend fuel (anthracite / petroleum-coke) could be fired in pulverized coal-fired furnace without causing flame stability problems. Using petroleum-coke as partly fuel for pulverized coal-fired furnace will be very beneficial in decreasing the expense of transportation of coal from far away than from petroleum chemical works. Blending petroleum-coke with coal could reduce the ash exportation to air and land. We know ash exportation is closed related ash content in fuel. Exportation of ash equal to $8A^f$ (ash content in coal). In our experiment A pulverized coal-fired furnace (670t/h) fired 150t/h high ash anthracite. If we reduce ash content in coal from 40% to 17%, the ash exportation will reduce 10.4 t/h and 72800 t/year (7000 hour operation time for furnace). This will be very beneficial both for environment and for apparatus of furnace.

Blending petroleum-coke with coal to be fired in pulverized coal-fired furnace has become an applicable technology. It will not bring environment and combustion problems to furnace and will improve the combustion behavior of high ash anthracite fired in pulverized coal-fired furnace. Which must be pay attention to is to control the S content to avoid the increasing of SOx emission. Control the ratio of petroleum-coke in blend fuel will keep the emission of pollution in the range of regulation. The S content in petroleum-coke depends on the oil property in chemical petroleum works.

APPENDIX
cpc 10% : coal 90% / petroleum-coke 10%=blend fuel 1;
cpc 30% : coal 70% / petroleum-coke 30%=blend fuel 2;
cpc 50% : coal 50% / petroleum-coke 50%=blend fuel 3;
CPC : coal /petroleum-coke;
PC :petroleum-coke;

REFERENCE

1. Jianyuan Chen. Study of coal ignition process and model; Ph.D thesis, 1989, HUST, WuHan. HuBei. P.R. of CHINA.
2. Hong Chen. Study of pore structure and combustion kinetics of char during raw coal burning. Ph.D thesis, 1994, HUST. WuHan HuBei. P.R. of CHINA
3. Richard W. Bryes. Fireside behavior of mineral impurities in fuels from marchwood 1963 to the Sheraton Palm coast 1992. Proceeding of Inorganic Transformation and Ash Deposition During Combustion. p3-71. March 10-15 1992. Palm coast. Florida. U.S.A.
4. T. Abbas, P. Costen and F. C. Lockwood. The effect of particle size on NO formation in a larger scale pulverized coal fired laboratory furnace: Measurement and Modelling. Combustion and Flame. p316-325. Vol. 93. May 1993.
5. J. R.Macdonald. Control of solid fuel slagging. Power Engineering. p 48-51. August. 1984.
6. Coal combustion and application. Progress in energy and combustion science. Vol.10 Number 2 1984.
7. Hans Beisswenger. Application of CFB technology to fired petroleum-coke of petroleum works. Journal of Petroleum Chemical Industry of U.S.A. 1985.
8. James Sahagian. Combustion of petroleum-coke in Pyroflow CFB. Journal of Petroleum Chemical Industry of U.S.A. 1985.
9. Toshizo Ito. Long-term Stable Operation of Petroleum Coke Firing boiler retrofitted from existing oil firing. Proceeding of the Engineering Foundation Conference On Inorganic Transformation and Ash Deposition During Combustion. March 10-15, 1991. Palm Coast Florida.

ALKALI SALT ASH FORMATION DURING BLACK LIQUOR COMBUSTION AT KRAFT RECOVERY BOILERS

P. Mikkanen[1], E.I. Kauppinen[1], J. Pyykönen[2], and J.K. Jokiniemi[2]
VTT Aerosel Technology Group, [1]VTT Chemical Engineering, [2]VTT Energy
P.O. Box 1401, FIN - 02044 VTT, Finland

M. Mäkinen
Finnish Meterological Institute
Sahaajakatu 22 E, FIN-00810 Helsinki, Finland

ABSTRACT

Recovery boiler is an essential part of paper pulping process, where waste sludge called black liquor is burned for chemical recovery and energy production. This study was carried out at an operating industrial recovery boiler in Finland.

Measurement of aerosol particles was carried out at bullnose level of furnace, at boiler exit, and at outlet of electrostatic precipitator (ESP). Aerosol mass size distributions in size range 0.02 - 50 μm were measured with Berner type low pressure impactor (BLPI) operated with precyclone. BLPI samples were further analysed with ion chromatography for water soluble Na, K, SO_4, and Cl. Particle morphology was studied with scanning electron microscopy (SEM). Phase composition of crystalline salts was measured with X-ray diffraction (XRD). Particles larger than 1 μm were analysed with computer controlled scanning electron microscopy (CCSEM) to derive particle composition classes.

At ESP inlet mass size distribution was bimodal with a major mode at about 1.2 μm and a minor mode at about 5 μm (aerodynamic diameter). At ESP outlet the mass size distribution showed only one peak at about 1.2 μm. Both submicron and supermicron particles were agglomerates formed from 0.3 to 0.5 μm spherical primary particles.

XRD analyses indicated that particles were crystalline with two phases of Na_2SO_4 (thenardite and sodium sulphate) and $K_3Na(SO_4)_2$. CCSEM results of individual particles larger than 1 μm showed that 79 to 88 volume percent of particles contained mainly Na and S, 7 to 10 volume percent Na, K, and S with minor amount of particles containing Na, S, and Ca.

Applications of Advanced Technology to Ash-Related Problems in Boilers
Edited by L. Baxter and R. DeSollar, Plenum Press, New York, 1996

INTRODUCTION

In recovery boilers waste sludge from pulping process called black liquor is used as fuel. Black liquor contains of about 20 % of sodium and less than 5 % of potassium. As about 10 % of Na and about 15 % of K is volatilized, the inorganic ash concentration in flue gases is up to 30 mg/Nm^3.

Black liquor is sprayed into the recovery boiler furnace. In ideal case the droplets dry and volatilize and the residue burns within the furnace. In reality some droplets entrain the furnace before complete burning and some deposit on the smelt bed at the bottom of the furnace before even drying. Fly ash particle formation is sensitive for temperatures and concentrations of species present, therefore each furnace atmosphere produces characteristic flue gas aerosol. In following some options for alkali salt fly ash formation are described.

In recovery boiler furnace burning particles reaches high enough temperatures to release small amounts of metals e.g. Mg and Ca. These metals forms oxides, that condense homogeneously forming tiny (30 nm) seed particles for alkali salt condensation. Volatilized alkali species (Na and K) form hydroxides in the boundary layer of the burning particles. If enough Na is present to capture all sulphur (e.g. in softwood combustion), excess forms carbonates and chlorides. Carbonates are formed on surfaces of particles and chlorides can condense on present particle surfaces or on heat exchanger surfaces. Particles also grow by coagulation and agglomeration of primary particles [Jokiniemi et al., 1995].

KOH forms sulphates, if SO_2 concentration in flue gases is high enough. If all sulphur is captured by Na, potassium forms chlorides. K_2SO_4 condenses on present particle surfaces and KCl condenses on particle surfaces or on heat exchanger surfaces. If chlorine is not present as chlorides i.e. when SO_2 is high, it forms HCl.

Flue gases formed at recovery boiler clearly differs from flue gases from pulverized coal fired boiler of similar capacity. Usual load of fly ash in flue gases at recovery boiler is 20-30 g/Nm^3 and at coal combustion 5-15 g/Nm^3. The particle mass size distribution shows that most of the fly ash from recovery boiler is in particles of about 1 µm in size, where as most of the mass of the fly ash from coal fired boiler is in particles larger than 5 µm [Joutsensaari et al., 1995]. At recovery boiler water content in flue gases is significantly higher than at coal fired boiler. NO_x content is 100 ppm at ESP inlet and SO_2 content can vary from 100 to 1500 ppm during one day.

EXPERIMENTAL

Measurements were carried out at an operating recovery boiler in Finland burning three different types of black liquor. Boiler operating conditions are shown in Table 1. During the experiments 10-20 tons/day of biosludge from the waste water treatment was burned with black liquor. Concentrated odour gases were burned in furnace and low concentrate odour gases were introduced with tertiary air.

Aerosol particle mass and composition size distributions were measured at boiler exit after the heat exchangers i.e. at ESP inlet. Individual particles from flue gases were collected at recovery boiler furnace at bullnose level and at boiler exit for characterisation and for comparison with particles collected from boiler bank, economiser and ESP ash hoppers. Total particle concentration was measured at boiler exit gas compositions for SO_2, O_2, CO_2, CO, NO_x, and H_2O were measured at ESP exit.

Stability of recovery boiler process was monitored by CO measurement (Fig. 1). The variation in gas composition shows that process was relatively stable during tests.

Table 1
Boiler operating conditions

black liquor flow	black liquor type	dry solids content
l/s		%
24.5	softwood	73
21.0	softwood	83
24.0	mixedwood	81

Particle Characterisation

Aerosol mass size distributions in size range 0.02 - 50 µm were measured with Berner type low pressure impactor (BLPI) [Kauppinen et al., 1990] operated with precyclone. Particles collected with precyclone were size classified with Wind Sieve method (WS). BLPI samples were further analysed with ion chromatography for water soluble Na, K, SO_4, and Cl.

Particle morphology was studied with scanning electron microscopy (SEM). Phase composition of crystalline salts was measured with X-ray diffraction (XRD) for samples of ESP hopper dust. Particles larger than 1 µm were analysed with computer controlled scanning electron microscopy (CCSEM) to derive particle composition classes.

RESULTS AND DISCUSSION

This recovery unit was chosen for experiments, because of the very high solids content of the black liquor. Also more common dry solids content 73 % was tested during experiments. Black liquor analyses results are shown in table 2.

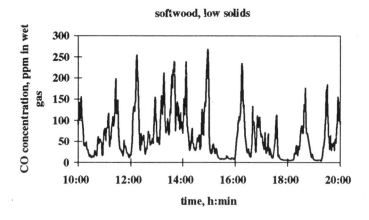

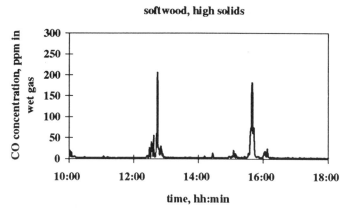

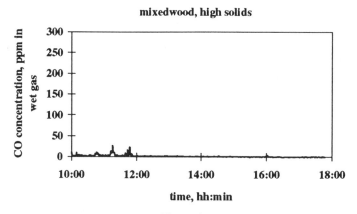

Figure 1
CO concentrations during recovery boiler experiments

Table 2
Black liquor analyses results

liquor type	softwood	softwood	mixedwood
dry solids	low solids	high solids	high solids
dry solids, %	73	83	81
C in dry solids, %	32.1	32.8	30.4
H in dry solids, %	3.1	3.3	3.1
N in dry solids, %	0.1	0.1	0.1
S in dry solids, %	6.2	5.9	6.0
Na in dry solids, %	22.7	22.1	22.9
K in dry solids, %	1.2	1.2	1.3
Cl in dry solids, %	0.2	0.1	0.2
calorific value, MJ/kg	12.7	12.8	12.4

Table 3
Gas composition and mass concentration for total particles and for particles smaller than 3 μm at boiler exit

liquor type	softwood, low solids	softwood, high solids	mixedwood, high solids
O_2, %	4.2	4.1	4.0
CO_2, %	14.5	15.1	15.7
SO_2, ppm	50	10	50
NO_x, ppm	80	85	100
H_2O, % vol	23	20	22
total mass concentration for particles at ESP inlet, mg/Nm^3	14	20	26
mass concentration for particles < 3 μm at ESP inlet, mg/Nm^3	9	11	8

Gas composition and mass concentration for total particles and for particles smaller than 3 μm at boiler exit are shown in Table 3.

Total particle concentrations were measured with different measurement techniques. In Fig. 2 results of measurements based on detection of salt particles collected onto water and analysed for Na ions and on opacity measurement are shown. During softwood experiments the data is consistent, because the opacity meter was calibrated for softwood black liquor combustion. During combustion of hardwood black liquor the optical characteristics of particles were significantly different causing 40 % difference between the results of these two methods.

413

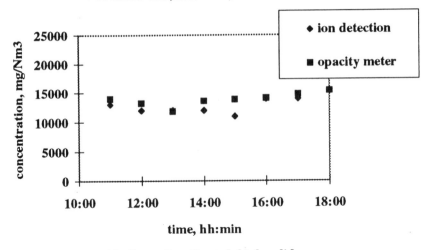

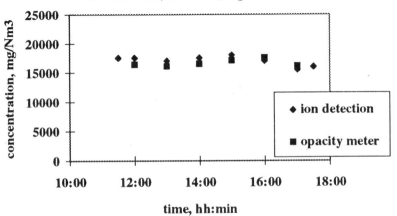

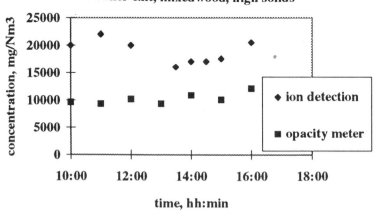

Figure 2
Total particle concentrations measured with different techniques

Fractions of Na, K, S, Cl, C, and N released from black liquor solids are calculated based on composition of particles smaller than 3 μm and flue gas composition (Table 4). Particles larger than 3 μm are not counted for, since their origin is unknown. Calculations of release rates for Cl and N are very sensitive to gas composition measurements. Only tens of ppm increase in HCl and NOx measurement results causes about 5 % increase in release rate results.

Increasing dry solids content in black liquor increases furnace temperature and therefore release rates are also increasing. Burning mixedwood black liquor (lower lignin content) decreases the furnace temperature and the release rates for Na and K are lower than during softwood combustion. Release rate for Cl is higher during mixedwood experiments than during softwood experiments.

Table 4

Release rates of Na, K, S, Cl, C, and N from black liquor solids

liquor type	softwood, low solids	softwood, high solids	mixedwood, high solids
Na, %	5.7	6.2	4.8
K, %	9.2	10.2	7.2
S, %	14.5	17.0	14.0
Cl, %	32	29	35
C, %	94	87	94
N, %	19	21	21

In Fig. 3 particle composition during recovery boiler experiments for particles collected from flue gases and for particles collected from the ESP ash hoppers. Ratio of K to Na for particles is higher for mixedwood than for softwood and ratio of Cl to Na is increasing with increasing solids content at ESP. The category other refers to the difference between the gravimetric mass and the water soluble ion masses detected with IC.

Aerosol mass size distributions at boiler exit are shown in Fig. 4. The differential mass size distribution is normalised with boiler load per hearth area. Mass size distribution is bimodal with a major mode at about 1.2 and a minor mode at 5 μm. The boiler temperature is highest during softwood combustion with high solids and therefore more alkaline is volatilised and the major mode is wider.

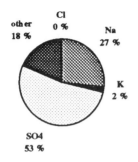

in particles, softwood, low solids

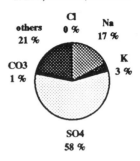

at ESP, softwood, low solids

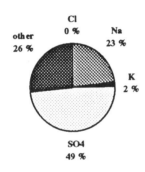

in particles, softwood, high solids

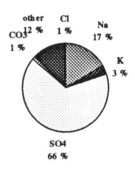

at ESP, softwood, high solids

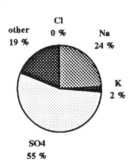

in particles, mixedwood, high solids

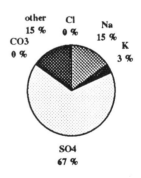

at ESP, mixedwood, high solids

Figure 3
Particle composition during recovery boiler experiments for particles collected from flue
gases and for particles collected from the ESP ash hoppers

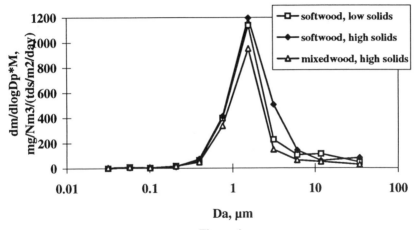

Figure 4
Aerosol mass size distribution at boiler exit measured with BLPI

Composition mass size distributions for water soluble Na ions at boiler exit are shown in Fig. 5, SO_4 in Fig. 6, K in Fig. 7, and Cl in Fig. 8. Na, K, and SO_4 composition size distribution are similar in shape as mass size distribution for all liquors. Chlorine is enriched into coarse particles during combustion of mixedwood black liquor.

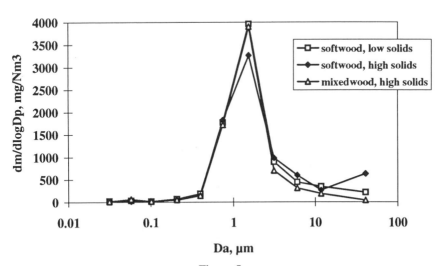

Figure 5
Differential composition size distributions for Na at boiler exit

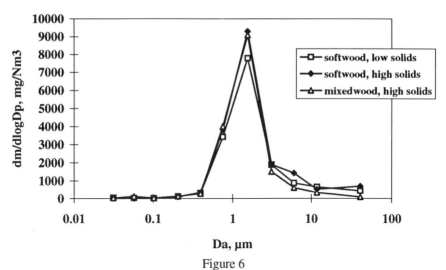

Figure 6
Differential composition size distributions for SO_4 at boiler exit

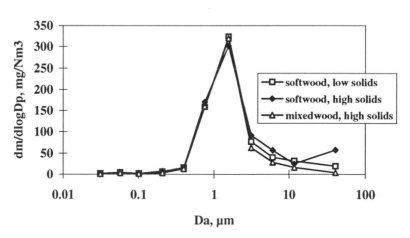

Figure 7
Differential composition size distributions for K at boiler exit

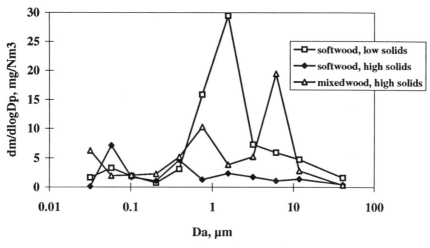

Figure 8
Differential composition size distributions for Cl at boiler exit

SEM micrographs of particles collected on BLPI substrates on aerodynamic size range 3.0-5.3 μm at boiler exit are shown in Figs. 8a., b. and c. Particles are agglomerates with primary particles of about 0.3 - 0.5 μm in size. Particles from softwood, low solids liquor combustion (Fig. 9a.) are less fused than particles from softwood, high solids liquor combustion (Fig. 9b.). In Fig. 9a. and 9b. size scale refers to the upper part of graphs and lower part is magnified by five. Particles are formed already in recovery boiler furnace as shown in Fig. 10. at about 900°C.

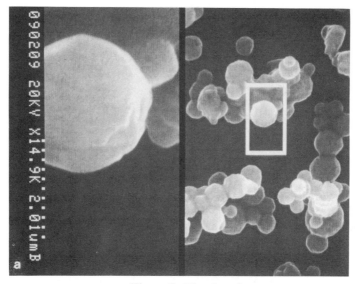

Figure 9. (Continued)

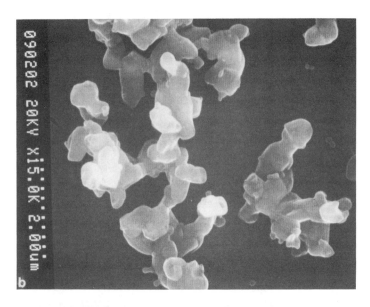

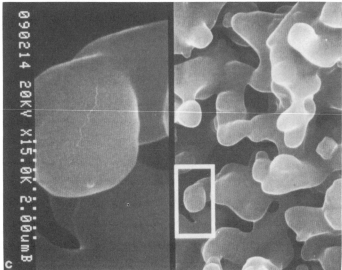

Figure 9
SEM micrographs of particles collected on BLPI stage six at boiler exit during softwood low solids (a), softwood high solids (b) and mixedwood high solids (c) black liquor combustion.

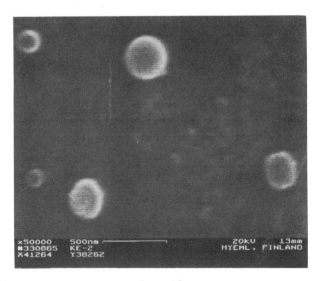

Figure 10
SEM micrograph of particles collected on copper grid from recovery boiler furnace at bullnose level at about 900°C during combustion of softwood, high solids black liquor.

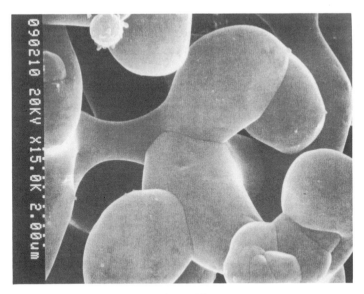

Figure 11
SEM micrograph of particles collected from ESP dust hopper during combustion of softwood, low solids black liquor.

In Fig. 11 SEM micrographs of particles collected from ESP dust hoppers are shown. Significant difference of morphology of particles is due to their hygroscopic nature.

Particles collected from ESP dust hopper and boiler bank dust hopper were analysed for crystalline salts. XRD showed segregated phases, i.e. two crystalline phases of Na_2SO_4 (thenardite and sodium sulphate) and $K_3Na(SO_4)_2$. Primary particles of one agglomerates had different contents.

CCSEM analyses were carried out on particles from ESP dust collected from another industrial recovery boiler (Table 5). The overall composition for both samples is quite similar, but there are significant differences in the abundance of various elemental categories. Volume percentage of Na-S particles is 11 % greater for high solids firing sample (H) than for low solids firing sample (L), whereas the volume percentage of Na-S-K particles is 12 % higher for sample L than for sample H. Na-S-Ca is found in particles in both samples in minor amounts.

Table 5
CCSEM elemental analyses for two different liquor types.

elemental categories	low solids (L)	high solids (H)
Na-S	77	88
Na-S-K	19	7
Na-S-Ca	2	1
others	2	4

CONCLUSIONS

Fly ash particles formed during black liquor combustion are agglomerates formed from primary particles of about 0.3-0.5 μm in size. Particle mass size distribution peaks at about 1.2 μm and at about 5 μm aerodynamic diameter at boiler exit conditions. Mass size distributions for all liquor types are similar, but total concentrations differ. Also composition size distributions for water soluble Na, K, and SO_4 ions for all liquors are similar in shape, but corresponding to the total concentration. Cl is enriched to the coarse particles during mixedwood, high solids combustion.

The overall composition for different combustion samples is quite similar, but there are significant differences in the abundance of various composition categories and in particle morphology. The phase segregation of particles showed differencies in primary particles of one agglomerates.

ACKNOWLEDGMENTS

We would like to acknowledge the researchers of the Aerosol Technology Group for help during the experiments, Q. Powell, A. S. Gurav, and T. T. Kodas from the University of New Mexico, A. D. Shah, F. E. Huggins, and G. P. Huffman from the University of Kentucky, for their co-operation on electron microscopy, and companies A. Ahlstrom, ABB Fläkt, AB Metsä-Botnia Oy, Tampella Power, and Veitsiluoto Oy for their advice and financial support. This work has been supported by the Finnish Ministry of Trade and Industry as part of the LIEKKI Research Programme.

REFERENCES

Jokiniemi, J. K., Pyykönen, J., Mikkanen, P. and Kauppinen, E. I. (1995) Modelling alkali salt deposition on kraft recovery boiler heat exchangers. In: Proceedings of International Chemical Recovery Conference 1995. Canada April 24-27, 1995.

Joutsensaari, J., Kauppinen, E. I., Jokiniemi, J. and Helble, J. (1994) Studies on ash vaporisation in power plant scale pulverized coal combustion. In: Williamson, J. and Fraser, W. (eds.) The Impact of Ash Deposition on Coal Fired Plant. The Engineering Foundation Conference Proceedings, Solihull, UK 20-25 June 1993. pp.613:624.

Kauppinen, E. I. and Pakkanen, T. A. (1990). Coal Combustion Aerosols: a field study. Environmental Science and Technology no 24 pp.1811:1818.

EFFECT OF ENVIRONMENT ATMOSPHERE ON THE SINTERING OF THAI LIGNITE FLY ASHES

C. Tangsathitkulchai[1] and M. Tangsathitkulchai[2]

Suranaree University of Technology
[1]School of Chemical Engineering and [2]School of Chemistry
Nakhon Ratchasima 30000, Thailand

ABSTRACT

Sintering of ash particles, related to deposit formation in a pulverized coal-fired boiler, was investigated for two lignite fly ashes obtained from Mae Moh and Bangpudum coal seams. The tests involved measuring the compressive strength of cold sintered pellets at varying sintering temperature, both under oxidizing (air) and non-oxidizing atmospheres (CO_2).

Under ambient air condition, Mae Moh fly ash which contained higher amount of glassy phase gave significantly higher sinter strength than Bangpudum fly ash. The role of glassy phase was confirmed by the lowering of sinter strength when HF-extracted fly ash was tested. Sintering under CO_2 environment resulted in larger strength development than sintering in air. Under this non-oxidizing condition, the pellet color turned black, indicating that most of the iron was in the reduced state and could form additional low melting-point glassy phase, hence facilitated sintering rate.

In addition, blending of the two ashes yielded intermediate maximum strength, under both air and CO_2 environments. This observation substantiates the important role of glassy phase in the sintering process and indicates the possibility of lowering deposit strength by judicious mixing of different raw coal feeds.

INTRODUCTION

Slagging and sintering tendencies of different coal ashes in pulverized coal-fired boilers have been studied by comparing the compressive strength of fly ash sinters (Tangsathitkulchai and Austin, 1985 and Tangsathitkulchai, 1986). Recently, this technique has also been used for rating sintering behavior of coal ashes in circulating fluidized bed boilers (Skrifvars, 1992). The slagging and sintering propensities from these laboratory tests seemed to correlate well with experiences found from full-scale and pilot-scale operation.

The sintering behavior of coal ash as a function of temperature could be explained by the variation in the amounts of amorphous glassy-phase material and the crystalline phase (Tangsathitkulchai, 1994 and Kuwarananchareon,

Applications of Advanced Technology to Ash-Related Problems in Boilers
Edited by L. Baxter and R. DeSollar, Plenum Press, New York, 1996

425

1994). Addition of alkali glasses in the test fly ash increased sintering rate (higher sinter strength). However, at temperature above 1000 °C alkali glass could lower the pellet strength, with calcium silicate showing stronger effect as compared to sodium silicate. Iron silicate, on the other hand, appeared to have negligible effect on the sinter strength.

Most sintering works reported previously were performed in an ambient atmosphere. Limited attempts have been made to study the role and influence of surrounding atmosphere on the sintering behavior. To gain further insight into the mechanism of fly ash sintering, tests were carried out in this work to investigate primarily the effects of oxidizing and non-oxidizing atmosphere (air versus CO_2), using Thai lignite fly ashes.

EXPERIMENTAL

The fly ashes used for sintering tests, designated as Mae Moh and Bangpudum samples, were supplied by the Electricity Generation Authority of Thailand (EGAT) from the Mae Moh and Krabi Power Plant, respectively. According to EGAT, these samples were taken from ESP dust collectors. Table 1 shows properties of the raw coals.

Both fly ash samples were heated in air for 2 hours at 700°C to burn off residual carbon. A pellet was made from 0.9424 gram of ash sample, mixed with a few drops of water and pressed in a cylindrical die, of 1 cm. dia. x 1 cm. high, to give pellet bulk-density of 1200 kg. m^{-3}.

The furnace used for sintering study consisted of an alumina tube (40 cm. long and 5 cm. internal dia.) which was externally heated by Kanthal A-1 heating wire. The heated section was insulated with a ceramic fiber sheath to reduce heat loss. Temperature inside the furnace was controlled by a PID temperature controller and a platinum-rhodium thermocouple. The maximum attainable temperature was 1200°C. The range of sintering temperature studied simulates the temperature gradient that exits in the ash deposit layer on the water-wall surface of the combustion zone. It should be noted that the sintered pellet was assumed to be in thermal equilibrium at the set temperature. Thus, the heat transfer flux through the pellet is immaterial.

Five pellets were placed in an alumina crucible and heated in the furnace at a heating rate of 5°C min^{-1} to the desired temperature, and held at that temperature for a period of one hour. The heating was then stopped and the furnace cooled down to room temperature.

The diameter of the cold pellets were measured and the compressive strength of the three pellets determined by an Instron Tester to obtain the average value. The other two pellets were kept for subsequent scanning electron microscopic (SEM) and X-ray diffraction (XRD) analyses.

For the tests performed in the non-oxidizing atmosphere, the furnace was continuously purged with CO_2 gas at a rate of 400 cm^3 min^{-1}. Test procedure under this condition followed that of the oxidizing atmosphere.

Some of the tests were also conducted with acid-extracted ashes. The test sample was prepared by mixing and agitating the raw fly ash with 10% hydrofluoric acid for 30 min., followed by filtering and drying at 110°C .

Table 1   Properties of raw coals.

	Coal Basin	
	Mae Moh	Bangpudum
Rank	lignite	lignite
Moisture, %	13.76	10.69
Volatile matter, %	41.58	40.48
Ash, %	25.94	16.92
Fixed carbon %	18.72	31.91

Table 2   Spectrochemical analyses of the test ashes.

	Weight percent of equivalent oxides	
Type of Oxides	Mae Moh fly ash	Bangpudum fly ash
SiO_2	35.56	45.68
Al_2O_3	19.56	21.52
Fe_2O_3	11.88	13.52
CaO	12.75	7.69
MgO	2.55	2.49
K_2O	1.82	2.10
Na_2O	0.56	0.09
SO_3	11.65	6.92

Table 3   Fusion temperature of ash sample in air

Fusion temp., °C	Mae Moh Fly Ash	Bangpudum Fly Ash
Initial temp.	1305	1268
Softening temp.	1317	1280
Hemispherical temp.	1327	1299
Fluid temp.	1357	1312

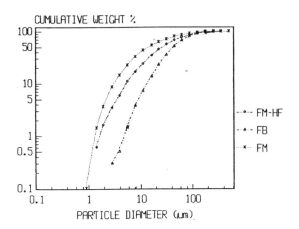

Figure 1 Size distribution of the test fly ashes, FM = Mae Moh fly ash,

FB = Bangpudum fly ash, FM-HF = HF-washed Mae Moh ash.

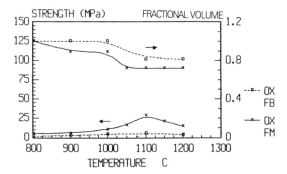

Figure 2 Compressive strength and the corresponding volume shrinkage as a function

of temperature for sintering in air, FM = Mae Moh, FB = Bangpudum.

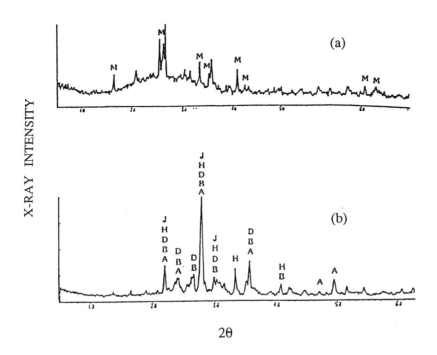

Figure 3 X-ray diffraction patterns of the as received Mae Moh fly ash (a), and

sintered at $1100^{\circ}C$ in air (b); only major peaks are labelled as

A = albite, B = albite calcium, D = anorthite sodium, H = margarite,

J = polyhalites, M = mullite

RESULTS AND DISCUSSION

OXIDIZING ATMOSPHERE (AIR)

The mean chemical composition analysis and the fusion temperature of the fly ashes are given in Table 2 and Table 3, respectively. Figure 1 shows the size distribution of the test ashes, as analyzed by the Microtrac Laser Diffractometer.

The relationship between the cold pellet strength and heat treatment temperature for the fly ashes tested in air is shown in Fig. 2. Mae Moh fly ash showed a rapid increase in the strength with temperature and passed through a maximum ($\approx$ 25 MPa) at 1100°C. Bangpudum fly ash, on the other hand, gave much lower strength development and exhibited no maximum strength. It was also noted from the same figure that the condition of maximum pellet strength corresponded to the maximum in shrinkage of pellet volume.

The sharp development of strength in Mae Moh fly ash was due to viscous flow sintering promoted by glassy materials, as evident from the results of SEM and XRD analyses. The presence of large amount of amorphous glassy phase was detected as the broad background of XRD pattern of the as-received fly ash, as shown in Fig. 3 (a). The only identified crystalline phase in this raw fly ash was mullite ($Al_6 Si_2 O_{13}$, m.p. 1810 °C). The morphology of ash particles from SEM showed the appearance of sinter necks and melting of the glassy phase. The amount of molten glass material increased with increasing temperature, giving rise to rapid sintering rate and hence sinter strength.

The decrease in strength of the Mae Moh sinter product at the higher temperature (>1100°C) was observed to be associated with the formation of crystalline phases. The content of crystalline phases appeared to increase with temperature, with corresponding reduction in the glass content, as observed in Fig. 3(b). The crystalline materials detected were polyhalite [$K_2Ca_2Mg(SO_4)_2$. $2H_2O$, m.p. 660 °C], albite [Na Al Si_3O_8, m.p. 1100°C], margarite [$CaAl_2(Si_2Al_2)O_{10}(OH)_2$, m.p. 1400°C] and anorthite [(Ca,Na) $(SiAl)_4O_8$, m.p. 1500-1600°C]. The occurrence of these crystalline materials with consequent reduction of the bonding glass phase would increase the viscosity of the residual liquid, hence lowering the rate of sintering.

Much smaller amount of glass material (amorphous phase) was detected in the XRD pattern of the as-received Bangpudum fly ash, as shown by the narrow base background of XRD pattern in Fig. 4(a). The major contents in the as-received Bangpudum fly ash was crystallines of quartz [SiO_2, m.p. 1710°C] and gismondine [Ca $Al_2Si_2O_8$. $4H_2O$, m.p. 1190-1450°C]. The transformation from glassy phases into new crystalline phases was rarely observed in the sintered pellets. This was inferred from similar XRD patterns between as-received fly ash and the heat treated pellets (Fig.4(a) versus 4(b)). It appeared then that the weak sinter strength of Bangpudum fly ash resulted from the slower rate of sintering of crystalline solid phase.

At the present stage, no attempt was made to identify and quantify the chemistry of the bonding glass phase due to the unavailability of the analytical instrument.

NON-OXIDIZING ATMOSPHERE (CO_2)

It was found that all the pellets that were heat treated in the non-oxidizing atmosphere of CO_2 gas turned into black color. This observation indicates that

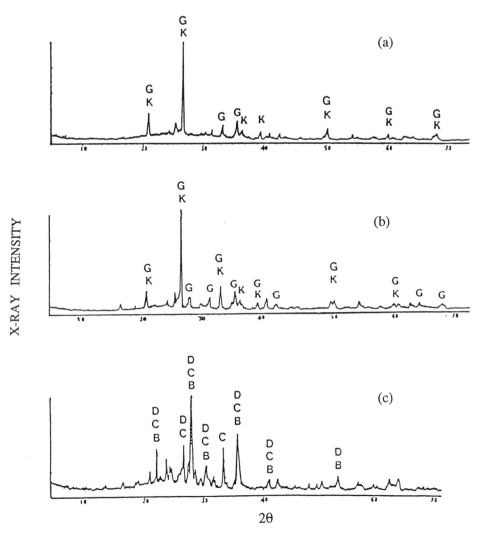

Figure 4 X-ray diffraction patterns of the as-received Bangpudum fly ash (a)

sintered at $1100^\circ C$ in air (b), and sintered at $1100^\circ C$ in CO_2(c),

major peaks : B = albite calcium, C = anorthite, D = anorthite sodium,

G = gismondine, K = quartz

Inside the figure (a):

X-RAY INTENSITY (vertical axis label)

2θ (horizontal axis label)

431

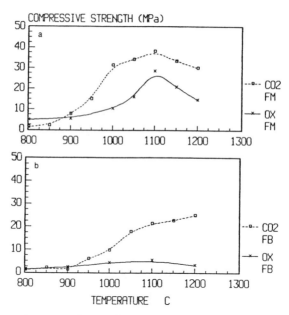

Figure 5 Compressive strength as a function of temperature of Mae Moh fly ash (a)

and Bangpudum fly ash (b), sintered in air (OX) and in CO_2

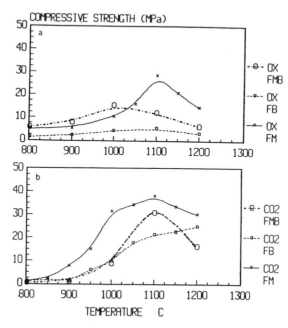

Figure 6 Compressive strength as a function of temperature of Mae Moh fly ash

(FM), Bangpudum fly ash (FB) and mixture of fly ashes (FMB), in air

(OX) (a) and in CO_2 (b)

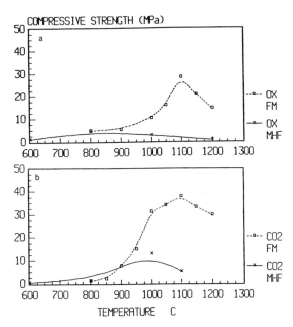

Figure 7 Compressive strength as a function of temperature of Mae Moh fly ash
(FM) and HF-washed Mae Moh fly ash (MHF) in air (OX) (a), and in
CO_2 (b)

most of the iron was largely present in the reduced state (Fe^0, Fe^{+2}). The iron in the reduced forms could readily enter into silicate and aluminosilicate glass and formed a lower melting-point iron glass. This newly formed melting glass could provide a liquid phase at lower temperatures than the glass material originally contained in the raw fly ash. As a result, the strength of both fly ashes in CO_2 gave higher strength than those in air, as shown in Fig. 5(a) and 5(b). As an example, at sintering temperature of 1100°C , a 36% and 300% increase in sinter strength were observed for Mae Moh fly ash and Bangpudum fly ash, respectively.

As to the role of CO_2 in changing the state of iron, it was hypothesized that CO_2 could dissociate into CO and O_2 according to, $2CO_2 \rightleftharpoons 2CO+O_2$, and the generated CO then acts as a reducing agent. Thermodynamic calculation indicates that at 1100°C, composition of CO is 0.04% in CO_2+CO+O_2 gas mixture at equilibrium condition. To check this hypothesis further, sintering tests under varying CO partial pressure could be studied.

Anorthite [$CaAl_2Si_2O_8$, m.p. 1550°C] and albite [$NaAlSi_3O_8$, m.p. 1100°C] were not detected in Bangpudum pellets sintered in air, but these crystallines did appear in the pellets sintered in CO_2 gas Fig. 4(c). It was possible that the presence of greater liquid content under the non-oxidizing condition could facilitate the interaction of sodium and calcium ions with aluminosilicate glass to form crystalline phase of albite and anorthite.

ADDITIONAL TESTS

To check further the important role of glassy phase material in the sintering process, tests were performed with mixture of fly ashes and fly ashes extracted with hydrofluoric acid.

Figure 6(a) and 6(b) compare the sinter strength of fly ash mixture (1:1 by wt.) and the original fly ash. As expected, the mixed fly ash gave intermediate maximum strength in both air and CO_2 environments, remembering that Mae Moh ash had higher glass contents than those of Bangpudum.

Results from spectrochemical analysis and the XRD of HF-washed fly ash indicated a major reduction in the amounts of glassy phase material as well as the alkali and iron constituents. That is, before extraction: CaO 12.75%, Fe_2O_3 11.88% and after extraction:CaO 6.07%, Fe_2O_3 3.43%

The strength of HF-washed pellet and the fly ash without acid extraction are compared in Fig. 7(a), for sintering in air. It is clear that the sinter strength of HF-washed fly ash was drastically reduced. This evidence strongly supports the role of glassy phase in promoting and affecting the extent of fly ash sintering.

Although the quantity of iron had been reduced in the HF-washed fly ash, it still gave higher strength in CO_2 as compared to the sintering in air (Fig. 7(a) vs 7(b)). Again, this emphasizes the contribution of iron in forming low melting-point glassy phase in the non-oxidizing atmosphere.

CONCLUSIONS

1. The larger glass content presented in the as-received Mae Moh fly ash promotes the more rapid sintering rate and gave higher strength than Bangpudum fly ash.

2. The role of glassy phase was confirmed by the reduction of sinter strength of the HF-extracted fly ash.

3. The lower strength at higher temperatures appeared in some of the sintered products, was the result of chemical transformation of glass materials into crystalline phase materials, thus removing the bonding glass and reducing the sintering rate.

4. Blending of the two fly ashes gave intermediate maximum strength under both air and CO_2 environment. This emphasizes the significant effect of chemical composition on the sintering strength of fly ash.

5. In the non-oxidizing atmosphere of CO_2, the formation of lower melting iron glass provides the more rapid sintering rate and gave significantly strength development than sintering in air.

6. Results from sinter strength measurement suggested that burning Bangpudum coal should give less slagging and fouling problem than burning Mae Moh coal. To date, this finding has not yet been verified with results from power plants burning these type of coals.

7. Some of the future work which require further investigation include :
- Sintering of synthetic mineral mixture of various combinations.
- Additional tests on a variety of fly ashes, particularly under the environment of flue gas composition.
- Distribution of crystalline and amorphous glassy phase of coal ash as a function of particle size.
- Sintering tests on ash samples collected from a drop-tube furnace for various size fractions.

REFERENCES

Tangsathitkulchai, M., and Austin, L.G. (1985), " Studies of Sintering of Coal Ash Relevant to Pulverized Coal Fired Boilers : 2. Preliminary Studies of Compressive Strengths of Fly Ash Sinter," *Fuel, 64*, 86-92.

Tangsathitkulchai, M.(1986), " Studies on Initiation, Growth and Sintering in the Formation of Utility Boiler Deposits," Ph.D. Thesis in Fuel Science, The Pennsylvania State University.

Tangsathitkulchai, M, and Tangsathitkulchai, C.(1994) " Sintering of Coal Ashes : Effects of Extraction and Addition of Alkali-Glass Materials," *Suranaree J. Sc. & Tech, 1*(2), 123-132.

Tangsathitkulchai, M, and Tangsathitkulchai, C.(1994) " Sintering of Coal Ashes : Effects of Temperature and Atmosphere, " 4th. *Chem.* Eng. Conf., Khon Kaen, Thailand.

Kuwarananchareon, J.(1994), " Sintering and Slag Deposit Formation of Coal Ashes," M.Eng. Thesis in Chemical Engineering, King Mongkut's Institute of Technology Thonburi, Thailand.

Tangsathitkulchai,C.,Kuwarananchareon, J.,and Tangsathitkulchai, M.(1994), "Sintering Study of Thai Lignite Ashes, " *2nd Engineering Conf.*, Faculty of Engineering, King Mongkut's Institute of Technology Thonburi, Thailand

Skrifvars, B.J., Hupa, M., and Hiltunen, M. (1992), "Sintering of Ash during Fluidized Bed Combustion, " *Ind. Eng. Chem. Res., 31*(4), 1026-1030.

ASH TRANSFORMATIONS IN THE REAL-SCALE PULVERIZED COAL COMBUSTION OF SOUTH AFRICAN AND COLOMBIAN COALS

Terttaliisa Lind, Esko I. Kauppinen, Tuomas Valmari,
Norbert Klippel[1] and Christer Mauritzson[2]

VTT Chemical Technology, Aerosol Technology Group
P.O.Box 1401, FIN-02044 VTT, Finland
[1]ABB Corporate Research, Baden, Switzerland
[2]ABB Fläkt Industri AB, Växjö, Sweden

ABSTRACT

In this work, the formation of ash particles in the combustion of South African Klein Kopie coal and a Colombian coal was studied by measuring the ash particle characteristics upstream of the electrostatic precipitator (ESP) at a 510 MW$_e$ pulverized coal fired power plant. We measured the ash particle mass size distributions in the size range 0.01 - 50 µm using low-pressure impactors and pre-cutter cyclones. Also, samples were collected for computer controlled scanning electron microscopy (CCSEM) with a cyclone with an aerodynamic cut-diameter of about 1 µm. The cyclone-collected samples were analyzed with standard CCSEM procedure by depositing the particles on a filter, and by embedding the particles in epoxy hence acquiring the cross-section analysis of the sample. All major mineral classes in both coals were found to undergo extensive coalescence during combustion. Iron, calcium and magnesium rich particles resulting from the decomposition of pyrite, calcite and dolomite were found to coalesce with quartz and aluminosilicate particles. The size distributions of the fly ash determined with CCSEM and low-pressure impactor-cyclone sampler were found to be similar.

INTRODUCTION

Extensive work has been carried out by many researchers during the past years to determine the mineral particle transformations during pulverized coal combustion because minerals have been found to play an important role in the boiler performance and maintenance. Large amount of the work has been conducted with laboratory and pilot reactors of different scales [e.g. Srinivasachar et al., 1990; Quann et al., 1990]. Modelling of the combustion process has been given an increasing interest [e.g. Wilemski and Srinivasachar, 1994; Jokiniemi et al., 1994]. However, to verify the laboratory tests and models, data from the operating, real-scale furnaces is needed for different types of coal, different boilers and different operating conditions.

Applications of Advanced Technology to Ash-Related Problems in Boilers
Edited by L. Baxter and R. DeSollar, Plenum Press, New York, 1996

Minerals are contained in pulverized coal as micrometer-sized inclusions and as excluded matter among the coal particles. In addition to minerals, coal contains other ash forming species, e.g. salts and organically bound inorganic species [Raask, 1985; Benson and Holm, 1985]. These constituents undergo transformations during combustion, and may interact with the mineral particles [Lindner and Wall, 1990]. Therefore it is essential that when we study the behaviour of the minerals in the combustion, we include the other ash forming constituents as a part of the whole picture. The interaction between minerals and other ash forming constituents may occur inside the coal particle during pyrolysis, during the char burning or after the combustion in the gas phase. In all cases, the interaction may significantly alter the characteristics of the ash derived from the mineral particles.

This work is part of a project where we study the behaviour of the given coals in real-scale boilers and in the laboratory laminar-flow reactor. We have determined the ash transformations in the operating plants with special attention on the extent of vaporization of the ash forming constituents and the interaction of the mineral particles with other ash forming species. The vaporization has been selected as a main interest because no reliable data from the laboratory reactors can be found in the literature, and its effect on the ash formation and the resulting ash characteristics is unquestionable. Also, large particle coalescence has been thoroughly studied. In this paper, we present data on the supermicron particles ($D_p > 1$ μm), and the ash vaporization and submicron particle formation is discussed elsewhere in this volume [Kauppinen et al., 1995]. The operating plant studies formed the first part of the research project, and the following step will be to compare the extensive data obtained from the real-scale plant for two different coals with the laminar-flow reactor test data. This work will be carried out in the near future and presented later.

In this work, we determined the fly ash composition and size distributions from an operating 510 MWe pulverized coal fired plant. In addition to supermicron ash measurements, we measured the size distributions of the submicron particles down to 10 nm in order to determine the extent of ash vaporization in the process. The measurements were carried out with two bituminous coals: South African Klein Kopie and Colombian El Dorado.

METHODS

Process Description and Coal Selection

The combustion experiments were carried out at a 510 MWe pulverized coal fired power plant in Staudinger, Germany, during two weeks in May 1994. The power plant is equipped with high-dust SCR NOx reduction and wet FGD units. The plant has a single wall-fired boiler, to which the coal is distributed from 4 coal mills each feeding coal to four burners. The ash formation in the combustion of two bituminous coals was studied. South African Klein Kopie coal was fired during the first week of experiments, and Colombian El Dorado coal was fired during the second week. The ash content of the Klein Kopie coal was 17 % and it was higher than the ash content of El Dorado which was 10 %. The boiler had low-NOx burners installed.

Coal minerals were analyzed with Computer Controlled Scanning Electron Microscopy (CCSEM) at Microbeam Technologies Inc., North Dakota [Huggins et al., 1980]. For the analysis, the coal particles were embedded in epoxy. The resulting pellets were cross-sectioned, polished and sputter-coated with carbon.

Size Distribution Determination

Low-pressure Impactor Sampling

The ash particle mass and elemental size distributions were determined by collecting size-classified coal combustion aerosol samples with an 11-stage, multijet compressible flow Berner-type low-pressure impactor [Kauppinen, 1992; Hillamo and Kauppinen, 1991; Kauppinen and Pakkanen, 1990]. A cyclone with Stokes cut-diameter of approximately 5 μm was used as a precutter to prevent overloading of the upper BLPI stages. The samples were collected from the flue gas channel upstream of the electrostatic precipitator (ESP) at the flue gas temperature of approximately 130-140°C. The cyclone-collected particles were size-classified with a Bahco sieve method. The mass size distribution of the ash was determined by weighing each BLPI-substrate prior to and after the sampling, and by weighing each Bahco-sieve size-classified fraction. The elemental compositions of the BLPI-collected particles were determined by analyzing each of the collection substrates with Inductively-Coupled Plasma Mass Spectrometry (ICP-MS) after the digestion in HNO_3.

CCSEM Sampling

Samples were also collected for the CCSEM analysis. For these samplings, a cyclone with a cut-diameter of about 1 μm was used. The samples were taken from the same sampling location as the BLPI samples. For the analysis, the samples were prepared with two different methods to obtain the particle surface and the inside compositions. For the inside composition determination (cross-section or x-s CCSEM) the samples were embedded in epoxy resin. The resulting pellets were ground and polished. For the surface composition determination (standard CCSEM), the samples were suspended in distilled acetone and deposited on Nuclepore filters. It has to be noted that the penetration depth of SEM beam is 2-3 μm, and hence for small particles the analysis volume becomes close to the particle volume, and for these particles no distinction between surface and interior composition obtained from CCSEM standard and x-s analyses should be made. In this study, we present the CCSEM data on all particle size classes. However, major fraction of the particle volume derives from particles larger than 2-3 μm. All the samples were sputter-coated with carbon to avoid charging effects in the SEM.

CCSEM Data Reduction

Particles were divided into composition classes by determining the maximum of three most abundant elements in each particle as obtained from EDX spectrum. The concentrations of all the elements together in a composition category had to be more that 80 %. The concentration of the second element had to be more than 10 % and the third element more than 5 %. The CCSEM analyzes the particles individually, and for each particle determines the composition and the area exposed to the beam. It is important to take into account the sample preparation while determining the resulting size distributions of the particles. While depositing the particles on a filter, the analysis gives a size distribution proportional to the area of the particles, whereas if the particles are embedded in epoxy, the resulting size distribution is the volume

Table 1. The coal ash composition in oxide weight % as determined with CCSEM and ash analysis.

Coal	Klein Kopie		El Dorado	
Oxide [w-%]	CCSEM	ash analysis	CCSEM	ash analysis
Na_2O	0.3	0.3	0.3	0.6
MgO	1.1	1.7	2.0	2.1
Al_2O_3	21	31	8.9	21
SiO_2	45	45	58	56
P_2O_5	0.6	1.2	0.6	0.3
SO_3	4.8	5.4	9.7	3.1
K_2O	1.3	1.5	1.2	2.1
CaO	17	7.3	12	4.1
Fe_2O_3	3.9	3.5	4.6	7.3
TiO_2	1.6	1.6	0.8	0.9

particles are embedded in epoxy, the resulting size distribution is the volume distribution of the particles. In this work, we used volume size distributions for all the results, and the data was converted accordingly, i.e. the area size distributions were converted into volume size distributions by multiplying the area of each particle with the particle's average diameter.

The particles were divided into six size classes according to their average diameter. The cut-diameters of the size classes were chosen to be close to those of the low-pressure impactor used to determine the particle mass size distributions from the flue gas. The size classes used were: 0.5-1.25 µm, 1.25-2.5 µm, 2.5-5.2 µm, 5.2-10.4 µm, 10.4-50 µm and particles larger than 50 µm.

RESULTS AND DISCUSSION

Coal Analysis

The ash compositions of the coals were analyzed with two methods: standard ash analysis and CCSEM. In Table 1 we compare elemental compositions of the ash converted to oxide weight percentages obtained with the two methods. For both coals, the concentration of calcium is significantly higher and aluminium lower when determined with CCSEM as compared to the bulk ash analysis. Similar results have been observed earlier when comparing the bulk ash analysis concentrations with CCSEM analysis results [Helble et al., 1994]. The difference in the analysis results with the different methods may be due to the attenuation of the EDX signal in the CCSEM for the lighter elements. The mineral matter analyzed with CCSEM was 58 % of the ash in the coal for Klein Kopie coal, and 40 % of the ash for El Dorado coal.

Table 2. The coal mineral and fly ash compositions as determined with CCSEM.

Composition Category	Klein Kopie			El Dorado		
	Coal	Fly ash, standard CCSEM	Fly ash, x-s CCSEM	Coal	Fly ash, standard CCSEM	Fly ash, x-s CCSEM
Si	7	6	10	31	21	31
Al-Si	59	53	44	23	37	26
Ca-Al-Si	5	24	31	1	7	7
Fe-Al-Si	1	3	2	1	8	12
K-Al-Si	8	7	6	18	26	22
Ca	6	1	2	3	1	2
Ca-Mg	5	-	-	5	-	-
Fe-S	2	-	-	7	-	-
others	5	6	5	11	-	-

The mineral transformations during the combustion were studied by comparing the compositions and size distributions of the coal minerals with the ash particles as determined with CCSEM. In Table 2 we present the concentrations of the major composition categories for the Klein Kopie and El Dorado coals and the fly ash particles collected from the flue gas channel into the sampling cyclone. The cyclone-collected particles were analyzed both with the x-s and standard CCSEM methods. According to CCSEM analysis, the only significant composition category in Klein Kopie coal is Al-Si, i.e. aluminosilicates, with minor amounts of quartz (Si), illite (K-Al-Si) and other aluminosilicates (Ca-Al-Si), pyrite (Fe-S), calcite (Ca) and dolomite (Ca-Mg). Calcium, iron and sulfur are more abundant in large particles (> 5 µm) and aluminium in small particles (< 5 µm).

The major mineral in El Dorado coal is quartz with 31 %, followed by aluminosilicates (23 %) and illite (18 %), Table 2. It also contains some pyrite (7 %), dolomite (5 %) and calcite (3 %). Compared to Klein Kopie coal, the mineral categories are more evenly distributed with several major species, when Klein Kopie mostly contains Al-Si particles. Also, the pyrite concentration in El Dorado is larger than in Klein Kopie.

Ash Mass Size Distributions

The comparison of the size distributions determined with low-pressure impactors and CCSEM method is presented in Fig. 1 for the ash samples of both coals. The CCSEM determined size distributions are volume distributions determined with the standard CCSEM from the samples collected in the cyclone and subsequently deposited on a filter. Impactor size distribution is based on the particle Stokes diameter calculated with an assumed density of 2.45 kg/dm^3. A precutter cyclone with a cut-diameter of about 5 µm was used with the low-pressure impactor and the particles collected in the cyclone were size-classified with a Bahco-sieve method. During the cyclone collection for Bahco-sieve with El Dorado coal the soot-blowing was on in the boiler, and the particle mass concentration in the sampling location was about double the usual concentration during the experiments resulting in the overestimation of the large particles in the size distribution. Hence probably for the El Dorado coal, the size distribution determined with CCSEM is the more accurate one. For the Klein Kopie coal, the ash particle size distributions determined with BLPI and CCSEM agree relatively well. The flue gas particle mass concentrations were clearly higher for Klein Kopie coal than for El Dorado coal, with averages of 18 and 8 g/Nm^3, respectively.

The mineral coalescence during combustion was studied by determining the number of mineral (N(min)) and ash particles (N(ash)) in several size classes from the CCSEM results. The relative amount of ash particles to mineral particles N(ash)/N(min) is 1 if every mineral particle forms 1 ash particle. In this study the ratio N(ash)/N(min) increases with increasing particle size, Fig.2. The ratio is <1 for particles smaller than approximately 4 - 7 µm and >1 for larger particles. This means that small mineral particles disappear and larger particles are generated during combustion due to extensive coalescence of the particles.

Mineral Transformations

Klein Kopie coal

When comparing the coal mineral and ash compositions in Table 2, it can be seen that Ca-Al-Si class that is absent in the coal appears in the ash. This is presumably due to the decomposition of calcite and dolomite during combustion, and subsequent interaction with Al-Si particles. Especially noticeable is the large amount of Ca-Al-Si particles in x-s analysis, i.e. in the particle interior. According to earlier field studies with Klein Kopie coal, the elemental size distribution for calcium shows a significant mode in the submicron size range, Fig. 3. This indicates that some calcium escapes from the mineral particles into the gas phase as vapour and condenses on the surfaces of submicron ash particles. The coexistence of the calcium in the interior parts of the large ash particles and in the submicron ash can be explained if we assume that the CaO particles resulting from the calcination of calcite and dolomite coalesce with aluminosilicate particles when the char is burning. The coalescence of CaO particles with aluminosilicates results in the formation of supermicron Ca-Al-Si particles. Subsequently, when the Ca-Al-Si particles are still in the high temperature region of the furnace, calcium is released from the surface of these particles to the gas phase where it condenses on the pre-existing submicron SiO_2-Al_2O_3-particles. The ternary diagrams of Ca-Al-Si particles for Klein Kopie coal and fly ash, Fig. 4, show wide variation in calcium concentration and no definite composition range for fly ash particles. This suggests that the interaction between calcium and aluminosilicates is mostly coalescence, not chemical reaction.

The mineral transformations and coalescence are further studied in Fig. 5, where we present the particle volume size distributions as determined with CCSEM for the coal minerals and standard and x-s fly ash samples for all particles and the major composition category Al-Si. Cross-section and standard CCSEM analyses give similar size distributions for the fly ash. Particles smaller than 5 µm make up 48 % of the particles by volume. The ash particles are larger with only 22 % of the particle volume in particles smaller than 5 µm. This indicates significant coalescence of the mineral particles during combustion. Al-Si particle size distributions are similar to those of all

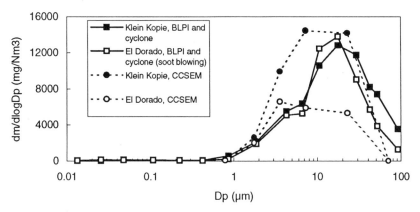

FIGURE 1. Particle size distributions at the ESP inlet at Staudinger power plant as determined with BLPI used with a precutter cyclone (cut size = 5 μm) and CCSEM from cyclone-collected particles (cut size = 1 μm).

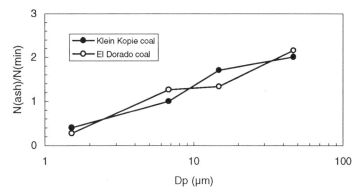

FIGURE 2. The ratio of ash particle number in each size class to the mineral particle number N(ash)/N(min).

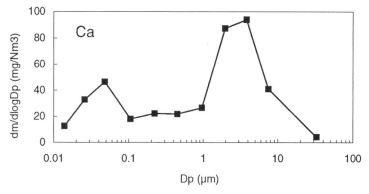

FIGURE 3. Calcium size distribution from the combustion of Klein Kopie coal determined with BLPI. Note that particles larger than 5 μm were collected with a precyclone sampler and not included in this figure.

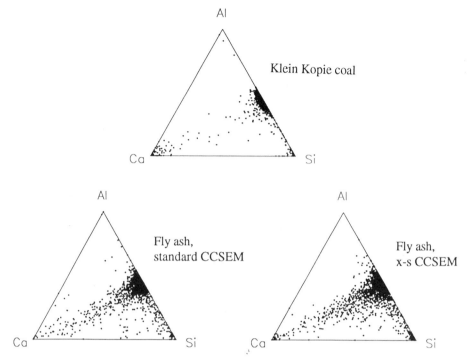

Figure 4. Ca-Al-Si ternary plot for Klein Kopie coal minerals, and fly ash analyzed with standard and x-s CCSEM methods.

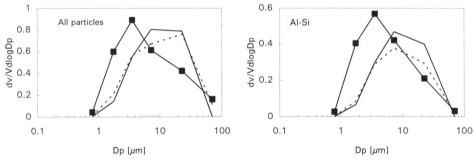

Figure 5. Volume size distributions of the Klein Kopie coal minerals and fly ash for all particles and Al-Si -composition category.

■— Klein Kopie coal

—— fly ash, standard CCSEM

- - - - - fly ash, x-s-CCSEM

particles, since Al-Si is the only major mineral class in the coal, but coalescence is apparent for all the particle composition categories.

El Dorado coal

The fly ash from El Dorado coal combustion contains mainly different aluminosilicates, and major composition classes are quartz, kaolinite and illite. The ternary plots for K-Al-Si particles in the coal and fly ash, Fig. 6, show two distinct composition categories along Al-Si axis, quartz and illite. In the fly ash, the distribution along the Al-Si axis is more even, probably due to the coalescence of the particles. The illite particles most abundant with potassium disappear from the fly ash probably due to coalescence of high potassium content illite particles with lower potassium content particles. The similar behaviour is seen for illite, kaolinite and quartz particles in Klein Kopie coal combustion.

Ca-Al-Si and Fe-Al-Si species appear in the fly ash during the combustion. This presumably happens when iron and calcium-rich particles resulting from the decomposition of pyrite, calcite and dolomite coalesce with aluminosilicates, Fig. 7. The coalescence is extensive for all the mineral classes, Fig. 8. Approximately 50 % by volume of the minerals are smaller than 5 μm. In the fly ash particle size distribution the fraction of small particles diminishes to 34 % smaller than 5 μm. This indicates extensive coalescence of all the major mineral classes. Illite particles are generally smaller than other minerals, both in coal and in the fly ash.

According to CCSEM analysis, quartz and Fe-Al-Si species are enriched in x-s analysis, i.e. in the particle interior, and Al-Si on the particle surface. The abundance of Fe-Al-Si species inside the particles can also be seen from Fig. 7, where more particles in this composition range are seen in the ternary plots for fly ash analyzed with x-s CCSEM than with standard CCSEM.

The magnesium content of El Dorado coal ash is approximately 2 %. The CCSEM and ash analysis give same concentrations, hence indicating that magnesium is mainly found in the mineral particles. The main magnesium containing mineral in the coal is dolomite, Fig. 9, which contains approximately 70 % calcium and 30 % magnesium in oxygen free basis. Also in the fly ash magnesium is mainly found in association with calcium. Mg can only be found in small concentrations in aluminosilicate particles in Mg-Al-Si ternary plot, Fig. 10, and no definite composition range can be seen. This is different behaviour than reported earlier for fluidized bed combustion of Venezuelan coal [Lind et al., 1995] and pulverized combustion of Polish coal [Joutsensaari et al., 1994]. In both studies Mg-rich particles were found in a certain, narrow composition area in Mg-Al-Si ternary plot for fly and bottom ash. In the coals studied, magnesium was mainly found in association with the coal organic structure, and it was concluded that the organically bound magnesium reacted with aluminosilicate particles after the decomposition. The results of this study show that magnesium originally found in dolomite does not chemically react with aluminosilicates. MgO and CaO particles resulting from the decomposition of dolomite coalesce with other minerals, mainly with quartz and aluminosilicates.

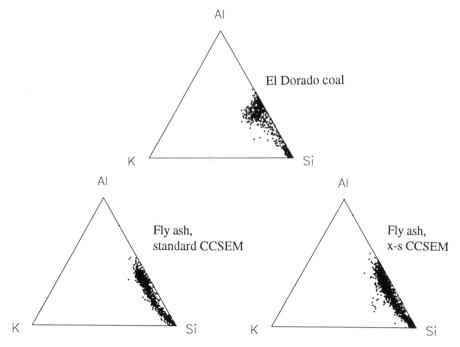

Figure 6. K-Al-Si ternary plot for El Dorado coal minerals, and fly ash analyzed with standard and x-s CCSEM methods.

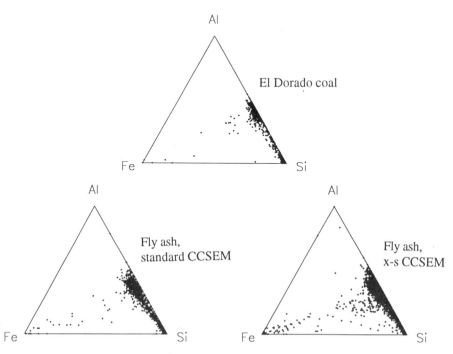

Figure 7. Fe-Al-Si ternary plot for El Dorado coal minerals, and fly ash analyzed with standard and x-s CCSEM methods.

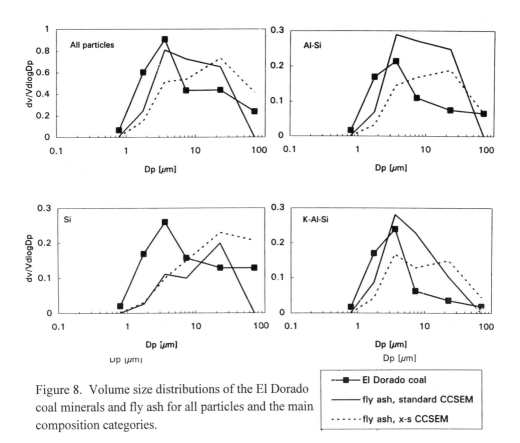

Figure 8. Volume size distributions of the El Dorado coal minerals and fly ash for all particles and the main composition categories.

- ■ El Dorado coal
— fly ash, standard CCSEM
---- fly ash, x-s CCSEM

CONCLUSIONS

The transformations of coal minerals were studied in real-scale pulverized coal combustion. Two bituminous coals were studied, South African Klein Kopie and Colombian El Dorado. The mineral particles were relatively fine, with approximately 50 % of the mineral particles smaller than 5 μm. The major mineral categories in El Dorado were illite, quartz and kaolinite, and in Klein Kopie kaolinite. All major mineral classes in both coals were found to undergo extensive coalescence during combustion. Iron, calcium and magnesium rich particles resulting from the decomposition of pyrite, calcite and dolomite were found to coalesce with quartz and aluminosilicate particles. The size distributions of the fly ash were determined with CCSEM and low-pressure impactor-cyclone sampler. The two methods were found to give similar size distributions.

ACKNOWLEDGMENTS

We would like to thank ABB Corporate Research, Baden, ABB Fläkt Industri AB, Växjö, and Ministry of Trade and Industry of Finland (SIHTI research program) for funding this study. We thank Mr. Sampo Ylätalo, Mr. Jorma Joutsensaari, the power plant personnel and all the others who participated in the experiments. Microbeam Technologies Inc. are gratefully acknowledged for carrying out the CCSEM analyses.

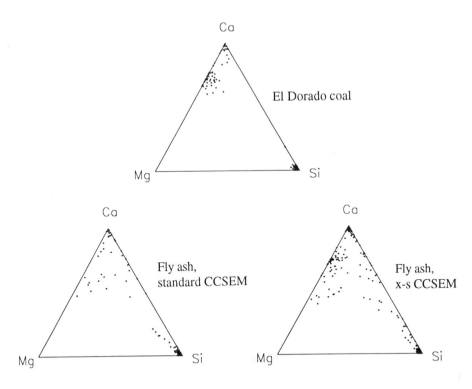

Figure 9. Mg-Ca-Si ternary plot for El Dorado coal minerals, and fly ash analyzed with standard and x-s CCSEM methods.

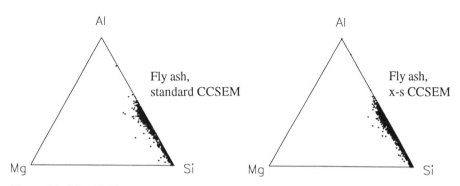

Figure 10. Mg-Al-Si ternary plot for El Dorado coal fly ash analyzed with standard and x-s CCSEM methods.

REFERENCES

Benson, S.A. and Holm, P.L. (1985). *Ind.Eng.Chem.Prod. Res.Dev.*, 24, 145-149.

Helble, J.J., Bool, L.E., Huffman, G.P., Huggins, F.E., Shah, N., Peterson, T.W., Gallien, D., Reschke, R., Sarofim, A.F. and Zeng, T. (1994) PSI Technology Company Quarterly Report No.5 for the program 'Fundamental study of ash formation and deposition: Effect of reducing stoichiometry', submitted to the U.S. DoE under contract DE-AC22-93PC92190.

Huggins, F.E., Kosmack, D.A., Huffman, G.P. and Lee, R.J. (1980). "Coal Mineralogies by SEM automatic image analysis" In O.Johari, (Ed.), *Scanning Electron Microscopy/1980/I*, SEM Inc., AMF O'Hare, Chicago, 531-540.

Jokiniemi, J.K., Lazaridis, M., Lehtinen, K.E.J. and Kauppinen, E.I. (1994). *J. Aerosol Sci.*, 25, 429-446.

Joutsensaari, J., Kauppinen, E.I., Jokiniemi, J.K. and Helble, J.J. (1994). In J. Williamson and F. Wigley, (Eds.) "The Impact of Ash Deposition on Coal Fired Plants": *Proceedings of the Engineering Foundation Conference,.,* Taylor&Francis, 613-624.

Kauppinen, E.I., Lind, T.M., Valmari, T., Ylätalo, S., Jokiniemi, J.K., Kodas, T.T., Powell, Q. and Mohr, M. (1995), "The Structure of Submicron Ash from Pulverized Combustion of South African and Colombian Coals." To be presented in the *Eng.Found.Conf. 'Application of Advanced Technology to Ash-related Problems in Boilers'* in Durham, New Hampshire, 16-22 July, 1995.

Kauppinen, E.I. (1992). *Aerosol Sci. Technol.,* 16, 171-197.

Kauppinen, E.I. and Pakkanen, T.A. (1990). *Environ. Sci. Technol.*, 24, 1811-1818.

Lind, T., Kauppinen, E.I., Maenhaut, W., Shah, A. and Huggins, F. (1995). "Ash Vaporization in Circulating Fluidized Bed Coal Combustion". Submitted to *Aerosol Sci. Technol.*

Lindner, E.R. and Wall, T.F. (1990). In R.W. Bryers and K.S. Vorres, K.S. (Eds.) *Mineral Matter and Ash Deposition from Coal,* United Eng. Trustees, Inc.

Quann, R.J., Neville, M. and Sarofim, A.F. (1990). *Combust. Sci. and Tech.*, 74, 245-265.

Raask, E. (1985). *Mineral Impurities in Coal Combustion*, Hemisphere Publishing Corporation.

Srinivasachar, S.S., Helble, J.J., Boni, A.A., Shah, N., Huffman, G.P. and Huggins, F.E. (1990). *Prog.Energy Combust.Sci.*, 16, 293-302.

Wilemski, G. and Srinivasachar, S.S. (1994). In J. Williamson and F. Wigley (Eds.) "The Impact of Ash Deposition on Coal Fired Plants": *Proceedings of the Engineering Foundation Conference,* Taylor&Francis, 151-164.

MODELING OF ASH DEPOSITION IN THE
CONVECTIVE PASS OF A COAL-FIRED BOILER

S.E. Allan, T.A. Erickson, and D.P. McCollor

Energy & Environmental Research Center, Grand Forks, ND

Phone (701) 777-5000, Fax (701) 777-5181

The Energy & Environmental Research Center (EERC) has developed a personal computer (PC)-based model, FOULER, to predict convective pass fouling deposit formation in coal-fired boilers. This program is used to evaluate the effects of coal quality and operational changes on both high- and low-temperature fouling. In addition, the effects of coal cleaning, blending, and switching options can be evaluated. FOULER will be incorporated in the Coal Quality Expert (CQE) software project. CQE is a comprehensive, PC-based program that can be used to evaluate various potential coal cleaning, blending, and switching options to reduce power plant emissions while minimizing generation costs.

The model is based on theory and a combination of laboratory-, pilot-, and field-scale test data. The code encompasses the hanging pendant, superheater, reheater, and economizer regions of the convective pass. The code predicts growth and removal of ash deposition through the interaction of several submodels: 1) Deposit Growth, 2) Deposit Strength Development, 3) Thermal Properties, 4) Deposit Removal, and 5) Sootblower Effectiveness. The deposit removal mechanisms included are thermal shock, gravity shedding, and sootblowing. The required inputs for the code include ash size and composition, boiler parameters, and operation conditions. Input parameters can be entered into the code directly or they can be predicted by other codes such as MMT (mineral matter transformation code) and CQE heat-transfer module. The submodels interact to produce outputs, based on a time basis, of the deposit mass, strength, resistivity, and removal rates. This report describes the fouling submodels, the rationale used in these submodels, and a description of how the experimental data were utilized to validate the algorithms.

INTRODUCTION

Convective pass ash-fouling deposits can decrease the efficiency of a utility boiler by reducing heat transfer from the hot gases to the tube banks. The tube banks contain steam used to generate electricity in the steam turbines. The extent of ash-related problems depends upon system conditions and the abundance and association of the inorganic components in the coal. Several key processes must be understood in order to develop a method to predict ash deposition. The first process is ash formation. During the

Applications of Advanced Technology to Ash-Related Problems in Boilers
Edited by L. Baxter and R. DeSollar, Plenum Press, New York, 1996

combustion of coal, the inorganics present in the coal combine or separate depending upon the composition, size, and physical location and association of the inorganics. The firing mode and conditions also greatly affect the resultant physical characteristics and composition of the ash. Solidification and condensation of the entrained ash species occur as the ash is transported by the gas stream to cooler regions of the convective pass.

The second process involves the transport of ash particles to the deposition surface and reactions that occur upon deposition. There are three primary modes of ash transport: inertial and eddy impaction, vapor-phase and small-particle diffusion, and thermophoresis/electrophoresis. These three modes of transport are responsible for the formation of the three general types of fouling deposits: inner layer, upstream, and downstream. The initial layer, 100–200 microns thick, forms both on the upstream and downstream sides of the heat-exchange tubes. This inner layer forms a base or anchoring site for both upstream and downstream deposition to occur. The inner upstream layers of deposition around a tube are generally deposited by vapor-phase and small-particle diffusion and thermophoretic/electrophoretic forces. The actual particles that deposit are dependent upon the flow characteristics around the heat-exchange tubes. Downstream deposits are formed by impaction from the recirculation eddies passing around the heat-exchange tubes. As the gas stream passes around the tube, those particles that do not inertially impact get caught in the recirculation eddies of the gas stream and are impacted into the downstream side of the tube surface.

The massive upstream deposits are formed by the inertial impaction of ash particles onto an already existing inner layer. The inner layer provides an attachment surface for the larger, less sticky ash particles that form the upstream layer. Not all particles that impact the deposit surface will necessarily stick. Kinetic energy, particle stickiness, particle size, and deposit surface conditions are all factors that affect whether a particle will stick or deflect off a deposit surface. Fouling deposit formation can be described as two interacting mechanisms: deposit growth and strength development. As the deposit grows, the temperature profile throughout the deposit changes, which affects the strength development and future deposit growth. Strength development is generally a result of one of two sintering mechanisms: silicate- or sulfate-based. Silicate-based sintering is controlled by viscous flow of amorphous material during and after deposition. Sulfate-based sintering is attributed to the filling of deposit pores by the sulfation of the alkaline and alkaline-earth components in the deposit. The effects of deposit growth and strength development can then be applied to the heat-exchange properties of the deposit and the deposit removability.

The last process involves the removal of the deposit from the tube surfaces. Three types of deposit removal mechanisms can occur in the convective pass: gravity shedding, thermal shedding, and sootblowing. Gravity shedding occurs when the deposit weight becomes greater than the deposit strength can support. Typical thermal shedding occurs when a utility boiler shifts from high- to low-load capacity during hours of low demand, which results in a temperature change. The deposit and the heat-exchange tubes have different expansion coefficients, which will result in cracks in the deposit, causing the deposit to shed. Thermal shedding can also occur during sootblowing because of the temperature difference of the deposit and the sootblowing media. Removal of the deposits by sootblowing is dependent upon the strength and mass of the deposit and the sootblower parameters. Sootblowing usually occurs at regular time intervals in the boiler. The number of sootblowers and configuration largely dictates the removal efficiency achieved.

FOULER

FOULER is a mechanistic model developed by the EERC to predict the fouling that occurs in the convective pass of a coal-fired utility boiler. FOULER receives the required input information, as shown in Fig. 1, from the CQE heat-transfer module, interface shell, and the MMT code. The heat-transfer module supplies the temperature and fluid flow properties of the system prior to deposition. The interface shell supplies the operational parameters such as sootblower configuration and mass loading as entered by the user. The MMT model supplies the necessary ash particle size and composition information. The FOULER code is separated into five submodels: 1) Deposit Growth, 2) Deposit Strength Development, 3) Thermal Properties, 4) Deposit Removal, and 5) Sootblower Effectiveness. The output from FOULER to the CQE model includes deposit resistivity, deposit growth rates, removal rates, and sootblowing effectiveness.

FOULER comprises 47 different subroutines, which are grouped together in the five submodels. The first submodel, Deposit Growth, separates the incoming MMT data into three separate ash categories. These ash categories and corresponding ash compositions are used in forming the initial downstream and upstream deposit layers. The second component of FOULER is the Deposit Strength Development submodel, which is separated into two sintering functions: silicate- and sulfate-based. Thermal Properties, the third submodel, supports the other models by supplying thermal property information such as deposit and gas thermal conductivities, heat-transfer values, and deposit densities. The fourth submodel, Deposit Removal, contains three deposit removal functions: sootblowing, gravity shedding, and thermal shock from load shifting. The final submodel, sootblower effectiveness, calculates the percent recovery of heat transfer for a given time interval, assuming no deposit removal has occurred until that time. Each of these submodels is described in greater detail below.

Deposit Growth

The first function of the deposit growth submodel is to organize the ash by size into three ash categories. The CQE code utilizes the MMT code created by PSI to predict the particle size and composition distribution (PSCD) of the entrained ash as a function of the original coal properties. The PSCD of the ash is divided into six size and seven composition bins, for a total of 42 different bins of ash particles. The first ash category includes ash particles less than 5 microns in diameter, which makes up the initial layer. All ash particles less than 10 microns are included in the second category and are used in forming the downstream deposit. Downstream deposits are formed by impaction from the recirculation eddies passing around the heat-exchange tubes. As the gas stream passes around the tube, those particles that do not inertially impact (generally less than 10 microns) get caught in the recirculation eddies of the gas stream and are impacted into the downstream side of the tube surface, as shown in Fig. 2. The last ash category forms the upstream deposit and includes all the size and composition bins.

As discussed previously, the three primary modes of fouling deposit growth are inertial and eddy impaction, vapor-phase and small-particle diffusion, and thermophoresis/ electrophoresis. The inner layer, formed by small-particle diffusion and thermophoretic/electrophoretic forces, comprises primarily vapors and particles less than 5 microns that traverse through the boundary layer surrounding the tube and deposit. Which particles actually deposit depends upon the flow characteristics around the heat-exchange tubes. At higher temperatures and faster gas velocities, the inner layer is

454

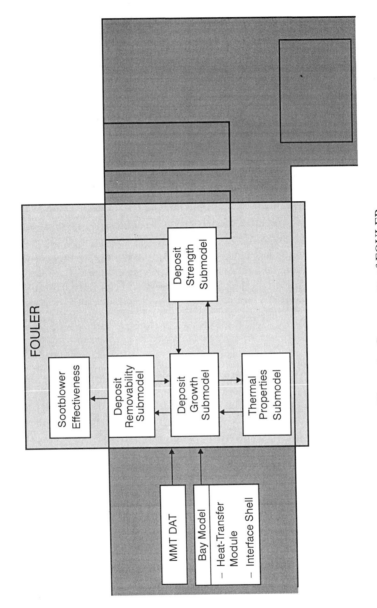

Figure 1. Components of FOULER.

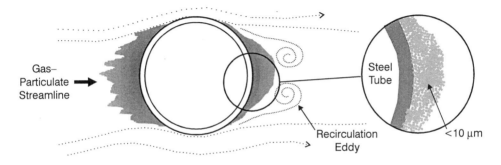

Figure 2. Downstream deposition.

enriched in vapor-phase species and remains loosely bound, while at lower temperatures (and lower velocities), the enrichment tends to shift to particles in the less-than-5-micron range. In both cases, the inner layer plays a role in the eventual formation of massive upstream deposits.

Both the downstream and inner layer deposition rates (DR_t), where t is the type of deposit—downstream ($_d$) or initial ($_l$) layer—are calculated from the following equation:

$$DR_t = \frac{V_g^2}{T_g} * R_r * M_{ash} * \frac{\delta}{A_d}$$ [1]

Potential deposition mass is the fraction of total mass in the respective ash categories described previously for initial and downstream deposition. Total ash is the total mass of ash passing through the convective pass.

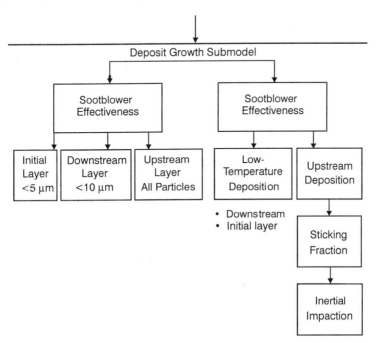

Figure 3. Components of deposit growth model.

The massive upstream deposits are formed primarily from inertial impaction onto the surface of the deposit. The deposit surface can be captive or noncaptive. The larger particles become separated from the gas stream as it flows around the tubes. The particles impact the surface and either stick or deflect off, depending upon their sticking ability and the captive surface of the tube. As massive deposits grow, the surface temperature of the deposit increases, developing a highly captive surface that will capture most of the impacting particles. As the deposit grows, it commonly becomes more aerodynamic, thus minimizing the amount of ash that impacts the surface.

FOULER's upstream deposition function is separated into three subfunctions, as shown in Fig. 3. The sticking fraction subfunction determines the deposit potential, which is the fraction of the total ash in the path of the heat-transfer surface that sticks [Srinivasachar et al., 1990]. A cut-off particle viscosity is calculated by using the values from the inertial impaction subfunction [Kalmanovitch and Frank, 1988]. If the particle viscosity is less than the cut-off viscosity, Equation 2, that particle is added to the sticking fraction.

$$DP = \alpha * \frac{M_p}{100} * IEF \qquad [2]$$

The impact efficiency value is a combination of two values: the impact efficiency strike (IES) and the impact efficiency velocity (IEV) [Wessel and Righi, 1988]. The IES value, (0 to 1) is the fraction of ash that will strike the tube, and the IEV value (0 to 1) is used to determine whether the particle has sufficient kinetic energy to rebound off or stick to the tube surface. The upstream deposition function goes through each ash particle-size fraction using the gas stream temperature and determines the impaction efficiency value for each size fraction and the cut-off particle viscosity. Summing all of the size bins, a cumulative sticking fraction (SF) value (0 to 1) can be determined. The cumulative sticking fraction is then used in Equation 3 to determine the upstream deposition rate.

$$DR_u = \frac{M_{ash} * A_d}{A_t} * \frac{SF}{A_d} \qquad [3]$$

Strength Development

Strength development is generally the result of one of two sintering mechanisms: silicate- or sulfate-based. The low viscosities necessary for silicate-based sintering are commonly attributed to higher temperatures and lower-melting-point phases such as sodium aluminosilicates. Some of the reduction in melting point happens after deposition, when the interaction of the deposited material and gas-phase species forms low melting point phases. Sintering develops over time as low-viscosity material flows and fills the pores of a deposit. Decreasing the numbers or size of the pores in the deposit increases strength. The larger the quantity of liquid phase, the higher the potential for sintering and generating hard deposits.

Sulfate-based sintering is attributed to the filling of deposit pores by the sulfation of the alkaline, alkaline-earth components in the deposit. Sulfates are generally unstable and decompose above 1850°F (1000°C), but form very rapidly at temperatures slightly below this decomposition temperature [Mulcahy et al., 1966]. The crossover temperature range from rapid sulfation to decomposition is narrow and can be crossed in some areas of the boiler as a result of load swings.

456

Both silicate- and sulfate-based strengths are determined for each deposit. The silicate strengths are a function of the viscosity and particle size of the deposited materials and the time duration of deposition. Equation 4 determines a silicate variable which is a function of the deposit potential, particle diameter, and particle viscosity.

$$Sil_v = DP * D_p * \mu_p \qquad [4]$$

The silicate variable is summed for all of the particle-size bins for that particular deposit type. For example, using Equation 5, the silicate variable for a downstream deposit strength would sum the ash size bins less than 10 microns.

$$\overline{Sil_v} = \frac{\Sigma Sil_v}{\Sigma DP} \qquad [5]$$

The cumulative silicate variable is divided by the total deposit potential, producing a weighted average silicate variable. The average silicate variable is the unitless time needed to achieve a completely sintered deposit. The silicate strength of the deposit, at that time interval, is determined using Equation 6.

$$STR_{sil} = 1 - (\frac{1}{(t+1)^{\overline{sil_v}}})^{S_{rc}} \qquad [6]$$

The silicate strength is summed over all of the deposit layers present. The longer a deposit has time to sinter, the higher the silicate strength. The weighted average silicate strength (STR_{sil}), Equation 7, is the sum of each deposit layer strength multiplied by that deposit layer's respective mass, divided by the total deposit mass. The weighted average silicate strength is an index value, from 0 to 1 with 1 being the maximum silicate strength.

$$\overline{STR_{sil}} = \frac{\Sigma (STR_{sil} * M_L)}{M_{dep}} \qquad [7]$$

The sulfate strength, Equation 8, is a function of composition and is assumed to reach maximum strength immediately. The PCfactor 1 and 2 in the equation are calibration numbers the EERC has obtained from previous projects [Wessel and Righi, 1988]. Sulfation strengths are set to zero if the temperature is above the sulfate decomposition temperature.

$$STR_{sulf} = \frac{\frac{Na_2O + MgO + K_2O + CaO}{Si_2O} * PCF_1 - PCF_2}{C_{sulf}} \qquad [8]$$

The sulfate strength is divided by a constant to produce a sulfate strength index number, from 0 to 1. Both sulfate and silicate strength mechanisms can occur at the same location. Figure 4 shows how the sulfate and silicate strength curves may intersect. The greatest strength determined from the two functions is chosen as the strength for that deposit layer

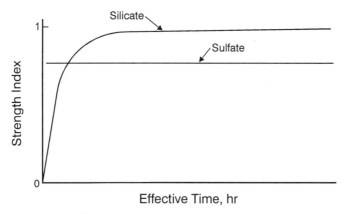

Figure 4. Strength development curve.

(STR$_{depL}$) at that time interval. An average deposit strength (STR$_{depavg}$) can be determined by summing the strength multiplied by the deposit layer mass and divided by the total deposit mass.

Thermal Properties

The thermal properties submodel is a collection of functions that support the other submodels. The thermal properties of the deposits depend upon the thickness, temperature, and physical sintered state of the deposit. Because of the changing temperature and sintered state throughout the thickness of a deposit, as well as its growth and removal, the thermal properties are not constant and require multiple iterations through the running of the FOULER model. Since the deposition rates and compositions of the upstream and downstream deposit are different, the thermal property calculations are considered separately for each half of the heat-exchange tube.

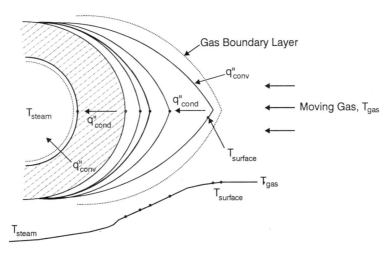

Figure 5. Heat transfer through deposit.

The main thermal property submodel is the heat-transfer function, which is responsible for the determination of the heat-transfer value through the deposit. The change in heat transfer is due to the convection and conduction through the gas stream, deposit, tube, and steam. Figure 5 shows the components of the heat-transfer function, along with a relative temperature drop through an upstream deposit. The forced convection heat-transfer rate, q''_{conv}, occurs at the boundary layer between the gas stream and the deposit surface. The forced convection value is dependent upon surface geometry and flow conditions and several fluid properties, which include density, viscosity, thermal conductivity, and specific heat. The thermal property submodel contains several fluid property functions, which are documented in the FOULER code. Once the fluid parameters are determined, a subfunction determines whether the flow at the boundary layer is turbulent or laminar. Depending upon the flow characteristics and average convection coefficient h_{conv}, q''_{conv} is determined. The conductive heat-transfer rate, q''_{cond}, through the deposit and tube is dependent upon thermal conductivity, temperature, and thickness of each layer. The convection heat-transfer rate through the steam is similar to solving the forced convection of the gas stream. The heat equation for the heat-transfer through the deposit is given in Equation 9.

$$q = \frac{T_{gas} - T_{steam}}{\dfrac{1}{2\pi Rad_{ti} Lh_{st}} + \dfrac{\ln(\dfrac{Rad_{t0}}{Rad_{ti}})}{2\pi L K_t} + \Sigma \dfrac{\ln(\dfrac{Rad_{dn}}{Rad_{dn-1}})}{2\pi L K_{dn}} + \dfrac{1}{2\pi Rad_{dt} Lh_g}} \qquad [9]$$

The deposit thermal conductivity is based on laboratory measurements of ash thermal conductivity by Mulcahy et al. [1966]. The conductivity of ash is nearly independent of the chemical composition, but is strongly dependent on the degree of sintering. Further, the ash conductivity exhibits hysteresis; once heated to a point where sintering occurs, the ash on cooling has a higher thermal conductivity than the original unsintered ash.

Deposit Removal

The deposit removal algorithm accounts for thermal shedding, sootblowing, and gravity shedding. The removal characteristics of the deposit are calculated based on the deposit growth and strength development. The removal indices have a value from 0 to 1, with 1 being complete removal. The sootblowing and load drop models are only applied at the time intervals supplied by the input data. The largest removal index is selected for each time interval.

Thermal shedding occurs when a utility drops load, which results in a temperature change in the boiler. Because of the different thermal expansion value of the ash deposit and the steam tube, the result is shear fracture in the deposit. The difference in thermal expansion can be correlated to density of the deposit. The change in temperature and the strength of the deposit are the main variables in determining the thermal shedding index.

Gravity shedding is common in the backpass regions of a utility boiler where strength development is low, but deposition is high. This form of deposit removal is correlated to the strength:mass ratio of the deposit. The gravity-shedding index is a function of the deposit strength and weight. Unlike load shedding and sootblowing, gravity shedding is possible at every time interval. Shedding occurs as the weight of the deposit becomes greater than the strength.

The amount of removal due to sootblowing is calculated by comparing the strength of the deposit with the shear stress applied to the deposit by a retractable sootblower as a function of the blowing media, pressure, nozzle angle, and spacing between sootblowers.

FOULER SUBMODEL INTERACTIONS

The fouling code utilizes the above-mentioned submodels to predict the heat-transfer effects of a particular coal on the convective pass of a boiler. The convective pass of a boiler can be divided into as many as twelve individual heat-exchange sections (within the primary superheater, reheater, economizer) for the fouling predictions. The code runs through the calculations for each section of the convective pass separately.

Figure 6 is a detailed diagram of the individual subfunctions within FOULER that has been numbered to show flow and interaction. These numbers are referred to throughout this section of the report. The FOULER code is separated into three main computer code sections. The first section contains functions that are not dependent upon time, and no computational iterations are needed. Once time intervals begin, the computer code enters a grouping of subfunctions that are dependent on time and may require several iterations as the deposit temperature, thickness, and other thermal properties change. The total test time of the analysis and the time interval between updates of the analysis are user-defined inputs. The time interval can range from ½ hour to 2 hours. The total time can be as long or short as the user desires, as long as the test time divided by the interval time is less than 200. For example, a ½-hour time interval would allow a maximum test time of 100 hours.

Input

FOULER receives as input four categories of information for each heat-exchange section or zone: 1) the boiler operational parameters, 2) gas temperature and composition distributions, 3) fly ash particle size and composition distribution, and 4) sootblowing and load drop parameters. The input values, which are provided by the CQE interface and the PSI MMT model, are loaded into classes before entering the main FOULER code (Boxes 1-4, Figure 6). If unit conversion of incoming data is necessary, it is performed in the data manipulation function (Box 5).

Flow of Submodels

For each zone, the ash particle sizes participating in the upstream, downstream, and inner layer deposition are identified for use in their appropriate growth and strength models, Boxes 6 through 19, as discussed previously. The removal index constants, as described earlier, are determined next in Boxes 20 through 23. The inner deposit layer is a special case because of the long times needed to develop; for this reason, a thickness of 200 microns is assumed (Box 28). The amount of mass needed to achieve the 200-micron thickness is determined using the deposition rate previously established. The initial layer is assumed to have the maximum strength of 1, to prevent future removal. The inner deposit layer thermal resistance is also determined (Box 29).

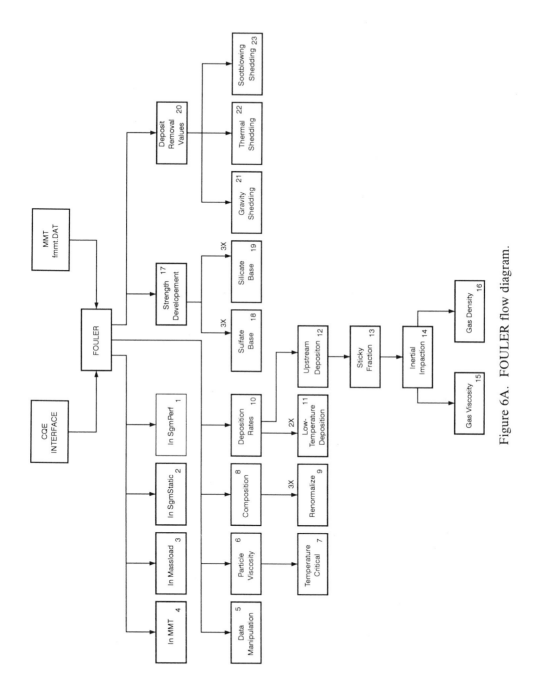

Figure 6A. FOULER flow diagram.

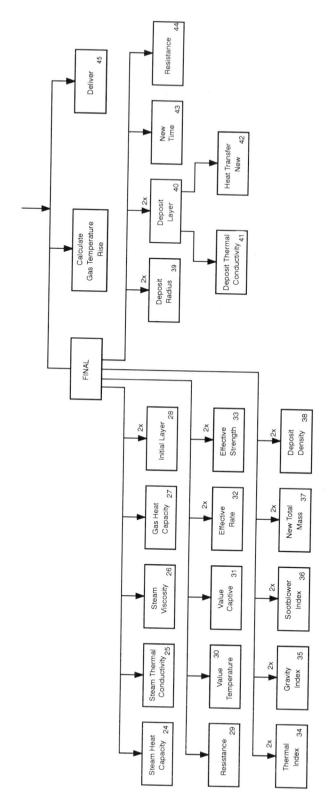

Figure 6B. FOULER flow diagram.

After all of the initial parameters have been calculated, the code begins the iterations to build the deposit layers. Figure 7 shows an example of how FOULER builds a deposit using layers. The initial layer (Layer 1), using a ½-hour time interval, would have been deposited on the heat-exchange surface for a total of 2 hours in this example. The outermost layer would have a deposition time of ½ hour. FOULER keeps track of each layer's mass, temperature, thermal properties, and chemical composition. These layers are the key to FOULER's methodology. Each layer may have different temperature, thickness, and thermal properties, which affect the other surrounding deposit layer properties.

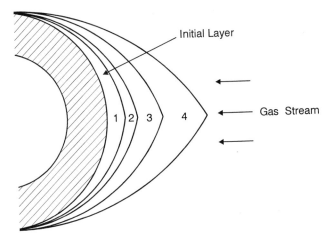

Figure 7. FOULER deposit layers.

As the deposit mass increases on the heat-exchange surface, the temperature of the gas stream increases as a result of less heat being transferred to the tube banks. The hotter gas stream and the insulating effect of the deposit layers will produce a higher outer deposit layer temperature than previous deposit layers has obtained. If the outer deposit layer becomes sticky because of the high surface temperature, a captive surface forms, thus increasing the amount of ash that is able to stick to the deposit surface. The temperature and captive values (Boxes 30 and 31, respectively) increase in value as the deposit grows. The temperature and captive variables are used in Equation 10 to determine the effective rate of downstream and upstream deposition. The effective rate uses the deposition rate determined previously and also takes into consideration the rising gas stream temperature and increasing stickiness of the outer deposit layer. Since the deposition rates and compositions of the upstream and downstream deposit are different, all of the models are considered separately for each half of the heat-exchange tube.

As a result of the interdependence of the deposit layer's temperature, thermal conductivities, and the heat transfer through the deposit, several computational iterations are necessary. Listed below are the steps taken in Box 40 to determine the deposit layer temperatures:

Step 1: Solve for each deposit layer temperature using Equation 11.
Step 2: Using the temperatures solved for in Step 1, determine the deposit layer's thermal conductivity (Box 41).
Step 3: Solve for the heat-transfer rate using Equation 9.
Step 4: Determine the outermost deposit layer temperature using Equation 12.
Step 5: Compare the temperature from Step 4 with the temperature determined in the previous iteration. If the difference is larger than 0.5 degrees, repeat Steps 1–5.

$$T_{dn-1} = T_{dn} - \frac{q * \log(\frac{Rad_{dn}}{Rad_{dn-1}})}{K_{dn} * 2 * \pi} \qquad [11]$$

$$T_{d_{outer}} = T_{gas} - \frac{q}{h_g * 2 * \pi * Rad_{dt}} \qquad [12]$$

The deposit thermal conductivity function, Box 41, assumes a characteristic thermal conductivity based on the degree of sintering and the average temperature throughout the deposit. With a known degree of sintering (based on the strength development) and a given temperature, a deposit thermal conductivity can be estimated.

Figure 8 shows that the area under the deposition rate curve is equal to the deposit's total mass. After the amount of mass loss due to the deposit removal models is subtracted, the "effective time" can be determined. The effective time represents the time state of the deposit and not the total time of deposition. This effective time is used in determining the next time interval's deposition rate. For this reason, it is essential to increment the time of the FOULER code in small enough segments to allow for the multiple interactions of the code.

The thermal resistance due to the deposition on the heat-exchange tubes is determined (Box 44) using Equation 13. When adding the upstream and downstream deposit resistances, the resistances must be added in parallel.

$$\frac{1}{R_t} = \frac{1}{\frac{\ln(\frac{R_{2up}}{R_{1up}})}{2\pi * K_{up}}} + \frac{1}{\frac{\ln(\frac{R_{2down}}{R_{1down}})}{2\pi * K_{down}}} \qquad [13]$$

After the thermal resistance is determined, the time is incremented by the time interval, and the model returns to Box 30. This process repeats until the maximum time has been

464

reached. The thermal resistance and deposit mass values are stored for each time interval and passed back to the CQE model through the CQE interface.

SUBMODEL TESTING

Experimental Approach

In order to study the wide range of variables involved in fouling and slagging of utility boilers, a large amount of experimentation, sampling, and analyses was required. Full- and pilot-scale tests have been conducted and advanced coal, ash, and deposit characterization performed in conjunction with this and other related projects.

Entrained ash, on-line deposits, and off-line deposits were sampled from the convective pass of full-scale utility boilers as part of a research program entitled "Project Calcium: Calcium-Based Deposition in Utility Boilers" (Hurley, 1992). Both tangentially fired pulverized coal (pc) boilers and cyclone-fired boilers were sampled. Entrained ash was collected using a three-stage source assessment sampling system (SASS) train with a final filter for collecting fine particulate. On-line deposits were collected on a fouling probe with a sacrificial insert for the examination of deposits from tube to tip. Data gathered from this testing has been used in the development of the downstream (low-temperature) deposition algorithms.

Pilot-scale tests have been designed to evaluate the fireside performance of test fuels in an environment where the unit-specific effects (such as boiler design, upper furnace convective pass tube spacing, and firing arrangement) can be eliminated, allowing an unbiased evaluation of fuel performance. Maintaining the same firing conditions, heat absorption, and temperature profiles in a full-scale unit to evaluate fuel performance while switching fuels is virtually impossible and very expensive. The pilot-scale testing allows for better control over the temperature profiles and heat fluxes and provides a means to

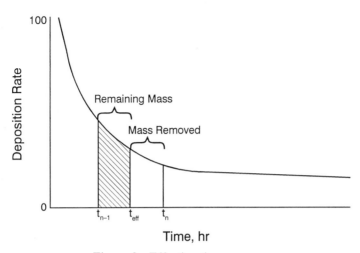

Figure 8. Effective time curve.

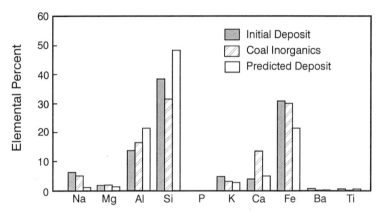

Figure 9. Comparison of measured, predicted, and original coal components for an upstream deposit collected at 2320°F in the ABB CE FPTF.

assess full-scale boiler phenomena in a controlled environment. Comparisons between pilot- and full-scale testing have shown good correlation in the simulation of fouling and slagging [Zygarlicke et al., 1992].

Comprehensive pilot-scale tests were conducted in the ABB CE Fireside Performance Test Facility (FPTF) to evaluate the combustion, furnace-slagging, and convective pass fouling characteristics of selected fuels. The FPTF is an upward-fired heat furnace including a complete fuel-handling system (for both solid and liquid fuels), pulverizer, and air preheater. The test furnace consists of a 5.5-m (18-ft)-high refractory-lined, 915-mm (36-in.)-diameter cylinder. The 6-in.-thick refractory lining minimizes the potential heat losses associated with the large surface-to-volume ratio inherent with small furnaces. Deposits were produced while firing an Illinois No. 2, 3, 5, and 5a blend at 4.0 MMBtu/hr and a Kentucky No. 11 coal at 3.6 MMBtu/hr. Excess air levels were maintained at 20%. Slag deposit samples were collected from Waterwall Panel No. 1 (WWP1), and a water-cooled sacrificial probe for collecting deposits was also used to collect slag deposits in the same waterwall region.

Analytical characterization of the original coal and resultant coal products (ash, slag, and deposits) was accomplished using various American Society for Testing and Materials (ASTM) techniques as well as computer-controlled scanning electron microscopy (CCSEM) and scanning electron microscopy point count (SEMPC) [Zygarlicke et al., 1992; Zygarlicke and Steadman, 1990]. The CCSEM technique allows for simultaneous determination of particle size and composition distributions for coal minerals or ash particles. This technique produces a database of particles of varying compositions and sizes for each analysis. The resultant data are easily manipulated through the use of Partchar® (particle characterization code) or by manual manipulation with a spreadsheet.

The SEMPC technique is highly effective for characterization of slags and deposits. The SEMPC technique uses the random point count routine in conjunction with the x-ray capabilities of a scanning electron microscope to quantitatively determine the highly variable compositions of slags and deposits. The SEMPC technique can be customized to determine the composition variations and porosity as a function of distance from the tube wall. Through experience in deposit and slag characterization, the state of the inorganic components can be inferred from their composition as crystalline or amorphous. The resultant data are easily manipulated through the use of Minclass® (mineral classification code) or by manual manipulation with a spreadsheet. Silicate-based viscosity profiles of the SEMPC data can also be obtained by using Viscalc (viscosity calculation code) using a modified Urban equation.

Deposit Composition Comparisons

The prediction of deposit compositions for high- and low-temperature deposits has been compared with pilot- and full-scale results, respectively.

Pilot-scale upstream deposits were collected on water-cooled sacrificial probes in the ABB CE FPTF firing a bituminous coal from the western Kentucky No. 11 seam. The deposits were collected at a gas temperature of 2320°F. The current fouling algorithms are designed to predict the potential for a given particle to impact and deposit on the leading edge of a heat-exchange surface in the absence of a captive surface. Since the deposit formed from the Kentucky No. 11 coal produced a highly liquid layer after significant deposition, the predicted results are only compared to the initial nonliquid layer. Input to the fouling code was generated from the MMT code as predicted from the initial coal properties. Figure 9 compares the deposit before the captive surface formation, predicted deposit, and the initial coal inorganic components. The predicted results compare well with the experimentally measured results, with the exception of the calcium content.

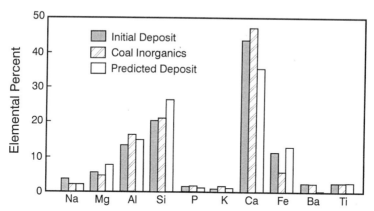

Figure 10. Comparison of measured, predicted, and original coal components for a downstream deposit collected at 1800°F in the Sherco Unit No.1.

Full-scale downstream deposits were sampled from Northern States Power Company's Sherco Unit No. 1 as part of Project Calcium. The feed coal was a Wyoming subbituminous. Input to the FOULER code was generated from analysis of entrained ash sampled from the same location as the deposit. The deposits were collected at a gas temperature of approximately 1800°F. Figure 10 compares the deposit, predicted deposit, and original coal inorganic components. The predicted values compare well with those measured from the full-scale sampling.

Deposit growth rates are being compared to the numerous full- and pilot-scale tests addressed previously to establish a correct rate development curve.

ACKNOWLEDGMENTS

The work presented here was supported by the Clean Coal Technology I program entitled Coal Quality Expert by the U.S. Department of Energy and the Electric Power Research Institute.

REFERENCES

Hurley, J.P. "Project Calcium," final report; EERC publication, Sept. 1992.

Kalmanovitch, D.P.; Frank, M. "An Effective Model of Viscosity for Ash Deposition Phenomena," Presented at Mineral Matters and Ash Deposition from Coal, Santa Barbara, CA, ASME Publication, 1988, 13 p.

Mulcahy, M.F.R.; Boow, J.; Goard, P.R.C. *J. Inst. Fuel* **1966**, *39* (308), 385–394.

Srinivasachar, S.; Helbe, J.J; Boni, A.A. "An Experimental Study of the Inertial Deposition of Ash Under Coal Combustion Conditions," *In* Proceedings of the 23rd International Symposium on Combustion; The Combustion Institute, 1990, pp 1305–1312.

Wessel, R.A.; Righi, J. "Generalized Correlations for Inertial Impaction of Particles on a Circular Cylinder," *Aerosol Science and Technology* **1988**, 9, 29–60.

Zygarlicke C.J.; Steadman E.N. "Advanced SEM Techniques to Characterize Coal Minerals," *Scanning Electron Microscopy* **1990**, *4* (3), 579–590.

Zygarlicke C.J.; Benson S.A.; Borio R.W.; Mehta A.K. "Examination of Ash Deposition in Full-, Pilot-, and Bench-Scale Testing," Presented at the Effects of Coal Quality on Power Plants (EPRI), La Jolla, CA, August 25–27, 1992.

APPENDIX A
Variable Definition List

A_d	=	Total cross-sectional area available for ash deposition, ft^2
A_t	=	Cross-sectional area of the convective pass, ft^2
B	=	Mass balance boundary condition variable
Baseline	=	Baseline resistance value
C	=	Minimum deposition variable
C_{sulf}	=	Sulfate constant
D	=	Deposit layer density
D_c	=	Density constant
D_p	=	Particle diameter, microns
DP	=	Deposit potential mass of ash that sticks to the tube, lb
DR_d	=	Downstream low-temperature deposition rate, $lb/(ft^2\ hr)$
DR_i	=	Initial layer low-temperature deposition rate, $lb/(ft^2\ hr)$
DR_t	=	Low-temperature deposition rate, $lb/(ft^2\ hr)$
DR_u	=	Upstream deposition rate, $lb/(ft^2\ hr)$
EDR	=	Effective deposition rate $lb/(ft^2\ hr)$
GS_{c1}	=	Gravity-shedding constant 1
GS_{c2}	=	Gravity-shedding constant 2
h_{conv}	=	Average convection coefficient, $Btu/(ft_2\ hr \cdot °F)$
h_g	=	Heat-transfer coefficient of the gas, $Btu/(ft_2\ hr \cdot °F)$
h_{st}	=	Heat-transfer coefficient of the steam, $Btu/(ft_2\ hr \cdot °F)$
IEF	=	Impact efficiency value – fraction of ash that will strike the tube
IES	=	Impact efficiency strike
IEV	=	Impact efficiency velocity
K_{down}	=	Average upstream deposit thermal conductivity
K_{dn}	=	Thermal conductivity of deposit layer, $Btu/(ft^2\ hr \cdot °F)$
K_t	=	Thermal conductivity of tube, $Btu\ (ft^2\ hr \cdot °F)$
K_{up}	=	Average upstream deposit thermal conductivity
L	=	Unit length of 1 ft
M_{ash}	=	Total ash, lb/hr
M_{dep}	=	Total deposit mass
M_{dep_r}	=	Total deposit mass after removal, lb/ft^2
M_L	=	Deposit layer mass
M_p	=	Mass percent for a specific particle size
PCF_1	=	PC sulfate factor
PCF_2	=	PC sulfate factor
PR	=	Percent sootblowing recovery
q	=	Heat-transfer rate, Btu/hr
q''_{cond}	=	Conductive heat-transfer rate
q''_{conv}	=	Forced convection heat-transfer rate
r	=	Rate degradation variable
Rad_{max}	=	Maximum radius of ellipse, in.

Rad_{min}	=	Minimum radius of ellipse, in.
Rad_{to}	=	Outside radius of tube, ft
Rad_{ti}	=	Inside radius of tube, ft
$\text{Rad}_{2\text{up}}$	=	Upstream outer deposit radius
$\text{Rad}_{1\text{up}}$	=	Upstream inner deposit radius
$\text{Rad}_{2\text{down}}$	=	Downstream outer deposit radius
$\text{Rad}_{1\text{down}}$	=	Downstream inner deposit radius
Res_{t}	=	Total deposit thermal resistance
R_{dn}	=	Deposit layer radius n, ft
$R_{\text{dn-1}}$	=	Next deposit layer radius n-1, ft
R_{dt}	=	Total deposit radius
R_{GS}	=	Gravity-shedding index, 0 to 1
R_{I}	=	Removal index 0 to 1, 1 being complete deposit removal
R_{r}	=	Empirical reference rate for downstream and initial layer
R_{TS}	=	Thermal shedding index, 0 to 1
R_{SB}	=	Sootblower index, 0 to 1
SF	=	Cumulative sticking fraction, 0 to 1
Sil_{v}	=	Silicate variable
S_{rc}	=	Silicate reference constant
STR_{avg}	=	Average deposit strength
STR_{depL}	=	Strength of deposit layer
STR_{sil}	=	Silicate strength
STR_{sulf}	=	Sulfate strength
t	=	Time
T_{g}	=	Gas temperature stream, R
T_{gas}	=	Temperature of the gas stream, °F
T_{dn}	=	Deposit layer temperature, °F
$T_{\text{dn-1}}$	=	Next deposit layer temperature, °F
T_{douter}	=	Temperature of outer most deposit layer
T_{steam}	=	Temperature of the steam, °F
TS_{c1}	=	Thermal-shedding constant
TS_{c2}	=	Thermal-shedding constant
V_{g}	=	Gas velocity, ft/s
Θ	=	Deposit resistance value before sootblowing
Φ	=	Deposit resistance value after sootblowing
ß	=	Ccaptive surface variable
τ	=	Temperature variable
δ	=	Potential deposition mass fraction
α	=	Angle factor
μ_{p}	=	Particle viscosity, poise

470

THE STRUCTURE OF SUBMICRON ASH FROM COMBUSTION OF PULVERIZED SOUTH AFRICAN AND COLOMBIAN COALS

E.I. Kauppinen[1], T.M. Lind[1], T. Valmari[1], S. Ylätalo[1] and J.K. Jokiniemi[2]
VTT Aerosol Technology Group
P.O.Box 1401, FIN-02044 VTT, FINLAND
[1]VTT Chemical Technology, [2]VTT Energy

Q. Powell, A.S. Gurav and T.T. Kodas
Center for Micro-Engineered Ceramics, Chemical Engineering Department
University of New Mexico, Albuquerque, NM 87131, USA

M. Mohr
ABB Corporate Research, Department CRBP.1
CH 5405 Baden-Dattwill, Switzerland

ABSTRACT

The formation of submicron ash particles during the utility-scale pulverized combustion of South African Klein Kopie and Colombian El Dorado coals was studied by measuring the ash particle number and mass size distributions in the size range 0.01 - 1 µm upstream of the electrostatic precipitator (ESP). Ash morphology, composition and microstructure were studied by high resolution scanning and transmission electron microscopes (SEM and TEM).

From 0.75 to 1.5 percent of the ash on a mass basis was in the particles smaller than 0.5 µm, i.e. had volatilized during combustion. Number mean sizes as determined by the differential electrical mobility method (DMA) varied from about 0.09 µm up to about 0.15 µm. Particles above 0.2 µm were spherical in shape, being formed due to transformations of the non-volatilized coal ash fractions. Volatilized ash species formed high number concentrations of submicron agglomerates having characteristic lengths up to 1 µm. They dominated the submicron number size distribution mode. The agglomerates were formed from a few to up to about one hundred partially fused primary particles. The size of the primary particles varied from 20 to 60 nanometers.

The amorphous primary particles from the combustion of Klein Kopie coal were not exactly spherical. They seemed to have formed from particles less than 10 nm in diameter. These tiny particles were almost completely fused together to form primary particles that appeared spherical in the lower magnification TEM and SEM micrographs. Similarly, the primary particles from the combustion of El Dorado coal seemed to have formed from the almost completely fused 20 nm spherical particles. Areas with clear crystalline fringes were observed within the amorphous primary

Applications of Advanced Technology to Ash-Related Problems in Boilers
Edited by L. Baxter and R. DeSollar, Plenum Press, New York, 1996

471

particles formed during the combustion of El Dorado coal, suggesting sufficiently high temperatures were reached for the crystallization of some of the condensed ash components.

We propose new mechanisms for the formation of submicron agglomerated ash particles in pulverized coal-fired boiler flames.

INTRODUCTION

Submicron ash particles are formed in pulverized coal combustion flames due to partial volatilization and condensation of ash components like Si, Al, Fe and Mg. The fraction of ash vaporized is typically only up to few mass percent of the total ash. Submicron ash particles contribute to deposit formation by controlling the surface area available for heterogeneous surface reactions and condensation of volatilized alkali species. Also, when deposited on heat exchanger surfaces, fine particles contribute to the structure of the surface, thus increasing the sticking coefficient for deposition of the bulk ash particles. The majority of the studies on ash vaporization and submicron particle formation has been carried at laboratory scales with laminar flow systems (e.g. Quann, 1982; Neville, 1982 and Helble, 1987). Flow conditions in boilers are highly turbulent, however, and subsequently the boundary layer characteristics surrounding the burning char particle as well as particle surface temperature may differ from those at the laminar flow conditions utilized in the detailed laboratory studies. Similarly, chemical reactions in the gas phase at turbulent furnace conditions due to enhanced mixing may significantly differ from those at laminar flow conditions.

Extensive work has been carried out by many researchers during the past years to determine the mineral particle transformations as well as coal ash volatilization during pulverized coal combustion. Large amount of the work has been conducted with laboratory and pilot reactors of different scales (e.g. Quann et al., 1990). Modeling of the vaporized ash species behavior during combustion processes has been given an increasing interest (Jokiniemi et al., 1994). However, to verify the laboratory tests and models, data from operating, real-scale boilers is needed for different types of coal, different boilers and different operating conditions.

A small fraction of ash, on the order of one percent on a mass basis, vaporizes during pulverized coal combustion. When the ash vapors condense homogeneously in the furnace, a high number concentration of tiny ash particles is formed via homogeneous nucleation. These small particles grow further when they collide due to their Brownian motion and when vapors condense heterogeneously on their surfaces. When the flue gases enter the ESP, a submicron fly ash mode at about 0.1 μm particle diameter has been observed during the earlier field studies. Detailed laboratory studies by Helble (1987) show that these submicron particles are agglomerates of from a few to less than 30 spherical primary particles from less than 10 to more than 100 nanometers in diameter, depending on the char surface temperature and fuel properties. Recent detailed studies by Joutsensaari et al. (1994) on the real scale combustion of Polish coal showed that submicron ash number size distributions can be bimodal with an ultrafine mode below 0.1 μm and an intermediate mode at about 0.3 μm. High-resolution scanning electron microscope studies revealed that the ultrafine particles were short agglomerates of only a few spherical primary particles, whereas particles in

472

the intermediate mode were spherical in shape. Ultrafine agglomerates were formed from the volatilized ash while the spherical particles were proposed to originate from the ash component that melt on the char surface during char combustion.

In this work we have studied the ash transformations in operating pulverized coal-fired boilers with special attention to the extent of vaporization of the ash-forming constituents and the interaction of the mineral particles with other ash-forming species. Especially, we have focused on the detailed determination of submicron ash characteristics by utilizing high resolution scanning and transmission electron microscopy to observe ultrafine ash particle morphology and microstructure down to the nanometer scale. Also, the coalescence of the large particles has been thoroughly studied as reported by Lind et al. (1995). We have determined the fly ash composition and size distributions from operating pulverized coal fired boilers. In addition to supermicron ash measurements, we measured the size distributions of the submicron particles down to 10 nm in order to determine the extent of ash vaporization in the process. The measurements were carried out with two bituminous coals: South African Klein Kopie and Colombian El Dorado.

METHODS

Process Description and Coal Selection

The combustion experiments were carried out at a 510 MWe pulverized coal fired power plant in Germany (boiler A) during two weeks in May 1994. The power plant is equipped with high-dust SCR NO_x reduction and wet FGD units. The plant has a single wall-fired boiler, to which the coal is distributed from 4 coal mills each feeding coal to four burners. The ash formation in the combustion of two bituminous coals was studied. South African Klein Kopie coal was fired during the first week of experiments, and Colombian El Dorado coal was fired during the second week. The ash content of the Klein Kopie coal was 17 % and was higher than the ash content of El Dorado which was 10 %. The boiler had low-NO_x burners installed, and the coal particle size was 8.5 % greater than 90 μm for Klein Kopie coal and 16.5 % greater than 90 μm for El Dorado coal. In addition, experiments were carried out at another power plant of 630 MW capacity (boiler B) in Denmark with Klein Kopie coal.

Size Distribution Determination

Number Size Distribution Measurements

The ash particle number size distributions at ESP inlet conditions were determined by the differential electrical mobility (DMA) method with the condensation nucleus counter (CNC) being used as a number concentration detector. DMA measurements covered the particle size range 0.02 - 0.8 μm. Flue gas was sampled in-stack through a cyclone precutter having a 3.5 μm Stokes cut diameter and diluted with clean, dry air at flue gas temperature within an ejector-based dilutor. Sample cooling and further dilution was carried out outside the stack within another identical diluter. A more detailed description of the DMA sampling procedure as well as data reduction methods is given by Joutsensaari et al. (1994).

Mass Size Distribution Measurements

The ash particle mass size distributions were determined by collecting size-classified coal combustion aerosol samples with an 11-stage, multijet compressible flow Berner-type low-pressure impactor [Kauppinen, 1992; Hillamo and Kauppinen, 1991; Kauppinen and Pakkanen, 1990]. A cyclone with Stokes cut-diameter of approximately 5 μm was used as a precutter to prevent overloading of the upper BLPI stages. The samples were collected from the flue gas channel upstream of the electrostatic precipitator (ESP) at the flue gas temperature of approximately 130-140°C. The cyclone-collected particles were size-classified by the Bahco sieve method. The mass size distribution of the ash was determined by weighing each BLPI-substrate prior to and after the sampling, and by weighing each Bahco-sieve size-classified fraction.

Particle bounce and re-entrainment is the major source of errors associated with inertial impactor method. Therefore, double stages were utilized during BLPI measurements at boiler A for the upper BLPI stages, in order to minimize the particle bounce. During tests at both sites the impactor substrates were carefully coated with a thin layer of Apiezon L vacuum grease using the method of Hillamo and Kauppinen (1992).

Samples for SEM and TEM analysis

Individual fly ash particles were collected for SEM analysis on planar Nuclepore filters made from polycarbonate films of about 15 μm in thickness. Sampling time, dilution ratio and flow rate were optimized based on number concentration measurements to deposit only individual particles on the filter surface. This is important as submicron particles will attach on the surface of the supermicron particles e.g. when collected on the ESP collection plate. When ESP collected dust is analyzed with SEM, it is difficult to determine the characteristics of particles while they are in the gas phase. Filter samples were coated with a thin layer of platinum and analyzed for both sub- and supermicron particle morphology with a high resolution scanning electron microscope (Hitachi S-800).

Fly ash particles were transferred from Nuclepore filter surfaces onto TEM grids by touching the TEM grid to the filter surface. Agglomerate structure as well as primary particle structure, composition and crystallinity were observed with a high resolution transmission electron microscope (JEOL JEM-2010) equipped with Energy Dispersive Spectroscopy (EDS) with light element (C,O,N) detection capability. The images were digitized and printed directly on a Kodak Colorcase printer.

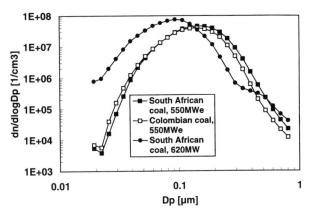

Figure 1. Submicron ash number size distributions as determined with DMA coupled to the two-stage ejector -based dilution system.

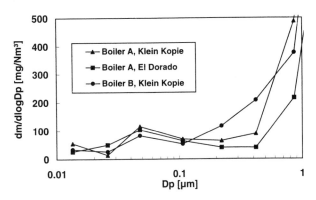

Figure 2. Ash mass size distributions for particles below 2 mm, as determined with the BLPI method.

RESULTS AND DISCUSSION

Ash Number Size Distributions

Submicron ash number size distributions, i.e. the number concentration of ash particles in the flue gas as a function of particle size are presented in Fig. 1. Both South African and Colombian coal fired at boiler B generate a broad submicron ash mode that peaks at about 0.15 μm in particle diameter. Interestingly, the combustion of South African coal at boiler B produces a submicron mode at about 0.09 μm particle diameter with evidence for another mode at about 0.4 μm. Intermediate mode was more pronounced in the flue gas after the combustion of Polish coal, which showed the ultrafine mode at about 0.04-0.06 μm (Joutsensaari et al., 1994). The mass fraction of ash vaporized in their study was only about 0.1 %, which can partially explain the small agglomerate average diameter.

These results indicate, that the size distribution of submicron ash depends both on coal properties as well as on the boiler operation conditions.

Ash Mass Size Distributions

The comparison of the overall mass size distributions determined with low-pressure impactors and with the computer controlled scanning electron microscope method (CCSEM) is presented by Lind et al. (1995) for the ash samples of both coals. Submicron ash mass size distributions are shown in Figure 2 for both coals and for both boilers. The impactor size distribution is based on the particle Stokes diameter calculated with an assumed density of 2.45 kg/dm^3. This is a reasonable estimate for the spherical ash particles that are formed from the non-volatilized ash fraction. However, for the fly ash particles formed during the condensation of volatilized ash, this value is too high, due to the agglomerate structure of particles, as discussed in more detail below. This is the reason why the submicron ash mass mode is found at about 50 nm, i.e. at significantly smaller size than the number size distributions show the major mode.

The fraction of ash volatilized can be estimated based on the measured mass size distributions. As a first estimate, the majority of the ash below 0.5 μm originates from the volatilized fraction. The fractions of ash below 0.5 μm are given in Table 1. The fraction of ash vaporized at boiler A during the combustion of El Dorado coal (1.5 %) is about the fraction volatilized for Klein Kopie coal at the same boiler. The fraction of Klein Kopie coal ash below 0.5 μm at boiler B is 1.2 %, being significantly higher than the corresponding value at boiler A. This is at least in part due to the slight particle bounce during boiler B test, which is evidenced by significant differences of mass size distributions in the size range 0.15 - 0.5 μm.

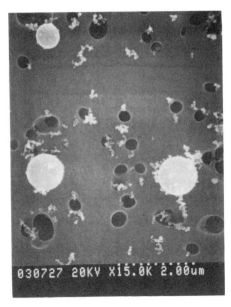

Figure 3. Morphology of the fine ash particles as determined with the scanning electron microscope (SEM) for the El Dorado coal at boiler A. Bimodal structure of the ash is clearly shown, with agglomerated particles having lengths up to about 1 μm, and spherical particles above 0.2 μm.

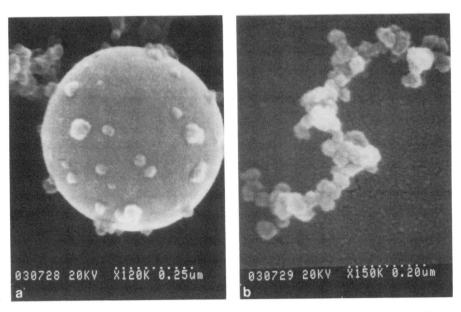

Figure 4. Structure of the submicron ash particles for the El Dorada coal at boiler A. a) Spherical, residual fly ash particles having gas-phase borne primary ash particles on the surface, b) about 0.8 μm long agglomerate composed of about 100 gas-condensed primary ash particles.

Table 1. The mass fraction of ash below 0.5 µm that is the first estimate for ash vaporized during the combustion process.

Coal	Boiler	Mass fraction below 0.5 µm (%)
El Dorado	Boiler A	1.5
Klein Kopie	Boiler A	0.70
Klein Kopie	Boiler B	1.2

Ash Morphology and Microstructure

Scanning Electron Microscopy (SEM)

The SEM micrograph in Figure 3 shows the two types of fly ash particles formed during the combustion of pulverized South African Coal at boiler A. Particles larger than about 0.2 µm were individual solid spheres. They were formed during the melting, coalescense and chemical reactions of coal minerals and other ash-forming constituents that are not volatilized during the combustion process of individual coal particles. Volatilized ash species form high number concentrations of submicron agglomerates having characteristic dimensions up to about 1 µm. They dominate the submicron mode observed both by the DMA in the number size distribution (Fig. 1) and by the BLPI in the mass size distribution (Fig. 2). Ultrafine primary particles have deposited onto the surfaces of the spherical residue particles via coagulation (Fig. 4a). The agglomerates are formed from few to up to about one hundred partially fused primary particles, as shown in more detail in Fig. 4b. Primary particle diameter varied from 20 to 60 nanometers.

The primary particle size in the agglomerated particles varied significantly from boiler A to boiler B for the combustion of Klein Kopie coal (Fig. 5 a and b). The agglomerates at boiler B were shorter, as also evidenced by the smaller number mean sizes measured with the DMA (note that DMA determines the electrical mobility equivalent diameter of the agglomerate in the gas phase, but neither it's physical size nor the size of the primary particle). The primary particle size for the El Dorado coal is almost equal to that for Klein Kopie coal when both coals are burned at boiler A, although the fraction of ash vaporized was significantly higher for El Dorado coal.

High Resolution Transmission Electron Microscopy (TEM)

The SEM results are confirmed by the TEM micrograph shown in Fig. 6, which shows the surface of 1 µm ash residue particle and several 20 to 60 nm primary gas-borne ash particles. TEM also indicated that the primary particles from the combustion of Klein Kopie coal were amorphous. EDS composition analysis carried out on the TEM on the areas marked with circles in Fig. 6 showed that the primary particles were composed mainly from oxides of silicon and aluminum. The primary particle surfaces were enriched in Al relative to bulk composition. EDS showed also Ca and Fe in the primary particles, Ca being enriched on the surfaces. The high-resolution TEM

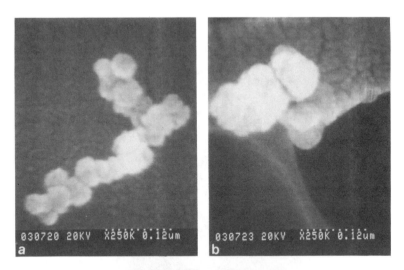

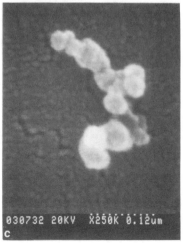

Figure 5. Comparison of the aglomerate structures for Klein Kopie and El Dorado coals: a) Klein Kopie at boiler A; b) Klein Kopie at boiler B; c) El Dorado at boiler A.

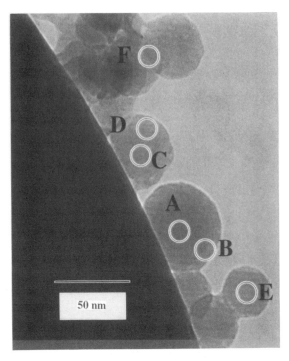

Figure 6. Comparison of the agglomerate and spherical ash particle as determined with TEM for Klein Kopie at boiler A. Al is enriched on the particle surface as compared to Si, as determined by EDS.

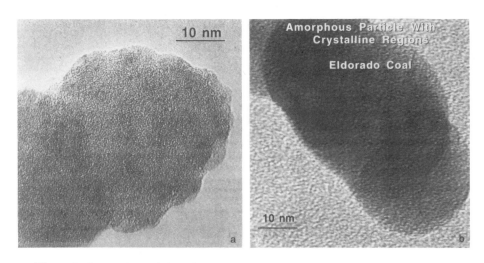

Figure 7. Comparison of the primary particle structures with high resolution TEM: a) Klein Kopie coal at boiler A: b) Eldorado at boiler A.

micrographs in Fig. 7 a and b show their structure down to atomic scale. Interestingly, the primary particles from the combustion of Klein Kopie coal are not exactly spherical. They seem to have formed from collisions of particles less than 10 nm in diameter. These small particles seem to have almost completely fused together to form primary particles that appear spherical in the lower magnification TEM and SEM micrographs. Similarly, the primary particle from the combustion of El Dorado coal seem to have formed from the almost completely fused 20 to 30 nm spherical particles. Areas with clear crystalline fringes were observed within the primary particle formed during the combustion of El Dorado coal (Fig. 7b). This indicated that some of the particles formed by gas-to-particle conversion reached sufficient temperatures for long enough residence time to crystallize.

The morphology of ash studied here differs significantly from that obtained at laminar flow laboratory combustion experiments (Helble, 1987). For comparable oxygen partial pressures (less than 0.2 atm, for the low NO_x burners probably about 0.1 atm) and gas temperatures (about 1600 K), the primary particles from laboratory combustion were smaller and the agglomerates were significantly shorter. At higher char surface temperatures during laboratory studies, significantly larger primary particles but similarly much shorter agglomerates were observed by Helble (1987). Interestingly, the fraction vaporized during this study was lower than that observed at laboratory studies at comparable conditions. The explanation for the longer agglomerates during real scale combustion studies could be the longer residence within the heat exchanger region of the boiler. The mechanism for the formation of both primary particles and agglomerates within the boundary layer of burning char particles via vapor nucleation and particle collisions, as proposed Helble (1987), cannot explain the observed gas-borne fly ash primary particle as well as agglomerate structure. The size of the smallest particles we have observed within primary particles agrees rather well with the size of the primary particles observed during laboratory studies at MIT (Helble, 1987; Neville, 1982).

We propose the mechanisms for the agglomerate ash particle formation as described schematically in Fig. 8. Aluminum and silicon are volatilized via the reduction reaction of their oxides at the char surface or within the char during the char combustion (Neville, 1982). When reduced metal oxide vapours oxidize at the boundary layer, they nucleate homogeneously forming high number concentration of very small spherical ash particles. Based on the work of Helble (1987), the particle size is 5 to 10 nm, when particles have diffused outside the boundary layer. Particle collisions continue after particles have been mixed into the turbulent gas flow in the flames. As the gas is cooling and particle size is increasing due to collisions, particle sintering rate is decreasing. As a result, particles are not any more exactly spherical, but show the structure of partially fused primary particles. When the gas temperature is further reduced at the upper furnace, the sintering rate of colliding particles is further reduced. Therefore close to spherical primary particles formed at flames do not sinter significantly any further. Instead, they continue to collide and accordingly form agglomerates that further grow when the combustion passes through the heat exchangers. Finally, at the ESP inlet conditions agglomerates have formed having few to up to one hundred 20 to 60 nm roughly spherical primary particles.

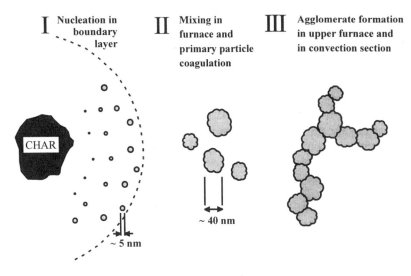

Figure 8. Proposed mechanisms for submicron agglomerate ash formation.

Clearly, more work is needed to characterize in more detail the microstructure of primary and agglomerated ash particles with respect to morphology and composition, in order to gain further understanding on submicron ash formation at realistic combustion conditions.

CONCLUSIONS

Ash vaporization and the characteristics of submicron ash have been studied during utility-scale pulverized coal combustion. Two bituminous coals were studied, South African Klein Kopie and Colombian El Dorado.

One to two percent of the ash was found in the particles smaller than 0.5 μm, i.e. had volatilized during the combustion. Submicron mode number mean sizes varied from about 0.05 μm up to about 0.15 μm. Particles above 0.2 μm were spherical in shape, being formed due to transformations of the non-volatilized coal ash fractions. Volatilized ash species formed high number concentrations of submicron agglomerates having characteristic dimensions up to about 1 μm. They dominated the submicron ash mode. Agglomerates were formed from few to up to about one hundred partially fused primary particles. The size of primary particles varied from 20 to 60 nanometers. The primary particles from the combustion of Klein Kopie coal were not exactly spherical. They seemed to have formed from less than 10 nm diameter particles. These tiny particles were almost completely fused together to form primary particles that appeared spherical in the lower magnification TEM and SEM micrographs. Similarly, the primary particle from the combustion of El Dorado coal seemed to have formed from the almost completely fused 20 nm spherical particles. Areas with clear crystalline fringes were observed within the primary particle formed during the combustion of El Dorado coal.

The microstructure of submicron ash differed significantly from that observed during laboratory combustion of pulverized coal particles at laminar flow conditions.

By comparing the microstructure of the ash agglomerates in our field scale study to those formed at laboratory scale studies, we propose new mechanisms for the agglomerated submicron ash formation, as described schematically in Fig. 8.

ACKNOWLEDGEMENTS

We would like to thank Ministry of Trade and Industry of Finland as well as ABB for funding this study via SIHTI research program. We acknowledge L. Lillieblad, C. Mauritzson and K. Porle at ABB Fläkt Industri AB, Växjö, Sweden, O. Riccius and N. Klippel at ABB Corporate Research, Baden, Switzerland, the members of the VTT Aerosol Technology Group as well as the personnel at the boilers for their support during this work.

REFERENCES

Helble, J.J (1987) PhD Thesis, MIT.

Hillamo, R.E. and Kauppinen, E.I. (1992) *Aerosol Sci. Technol.* 14, 33-47.

Jokiniemi, J.K., Lazaridis, M., Lehtinen, K.E.J. and Kauppinen, E.I. (1994). *J. Aerosol Sci.*, 25, 429-446.

Joutsensaari, J., Kauppinen, E.I., Jokiniemi, J.K. and Helble, J.J. (1994). In J. Williamson and F. Wigley, (Eds.) "The Impact of Ash Deposition on Coal Fired Plants": *Proceedings of the Engineering Foundation Conference,.,* Taylor&Francis, 613-624.

Lind, T.M., Kauppinen, E.I., Valmari, T. Klippel, N. and Mauritzson, C. (1995) " Ash Transformations in the Real-Scale PCC of South African and Colombian Coals." Proceedings of the *Eng.Found.Conf. 'Application of Advanced Technology to Ash-related Problems in Boilers',* Durham, New Hampshire, USA, 16-22 July, 1995.

Kauppinen, E.I. (1992). *Aerosol Sci. Technol.,* 16, 171-197.

Kauppinen, E.I. and Pakkanen, T.A. (1990). *Environ. Sci. Technol.,* 24, 1811-1818.

Neville, M. (1982) PhD Thesis, MIT.

Quann, R.J. (1982) PhD Thesis, MIT.

Quann, R.J., Neville, M. and Sarofim, A.F. (1990). *Combust. Sci. and Tech.,* 74, 245-265.

THE EFFECT OF OPERATION CONDITIONS ON SLAGGING

A.N. Alekhnovich and V.V. Bogomolov

Ural Heat Engineering Institute (UralVTI)
Chelyabinsk, Russia

ABSTRACT

Operation conditions have effect on slagging as due to changes of the flue gas temperature and the heat flux so due to changes of slagging properties of the fly ash. The effect of the same factor for various coal types and boiler zones may differ not only in degree but has opposite sign. Besides, the effect of the factor may be various depending on a combination of the other factors.

The experimental results confirmed the above idea. It also presents examples of a successful decrease of a slagging with the change of operating conditions.

1. INTRODUCTION

Operation conditions effect the slagging in boilers in two ways. First of all it changes the characteristics of the flue gas such as temperature, heat flux, composition and velocity pattern. At the same time the slagging properties of the fly ash can be changed. The effect of these two factors on slagging can be as the same so with an opposite sign. Thus, the slagging in some boilers decreased when they were overhauled with stage burning system though the slagging properties of the fly ash worsened (fly ash became more slagging).

The change of the flue gas characteristics, as a rule, can be calculated or estimated, but the change of the ash slagging properties is established by experiments at that time.

It is also necessary to take into consideration that the operation conditions do not have the same effect on various surfaces of the boiler and the effect of the same factor depends on combination and level of others. For example, the effect of the flue gas speed on the slagging rate depends on the temperature level (Alekhnovich et al., 1995).

Below there are certain results which have been obtained by UralVTI in full scale boilers. Some of them have a particular case and they don't pretend to be hold for wide range of coal types and boiler designs.

2. AIR EXCESS

The fuel burning with low air excess is one of the mean factors that increases the slagging on boilers apparently for all coal types. On the contrary, an increase of the air excess is used as an effective method to maintain a slag-free

Applications of Advanced Technology to Ash-Related Problems in Boilers
Edited by L. Baxter and R. DeSollar, Plenum Press, New York, 1996

485

operation. First of all, it is caused by the change of the flue gas volume and the temperature correspondingly. As a result of only this factor the temperature of uncooled probes in a superheater zone reduced down to 12 - 55 C when the air excess went up by 0.1. In reality the air excess effect is large if the change in slagging of the low part of the furnace is taken into consideration.

The increase of slagging properties is another important factor when burning certain coals in conditions of a low air excess. In particular, when burning low slagging Ekibastuz coal that operation causes the reduction of the temperature of the start of slagging, t_{sl}, by 75 C, the increase of slag sintering and the formation of iron-enriched deposits, which are base deposits in slagging. The change of the t_{sl} temperature has the same effect on slagging as the change of a boiler load by approximately 30%. The results of measurements of the slagging temperature for Ekibastuz coal and for two Siberian coals with a modified acid/base ratio $K/O = (SiO_2+Al_2O_3+TiO_2) / /(CaO+MgO+K_2O+Na_2O) = 10-12$ are shown in Fig.1. The change of the temperature of the start of slagging take place both in the burner zone and at the furnace exit and, according to data of Ekibastuz coal have effect below the critical value of the air excess.

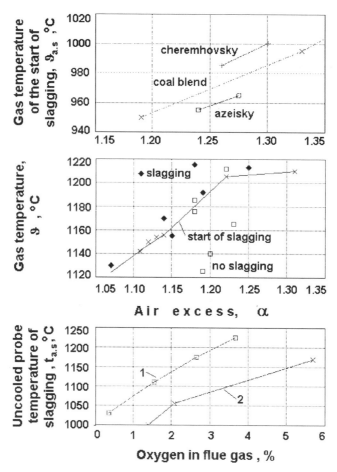

Figure 1. Temperature of the start of slagging in the upper part of the furnace depending on the air excess for certain coals and the uncooled probe temperature of the start of slagging in the burning zone depending on oxygen percentage for Ekibastuz coal. 1) - all burners operating, 2) - three burners off.

When the percentage of oxygen in the flue gas was very low and the carbon content in fly ash was high it again has been observed absence of slagging.

The measurements showed that the temperature of the start of slagging did not changed or it altered marginally depending on the air excess for other more slag forming coals such as Kuznetsk, Kyzyl-Kija whose modified acid/base ratio was less then 10 and for Anju coal (K/O = 10 -12).

Chemical composition of the mineral matter of studied coals is changed in rather wide limits. Average data of ash analyses of mentioned coals are cited in Table 1.

Table 1. Ash analyses (wt %) of coals.

	Ekibastuz	Azejsky	Anju	Kuznetsk	Donetsk	Kyzyl-Kija
SiO_2	61.45	55.89	59.96	60.93	54.06	54.75
Al_2O_3	28.38	29.8	24.99	22.07	24.8	17.02
TiO_2	1.22	0.95	0.92	0.89	0.99	1.11
Fe_2O_3	6.36	4.98	5.6	4.81	12.62	14.32
CaO	1.08	5.64	1.73	4.61	2.24	7.46
MgO	0.6	1.66	2.0	2.56	1.45	2.76
K_2O	0.63	0.66	3.9	2.79	2.84	1.74
Na_2O	0.28	0.42	0.9	1.34	0.99	0.84
Total	100	100	100	100	100	100
K	91.05	86.64	85.87	83.89	79.86	72.88
O	2.59	8.38	8.53	11.3	7.52	12.8
K/O	35.15	10.34	10.07	7.42	10.62	5.69
Ash, A^d	40	19	67.5	19.75	28	18
t_{sl} C	1205	995	1060	1010	1020	970

$K = A = SiO_2 + Al_2O_3 + TiO_2$.
$O = B - Fe_2O_3 = CaO + MgO + K_2O + Na_2O$.
t_{sl} - temperature of the start of slagging.

For the coals we deal with here it can be explained by the following factors. The first one is the content of the unburned carbon in the fly ash. On the whole, the fly ash is not sticky and it sinteres less, when the carbon content is high. The temperature of the start of slagging is higher in the burning zone as compared with the furnace exit zone due to that factor.

The second factor is the change of the iron quantity in the ash melt (in ash glass when cooling) and the degree of iron oxidation in glass. The fraction of iron in glass and the degree of its oxidation was measured by means of Mossbauer spectroscope. It has been found out that the fraction of Fe^{2+} in the ash glass increases when the air excess becomes less. The results are shown in Figure 2.

There are good reasons to take the fraction of Fe^{2+} in ash glass as the base constituent and Fe^{3+} - as an acid one. Then the empirical equation for t_{sl} estimates (Alekhnovich et al., 1988) has the form of :

$$t_{sl} = a+b* (K+Fe^{3+} *Fe_2O_3) / (O+Fe^{2+} *Fe_2O_3).$$

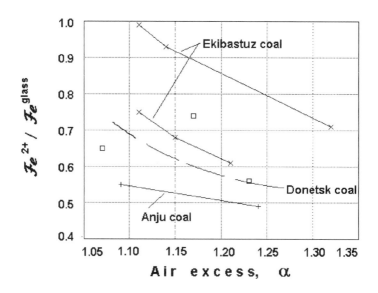

FIGURE 2. Ratio of iron as Fe $^{2+}$ against the iron in ash glass depending on the air excess.

According to the equation the effect of the change of the iron oxidation degree is stronger for ashes with large K/O radio. This corresponds with the above mentioned experimental data and equation of that form can be used for the evaluation of the effect both of the ash chemistry and operating conditions (Figure 3).

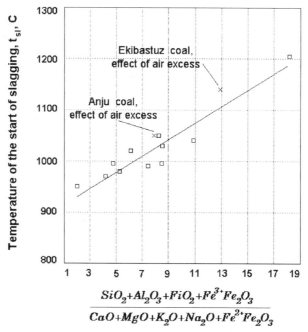

$$\frac{SiO_2+Al_2O_3+FiO_2+Fe^{3+}Fe_2O_3}{CaO+MgO+K_2O+Na_2O+Fe^{2+}Fe_2O_3}$$

FIGURE 3. The temperature of the start of slagging depending the ash chemistry.

3. REBURNING

Various methods of stage burning including the reburning of a part of fuel are introduced in boilers in order to reduce NO_x emission at last years. The change of burning and slagging conditions play their part in that process. The slagging problems in furnaces and superheaters have been reduced or excluded completely on all three boilers tested by UralVTI where the reburning method was used.

However the experiments have shown that the slagging properties of the fly ash become worse not only in the burning zone with low oxygen percentage but also at the furnace exit. It means that the slagging rate increases at the same temperature and the results correspond to data obtained by Mossbauer spectroscope, sintering tests and investigations of the ash glass structure. The fly ash sampled in boilers using the method of reburning has higher content of iron in glass, larger fraction of Fe^{2+}, more strength sintered samples and more even structure of the glass. Thus, the changes, appearing in the structure and composition of the fly ash with low oxygen percentage do not disappear completely during the next stages of burning accompanied by a high air excess. That is caused by the slow oxidation of iron in the glass, that we measured in air and, by PSITechnologies - in flue gas (Bool et al., 1995)

The tests have been carried out only on low slagging coals (Ekibastuz, Azejsky) which show a dependence of ash slagging properties on air excess and, apriori, cannot be extended to other high slagging coals. But, in any case, this phenomenon can take place and it should be taken into consideration.

4. BURNING TEMPERATURE, FINENESS OF PULVERIZED COAL

The temperature of the start of slagging has been measured in the boilers equipped with wet-bottom furnace (burning temperature above 1650 ° C) and in boilers equipped with dry-bottom furnace. The tests were made on high quality Kuznetsk coal with a medium slagging potential. The resulting values in these conditions were the same though it might not be common for other coal types.

It was also found that there was not a large difference depending on coal dust size for some tested coals. But when these two factors were changed within a wider range (the swirling low temperature furnace operating on crushed Azejsky coal) there was no slagging in the furnace and the temperature of the start of slagging was above 1100 ° C and could not be determinated, though the iron-enriched deposits formed actively.

When we said above that t_{sl} depended slowly on coal fineness it did not mean that the did not depend on that factor. In reality this factor can have a strong effect on slagging and this effect can be opposite for various boiler surfaces and conditions. The furnace exit temperature decreases when it is used more rough coal dust in spite of the delay of burning if iron-enriched deposits do not form. Such result were received repeatedly in boilers operating on the Kuznetsk coal and in some of the boilers using the Ekibastuz coal and they used to reduce the slagging of the superheater and the lower part of the furnace.

On the contrary our recommendation to use finer coal dust in order to reduce the slagging of the upper part of the furnace and the superheater of boiler type P-57 has been successful. In that case iron-enriched and slug deposits were formed. Burning of fine dust increases the rate of friable deposits and the forming of iron-enriched ones and slagging are decreases or it is ruled out.

5. DIRECTION OF THE AIR SWIRL IN BURNERS

The PK-39 boiler of an 300 MW unit equipped with turbulent burners arranged in two opposite rows. The problems of slagging concerned with the slagging of the low part of the furnace and the upper part of the bottom ash hopper due to a short distance between the burners and the hopper, between the burners and adjoining walls.

The temperature and the composition of the flue gas and the slagging rate were studied in this zone under various operating conditions which included the change the value and the direction of the air swirl of the burners. Seven combinations of the swirl direction were tested and they can be divided in two groups ("swirl up" and "swirl down") according to the direction of the air swirl in the outer low burner (Figure 4). The tests has been carried out in the zone of that burner with the help of a deposition probe. The water-cooled probe was equipped with a non-cooled test specimen. The probe was moved along the front wall of the furnace with the help of a hand trolley. As there was a large temperature gradient near the water wall the special fasteners were used to clamp the probe in positions.

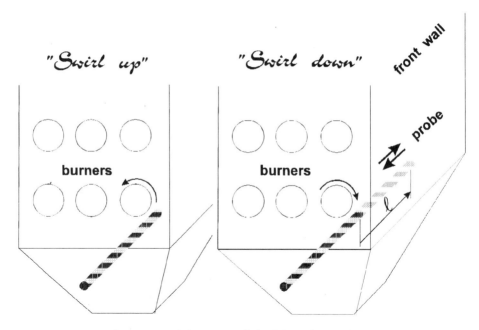

FIGURE 4. A low part of the PK-39 boiler furnace.

It has been measured that the temperature and the slagging rate decreased when the swirl direction pointed down and all burners were operating (Figure 5). But when a part of the burners were cut off from fueling or when the burner of interest is overload the result can be opposite. The effect of the swirl value is also ambiguous. Under more or less even load of the burners the temperature and the slagging rate increase for "swirl down" when the swirl value decreases and change little", depending on that factor for "swirl up".

These results can be explained by interaction between the burner flow and the upward flow of a furnace and the change of a distribution of local draft losses.

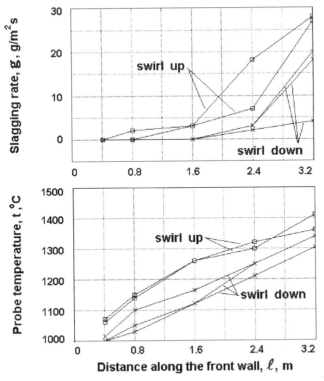

FIGURE 5. Temperature and slagging rate depending on the direction of the swirl.

6. DISTRIBUTION OF BURNERS LOAD

Uneven distribution of loads between burners often causes the slagging of the furnace. However, this factor sometimes can be used successfully to maintain slag-free operation of the superheater.

The study of temperature patterns of the superheater zone was carried out on four boilers equipped with turbulent burners. It is obvious that the reduction of the fuel supply in central burners leads to a more even temperature pattern in the superheater zone if there were no problems with the slagging of the lower part of the furnace and if the burners were located on water walls under and opposite the furnace exit window. The change of the fuel distribution for this purpose is used less when the burners are located on adjoining walls like it is shown in Fig. 6. The reduction of the temperature in a superheater zone was achieved when the burners located near the front wall had more load than those located near the rear wall and when the burners were located on the opposite walls in one row. The temperature was measured with the help of a set of non-cooled probes equipped with several thermocouples. If the burners are located in two rows (a position 2 ' of the second burner in Fig. 6) the results also depends on the level of the burner and it is better, when the low central burner is loaded more, than others.

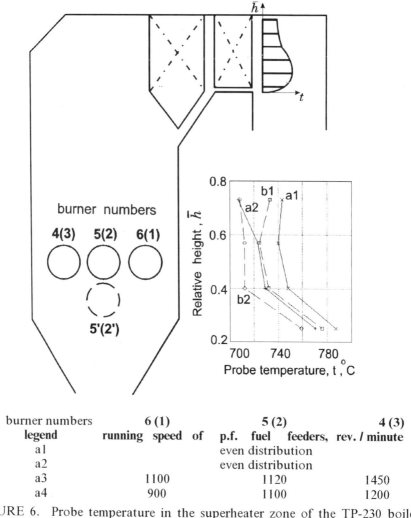

burner numbers	6 (1)	5 (2)	4 (3)
legend	running speed of	p.f. fuel feeders, rev. / minute	
a1		even distribution	
a2		even distribution	
a3	1100	1120	1450
a4	900	1100	1200

FIGURE 6. Probe temperature in the superheater zone of the TP-230 boiler depending on coal distribution between burners.

7. CONCLUSION

It has been found that the slagging properties of the fly ash can be changed depending on operating conditions and the effect is not the same for various coal types. The result can be explained by the change of the iron fraction in the ash glass and degree of its oxidation.

Operation conditions have their effect on slagging both as a result of change of slagging properties of the fly ash and of the change of flue gas characteristics. The examples presented here show that the same operating factor can have, in different cases, the opposite effect and may depend on a combination of others.

The deposits of several types are formed on the boiler surfaces and it is possible to effect on the deposits of other types and the process of slagging as a whole if it is effected the deposition rate of one of the deposit types.

REFERENCE

Alekhnovich, A. N. and Bogomolov, V. V. (1988). "Temperature Conditions of the Start of Slagging when Burning Coals with Ash of Acidic Composition". **Thermal Engineering, 35** (1), 20-24.

Alekhnovich, A. N., Bogomolov, V. V. and Artemjeva, N. V. (1995). "The Investigation of some Slagging Problems at the Rigs". Preprint of the Conference "Application of Advanced Technology to Ash-Related Problems in Boilers", Watervilly Valley, July 16-21.

Bool, L. E. and Helble J. J. (1995). "Iron Oxidation State and its Effect on Ash Particle Stickiness". Preprint of the Conference "Application of Advanced Technology to Ash-Related Problems in Boilers", Watervilly Valley, July 16-21.

NEURAL NETWORK PREDICTIONS OF SLAGGING AND FOULING IN PULVERIZED COAL-FIRED UTILITY BOILERS

David Wildman, Scott Smouse, and Richard Chi

Pittsburgh Energy Technology Center
P.O. Box 10940
Pittsburgh, PA

ABSTRACT

Feed-forward back-propagation neural networks were trained to relate the occurrence and characteristics of troublesome slagging and fouling deposits in utility boilers to coal properties, boiler design features, and boiler operating conditions. The data used in this effort were from a survey of utility boilers conducted by Battelle Columbus Laboratories in an Electric Power Research Institute project. This data base is the largest known collection of public information on slagging and fouling in utility boilers. Although the data base was large, the actual number of data sets with adequate information to use as inputs into the networks was relatively small. This situation limits the usefulness of the algorithms developed in this study. However, the data were adequate to demonstrate that neural networks can be used to reliably predict slagging and fouling in utility boilers using simple, readily available input parameters. This suggests that others can use a similar approach to develop either a general tool that is applicable to all boilers or a specialized tool that is applicable to a limited set of boilers, such as those of one utility or boiler manufacturer.

Two networks were developed in this study, one for slagging and one for fouling, to predict ash deposition in various types of boilers (wall-, opposed wall-, tangentially, and cyclone-fired) that fire bituminous and sub-bituminous coals. Both networks predicted the frequency of deposition problems, physical nature (or state) of the deposit, and the thickness of the deposit. Since deposit characteristics vary with boiler location and operating conditions, the worst documented cases of ash deposition were used to train the neural networks. Comparison of actual and predicted deposition showed very good agreement in general. The relative importance of some of the input variables on the predicted deposit characteristics were assessed in a sensitivity analysis. Also, the slagging and fouling characteristics of a blend of two coals with significant different deposition characteristics were predicted to demonstrate a practical application of developed neural networks.

INTRODUCTION

The degree of ash deposition (i.e., slagging and fouling) that will occur is governed by boiler design and operation and by the characteristics of the fuel being fired. The interdependence of these factors makes it difficult to predict quantitatively the extent of ash deposition throughout a boiler on any practical time scale. The ash deposition

Applications of Advanced Technology to Ash-Related Problems in Boilers
Edited by L. Baxter and R. DeSollar, Plenum Press, New York, 1996

propensity of a coal can be assessed in general terms by empirical relationships that have been established between various fuel properties and actual field or test performance. The most widely used of these slagging and fouling indices are based on coal ash content, ash chemistry, and ash-melting behavior (Weingartner, 1974).

Although the literature is replete with observations on the inadequacies of these indices (Bryers and Vorres, 1988; EPRI, 1987a and 1987b; Hough, 1988; Raask, 1985), they continue to be the only generally accepted means to predict the ash deposition tendencies of coals in boilers, and, in fact, are part of some coal purchasing contracts. In recent years, however, most researchers have realized that a more fundamental approach towards understanding and predicting ash deposition in utility and industrial boilers is required. Several examples of situations creating these problems follow. Stricter environmental legislation (e.g., the Clean Air Act Amendments of 1990) has increased the interest in coal beneficiation, coal switching (particularly to lower sulfur western coals), coal blending (including blends of eastern and western coals), and low NO_x combustion technologies (which can exacerbate furnace slagging). With the price of many fuels uncertain and their long-term availability in question, the increasing domestic demand for electricity has resulted in numerous life extension, repowering, and boiler conversion projects. The decline of steel manufacturing in the U.S. has made low-sulfur metallurgical coals available on the steam coal market. The increasing competitiveness of the world's coal exporting countries has introduced foreign coals into the U.S. market. These situations have highlighted the inadequacies of the widely used empirical ash deposition indices and correlations, which were developed using limited data bases of U.S. or other coals. However, advanced analytical techniques, e.g., scanning electron microscopy and chemical fractionation, have been developed that are shedding light on the fundamental mechanisms involved in ash formation and deposition. Also, numerical models of the chemical and physical processes of ash formation and deposition are being advanced by several organizations and have been used to solve real boiler problems. Although significant progress has been made in these areas, there still is a need for a simple, reliable means to assess the deposition propensity of coals because of the cost or time associated with these approaches.

As part of an ongoing effort at the Department of Energy Pittsburgh Energy Technology Center (PETC) to demonstrate new techniques for the improved operation and control of coal-fired utility boilers, advanced approaches for data analysis, such as neural networks, are being explored. The authors have previously shown neural networks to be superior to traditional data analysis techniques, such as multiple linear regression analysis, for predicting ash fouling in a pilot-scale combustor and NO_x emissions from utility boilers (Wildman et al., 1993; Smouse et al., 1993; Wildman and Smouse, 1995).

This paper summarizes the development of two neural-network-based algorithms to predict slagging and fouling in utility boilers, including a review of the data bases used in the effort and a sensitivity study of the major input parameters. The neural network training data were based on complete sets of responses to an Electric Power Research Institute (EPRI) sponsored survey of utility companies conducted by Battelle Columbus Laboratories (EPRI, 1987b). This data base, which was obtained as an electronic file from Battelle with EPRI's permission, is the largest known collection of public information on slagging and fouling in utility boilers. Although the data base was large, the actual number of data sets with adequate information to use as inputs into the networks was relatively small.

NEURAL NETWORK APPROACH FOR ESTIMATING DEPOSITION CHARACTERISTICS

Neural networks became popular in the 1980s for analyzing complex, interrelated data sets. Neural networks do not assume that a relationship is known between process input and output but rather try to determine the relationship by analyzing sets of input and output data. Neural networks, which can be viewed as nonlinear data analysis techniques, are computational systems that use the organizational principles of biological nervous systems. Computer scientists imitate these principles because biological systems easily outperform all current approaches for pattern recognition. The primary requirement for developing a neural network to predict any value is a set of data that fully describes the parameters affecting the value to be predicted. Also, the proper representation of the input (e.g., a ratio of two variables may be the best representation) hastens the training process. The principal aim of the technology is to mimic nature's approach for processing data and information. The back-propagation algorithm developed independently by at least three groups (Rummelhart et al., 1986; Parker, 1987; Werbos, 1974) is the most widely used neural network approach.

Neural networks are being used to solve real problems. For example, they have been shown to be effective in estimating the fatigue life of mechanical parts (Troudet and Werrill, 1990), performing fault detection in chemical plants (Hoskins et al., 1990), diagnosing automobile malfunctions (Marko et al.,1990), and recognizing human speech (Hampshire and Waibel, 1990). Neural networks are also being widely applied within the power industry. During the last five years, more than 200 technical papers discussing artificial intelligence applications in the power industry have been published; more than one-half of these applications involved the back-propagation algorithm (Niebur, 1993). All of these applications involved noisy data, in which the relationships between input-output pairs were complicated and poorly understood. Neural networks generally excel over traditional data analysis techniques, such as multiple linear regression techniques, in analyzing noisy data sets.

Commercial neural network software, Brainmaker Professional (Version 2.03) from California Scientific Software, was used in this study. During the network development process, the user defines the problem and collects the input-output pairs. If the user neglects key input parameters for a particular application, the predictive capabilities of the network will suffer. A neural network is not programmed with rules like an expert system, but rather it "learns" in much the same way that people do, i.e., by example and repetition.

After the input-output pairs are collected, the learning stage commences with the network associating output data with input data. Each time an input is presented, the network sends back an answer of what it thinks the output should be. When it is wrong, the network corrects itself by changing the weight matrices that are used to adjust the input-output relationships. Brainmaker Professional uses a back-propagation algorithm to adjust these weight matrices. The training process is repeated until the network derives answers that are within the user-specified tolerance for all inputs. This learning process can take considerable time, possibly up to several days or weeks, to complete depending on the complexity of the problem and the processing speed of the computer used. The final weight matrices are saved to file and can be used external to Brainmaker Professional to predict outputs for other sets of input parameters.

Table 1. Boiler data base information as a function of firing mode.

Boiler Information	Firing Mode			
	Wall	Opposed-wall	Tangential	Cyclone
Slagging Data				
Number of Boilers	6	18	16	6
Boiler Capacity Range (Gross MW$_e$)	417-500	330-1325	333-919	330-1150
Fouling Data				
Number of Boilers	1	13	18	9
Boiler Capacity Range (Gross MW$_e$)	500	330-1325	333-919	330-1150

Key to the training process in a neural network is the transfer function. Output is determined by the network's nodes, which combine signals sent to them by lower-layered nodes and transfer these adjusted signals to higher-layered nodes. Neural network nodes are designed to process signals with values between 0 and 1; therefore, all input variables must be scaled to the unit interval. Nonlinear transfer functions send signals from node to node within the network. The default transfer function used by Brainmaker Professional is the sigmoid function:

$$f(x) = \frac{1}{1 + \exp(-x)}$$

It is often called a "squashing function" because very large positive values are asymptotically mapped to 1 and very large negative values are asymptotically mapped to zero. The squashing function forces the network to focus on the bulk of the data and place less emphasis on the data at the extreme lower and upper ends of a data set. Hence, network predictions are generally less accurate at the data extremes.

A network stops training when its answers are within a user-specified tolerance. This tolerance specification is applied to the data after it has been subjected to the squashing function. The nonlinear nature of the squashing function often causes extreme data values to deviate more from actual values than nonextreme data. Typically, training is conducted at progressively smaller tolerance levels until the deviation between the predicted and actual values is acceptable to the user. The value of acceptable deviation between the network answers and the actual data depends upon the accuracy of the available input-output pairs. For example, if inherent errors in the available input-output pairs limit their repeatability to 5%, then training to a tolerance less than 0.1 is unwise. Although decreasing the tolerance will force the network answers to more closely approximate the training data, the network will not necessarily perform better on new data sets. Usually, continued training at progressively smaller tolerances improves the network's performance to a point beyond which it deteriorates. Once confidence is obtained in the predictive capabilities of the neural network, it is used to generate answers for input sets with unknown outputs.

Table 2. Neural network input parameters.

Input Parameter	Data Ranges	
	Slagging	Fouling
Firing Mode: 1-wall, 2-opposed wall, 3-tangential, 4-cyclone	1, 2, 3, 4	2, 3, 4
Steam Flow Rate/Plan Area, $kg/m^2/s$ or $(MMlb/ft^2/hr)$	747.5 - 3411.3 (0.55 - 2.51)	747.5 - 3411.3 (0.55 - 2.51)
Steam Flow Rate/Furnace Volume, $kg/m^3/s$ or $(lb/ft^3/hr)$	0.0313 - 0.0691 (7.02 - 15.50)	0.0319 - 0.0941 (7.15 - 21.10)
First Convection Section Tube Spacing, m or (in)	—	0.076 - 0.648 (3.0 - 25.5)
Soot Blow Frequency, cycles/shift	—	0 - 5
Ash Burden, wt%/J/kg or (wt%/kBtu/lb)	2.0e-6 - 9.04e-6 (4.77 - 21.00)	1.79e-6 - 8.05e-6 (4.16 - 18.70)
Sulfur, wt%, dry basis	0.41 - 4.51	0.32 - 4.51
Base-to-Acid Ratio	0.16 - 0.96	0.19 - 0.96
$SiO_2/(SiO_2+Fe_2O_3+CaO+MgO)$	0.42 - 0.81	0.42 - 0.81
SiO_2/Al_2O_3	1.52 - 3.11	1.66 - 3.11
$(SiO_2+Al_2O_3)/(Fe_2O_3+CaO)$	1.12 - 7.60	—
Fe_2O_3/CaO	0.19 - 24.49	—
Alkali Ratio	—	0.08 - 30.00
Ash Initial Deformation Temperature, reducing atmosphere, °C or (°F)	—	1060 - 1253 (1940 - 2288)
Ash Fluid Temperature, reducing atmosphere, °C or (°F)	1190 - 1538 (2118 - 2800)	—
Ash Fluid Temperature Minus Ash Initial Deformation Temperature, reducing atmosphere, °C or (°F)	—	37 - 308 (66 - 555)

DEFINITION AND SCOPE OF STUDY

Although the Battelle data base is the largest known collection of utility data on slagging and fouling, many of the data sets lacked key inputs. Therefore, limited data were available to train the networks. A number of possible input parameters, based on boiler design features and operating conditions and on fuel properties, were considered during development of the algorithms. The objective of this study was not necessarily to develop a final algorithm for general use but to demonstrate the use of neural networks for predicting ash deposition in utility boilers. Therefore, it was desirable to use a limited number of simple, readily available parameters as inputs as long as the predicted deposit characteristics agreed reasonably with actual data for a range of boiler types, boiler operating conditions, and coal properties. Some of the input parameters that were identified by Battelle as being statistically significant were selected for this study. Additional coal properties were selected based on the fact that they are commonly used to assess ash deposition propensities of coals; additional boiler design and operating parameters were also selected because they are generally believed to significantly influence ash deposition in utility boilers.

Table 1 indicates, as a function of firing mode, the number and size range of those boilers in the Battelle survey that had sufficient information for this study. Forty-six complete slagging data sets and 41 fouling data sets were used. For the data sets with complete information, about equal numbers of surveys were available from boilers firing bituminous coals and subbituminous coal-fired boilers. While the Battelle survey did ask whether low-NO_x firing systems, which can affect furnace ash deposition, were installed, this information was not included in the data base provided by Battelle. No attempt was made to collect additional information because the objective of this effort was to demonstrate the feasibility of using neural networks to predict ash deposition in utility boilers, not to develop an algorithm for general use. Table 1 indicates that fouling data was obtained for one wall-fired unit.

Table 2 summarizes the parameters that were eventually selected as inputs to the neural networks that were subsequently developed to predict slagging and fouling in utility boilers. The neural network generated for predicting fouling deposit characteristics is not recommended for wall-fired units because of a scarcity of available data. Steam flow rate was used as an indicator of boiler load because it was readily available and other indicators, such as fuel flow rate or electrical output, were not. Furnace height, which is used to calculate the steam flow-rate-to-furnace-volume ratio, was defined as the distance from the bottom of the ash hopper to the furnace roof (i.e., the overall boiler height). The more standard definition for boiler height, which is the distance from the furnace hopper knuckle to the furnace nose, was not used because detailed furnace dimensions were not available for all units. Tube spacing refers to the distance between tubes in a direction normal to the gas flow in the first convection section not exposed to direct radiation from the furnace (i.e., not radiant or platen superheaters). The various ratios based on the elemental oxide composition of the coal ash were selected because they are widely used in empirical indices to assess the ash deposition propensity of coals (Weingartner, 1974).

The neural networks were trained to predict the frequency of troublesome deposits, the physical nature or state of the deposit, and the deposit thickness. Because deposit characteristics vary with boiler location, the worst documented conditions were used for training. Figure 1 indicates the discrete nature of the output parameters. The frequency of slagging problems was classified either as daily, several times per month, occasionally, or rarely. Deposit thickness was classified as either less than 1 inch, between 1 inch and 3 inches, or greater than 3 inches. The physical nature of the slagging deposits was classified either as very fluid, molten, fused, or dry and the nature of the fouling deposits was classified either as molten, dry sintered, or dry unsintered.

The parameters to be predicted are not continuous variables; therefore, they were modeled as integer variables in this study. However, neural network algorithms treat output parameters as continuous variables. Therefore, the network predictions described in this study were rounded to the nearest integer before being reported.

NEURAL NETWORK RESULTS

Figures 1 through 6 compare the actual deposit characteristics with the predicted results for the training data. Figures 1-3 compare the actual frequency of slagging problems, deposit thickness, and deposit nature with the predicted values; Figures 4-6 make the same comparisons for fouling. The integer beside each point in these figures indicates the number of data sets that fall at that point. If the networks predictions matched the actual data, a line of unit slope passing through the origin would result. Unfortunately, this is not the case as shown in Fig. 2, 3, 5, and 6. These graphs have points that do not fall on the line of unit slope. All of the points that were poorly predicted are points where there were only a few data sets. For example, Fig. 5 depicts the predicted and actual fouling deposit thickness. Of the 37 data sets used to train the algorithm, only one data set had a deposit thickness of less than 1 inch. In all, there are 48 possible predicted outcomes (3 levels of thickness, 4 types of deposit nature, and 4 frequencies of soot blowing). Most of the training data lies in the middle of the range of results, indicating there are many possible output combinations not represented. Until the training data set is representative of the boiler population, conclusions can only be regarded as preliminary.

Because each network predicts three parameters (i.e., the frequency of troublesome deposits, the deposit's thickness, and the deposit's nature), it is difficult to compare the actual and predicted data with one graph. Figure 7 is a key to Fig. 8-11, which compare actual data with values estimated by the neural networks for four test cases using data that were not used to train the network. Figure 7 shows that the nature of the deposit is represented by the pattern of the block, the length of the block corresponds to the frequency of the troublesome deposits, and the height of the block corresponds to the thickness of the deposit. The four blocks on the left of Fig. 8 and 9 represents the actual data and the blocks on the right represent the network predictions.

Figure 8 shows that the neural network for slagging accurately predicted all three parameters for two of the four test cases, i.e., cases one and four. The frequency of deposition problems and the deposit thickness were correctly predicted for the second and third test cases but the physical nature of the deposit was incorrectly predicted. In the second case, the slagging deposit was actually molten but was predicted to be fused, which is a difference of one adjectival category. In the third case, the slagging deposit was actually fused but was predicted to be runny, which is a difference of two adjectival categories.

Figure 9 shows similar data for fouling. The network accurately predicted the nature and

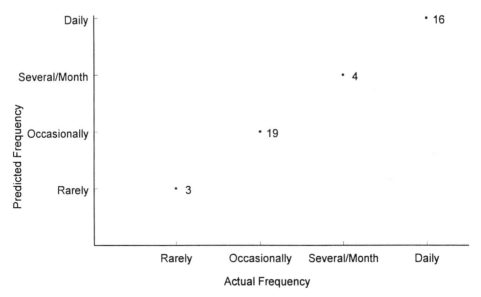

Figure 1. Comparison of actual and predicted frequency of slagging problems.

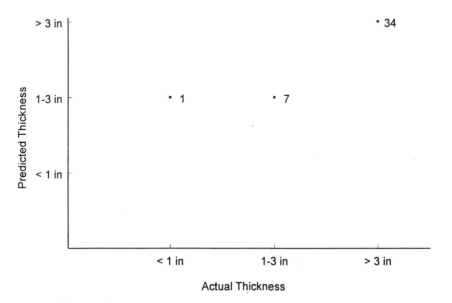

Figure 2. Comparison of actual and predicted thickness of slagging deposits.

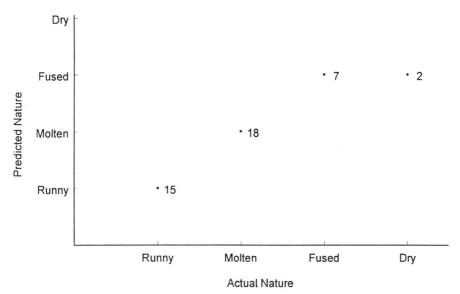

Figure 3. Comparison of actual and predicted physical nature of slagging deposits.

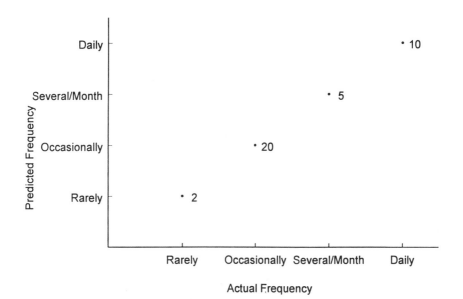

Figure 4. Comparison of actual and predicted frequency of fouling problems.

thickness of fouling deposits for the first and fourth test cases but over predicted the frequency of deposition problems. The network predictions of the frequency of fouling problems and deposit nature were correct for the second test case but slag thickness was over predicted. All three parameters were correctly predicted for the third test case.

To summarize the performance of the slagging and fouling predictions using the PETC-developed neural networks, it can be said that they were encouraging. In most cases, at least two of the three predicted parameters agreed with actual data; in three cases, all three parameters agreed. This approach of predicting the frequency of troublesome deposits, the physical nature of the deposits, and the deposit thickness represents a significant gain in practical information over that obtainable by using the traditional empirical ash deposition indices and correlations, which generally just classify ash deposition adjectivally (e.g., low, medium, high, or severe). Additional training and testing data are required before this approach can be applied to practical problems.

SENSITIVITY OF INPUT VARIABLES

The sensitivity of PETC's neural network-based slagging and fouling algorithms was investigated to assess their reliability. While these experiments do not fully evaluate the accuracy of the developed algorithms because actual data were not available for comparison in all cases, they do reveal trends that can be compared with conventional wisdom and practical experience. These sensitivity analyses are not meant to describe trends that are universally applicable to all coal-fired boilers but rather to show that, while the trends are generally reasonable, specific analyses can yield results that require further explanation. To simplify these analyses, the input parameters were varied individually except when fuel analysis effects were being assessed. Individual parameters were varied over a range of values up to 20% of their actual value. Greater variations were not investigated because of a concern for exceeding realistic operational limits.

For example, the sensitivity of the predictions to the boiler-steam flow rate (i.e., boiler load) was investigated for both slagging and fouling. Increasing the steam flow-rate-to-furnace volume or the steam flow-rate-to-plan area ratios resulted in more frequent deposition problems and thicker fouling deposits but the nature of the deposits remained constant within the limits explored in this study. Increasing these ratios is equivalent to increasing the fuel flow rate, which will increase the furnace exit-gas temperature.

Therefore, more frequent deposition and thicker deposits are reasonable expectations for 20% increases in fuel flow rate. The deposit nature is not likely to change for most coals until the furnace exit-gas temperature exceeds some critical temperature that causes the deposit to fuse or melt. Varying these ratios had little effect on the slagging predictions. Other sensitivity analyses were conducted that yielded similar predictions but are not described here. In general, the predicted ash deposition agreed very well with the actual deposition in the utility boilers as reported in Battelle's survey where data was available for comparison.

Using two coals with significantly different deposition characteristics, the effects of coal blending on slagging and fouling are demonstrated in Fig. 10 and 11. Input parameters used to generate Fig. 10 and 11 are found in Table 3. As shown in Fig. 10, coal A's slagging deposits were characterized as occurring occasionally, having a thickness of greater than 3 inches, and having a fused nature while coal B's deposits were characterized as occurring daily, having a thickness of greater than 3 inches, and having

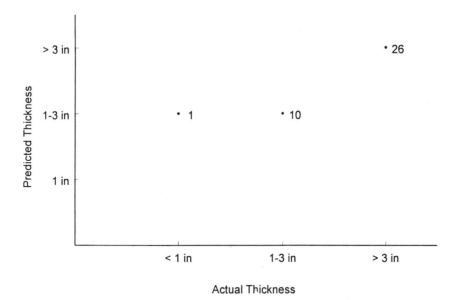

Figure 5. Comparison of actual and predicted thickness of fouling deposits.

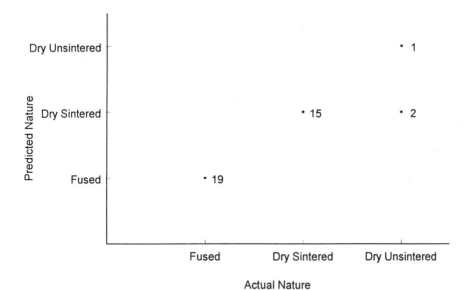

Figure 6. Comparison of actual and predicted physical nature of fouling deposits.

a molten nature. This figure shows the predicted frequency of troublesome fouling, the deposit thickness, and deposit nature of various hypothetical blends. Figure 11 depicts the same information for the fouling deposits. Both figures indicate that the deposition characteristics of the blends are the same as the characteristics of the blend's dominating coal. It is not known whether these predictions are correct because no actual data were available for comparison. However, the accuracy of such predictions could easily be assessed by a utility that is blending two or more coals with different ash deposition characteristics (e.g., a blend of eastern and western coals as is currently being fired in some midwestern U.S. boilers). The purpose of this exercise was to demonstrate a potential use for neural-network-based algorithms, such as those developed in this study, to predict ash deposition in utility boilers.

CONCLUSIONS/SUMMARY

Neural networks were developed to predict the slagging and fouling propensities of bituminous and subbituminous coals when fired in wall-, opposed wall-, tangentially, and cyclone-fired boilers. The data base used in this effort was gleaned from the responses to a survey of utility boilers conducted by Battelle Columbus Laboratories for the Electric Power Research Institute. Too little data were available for the wall-fired boilers to expect the algorithm developed for fouling to make accurate predictions. Simple, readily available parameters were selected as input parameters for the slagging and fouling networks. The selected parameters (11 for slagging and 13 for fouling) included coal properties, boiler design features, and boiler operating data. Each network predicts three practical parameters related to ash deposition that are easily understood by a boiler manufacturer or operator: the frequency of troublesome deposits, the deposit's thickness, and the deposit's nature. In most cases, at least two of the three predicted parameters agreed with actual data; in three cases, all three agreed. This approach of predicting the frequency of troublesome deposits, the physical nature of the deposits, and the deposit thickness represents a significant gain in practical information over that obtainable by using the traditional empirical ash deposition indices and correlations, which generally just classify ash deposition adjectivally (e.g., low, medium, high, or severe). Additional training and testing data are required before this approach can be applied to practical problems. A sensitivity analysis indicated that increasing the steam flow rate-to-furnace volume or the steam flow rate-to-plan area ratios resulted in more frequent deposition problems and thicker fouling deposits but the nature of the deposits remained constant within the limits explored in this study. Increasing these ratios would be equivalent to increasing the fuel flow rate, which would increase the furnace exit-gas temperature. More frequent deposition and thicker deposits are reasonable expectations. Using two coals with significantly different deposition characteristics, the possible effects of coal blending were demonstrated. Deposition characteristics of the blends were the same as the characteristics of the blend's dominating coal.

DISCLAIMER

Reference herein to any specific commercial product by trade name, trademark, manufacturer, or otherwise does not necessarily constitute or imply its endorsement, recommendation, or favoring by the United States Government or any agency thereof.

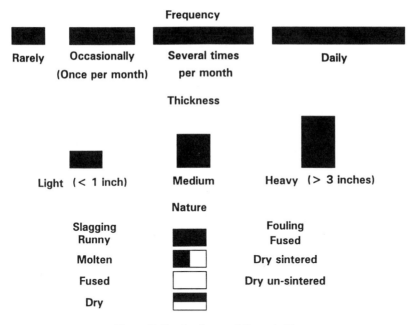

Figure 7. Key for figures 8 through 11

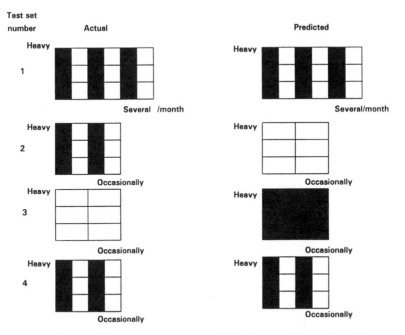

Figure 8. Comparison between actual and predicted slagging data.

507

Table 3. Input parameters used in the coal-blending studies.

Input Parameter	Data Ranges			
	Slagging		Fouling	
	Coal A	Coal B	Coal A	Coal B
Ash Burden, wt%/J/kg or (wt%/kBtu/lb)	4.86e-6 11.3	3.87e-6 9.0	4.91e-6 11.4	3.74e-6 8.7
Sulfur, wt%, dry basis	0.9	2.1	0.6	1.8
Base-to-Acid Ratio	0.20	0.42	0.22	0.27
$SiO_2/(SiO_2 + Fe_2O_3 + CaO + MgO)$	0.80	0.67	0.81	0.74
SiO_2/Al_2O_3	1.76	2.84	2.7	2.0
$(SiO_2 + Al_2O_3)/(Fe_2O_3 + CaO)$	6.8	3.3	-	-
Fe_2O_3/CaO	4.5	0.9	-	-
Alkali Ratio	-	-	3.0	11.2
Ash Initial Deformation Temperature, reducing atmosphere, °C or (°F)	-	-	1235 (2255)	1211 (2212)
Ash Fluid Temperature, reducing atmosphere, °C or (°F)	1478 (2692)	1254 (2290)	-	-
Ash Fluid Temperature Minus Ash Initial Deformation Temperature, reducing atmosphere, °C or (°F)	-	-	114 (205)	217 (390)
Firing Mode	Opposed wall		Tangential	
Steam Flow Rate/Plan Area, kg/m²/s or (MMlb/ft²/hr)	2202 (1.62)		1794 (1.32)	
Steam Flow Rate/Furnace Volume, kg/m³/s or (lb/ft³/hr)	0.0495 (11.1)		0.0424 (9.5)	
First Convection Section Tube Spacing, m or (in)	-		0.254 (10)	
Soot Blow Frequency, cycles/shift	-		Daily	

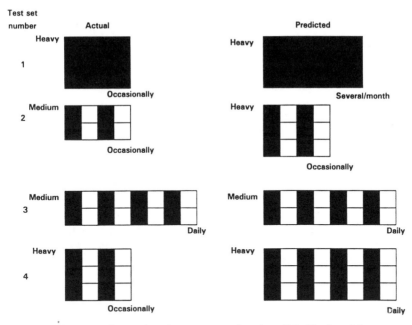

Figure 9. Comparison between actual and predicted fouling data.

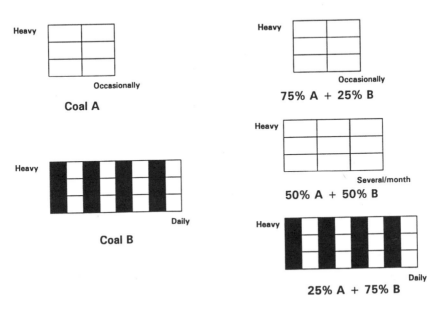

Figure 10. Effect of coal blending on slag charactistics.
(opposed wall-fired unit)

509

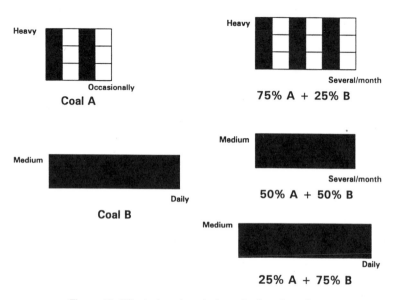

Figure 11. Effect of coal analysis on fouling deposits.
(tangentially-fired unit)

510

ACKNOWLEDGMENT

Dr. Arun Mehta of the Electric Power Research Institute is gratefully acknowledged for allowing Battelle Columbus Laboratories to release their survey data to the authors on electronic media.

REFERENCES

Bryers, R. and Vorres, K. (Ed), Proceedings of the Engineering Foundation Conference on Mineral Matter and Ash in Coal, Hemisphere Publishing, New York, NY, 1988.

EPRI, Final Report No. CS-5523, "Slagging and Fouling in Pulverized Coal-Fired Utility Boilers," 1987a.

EPRI, Research Project 2256, Proceedings of the EPRI Conference on Effects of Coal Quality on Power Plants, Atlanta, GA, 1987b.

Hampshire, J.B., and Waibel, A., "Connectionist Architectures for Multi-speaker Phoneme Recognition," Advances in Neural Information Processing Systems 2, 1990.

Hoskins, J.C., Kaliyur, K.M., and Himmelblau, D.M., "Incipient Fault Detection and Diagnosis Using Neural Networks," Proceedings of the International Joint Conference on Neural Networks, 1990.

Hough, D.C., "ASME Ash Fusion Research Project," prepared for the American Society of Mechanical Engineers Research Committee on Corrosion and Deposits from Combustion Gases, Babcock Power Fuel Consultancy Report No. FT/88/01, 1988.

Marko, K.A., Feldkamp, L.A., and Puskorius, G.V., "Automotive Diagnostics Using Trainable Classifiers: Statistical Testing and Paradigm Selection," Proceedings of the International Joint Conference on Neural Networks, 1990.

Naser, J., "Exploring Neural Network Technology," EPRI Journal, December, 1992.

Niebur, D., "Artificial Neural Networks for Power Systems - A Literature Survey," Proceedings of the EPRI Conference on Expert System Applications for the Electric Power Industry, Phoenix, AZ, 1993.

Parker, D.B., Optimal Algorithms for Adaptive Networks: Second Order Back Propagation, Second Order Direct Propagation, Second Order Hebbian Learning," Proceedings of IEEE First International Conference on Neural Networks, 1987.

Raask, E., Mineral Impurities in Coal Combustion, Hemisphere Publishing Corporation, New York, NY, 1985.

Rumelhart, D.E., Hinton, G., and Williams, R.J., "Learning Internal Representations by Error Propagation," Parallel Distributed Processing, Vol. 1, MIT Press, 1986.

Smouse, S.M., Wildman, D.J., McIlvried, T.S., and Harding, N.S., "Estimation of NO_x Emissions from Pulverized Coal-Fired Utility Boilers," EPRI/EPA Joint Symposium on Stationary Combustion NO_x Control, Bal Harbour, FL, 1993..

Troudet, T., and Werrill, W., "A Real Time Neural Net Estimator of Fatigue Life," Proceedings of the International Joint Conference on Neural Networks, 1990.

Werbos, P.J., "Beyond Regression: New Tools for Prediction and Analysis in Behavior Sciences," PhD Thesis, Harvard University, Cambridge MA, 1974.

Weingartner, E.C., "Coal Fouling and Slagging Parameters," ASME Research Committee on Corrosion and Deposits from Combustion Gases, 1974.

Wildman, D.J. and Smouse, S.M., "Estimation of NO_x Emissions from Pulverized Coal-Fired Utility Boilers," EPRI/EPA Joint Symposium on Stationary Combustion NO_x Control, Kansas City, MO, 1995.

Wildman, D.J., Ekmann, J.M., and Smouse, S.M., "Prediction of Pilot-Scale Coal Ash Deposition: Comparison of Neural Network and Multiple Linear Regression Techniques," Engineering Foundation Conference on the Impact of Ash Deposition on Coal-Fired Plants, Solihull, England, 1993.

SLAGGING AND FOULING PROPENSITY:
FULL-SCALE TESTS AT TWO POWER STATIONS IN WESTERN DENMARK

Karin Laursen

Geological Survey of Denmark and Greenland
DK-2400 Copenhagen NV, Denmark

Flemming Frandsen

Department of Chemical Engineering
Technical University of Denmark
DK-2800 Lyngby, Denmark

Ole Hede Larsen

Faelleskemikerne, Fynsvaerket
DK-5000 Odense C, Denmark

ABSTRACT

As part of a project on slagging and fouling in pulverized coal-fired boilers, four full-scale trials were performed at two power stations in Denmark. One took place at the Ensted power station and three at the Funen power station. The coals burned during the tests were from four different continents and varied in rank from subbituminous to bituminous. Chemical analyses of the ashes showed that one coal had a lignitic type of ash while the three others had bituminous ashes. Computer controlled scanning electron microscope analyses of the minerals in the coal and the ashes indicated that the internal locations of the coal minerals influenced the chemical composition of the fly ash particles. Based on empirical predictions of deposition propensities none of the coal types tested would be expected to show major slagging or fouling problems. Deposits collected on probes exposed in the furnace were highly enriched in iron compared to the minerals present in the coal and fly ash. Deposits from the convective pass were also enriched in iron, especially in the initial layer. The extent of the deposits agreed well with the empirical predictions.

Applications of Advanced Technology to Ash-Related Problems in Boilers
Edited by L. Baxter and R. DeSollar, Plenum Press, New York, 1996

513

INTRODUCTION

During combustion of coal the inorganic constituents transform into fly ash particles. These particles may deposit on the heat transfer surfaces of the boiler. Slagging and fouling causes a decrease in the heat transfer from the combustion gas to the water-steam system. Severe slagging and fouling may also lead to damages in the boiler. Traditionally, the slagging and fouling tendencies of coal has been predicted using empirical indices based on the bulk chemical composition of the ash (Winegartner, 1974).

A three year collaborative project on mineral transformations and ash deposition in pulverized coal fired systems was initiated in Denmark in 1993. The study is a Research & Development Project supported by ELSAM, the Jutland - Funen Electricity Consortium. The participants in the project are ELSAM (project coordinator and full-scale trials), the Geological Survey of Denmark and Greenland (coal, ash, and deposit analyses), and the Combustion and Harmful Emission Control (CHEC) research programme at the Technical University of Denmark (modeling of ash deposition propensities).

The project focuses on existing boilers and the new USC boilers (steam parameters 290 bar / 580 °C) under construction in western Denmark, but also the next generations pf-boilers (325 bar / 620 °C - 700 °C) are considered. Many of the existing boilers are equipped with low NO_x-burners. Because coals of worldwide origins are used in the power stations, it is essential to increase the knowledge of combustion behavior of different coal types. The objective is to increase the knowledge of the mechanisms controlling ash deposition in boilers in western Denmark. The project involves improvements of the characterization of the inorganics in coal by use of computer controlled scanning electron microscopy (CCSEM) . An empirical deposition model, based on the work of Barrett et al. (1984) and EPRI (1987 a, b) is currently being developed (Frandsen et al., 1995). This model involves not only the influence of the ash composition on deposition tendencies, but also the boiler design and operational parameters for the plant. In relation to the project, full-scale trials are carried out at different power stations in western Denmark.

FULL-SCALE TRIALS

As part of the project four full-scale plant trials were conducted: one at the Ensted power station (Enstedvaerket) and three at the Funen power station (Fynsvaerket). The Ensted power station, Unit 3, is a 630 MW pf-fired unit. The boiler is fired with 36 low-NO_x combination burners situated in tree levels in a boxer arrangement (Fig. 1). The Funen power station, Unit 7, is a 350 MW pf-fired unit with 16 tangential arranged low- NO_x burners in 4 levels. The major plant and operational informations for the two units are listed in Table 1 and boiler profiles are shown in Figure 1.

The coal types burned were from four continents and varied in rank from subbituminous to bituminous. These four coal types are very representative for the actual coal consumption in Danish power stations. During all four tests, samples of pulverized coal, fly ash (from

Table 1. Major plant and operational informations on the Ensted Power Station, Unit 3 and the Funen Power Station, Unit 7.

Parameter	EV-3	FV-7
Unit capacity	630 MW	350 MW
Firing method	opposed wall	tangential
Burners	36 Low NOx	16 Low NOx
Bottom ash condition	dry	dry
Superheater steam	186 bar / 808 K	250 bar / 818 K
Reheater steam	40 bar / 808 K	58 bar / 813 K
Steam flow	513 kg/s	328 kg/s
Furnace height	48.6 m	54.62 m
Furnace width	22.34 m	15.11 m
Furnace depth	12.24 m	15.11 m
Height of burner belt	5.88 m	12.74 m

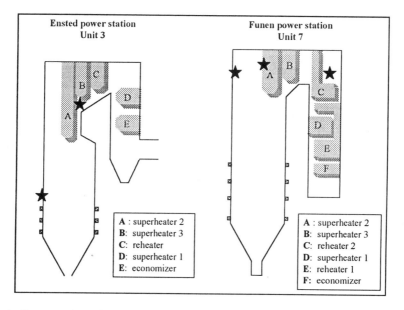

Figure 1. Cross section of the Ensted power station, Unit 3, and the Funen power station, Unit 7. The placement of the deposition probes in the furnace and in the convective pass are indicated by stars. The star in the convective pass of Ensted Unit 3, indicates the locations of two probes; one from each side of the boiler.

the electric precipitator), and bottom ashes were collected. Deposition probes were inserted in the furnace and the convective pass for collection of deposits. Additionally, during the test at the Funen power station, char and fly ash particles were collected from the combustion zone.

Sampling methods

Two types of deposition probes were used for collection of deposits from the boilers. For collection of slags in the furnace, a water-air cooled probe was used. For collection of fouling deposits, an air-cooled probe was inserted. The probes were designed and built for testing of corrosion by ELSAMPROJEKT (the engineering division of ELSAM), and modified for deposition tests. The probes consist of concentric tubes that allow the passage of cooling water and pressurized air. The outer part of the probes consist of an exchangeable metal tube (test element) for collection of deposits. Metal temperatures are measured with two thermocouples on the up- and downstream side of the tube adjacent to the test element. Temperatures are registered continuously during exposure.

During the full-scale test at the Funen power station char and fly ash particles were collected from the burner zone of one burner. The particles were removed from the combustion zone with a suction pyrometer and collected on a filter in a sampling box. The morphology of the chars were determined using an eight group classification system. The composition of the minerals in the chars was determined using computer controlled scanning electron microscopy. The relationship between the char morphology and mineral composition is under investigation.

Course of events during the full-scale trials

The full-scale trial at the Ensted power station was carried out during a three-day test burn of a subbituminous coal (Coal A) in March 1994. A decrease of the excess air supply after the first test day was the only operational variation. The three full-scale trials at the Funen power station were conducted during January and February 1995. These trials involved direct collaboration with an ELSAM Research and Development Project on "Coal Quality and NO_x-emission". Therefore, the coals were selected because of their general combustion behavior rather than for their ash deposition propensities. The three coals (Coal B, Coal C and Coal D) burned during the three trials were all bituminous, and had origins from three continents. Trials for each of the three coal types lasted three days. The operation of the plant and the excess air was constant on a given day for each coal type, but varied from day 1 to day 3 of the test as follows: day 1, (1.17 %, normal load); day 2, (1.15 %, normal load); and day 3, (1.17 %, and higher coal load plus lower excess air on the lowest burner level). For collection of deposits probes were inserted at different locations in the two boilers. Probe locations are shown in Figure 1. At the Ensted power station the probe surface temperatures were evaluated in the temperature ranges 450 - 500 °C in the furnace and, 550 - 700 °C in the convective pass. Exposure times were 2-20 hours. At the Funen power station probe surface temperatures were maintained at 570 °C for all tests. Exposure times were 8 hours for each coal / operational condition for two of the probes, whereas the third probe near superheater 2 (Fig. 1) was exposed for 36 hours for each coal. Samples of pulverized coal, fly ash and bottom ashes were collected during the trials; additionally, at the Funen power station, char and fly ash particles were sampled from one burner.

516

Table 2. Proximate analyses for the coal burned at Ensted (Coal A) and the three coals burned at Funen (Coal B - Coal D). All the parameters are given on an "as received" basis.

Parameter	Coal A	Coal B	Coal C	Coal D
Moisture (%)	15.5	12.6	8.4	11.7
Ash (%)	6.5	11.4	14.3	11.8
Volatile matter (%)	35.0	30.3	22.2	27.5
Total sulfur (%)	0.81	0.77	0.47	0.80
Heating value (MJ/kg)	25.16	25.50	25.80	25.88

Table 3. Bulk ash analyses for the coal burned at Ensted (Coal A) and the three coal burned at Funen (Coal B, Coal C, Coal D). All data are on a percent by weight basis.

Component	Coal A	Coal B	Coal C	Coal D
SiO_2	48.5	57.4	46.7	47.4
Al_2O_3	25.4	21.4	30.8	26.4
Fe_2O_3	11.1	7.53	3.35	9.04
CaO	2.13	3.01	6.55	4.67
MgO	2.24	1.61	1.62	2.77
Na_2O	0.41	0.58	0.15	1.01
K_2O	2.13	1.98	0.69	2.16
TiO_2	1.12	0.97	1.74	1.16
P_2O_5	0.52	0.30	1.33	0.66
SO_3	2.40	2.60	4.30	3.85

Coal Analyses

The proximate analyses for the coal burned at the Ensted power station (Coal A) and the three coals burned at the Funen power station (Coal B, Coal C and Coal D) are given in Table 2. All four coal types have a relatively low ash content, especially Coal A. The chemical composition of the coal ashes are shown in Table 3. The coals have similar ash compositions. All are high in SiO_2 and Al_2O_3 and variable but relatively low in Fe_2O_3 and CaO.

THEORETICAL PREDICTIONS

As part of the project is was decided to develop tools for prediction of ash deposition propensities. A PC-programme based on the concept of pairing a bulk ash chemistry index with a plant parameter (Barrett et al., 1984) was developed. The ash chemistry index accounts for the potential chemistry of the coal ash (e.g., a high S or Na content), while the plant parameter accounts for the configuration (e.g., furnace, boiler) and the operation (e.g., heat input, steam generation rate) of the system. Comprehensive reviews of ash indices and plant parameters are given by Winegartner (1974), Barrett et al. (1984), Badin (1984). EPRI (1987a, b), Singer (1991), Skorupska (1993) and Skorupska & Couch (1994). In the program used in this work (Frandsen et al., 1995), 22 conventional bulk ash chemistry indices and 9 plant parameters are calculated and combined according to the guidelines given in EPRI (1987a, b) for each combination of coal, system configuration and operational data.

An example of an ash chemistry index is the Attig & Duzy slagging factor R_s, defined as:

$$R_s = \frac{\%Fe_2O_3 + \%CaO + \%MgO + \%Na_2O + \%K_2O}{\%SiO_2 + \%Al_2O_3 + \%TiO_2} \cdot \%S$$

where %S is the sulfur content on coal dry basis. The steam flow per unit plant area of the boiler, STMFLPA, is an example af a plant parameter (Barrett et al., 1984). STMFLPA is defined as:

$$STMFLPA = \frac{\text{steam generation rate (kg/s)}}{\text{furnace plan area (m}^2)}$$

In Figure 2, the Attig & Duzy slagging factor, R_s, is plotted against the steam flow per unit plan area (STMFLPA).

Based on empirical data (EPRI, 1987a, b) and a statistical regression analysis it is possible to define the rare and the frequent slagging zone in Figure 2. The operational points for the two power stations and the coal considered are clearly placed in the rare slagging zone of the Figure 2.

RESULTS OF FULL-SCALE TRIALS

Fly ash and bottom ashes collected during thetrials were analyzed chemically by inductively coupled plasma atomic emission spectrometry (ICP-AES) and the mineralogy

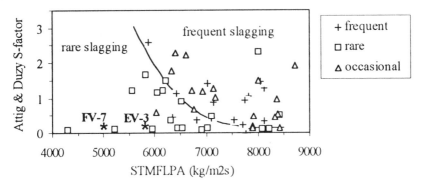

Figure 2. The Attig & Duzy slagging factor, R_s, plotted against the steam flow per unit plan area. FV-7 and EV-3 represents the operational points af the Funen poser station and the Ensted power station. Empirical data source: EPRI, 1987a, b.

was determined by x-ray diffraction analysis (XRD). Minerals in the pulverized coal and fly ash were analyzed by computer controlled scanning electron microscopy (CCSEM). The analyses were performed using a Phillips XL 40 scanning electron microscope and a Tracor Voyager analytical system. The technique used is very similar to the technique previously reported by Wigley and Williamson (1994). The main discrepancy between this technique and the other techniques (e.g., Jones et al., 1992) is that the chemical composition of the minerals located in the sample is represented by an average composition of the grain and not the composition of the center of the grains. Thus, if the mineral particles are adjacent to one another they will be represented by an area-weighted composition of the mineral end-members. The test elements from the deposition probes were sliced and cut into four pieces representing the upstream, downstream and two middle pieces of the tube. Each piece was embedded in epoxy and polished. The individual tubes were analyzed by scanning electron microscopy and energy dispersive x-ray analyses (SEM-EDX).

Table 4. Result of bulk chemical analyses of fly ash (electric precipetator) and bottom slag. All data are given on a percent by weight basis.

Component	fly ash A	slag A	fly ash B	slag B	fly ash C	slag C	fly ash D	slag D
SiO_2	50.68	52.25	58.50	64.80	49.50	42.00	50.00	51.70
Al_2O_3	25.97	26.33	20.70	17.10	33.00	27.80	27.00	23.30
Fe_2O_3	10.18	11.40	7.35	8.08	3.29	3.93	8.70	11.20
CaO	2.42	2.84	2.98	3.66	6.78	7.42	4.18	4.32
MgO	2.05	2.05	1.61	1.66	1.71	1.71	2.71	2.75
K_2O	1.92	1.92	1.97	1.53	0.69	0.54	2.29	2.00
Na_2O	0.36	0.36	0.61	0.58	0.14	0.07	1.14	0.73
TiO_2	1.30	1.30	0.99	0.87	1.76	1.51	1.14	0.99
P_2O_5	0.59	0.59	0.26	0.25	1.31	1.23	0.60	0.45
SO_3	0.32	0.32	0.54	0.15	0.40	0.21	0.81	0.16

Analyses of coal and ashes

The results of bulk chemical analyses of the fly ash and the bottom ashes are given in Table 4.

CCSEM analyses of Coal A burned at the Ensted power station shows that this coal is very rich in clays, with kaolinite and K-Al silicate comprising nearly 60 % of the minerals present in the coal. In a SiO_2-Al_2O_3-Fe_2O_3 ternary diagram these are seen as a peak along the SiO_2-Al_2O_3 axis (Fig. 2). Other significant minerals are quartz and pyrite with lesser amounts of iron-oxide, which are represented by the peaks in the SiO_2 and the Fe_2O_3 corners. There are limited particles located on the tie-line between pyrite/iron oxide and the clay peak (Fig. 3.a). Likewise, there are very few minerals with compositions intermediate between clay and quartz. This diagram indicates that the associations between pyrite/iron oxide and clay and between pyrite/iron-oxide and quartz are weak. Analyses of the precipitator ash (Ash A) from the burning of Coal A show that the majority of the fly ash particles are aluminosilicates (75 %) with compositions very similar to the clay minerals present in the raw coal. The quartz content of the fly ash is nearly identical with the content in the coal. Compared to the coal the fly ash contains few iron rich particles (< 2%). A portion of the aluminosilicates are enriched in Fe_2O_3, and are located on the tie-line between the aluminosilicates and the Fe_2O_3 corner (Fig. 3.b). The fly ash is not significantly depleted in iron, but the iron is now present in the aluminosilicates instead.

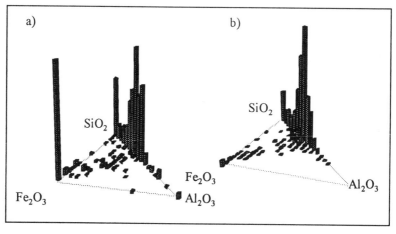

Figure 3. Fe_2O_3-SiO_2-Al_2O_3 ternary diagrams. a) pulverized coal, Coal A. b) electric precipitator ash, Ash A. The heights of the columns show the relative weight percentages present in the sample with the composition indicated by the location of the column.

Coal B consist of an equal amount of K-Al-silicate and kaolinite, comprising 50 % of the minerals. Quartz is another significant mineral in the coal (15 %), followed in importance by pyrite (7%). Ternary diagrams (not shown) reveal that both clays and quartz are likely to be associated with the Fe_2O_3-rich particles. The fly ash (Ash B) has a very low content of Fe_2O_3 rich particles (< 0.5 %). Particles with intermediate composition between the Fe_2O_3 corner and the aluminosilicates, and between the Fe_2O_3 corner and the SiO_2 are very abundant.

Coal C has a high content of kaolinite, constituting 62 % of the minerals. The analysis indicates that the relatively high content of calcium in the coal (Table 3) is present as calcite and Ca-Al-silicates, which probably are mixtures of clays and calcite. Ternary diagrams (not shown) indicate that the association between the quartz and the clay minerals is limited. Such a limited association also applies for the clay and the Fe_2O_3 rich particles. The fly ash (Ash C) is rich in minerals with compositions similar to the clays in the coal. The fly ash is markedly enriched in Ca-Al-silicates comprising nearly 30 % of the minerals.

Coal D has a high content of clay minerals, with kaolinite and K-Al silicates constituting 70 % of the minerals. Other significant minerals are pyrite, quartz and iron oxide. Fe-Al silicates are more abundant in this coal than in any of the other three coals. In the SiO_2-Al_2O_3-Fe_2O_3 ternary diagram these particles are located on the tie-line between the clay peak and the Fe_2O_3 corner (Fig. 4.a). The particles with intermediate compositions indicate that clay minerals and iron-rich particles (pyrite and iron oxide) are closely associated in the coal. The fly ash, Ash D, sampled at the Funen power station has a large amount of particles with intermediate compositions between the aluminosilicates and the iron-rich particles (Fig. 4.b).

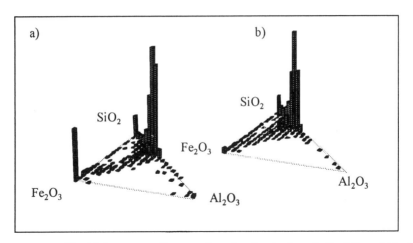

Figure 4. Fe_2O_3-SiO_2-Al_2O_3 ternary diagrams for: a) pulverized coal, Coal D and b) electric precipitator ash, Ash D. The heights of the columns show the relative weight percentages present in the sample with the composition indicated by the location of the column.

Ash deposits from the furnace

Deposits collected on probes in the radiant zone of the boiler at the Ensted and the Funen power stations are all very similar in morphology and composition. Based on texture and morphology the deposits can be separated into three main zones: oxide zone, inner deposit zone, and outer deposit zone (Fig. 5). The oxide zone consist of an inner zone with radiantly orientated crystals and an outer more homogeneous zone (Fig. 5). The inner deposit layer consist of fly ash particles embedded in a matrix with low porosity and a relatively homogeneously texture (Fig. 5). Deposits collected at the Ensted power station did not show any visible differences in the upstream and the downstream side of the probe. However, the deposits from all three trials at the Funen power station showed markedly visible differences between the two sides. The upstream side is characterized by relatively small fly ash particles (0.25 - 2 μm), whereas the downstream side has fewer and generally larger fly ash particles (0.25 - 5 μm) (Fig. 5). The matrix is not always totally homogeneous, locally remnants of original fly ash particles can be distinguished. The fly ash particles are characteristically round, but angular (typically quartz) particles are also seen. On the upstream side of the tubes the contact between the fly ash particles and the matrix is less distinct than on the downstream side. These diffuse contact zones could be the reaction rims indicating dissolution of the fly ash particles. Locally on the upstream side the deposits appear more texturally homogeneous. The composition of these layers is very similar to the bulk chemistry of the fly ash-rich layers, indicating that the former areas result from a local in-situ sintering of the fly ash and the matrix. The outer deposit consist of a thin layer of fly ash particles on the surface of the inner deposit. Usually only one layer of particles are seen, however, isolated regimes of multi-layer particles are seen.

SEM-EDX analyses of the deposits from the furnace reveal that both the oxide and the matrix of the inner deposit consists mainly of iron. The fly ash particles are mainly aluminium silicates and fewer are quartz. The Al-Si rich particles are characteristically spherical. In contrast to these spheres the Fe-rich particles are more irregular in shape, and their shape seems to be controlled by the particles around them. Large Fe-rich particles are rather common as islands in the deposits from the trial at the Ensted power station. The contact relations between the inner deposit and these large iron-rich particles indicate that the fly ash particles had a very low viscosity during impact.

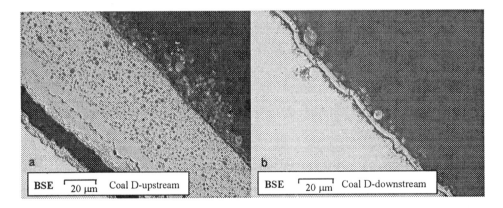

Figure 5. Upstream (a) and downstream (b) sides of a slagging deposit collected on a probe inserted in the top of the furnace at Funen power station during the combustion of Coal D.

The deposits collected in the furnace at the Funen power station were significantly thicker on the upstream side (10 - 80 μm) of the probes than on the downstream side (2.7 - 12 μm). Deposits from Ensted shows no such differences. The differences in the thickness between the two sides were greater for Coal C and Coal D, than for the Coal B. The deposits on the probes from burning of Coal B were much thinner than for the two other coal types. Coal B did not give any difference in slagging behavior during the three tests days. The deposits from the burning of Coal C, were thinnest on day 1(excess air 1.17 %, normal load), and were similar on day 2 (excess air 1.15 %, normal load) and day 3 (excess air 1.17 %, higher load at the lowest burner level). Deposits from burning of Coal D were more extensive on day 1, compared to the following two days.

Ash deposits from the convective pass

The probes inserted in the convective pass were all covered with a layer of un-consolidated fly ash particles. This layer was much thicker on the upstream side than on the downstream side of the tube. However, it is possible that additionally loose deposits were blown off as the probes were removed from the boiler. On the upstream side the layer was shaped like a flattened cone whereas on the sides and the downstream side the ash was equally distributed. The loose deposit was gently removed and analyzed separately. The adhering deposits from the convective pass of both power stations can, based on texture and morphology, be separated into the same three main zones as the deposits from the furnace (oxide zone, inner deposit zone and outer deposit zone). The morphology and the texture of the zones are very similar to the deposits from the furnace. All deposits showed distinct differences between the upstream and downstream side of the tubes. The thickness of the adhering deposit on the upstream side varied between 5 - 20 μm and on the downstream side between 2 - 10 μm. Similar to the deposition pattern in the radiant zone, the upstream side consisted of many but very small fly ash particles (0.25 - 2.5 μm) and the downstream side contained few but generally larger fly ash particles (0.25 - 10 μm). The matrix on the downstream side is very homogeneous; but on the upstream side remnants of individual particles can be identified. The outer deposit consists of a single layer of fly ash particles which are the remnants of the loose deposits that were removed from the probe.

Of the three coal types burned at the Funen power station the deposits from the burning of Coal B, were much thinner than for the two other coal types. The empirical predictions showed that Coal C was the most prone to cause fouling problems. Of the three test days, day 1, created the thinnest deposits for all three coal types, and day 3, gave the thickest deposits.

Probes from the Ensted power station exposed with the same metal temperature but for different times showed deposits thickness is proportional to exposure time. Exposure time does not appear to affect the chemical composition of the deposits. Probes exposed for the same time but with various metal temperatures showed that a rise of temperature from 500 °C to 600 °C increased in the thickness of the deposit approximately 2-fold.

The matrix of all the deposits are iron-rich. As with the probes from the furnace, the inner zone of the oxide is enriched in chromium relatively to the tube and the outer part is depleted. The fly ash particles in the deposits are mainly aluminosilicates with very few iron-rich particles.

Mechanisms controlling the ash deposition

Boiler deposits from the radiant zone enriched in iron relative to the iron content in the coal and the fly ash have been identified (Wigley & Williamson, 1994). It is generally accepted that iron, especially low melting FeS-FeO intermediates from pyrite oxidation, is an important element affecting the formation of deposits in the radiant zone (Couch, 1994). The presence of sulfur in the iron-rich matrix indicates that intermediates from pyrite oxidation were present. However, the sulfur could also have its origin in the flue gas that could react with particles or the surface of the tubes creating iron sulfates. Of the four principal slagging types suggested by Hatt (1990), the deposits from both the Ensted and the Funen power stations would be classified as sintered slags, which typically are composed of partially fused particles.

Traditionally fouling deposits are classified based on bonding mechanisms: 1) silicate liquid phases (high temperature fouling) and 2) sulfates (low-temperature fouling)(Hurley et al., 1994; Benson et al., 1993). High temperature fouling normally occurs above 1050 °C (for high-calcium subbituminous coal), whereas low temperature fouling occurs below 1050 °C (Hurley et al., 1994). The deposition probes at Ensted were situated in the convective pass where the flue gas temperature was estimated to be 985 °C. Fouling in this area would therefore fall in the class of low temperature fouling. The probes at the Funen power station were located at positions where the temperatures are estimated to 1150 °C in the region of superheater 2, which would be in a regime of high temperature fouling and 850 °C in the region of reheater 2, a low temperature fouling regime. However, morphological and chemical analyses of the deposits from both boilers did not show any sign that any of these expected mechanisms were active. Iron was clearly the most dominant bonding medium in all the fouling deposits. One reason for iron being so dominant in the deposits could be that all the deposits are very thin and thus deposition was just initiated. However, the long time exposed probes from both Ensted and Funen gave no indications of alternative bonding mechanisms. Currently, it is not possible to determine the bonding mechanism for the loose deposits on the probes. Further investigations of these ash particles may give a better understanding of the mechanisms controlling bonding of loose ash particles.

Mechanisms involved in ash deposition may be divided into four types: 1) inertial deposition, 2) condensation, 3) thermophoresis, 4) chemical reaction (Baxter & DeSollar, 1993). Particles transported to a heat transfer surface will either adhere to or bounce off the surface. Whether or not a particle will stick to the surface is controlled by the particle viscosity, surface tension, velocity, and angle of impact, but also by the chemical and physical state of the surface of impacttion (Richards et al, 1993). There are no indications that the impacted Si-Al rich particles had low viscosity's, because most of these are spherical. On the other hand, a thin low viscosity coating on the fly ash particles could cause some "stickiness". However, it seems more likely that the Si-Al rich fly ash particles would adhere because the surface was "sticky". Characteristic of the deposits is that the upstream side consist of many small fly ash particles, and the downstream side of fewer but relatively larger fly ash particles. A similar size distribution has been reported by Hurley et al., 1994. The reasons for the particle size distribution found in this work is not clear, but several mechanisms could be involved. The deposition of the small particles (< 4 μm)on the upstream side is probably mainly controlled by thermophoresis. Whereas the larger particles (4 - 10 μm) on the downstream side probably are controlled by eddy deposition. On the upstream side impacting larger particles will bounce off due to their high velocity. These particles will on the downstream side be caught by eddies behind the tubes. The particles will have lower velocity and are more likely to stick on impact. Larger fly ash particles (> 10 μm), which comprise the majority of the fly ash, will not be caught by the eddies, but follow the main gas flow.

The iron-rich matrix of both the fouling and the slagging deposits can have two origins: 1) iron-rich fly ash particles; and 2) diffusion of iron from the tube due to oxidation (Cutler et al, 1980). Remnants of iron-rich particles can locally be identified in the deposits, also in the convective pass, indicating that impacting iron-rich particles contributes to the matrix. These remnants of particles indicates that reduced iron-rich particles were present even in the back of the convective pass. However, so far it is not possible to determine if iron-rich particles were the only source to the iron enrichment in the deposits.

FUTURE WORK

The analyses of the deposits from the Ensted and Funen power stations presented in this paper are the preliminary results. Further analyses of these deposits will probably contribute to a better understanding of the mechanisms controlling the deposition. The next step in improving the empirically based ash deposition propensity model will be to include CCSEM analyses of coal minerals in an ACI (ash chemistry index) instead of bulk-chemical ash analyses. The final step in the modeling is to develop a mechanistic model for prediction of fly ash composition based on CCSEM analyses of the coal. Within the time duration of this project at least one more full-scale trial will be performed at a power station in western Denmark. Combustion of biomass (straw and wood) is contributing to an increasing fraction of the energy produced in Danish power stations. The biomass used for energy production will be burned alone or together with coal. Problems with slagging and fouling during co-combustion of coal and straw will also be considered in this project.

CONCLUSIONS

* As part of a Danish project on slagging and fouling four full-scale trials have been completed at two power stations. Four coals from four continents were burned during the trials.

* CCSEM analyses of the coals showed that they were all rich in clays, quartz and various amounts of pyrite. The fly ash were all rich in aluminosilicates and quartz. The internal association between the minerals in the coals were important for the chemical compositions of the fly ash particles.

* A program developed for empirical predictions of deposition propensities, showed that none of the coal burned during the full-scale were expected to cause problems with slagging or fouling.

* As predicted by the empirical prediction program the deposits collected on probes in the furnace and the convective pass were all of limited extent. All deposits were significantly enriched in iron relative to the coal and the fly ash.

* The samples collected during the two full-scale trials will be further analyzed. Additionally, more full-scale trials will be conducted in power stations in Denmark. The deposition propensity program will be extended to include CCSEM analyses of coal.

ACKNOWLEDGMENTS

The work presented in this paper has been financially supported by ELSAM - the Jutland-Funen Electricity Consortium. The authors wish to thank the Ensted power station and the Funen power station for support during the full-scale trials. The authors also wish to thank University of North Dakota, Energy and Environmental Research Center, for much support during the development of CCSEM in Denmark and empirical prediction methods. Thanks is also extended to Fraser Wigley, Imperial College, London, for helping with establishment of the CCSEM technique in Denmark.

REFERENCES

Badin, J.E. (1984). "Coal combustion chemistry-correlation aspects". *Coal Science and Technology*. Volume 6. Elsevier, Amsterdam

Barrett, R.E., Murrin, J.M., Dimmer, J.P. & Mehta, A.K. (1984). *"Slagging and fouling as related to coal and boiler parameters"*. Proceeding of the Engineering Foundation Conference. *"Slagging and fouling due to impurities in combustion gases"*, Copper Mountain Resort, Colorado, 1984.

Baxter, L.L. & DeSollar, R.W. (1993) *"A mechanistic description of ash deposition during pulverized coal combustion: prediction compared with observations"*. *Fuel*, 72 (10), 1411-1418.

Benson, S.A., Jones, M.L. & Harb, J.N. (1993) "Ash Formation and deposition". In D.L. Smoot (Ed.), *Fundamentals of coal combustion for clean and efficient use*. New York, Elsevier, 299-371.

Cutler, A.J.B., Flatley, T. & Hay, K.A. (1980) *"Fire-side coorosion in power-station boilers"*. **Combustion**. December, 1980.

Couch, G. (1994). *Understanding slagging and fouling during pf combustion*. IEA Coal Research Report No. IEACR/72, London.

Electric Power Research Institute (EPRI) (1987a). *"Slagging and fouling-coal-fired utility boilers. Volume 1: A survey and analysis of utility data"*. EPRI Research Project No. 1891-1, Final Report.

Electric Power Research Institute (EPRI) (1987b). *"Slagging and fouling-coal-fired utility boilers. Volume 2: A survey of boiler design practices for avoiding slagging and fouling."* EPRI Research Project No. 1891-1, Final Report.

Frandsen, F., Dam-Johansen, K., Laursen K. & Larsen, O.H.(1995). *"Prediction of ash deposition propensities at two coal-fired power stations in Denmark"*. Paper presented at the 12th Annual International Pittburg Coal Conference, Pittsburg.

Hatt, R.M. (1990). *"Fireside deposits in coal-fired utility boilers"*. **Prog. Energy Combust Sci.**, 16, 235-241.

Hurley, J.P., Benson, S.A. & Mehta, A.K. (1994). "Ash deposition at low temperatures in boilers burning high calcium coals". In J. Williamson and F. Wigley (Eds.) *The Impact of ash deposition on coal fired plants.* Proceeding of the Engineering Foundation Conference, Solihull, England, June 20- 25, 1993, 19-30. Taylor & Francis.

Jones, M.L., Kalmanovitch, D.P., Steadman, E.N., Zygarlicke, C.J. & Benson, S.A. (1992). "Application of SEM techniques to the characterization of coal and ash products." In H.L.C. Meuzelaar (Ed.) *Advances in coal spectroscopy*, Plenum,1-25.

Richards, H.R., Slater, P.N. & Harb, J.N. (1993) "Simulation of ash deposit growth in pulverized coal-fired pilot scale reactor." *Energy and Fuels,* 7, 774-781.

Singer, J. (1991). *"Combustion - fossil power systems."* Combustion Engineering Inc. Windsor, CT.

Skorupska, N.M. (1993). *"Coal specification - impact on power station performance"*. IEA Coal Research Report No. IEACR/52, London.

Skorupska, N.M. & Couch, G. (1994).*"Coal characterization for predicting ash deposition: An international perspective."* In J. Williamson and F. Wigley (Eds.) *The Impact of ash deposition on coal fired plants.* Proceeding of the Engineering Foundation Conference, Solihull, England, June 20- 25, 1993, 137-150. Taylor & Francis.

Wigley, F. & Williamson, J. (1994) "Fly ashes, coal and deposits from boiler trials." In J. Williamson and F. Wigley (Eds.) *The Impact of ash deposition on coal fired plants.* Proceeding of the Engineering Foundation Conference, Solihull, England, June 20-25, 1993, 385-407. Taylor & Francis.

Winegartner, E.C. (1974). *"Coal fouling and slagging parameters"*. **The Am. Soc. Mech. Eng.**

INVESTIGATION OF SOME SLAGGING PROBLEMS AT THE RIGS

A.N. Alekhnovich, V.V. Bogomolov, N.V. Artemjeva

Ural Heat Engineering Institute (UralVTI)

Cheljabinsk, Russia

ABSTRACT

It is not possible to simulate deposition processes exactly on small rigs. However, slagging simulation and imitation have sense. It helps to pick out separate aspects, to extend the range of variables and so to achieve the exaggerated results that help to understand the process. It is also possible to achieve reliable results to solve certain practical problems. The different rigs are used by UralVTI for the different problems.

The investigations of the slagging of the upper part of a furnace have been carried out on an isothermal air rig with captive transparent walls. The slagging of the lower part of the furnace has been studied on a gas fired rig. The fly ash here is synthetic (a mixture of real fly ash and pane glass powder) and it is injected into a furnace through burners. The effect of a gas velocity, the viscosity of the sticking agent and the percentage of the agent has been studied on the special rig. The low heated air and the mixture of colophony, a common salt, a fine silicate powder were used on the rig. Such a composition of the mixture helps to disclose the effect of various ash particles and it offers to analyse the deposits easy.

Slagging and fouling properties of coals are explored on the coal fired rig with 50 kg/h input.

1. INTRODUCTION

In investigations of slagging problems it is usually concerned with the composition of the coal mineral matter, mineral matter transformation during the burning of coal, deposits properties and thus we deal with a chemical and physical-chemical aspects of the problem. The rigs and facilities of a different scale and design are used for the such investigations. In these circumstances the real coal is burned in the rigs and it is tried, as far as it possible, to have the same operating conditions as in utility boilers (temperature, burning time, atmosphere). The problems of dynamic interaction between ash particles and the slag surface, slagging as a mechanical process, the studies of slagging in geometrically similar models attracted less attention, at least in Russia. The mathematical models of a slagging known to us are not an exception and the trapping of ash particles

Applications of Advanced Technology to Ash-Related Problems in Boilers
Edited by L. Baxter and R. DeSollar, Plenum Press, New York, 1996

529

in the deposits are considered on the background of their material properties (the viscosity, the part of liquid particles).

It is impossible to simulate exactly the deposition process and various types of the deposits owing to the incompatibility of dimentionless numbers when we use synthetic material and small rigs. However the simulation and imitation of slagging have a sense. It allows to pick out separate aspects, to extend the range of variables and so to get exaggerated results which help to understand the slagging process. It is also possible to get reliable results for certain practical problems.

UralVTI used various rigs, including isothermal and coal fired ones , depending on boiler zones and goals. Brief information about used rigs is cited in Table 1.

Table 1. Brief Information about Used Rigs.

The rig	A material for slagging imitation or simulation	The goal of investigations	The salient features of experiments
The isothermal model	A size fraction of fly ash	An optimisation of a design of a furnace upper part and gas ducts	Using of sticky transparent surface, a measurement of a luminous flux
The rig operating on a low heated air	A mixture of - colophany, -common salt, -fine silicate powder	A study of a slagging mechanism. Slagging/ fouling rate and deposit composition in dependence on gas velocity, a sticky agent viscosity, the agent percentage.	- Design of an experiment. - Deposit analysis by leaching in a water and in an alcohol
The gas fired rig equipped with a system for powder supplying	A mixture of - fly ash, - glass powder	- An optimisation of a design of the furnace lower part. - A study of slagging mechanism.	- Simulation of a sticking process. - Measurement of a slagging pattern of furnace walls.
The coal fired rig	A real pulverised coal	- Determination of slagging/ fouling properties of coals and blends (dependence of slagging rate and deposit strength on a temperature, the temperature of the start of slagging, fouling index). - A study of slagging properties depending on operation conditions and burning arrangements.	- Measurement of a tensile strength of deposits "in situ". - Measurement of a heat-absorption efficiency during fouling.

2. THE GAS FIRED RIG

2.1. Simulation of Slagging

When the approximate simulation of deposition in the furnace and the flue duct is used it is necessary to maintain the geometrically simulation, the simulation of a two-phase flow and temperature patterns, the simulation of a particles trapping in the deposits. The processes in the burning zone of a furnace are substantially non-isothermal and reliable results can be obtained in fuel fired models. The effect of the burning schemes on slagging properties of fly ash has been studied by us on a coal fired model. The model is geometrically similar the PK-39 boiler of an 300MW unit and belong to the institute KazNIIE (Kazakhstan). Additionally, there was an idea to use a gas fired rig to study deposition in the burning zone and the experimental procedure has been worked out (Alekhnovich et al., 1988). The rig was equipped with a system of powdered material supply through burners and the waterwalls were covered with local demountable shields of a refractory set.

Considerations on the approximate simulation of flow and temperature patterns have been published by a number of researches and we do not intend to discuss them in this article. We tried to formulate the ideas concerning the simulation of slag formation based on several points. The fly ash particles are different according to their composition, size, state of aggregation of matter and so by the part they play in the deposition. The slag deposit growth is the result of opposing processes. There is the viscous, tough deformation and an adhesion of particles on the one hand, and the destruction of the bonds by more coarse and less sticky particles on the other hand. Thus it was supposed that the material for the slagging simulation should include different fractions and their dimentionless numbers should differ. The elasticity is vital for the destruction process and the velocity scale $M_w = 1$ is a criterion of an approximate simulation. For sticky material such criterion is $M_w = M_d$, where M_d is the particle size scale. The simulation criterion of material characteristic reduces to uniform viscosity when we use resembling materials in a boiler and in a rig. The problem of the choice of the material can be solved if it is used certain simulation materials whose stickiness is higher or lower, than it is necessary as stipulated by the simulation conditions. It is evident, that the result is reliable if the change of material has no substantial effect on it. For the simulation of the Kuznetsk fly ash in a gas fired rig the mixture of a high-melting fly ash and a glass powder was used.

2.2. Study of Slagging of the Burning Zone

The study of slagging of the burning zone has been carried out on the model of the TPP-804 boiler of an 800 MW power-generating unit. This boiler is the most powerful in the country and it is equipped with new type flat-flame burners which was studied deficiently in terms of slagging. The experimental procedure made it possible to measure the distribution of the slagging on the walls with different position of the flames. The schematic diagram of the gas-fired model and the example of the slagging distribution with the horizontal position of the flames are shown in Fig.1.

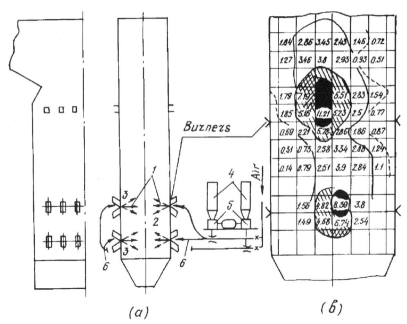

FIGURE 1. Schematic diagram of (a) the gas fired model of the boiler furnace type TPP-804 and (b) an example of slagging distribution.
1) horizontal position of flames; 2) close-flame arrangement; 3) open-flame arrangement; 4) bins containing powder material; 5) feeder; 6) powder to burners. The numbers in Fig. 1(b) mean a relative slagging rate.

It was found out that the burners of this type help to vary the slagging pattern over a wide range depending on the arrangement of flame and can be used for an overhauling of boilers which have slagging problems. At the same time intensive spread in the horizontal plane of the jets from burners results in the appearance of areas of excessive slagging on the side walls, so that it is advisable to turn the end burners towards to the centre or to increase the distance between the burners and the wall as compared with other burner types.

2.3. Study of Slagging Mechanism

The tests in the gas-fired rig also yielded useful results concerning the slagging mechanism. It was found out that the rate, in general, is not a linear one with respect to the fraction mass of sticky particles which reach the deposit surface. This is shown in Fig. 2. A part of particles which have not necessary sticky properties is bonded in the deposits due to the stickiness of other ones in certain conditions. And on the contrary, there are sticky particles in fly ash at temperatures below the temperature of the start of slagging. These results verify experimentally the matrix hypothesis of slagging. The observation on another rig with an open flame of a single burner has validated the destruction of deposits simultaneously with their growth.

The tests in the gas-fired rig using synthetic materials also showed clearly a possibility of a considerable error while using an average ash chemistry or viscosity for slagging prediction.

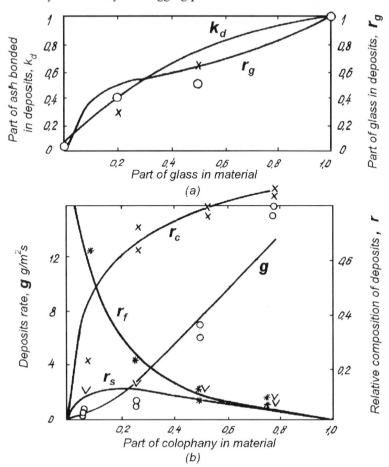

FIGURE 2. Results of (a) a deposition study in the gas fired rig and (b) in a rig operating on a low heated air. Points - experimental data, lines - calculated data based on a version of "probability" deposit growth model. r_c - fraction of colophany in deposits, r_f - fraction of fine silicate powder, r_s - fraction of salt.

On the basis of said observations the "probability" deposit growth model has been proposed. The features of the model are the following. As a result of a contact between fly ash particles and particles located at a deposit surface it can be a strong bond, a failed bond during the next phase of the process and a rebound. The particles are grouped into several types according to the part they play in slagging and probability of contact of particles of different types on the unstable surface deposits and the result of those contacts are calculated. More detail information about the "probability" model and its verification is in another our presentation at this Conference.

3.   SLAGGING IMITATION IN THE RIG USING MULTICOMPONENT SYNTHETIC MATERIAL

As was mentioned above, the deposition is a set of processes and it involves both the bonding of sticky and fine particles and the destruction of bonds by coarse non-sticky particles. This insight was realised on a special rig which operates on a low heated air (60 - 130 degrees C) with synthetic material. The material contains colophony as a sticky constituent, common salt as a coarse one and fine silicate powder. Such composition of the material helps to analyse the deposits easily by leaching them in water and then in alcohol. The schematic diagram of the rig is shown in Fig. 3. The deposits formed in the rig had a similar structure and look like the deposits formed in full scale boilers, that is shown in Fig. 4. Together with an agreement in results of investigations in the rig and in boilers this corroborates accepted preconditions.

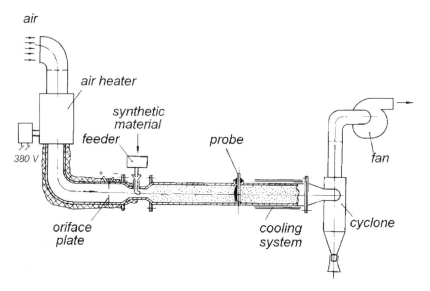

FIGURE 3. Schematic diagram of a rig operating on a low heated air.

The effect of the gas velocity, the viscosity of the sticky agent, the percentage of the agent and the proportion of fine and coarse agents were studied on the rig. In our research we used the design of an experiment. First of all, the data verify the salient points of the deposit growth model and the results obtained in the gas fired rig. This is shown in Fig. 2. Besides, it showed, that the temperature of the start of slagging depends on external factors such as a flow velocity.

FIGURE 4. External appearance of a deposit probed in the rig using synthetic material (colophany - fine silicate powder - salt).

It was also found out that the effect of individual factors depends on the combination and the level of other factors. Thus, the slagging rate increases with the increase of kinetic energy at high temperatures and reduces at low temperatures with other factors being equal. This is shown in Fig. 5. The data are in qualitative correspondence with the data obtained in boilers by us and other researches but are of a more general character. In particular, the measured rate of slagging shows a good agreement with calculated slagging rate based on an empirical equation that has been formulated by Tallinn Technical University (Apik, 1966).

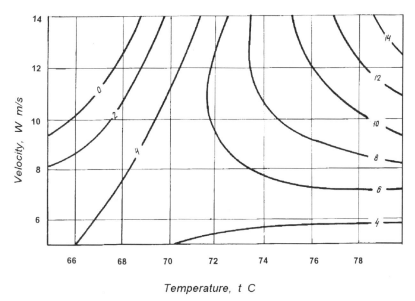

FIGURE 5. Graph showing the dependence of the deposition rate on "gas" velocity and temperature for the rig operating on low heated air.

4. STUDY OF SLAGGING OF FURNACE ROOF

Boilers with two symmetrical downdraft ducts (T - type arrangement) are manufactured in Russia on a large scale. Unlike conventional designs the operation of boilers with T - type arrangement sometimes presents difficulties resulting from the slagging of the furnace roof.

The study of this phenomenon has been carried out on an isothermal model of the upper part of the furnace. The model is shown in Fig. 6.

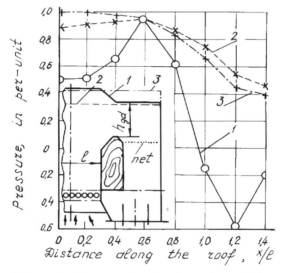

FIGURE 6. Schematic diagram of an isothermal model (its right part) and a graph showing the distribution of the relative pressure along a furnace roof.
1) a new design of the roof; 2) a planar roof, 2*h / l =1.48; 3) a planar roof, 2*h / l =1.91.

The model design allows to change easily the air-foil form and sizes of the ducts, patterns of the air velocity and the solid phase distribution. The possibility to use an isothermal model was due to minor changes of the flue gas temperature in the upper part of the furnace. The conditions for an approximate simulation of a two-phase flow have been folded by the choice of the velocity scale (Fr dimensionless number), the particle size scale (St dimensionless number) and net arrangement in the ducts (Eu dimensionless number). The conditions of the deposition were not simulated. The mass of particles reaching the surface "g" and the dynamic interaction between the flow and the surface of the furnace roof "p" have been used as slagging indexes. The mass of particles was estimated by the measurement of the attenuation of a luminous flux passing through the sticky transparent surface of the roof. The dynamic interaction was measured as the difference of the pressure on the surface and outside of the boundary layer. The results were treated in the following way. The growth of the pressure difference defined as the increase of the plastic deformation and deposits strength, whereas the growth of the mass at the low pressure difference is taken as an increase of the rate of friable deposits.

The experiments have shown that it is possible to reduce the slagging of the roof at the same temperature by means of optimisation of proportion of the ducts and by using new designs. In particular, it was proposed to locate the furnace roof above the roof of horizontal flue-gas ducts, as it is shown in Fig. 6 (Alekhnovich et al., 1984) or to direct the recirculation gas through the roof (Alekhnovich et al., 1982).

5. COAL FIRED RIG

The study of the slagging properties while burning of various coals has shown that these properties can not stable even within one coal field. Thus it is taken the following procedure of tests of slagging. Tests with a typical coal in full scale boilers were accompanied by the tests on a coal fired rig with various batches of the same coal type. The results obtained in the boiler and in the rig differ quantitatively, but the correlation was defined as shown in Fig. 7.

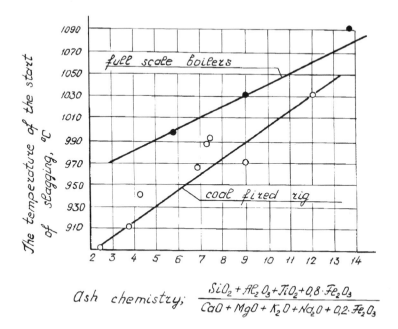

FIGURE 7. Temperature of the start of slagging depending on the ash chemistry.

The coal fired rig of UralVTI contains the equipment for coal preparation, a system for the supply of pulverised fuel and secondary air, a pulverised - fuel combustor (furnace), a horizontal duct for studying of slagging and fouling, an air heater, several flue-gas desulfurizasion systems and dust precipitators. The coal input is 50 kg/h.

The schematic diagram of the combustor and the ducts are shown in Fig. 8. The cylindrical refractory lined combustor is equipped with a turbulent burner which has a variable swirling of the stream and is located on a combustor bottom. The flue gas temperature in the horizontal "experimental" duct reduces evenly from 1200 C down to 700 C which allows to study both the slagging and the fouling problems.

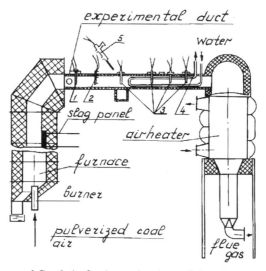

FIGURE 8. The coal fired rig for investigation of slagging properties.
1) tensile strength device, 2) probe with an adjustable temperature of a wall, 3) non-cooled probes, 4) fouling probe, 5) thermocouple with gas exhausting.

The "experimental" duct is fitted with a complement of probes, temperature and heat flux sensors, flue gas analyser. The tensile strength of slag deposits is measured in hot state ("in situ") by a specially designed device (Alekhnovich et al., 1989). During the studies of the slagging properties of coals in the rig we defined the temperature of the start of slagging, the slagging rate and deposit strength depending on the temperature, fouling index and extracted samples of deposits, fly ash for further researches in the laboratory.

6. CONCLUSION

UralVTI has developed, constructed and used various rigs for the studying of slagging problems. New techniques of slagging imitation and investigation have been worked out.

With the help of the rigs a specific information about the slagging mechanism was obtained, the designs of furnaces and their parts have been optimised, and the range of coals types studied as to their slagging properties has been extended.

Besides extending the range of investigated coals the experiments in the coal fired rig increase accuracy of an evaluation of coal slagging properties. It is due to the measurements of important additional indexes (deposit strength) and mainly due to study of several controlled batches of the coal. The correlation between results obtained in the rig and in full scale boilers was get by comparative studying of the same coals.

The goal of investigations in the rig using multicomponent synthetic material was to elucidate the slagging mechanism. The investigations have qualitative effect. At the same time, correspondence of the results with data obtained in full scale boilers corroborates the accepted preconditions.

Study of slagging of a furnace roof on isothermal model and a lower furnace part on gas fired one has been done based on approximate simulation rules. Although the results are qualitative they offer to prevent serious design mistakes and to evaluate expediency of new designs.

REFERENCES

Alekhnovich, A.N., Brykov, V.Y., Orenbakh, M.S. (1989).The device for measurement of deposit strength (Ustrojstvo dlja Izmerenija Prochnosti Otlojeniy). Patent USSR. no. 1449762

Alekhnovich, A.N., Lisovoj, V.G., Petrov, E.V. and Sotnikov, I.A. (1984). Boiler (Kotel). Patent USSR. no. 1065656.

Alekhnovich, A.N., Osintsev, V.V. and Lisovoj, V.G. (1982). The Vertical Prismatic Furnace of the Boiler with T-type Arrangement (Vertikalnaja Prizmaticheskaja Topka T-obraznogo Kotla). Patent USSR. no. 974035.

Alekhnovich, A.N., Verbovetsky, E. Kh. and Abrosimov, A.A. (1988). A Study of the Position of the Zones of Slagging on a Model of the TPP-804 Boiler of an 800 MW Power-generating Unit. **Thermal Engineering, 35** (8), 483-484 (54-57 in Russian).

Apik, I.P. (1966). An Effect of Oil Shale mineral Matter on the Boiler Operation Conditions (Vlijanie Mineralnoj Chasti Slantsev na Rabotu Kotloagregata). Estonian State Publishing House. Tallinn. 129-130 (in Russian).

COAL ASH FUSION TEMPERATURES - NEW CHARACTERISATION TECHNIQUES, AND ASSOCIATIONS WITH PHASE EQUILIBRIA

T F Wall[1], R A Creelman[2], R P Gupta[1], S Gupta[1] C Coin[3] and A Lowe[4]

1. Department of Chemical Engineering, University of Newcastle, NSW 2308
2. R A Creelman and Associates, 108 Midson Road, Epping, NSW 2121
3. ACIRL Ipswich, PO Box 242, Booval Qld 4304
4. Pacific Power, GPO Box 5257, Sydney, NSW 2001

Abstract

The well-documented shortcomings of the standard technique for estimating the fusion temperature of coal ash are its subjective nature and poor accuracy. Alternative measurements based on the shrinkage and electrical conductivity of heating samples are therefore examined with laboratory ash prepared at about 800°C in crucibles, as well as combustion ash sampled from power stations. Sensitive shrinkage measurements indicate temperatures of rapid change which correspond to the formation of liquid phases that can be identified on ternary phase diagrams. The existence and extent of formation of these phases, as quantified by the magnitude of 'peaks' in the test, provide alternative ash fusion temperatures. The peaks from laboratory ashes and corresponding combustion ashes derived from the same coals show clear differences which may be related to the evaporation of potassium during combustion and the reactions of the mineral residues to form combustion ash.

1.   INTRODUCTION

A number of tests and analytical techniques are used to evaluate the potential of coals to foul and slag furnace surfaces (Couch, 1994). There are tests and measurements that attempt to predict slagging propensity amongst which are the ash fusion temperatures (AFT) test, oxide analysis and the estimation of indices based on ratios of elemental oxides, and measurements of the viscosity of ash. A more direct approach is to conduct deposition experiments in test furnaces or an operating plant. The first approach has proven to produce results that are of doubtful accuracy (Juniper, 1995), and deposition experiments in furnaces are prohibitively expensive.

The ash fusion test has become the most accepted method of assessing propensity of coal ash to slag in a furnace. The AFT measures the softening and melting behaviour of ash prepared in a laboratory at temperatures of about 800ºC. Coal specifications normally define limits for softening and melting in the belief that AFT behaviour is a predictor of ash behaviour in the furnace. There are well-documented shortcomings of these approaches

Applications of Advanced Technology to Ash-Related Problems in Boilers
Edited by L. Baxter and R. DeSollar, Plenum Press, New York, 1996

relates to their uncertainties as predictive tools for plant performance, and the reproducibility of ash fusion measurements.

Of particular concern is the estimation of the initial deformation temperature (IDT). IDT is the temperature at which the rounding of the tip of an ash cone is noted, which has been accepted as the temperature where the ash first softens and may become sticky. Gas temperatures entering tube banks are designed to be below the IDT to avoid fouling problems. However, the repeatability and reproducibility of the IDT is poor, and because it is a prime determinant for the heat transfer area which must be provided in furnace design, there are uncertainties when a furnace is being designed or modified if IDT is the sole predictor.

AFT tests are not really measurements, they are observations. There is a need for a better understanding of processes of ash fusion, and for tests that are more objective. The present paper is concerned with alternative procedures for determining ash fusibility and is accompanied by investigations into the mechanisms involved in the associated chemical and physical processes and the differences in results when using either laboratory ash or ash generated in the course of PF combustion from a given coal.

2.   CHARACTERISATION OF ASH FUSION

As ash heats, particles sinter and fuse and eventually form a liquid slag, these changes providing bases for characterising the fusion processes. The AFT operator observes changes in a standard ash cone as it is heated through the temperature range expected in a furnace. The temperature at which the tip of the cone rounds is the Initial Deformation Temperature. Softening Temperature (ST) is taken to be when the height of the cone equals the width of the cone. Hemisphere temperature (HT) is when cone height is a half the cone width. Flow temperature (FT) is when sample height is about 1.5mm. The temperature range involved is 1000 - 1600 ^{o}C. Despite the best efforts of the operators, and the provision of auxiliary devices to assist, the AFT is still an observation.

The need is for a measurement. Shrinkage of a sample ash as it heats should characterise the reduction of porosity associated with the fusion of particles or the flow of liquids under gravity. Electrical resistance should characterise the conduction path through the layer, and therefore particle/particle contact and fusion. Both these features can be measured, therefore they provide at least two avenues by which measurements can be substituted for observations, as suggested by Raask (1979).

Many previous studies have described experiments of these types, (Sanyal and Cumming (1981), Gibson and Livingston (1991), Lee et al (1991), Khan and Williford (1989). However, a systematic study comprising several measurement types, laboratory ash and combustion ash with investigations of the fusion mechanisms has been lacking.

3.   COAL AND ASH SAMPLES

The ash samples used in this study comprised a range of coals from those used in domestic power stations, to those exported from Australia as steaming coal. A sample set of 9 export coals and a further set of 9 made up of pulverised fuel and the corresponding precipitator (combustion) ash sampled from power stations were selected. Laboratory ash was prepared from the power station coals and the export coals giving a total of 27 ash samples. The chemical compositions of the laboratory ash samples are compared on Fig. 1.

Several of the export coals were selected on the basis of the difficulty of the estimation of their IDT's. This is apparent on Fig. 2 which compares measurements of ash fusion temperatures from three commercial laboratories. Laboratory 1 generally reported the lowest estimates, Laboratory 3 the greatest with differences as great as 400°C, indicating obvious differences in the interpretation of the initial rounding of the cone tip by operators. Fig. 2 also indicates that there is a general difference between the export coal ash samples and the power station ash samples, the former having, on average, IDTs some 100°C lower than the latter. The samples are, therefore, not strictly comparable.

4.   EXPERIMENTAL PROCEDURES

The study used several techniques with the same ash samples. Fig.3 and Table 1 providing comparisons between the equipment and techniques.

4.1   The ACIRL test, also called the Improved Ash Fusion test (Coin et al, 1995)

The ACIRL test measures the shrinkage of ash pellets that are placed between two tiles. The procedure was designed by the Australian Coal Industry Research Laboratories to provide a more precise measure of ash fusion temperatures than the standard procedure. Four ash cylinders (2 mm x 2 mm dia) are used as pillars to separate two alumina tiles. The distance between the tiles is measured as the assembly is heated, either continuously or at temperature intervals of up to 20°C, from 1000°C to 1600°C. The present data is for 20°C intervals.

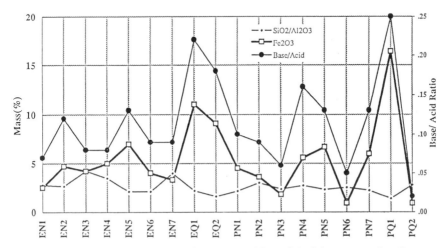

Figure 1.   Comparison of the chemical composition of the laboratory ashes for export (E..) and power station (P..) coals.

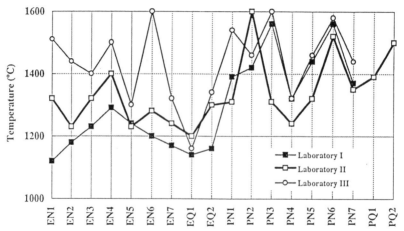

Figure 2. Comparisons of the initial deformation temperatures for the laboratory ashes from three commercial laboratories.

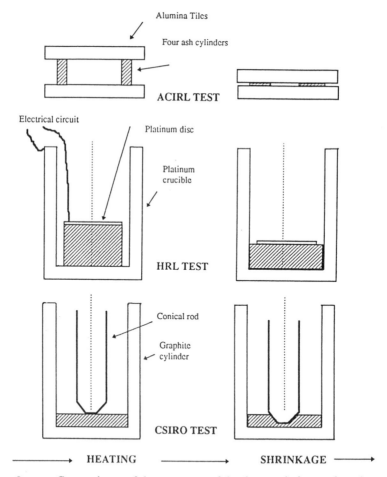

Figure 3. Comparisons of the geometry of the three techniques, the ash samples shown shaded. Left - the initial sample, right - deformed sample after heating.

Table 1.   Comparison of the three tests used

Common Aspects

Heating rate, $8°C$/min.

Gas atmospheres, reducing for HRL and ACIRL, inert gas for CSIRO

Differences

Amount of ash used : HRL ~ 0.3g, ACIRL~ 0.1g, CSIRO ~ 35 mg

Ash compaction : CSIRO, loose ash, others use pellets with dextrin.

Initial temperatures : ACIRL, $1000°C$, others room temperature for data but $900°C$ selected as initial condition for shrinkage analysis (i.e. temperature for initial dimension of ash sample).

Maximum temperature of ~ $1350°C$ for HRL

Data at temperature intervals of $20°C$ for ACIRL, continuous for CSIRO, intervals of $4°C$ for HRL.

All tests monitor shrinkage, HRL also monitors electrical conductivity.

Table 2.   Correlations between primary fusion temperatures and eutectic points from phase diagrams

Primary Fusion Temperature $°C$	System	Eutectic Temperatures $°C$	Position of System on Phase Diagram
930	SiO_2 - Al_2O_3 - K_2O	923	K-rich components; in the lower leucite field
1005	SiO_2 - Al_2O_3 - K_2O	985	K-poor components; eutectic at join of mullite, tridymite and potash feldspar fields.
1080	SiO_2 - FeO - CaO or SiO_2 - Al_2O_3 - FeO	1093 1088	Lowest point in this system; olivine field. Lowest point in system; start of iron cordierite field.
1220	SiO_2 - Al_2O_3 - FeO	1205	End of iron cordierite field; join with mullite field.
1320	SiO_2 - Al_2O_3 - CaO	1310, 1315, 1335, 1345	Join of gehlenite, pseudowollastonite & rankinite fields for low Al_2O_3 systems; $1335°C$ refers to higher Al_2O_3 composition fields.
1440-1480	SiO_2 - FeO - CaO	1470	Join of crystobalite, mullite and anorthite field.
1590	Al_2O_3 - SiO_2 binary	1590	Last melt in silica - alumina system.

4.2   The HRL test (Ellis, 1989)

The HRL test measures concurrently electrical resistance and shrinkage of ash pellets. The procedure was developed by HRL Technology Pty Ltd. Ash pellets (7mm high x approx. 7mm dia, ~0.3gm) with 10% dextrin were pressed and located in a platinum crucible 10mm dia, with a platinum disc located on the top face. A circuit to measure electrical conductance/resistivity established a pulsed voltage across the pellet and a standard resistance in series with it. A square waveform with a period of approx 90ms and amplitudes of approx +2V / -2V was used. The voltage drop across the known resistance was monitored with a datataker concurrently with the pellet height. Measurements at the times of constant applied voltage (+2V and -2V) were measured, as the samples were heated from low temperatures (30°C) to above 1350°C or when the sample contracted and the top disc contacted the crucible.

4.3   The CSIRO test

The CSIRO test measures the shrinkage of loose ash. The procedure was developed by the CSIRO Division of Coal and Energy Technology. A special sample holder has been designed for a Thermomechanical Analysis machine (Setram TMA92). Loose ash (~35mg) is located between container with a flat bottom and a rod with a tapered end. As the assembly (and ash) is heated (from room temperature to 1600°C) the rod sinks into the ash and eventually flows as slag into the annular gap between rod and container.

5.   CORRELATION BETWEEN SHRINKAGE MEASUREMENT TECHNIQUES AND RESULTS

It has been noted that the samples can be divided into three groups.   For the purposes of discussion the three groups are represented by:

- EN3, on ash with a high SiO_2/Al_2O_3 ratio, low Fe_2O_3 and poor reproducibility for the IDT test.

- EN6 - an ash with a SiO_2/Al_2O_3 ratio within the norm of Australian coals, low Fe_2O_3, and very poor reproducibility for the IDT test.

- EQ1 - an ash with the highest Fe_2O_3 content and good reproducibility for the IDT test.

The three ashes have low CaO levels (0.8 to 3.1%) with the Fe_2O_3 being primarily derived from sideritic minerals in the coal.

Figs. 4-6 compare the techniques with results for (top plot) electrical resistance and (bottom three plots) the three shrinkage tests. Also indicated are the ash fusion temperatures from laboratory 2, as given on Fig. 2. The figures also indicate as 'peaks' the temperatures of rapid shrinkage, with the height of the peak being proportional to the maximum change in shrinkage with change in temperature.

There are a number of observations that can be made from the data. Significant shrinkage begins at around 900°C whereas the ash fusion measurement commences at 1000°C. The magnitude of the shrinkage in some samples is high, suggesting that the use of the 1000°C start point misses a series of what may be significant events.

546

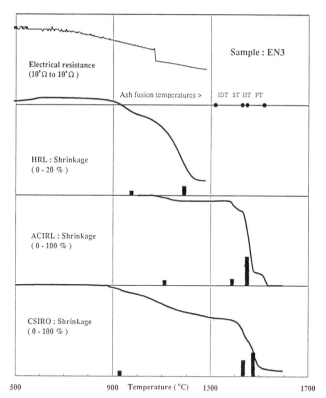

Figure 4.   Sample EN3, comparisons of the results for the techniques. Top plot - electrical resistance data from HRL test, bottom three plots - shrinkage experiments reported as progressive ash sample dimension on heating. The bars indicate 'peak' temperatures of rapid change in ash sample dimension with temperature.

Comparing the three measurements of shrinkage, some common peaks are evident on Figs. 4 to 6 for a given sample. The temperatures of the peaks for the CSIRO test appear to be reproducible to ± 5°C, and the ACIRL test has similar sensitivity when continuous recording is used. Further work is necessary to quantify this aspect. The magnitude of the peaks and the extent of the initial shrinkage observed at low temperature (<1100°C) is also seen to depend on the state of the sample. The CSIRO test uses loose ash, rather than the ash being compacted in the other two tests which results in an amplified initial shrinkage compared to the other two procedures. As illustrated on Fig. 3, during melting the ash is penetrated by a rod during the CSIRO test. The CSIRO test therefore provides the most sensitive data over the complete temperature range for the present sets of data.

In Fig. 6, the HRL test is seen to give an expansion of the sample at temperatures above 1200°C when the other tests are indicating rapid shrinkage. This event, we believe is a bloat associated with the escape of gases from the sample, and was observed for only two samples.

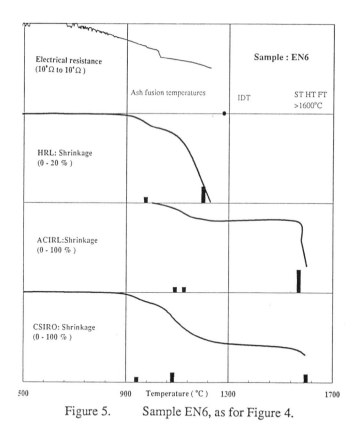

Figure 5.   Sample EN6, as for Figure 4.

Figs. 4 to 6 provide insights into the relationships between the AFTs and shrinkage and the reproducibility of the IDT is quantified by the range in reported results on Fig. 1. For sample EN3, as seen on Fig. 4, the IDT is within a temperature region without substantial change in measured shrinkage. The poor reproducibility of the IDT procedure is therefore expected. In contrast the ST and HT are in a region of rapid shrinkage and should therefore be determined with good reproducibility. For sample EN6, as seen on Fig. 5, the IDT is again within a temperature region of little shrinkage. In fact, little shrinkage is measured from 1100-1550°C. This results in extremely poor reproducibility for the IDT. For sample EQ1, as seen on Fig. 6, the IDT occurs at a temperature of rapid shrinkage, hence its good reproducibility.

6.   ELECTRICAL RESISTANCE RESULTS

Referring to the raw data presented on Figs. 4 to 6, the noise in the signal reported at low temperatures (<800°C) are due to the square waveform of the applied voltage and poor contact between the platinum electrodes and the ash, the contact obtained at higher temperatures resulting in a smooth trace.

The common technique used to interpret data for electrical resistance (R) against temperature (T) is to plot Log R against T. A change in slope on such a plot has been called (Sanyal and Mehta, 1994) a 'breakpoint' and appears to be related to a change of phase or melting of the ash. Few obvious 'breakpoints' were observed in the present data. The data is therefore being reanalysed in terms of the temperature required for changes in electrical conductance levels, (Wall et al, 1995).

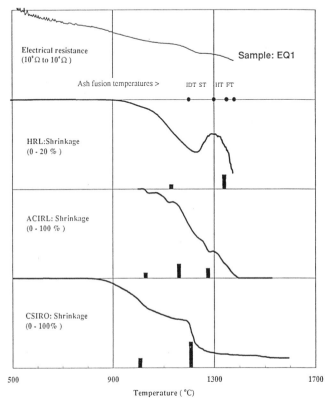

Figure 6.   Sample EQ1, as for Figure 4.

7.   MECHANISM OF FUSION AND SHRINKAGE

Additional pellets of crucible ash and combustion ash were prepared, at the same heating rate and atmosphere, to a number of determined temperatures and then cooled (in about 5 to 10 minutes to a quenched temperature of around 800°C). The nominal temperatures selected were 1200, 1300, 1400 and 1500°C, and were based on the ranges of AFTs reported on Fig. 2. The pellets were prepared as polished sections and examined by SEM.

The observed processes associated with ash fusion are illustrated on the SEM images given on Fig. 7. The sequence of events observed as temperature increases is :

1.   Melting and reaction to form the first melt (liquid) phase. The most significant melt phase is iron rich and often has a composition close to iron cordierite, (Fig. 7B).

2.   The reaction of the liquid with specific solids (SiO_2 rich) until the progressive depletion of the solids, (Fig. 7C).

3.   As the proportion of liquid increases the pellet slumps until it is supported by the remaining solids, that is, the height decreases, but diameter increases.

4.   The voids (or gas bubbles) increase in size as they coalesce. The final state is a homogeneous liquid containing large bubbles, (Fig. 7D).

The lowest temperatures at which the ash pellets were prepared was approximately 1200°C. At this temperature the ratio of liquid to solid in some cases exceeded 30%. It can be concluded that at the IDT there has been considerable melting. The CSIRO shrinkage tests have shown that there are significant shrinkage events as low as 930°C and later at around 1000°C. As the ash fusion test begins at 1000°C there has been a considerable loss of information if no observations or measurements are taken below 1000°C. We consider that these lower temperature events represent the melting of alumino silicate phases that have significant potassium content.

AFT and shrinking methods are essentially different approaches to the problems identifying liquid phases in an ash pellet. Huffman et al (1981) concluded from their investigations that AFT data corresponds approximately to the liquidus surface of the pseudo-ternary system $A\ell_2O_3$-SiO_2-Base where the drivers are FeO, CaO and K_2O with modifying effects from MgO and Na_2O.

Shrinkage methods measure dimensional changes in sample. The causes of the dimensional changes are numerous but the most rapid changes coincide with liquid formation. Heating and cooling a mass of ash results in three processes; fusion (liquid formation), adsorption (liquid solid interaction to form more liquid) and crystallisation on cooling or after supersaturation.

The CSIRO and ACIRL shrinkage tests generate an output of the rate of shrinkage as the sample temperature increases. Rapid fusion is the stage where liquid is being formed and when the first derivative of the shrinkage curve is taken rapid fusion events are seen as peaks. Illustrative traces are given on Fig. 8 for samples EN3, EN6, and EQ1. Figure 9 presents the temperatures of these peaks plotted in order of increasing temperature for all samples. Figure 10 shows that there is a general repeatability for certain temperatures and these temperatures are common to many samples. These may be termed *Primary Fusion Temperatures* (PFT). It is postulated that there is a second set of peak temperatures are the result of slumping of the sample as minerals are adsorbed which may be termed *Derived Fusion Temperatures*, and will occur at temperatures higher than a PFT. Overlapping peaks associated with two PFTs may also result in derived fusion temperatures, to values either higher or lower than the PFT.

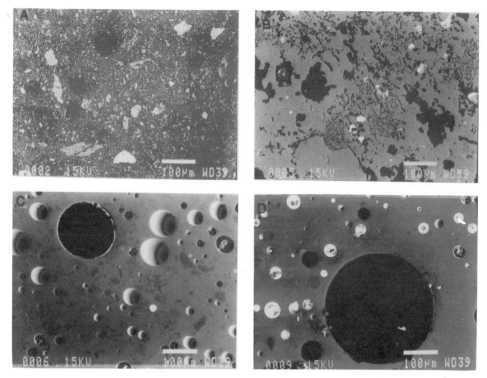

Figure 7.   Scanning electron microscope images of ash pellets at various temperatures, Sample E3 (Export Coal).

A: Crucible ash (800°C).   Note minerals are discrete particles. Carbonates and sulphates are thermally degraded, but as yet there has been no fusion.

B: Ash pellet, 1350°C.   Melt phase is dominant, but quartz (silica) particles are discrete inclusions within the melt. Alumino-silicates are in the process of reacting with the melt.

C: Ash pellet, 1460°C.   At this temperature most alumino silicates have been injested, and silica is being absorbed into the melt.   Note the appearance of gas bubbles.

D: Ash pellet, 1500°C.   Silica now small particles in melt. EMPA analyses shows 5-10% Al_2O_3 in silica particles indicating high level of diffusion between melt and remaining silica particles.

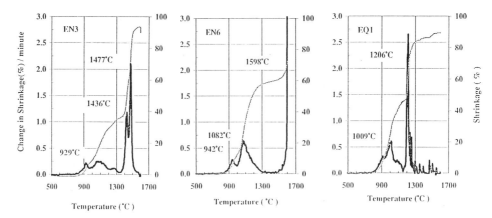

Figure 8.   Outputs from the CSIRO test giving shrinkage and rate of change with temperature, indicating peak temperatures corresponding to rapid shrinkage, for laboratory ash samples EN3, EN6 and EQ1. To convert value to Change in Shrinkage (%)/°C, multiply by 0.125.

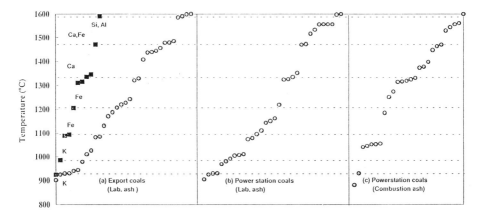

Figure 9.   Peak temperatures plotting in order of increasing temperature for the three ash types. The points shown are the eutectic temperatures, with the associated elements being indicated, the dashed lines are the primary fusion temperatures identified in Table 2.

Shrinkage data require more holistic use of phase diagrams whereas AFT observations require the identification of the effects of liquid formation, shrinkage identifies regions of rapid fusion that correspond to discrete liquid formation events. An important conclusion reached by Huffman et al (1981) was that IDT recognition required extensive liquid formation. With shrinkage the PFT event is recognised and the magnitude of the peak is a measure of the importance of that event.

Huggins et al (1981) have used phase diagrams to interpret ash fusion temperatures and addressed the question of the appropriate ternary system to use in interpreting such data. The authors contended that Mossbauer studies of ashes used in their experiments showed all iron to be ferrous under standard ASTM conditions.

In the course of the investigation Callide B power station in Queensland, (ACIRL 1993) Mossbauer studies of boiler deposits showed that magnetite was the predominate spinel present in the glasses, with a iron silicate phase being ferrous iron silicate. Typical ratios were 65:25:10 Magnetite: Ferrous iron silicate: Ferric iron silicate. One class of deposit, minor to rare in abundance, consisted of haematite and ferrous iron silicate in the ratio typically 70:30.

Theoretically it could be expected that the environment of the PF furnace would favour ferric iron compounds, but the experimental evidence shows that the compounds are dominated by ferrous forms. The use of $SiO_2-Al_2O_3$ - FeO ternary diagram as distinct from an equivalent of $SiO_2-Al_2O_2-Fe_2O_3$ is therefore justified. Fe_2O_3 is present, but at the detail the ternary diagram are being used, the amount of ferric present would not change greatly the stability fields depicted in the $SiO_2-Al_2O_3-FeO$ ternary.

The total ash system is the sum total of a number of subsystems. Huffman et al (1981) recognised the importance of the pseudoternary system $SiO_2-Al_2O_3$-Base (FeO + CaO + K_2O) and this study confirms that the important ternary subsystems are SiO_2 - Al_2O_3 - FeO, SiO_2 - Al_2O_3 - CaO (modified by MgO), SiO_2 - Al_2O_3 - K_2O (modified by Na_2O), SiO_2 - FeO - CaO (modified by MgO), Al_2O_3 - FeO - CaO (modified by MgO).

The temperatures appear to be correlatable with specific eutectic points within the subsystem. The eutectics are in positions on the ternary diagrams that are close to the plots on those diagrams of the bulk composition of the ash. Deviations of the eutectics from the primary fusion temperatures are due to modification by other oxides that modifies the system to other temperatures. The comparison between the measured Primary Fusion Temperatures as shown in Fig. 9 and the important eutectic points within the various systems are given on Table 2.

The peaks in the range 1080°C - 1220°C appear to be related to two systems, the SiO - FeO CaO and SiO_2 - Al_2O_3 - FeO. In practice it is observed that the high iron ashes have strong peaks at 1200°C. This is related to the iron cordierite field and as such is a signature of the iron in the system. The peaks in the range 920°C - 1010°C appear to be potassium signatures. The peaks in the range 1100°C through to 1480°C are all parts of the FeO - CaO bases reacting with various proportions of SiO_2 and Al_2O_3.

Therefore, in general, both the ACIRL and CSIRO shrinkage results show peaks in three distinct groups. The lower temperature peaks up to 1100°C are related to potassium content, the 1200°C peak to the iron content and the range 1220°C to 1440°C to FeO and CaO related reactions.

553

8. COMPARISONS OF RESULTS FOR LABORATORY AND COMBUSTION ASH

The AFTs for the laboratory ash and combustion ash were generally within 100°C. Comparison of the shrinkage results from the HRL tests in the range 0-20% indicates that temperatures for an equivalent shrinkage were higher for seven samples and lower for two.

Figure 9 presents in ascending order the peak temperatures from the CSIRO test for the three ash types together with the eutectic temperatures identified on Table 2 as derived from the phase diagrams with horizontal lines indicating temperatures of common peaks. The laboratory ash is mostly comprised of only slightly modified coal minerals, the combustion ash is made up of the reaction products of mixtures of these minerals which will be homogenised at the higher temperatures in the furnace. The apparent loss of potassium peaks at 930°C and 980°C is expected due to the loss of potassium experienced during combustion. The same trends are evident in the ACIRL data. The appearance of iron related peaks between 1100°C and 1200°C with a very strong peak in any iron bearing ash at 1200°C can be explained by the change in composition as melt products form from iron particles. In general the combustion ashes have a more uniform Fe and Ca composition. The iron related peaks in the range 1080°C -1220°C are not apparent in the power station ashes. The range 1310°C to 1345°C appears to represent various calcium peaks. At approximately 1320°C a strong peak occurs that is related to calcium iron silicate compounds and the 1590°C peak is silicon/aluminium related and represents the last melts which are rich in silica. The silicate aluminium peaks are not apparent in the power station coals.

9. CONCLUSIONS

Measurements of the shrinkage and electrical conductivity of heating ash samples have been previously suggested as alternative characterisation techniques for ash fusion. Results from these alternative procedures reveal that ash samples have reacted at temperatures well below the initial deformation temperature (IDT) and at the IDT a substantial proportion of the sample is liquid. The IDT is clearly not the lowest temperature for ash to soften.

The present electrical resistance data has not provided breakpoint temperature which relate to ash fusion temperatures (AFT) as some previous studies have done. However the sensitive shrinkage measurements may be interpreted to provide ash fusion temperatures. The CSIRO shrinkage test, based on a TMA system, provides peak temperatures identifying rapid shrinkage events during ash heating which are related to the formation of eutectics identified on phase diagrams. The existence and magnitude of these peaks provide alternative ash fusion temperatures. The ACIRL shrinkage test may also provide these peak temperatures when continuous measurements are obtained.

The CSIRO and ACIRL tests have also revealed differences between laboratory and combustion ash from the same coals which are related to the loss of potassium and reactions between the mineral residues at the higher temperatures experienced by combustion ash.

10. ACKNOWLEDGEMENTS

We are most grateful to Mr Dick Sanders of Quality Coal Consulting Pty. Ltd. for managing the project, selecting coal samples and providing insights into the AFT procedure. The study was supported by the Australian Coal Association Research Program with Mr Grant Quinn of BHP Australia Coal as project monitor. The assistance of Dr Alf Ottrey at HRL and Dr John Saxby at CSIRO is greatly appreciated.

11. REFERENCES

ACIRL (1993), Investigations of Slagging at Callide B Power Station. Unpublished Report to Queensland Electricity Commission.

Coin, C., Reifenstein, A.P., Kahraman H. (1995), Improved Ash Fusion Test, this symposium.

Couch, G. (1994), Understanding Slagging and Fouling in PF Combustion IBACR/TJ2, IEA Coal Research Laboratories, August.

Ellis, G.C. (1989), The Thermomechanical, Electrical Conductance and Chemical Characteristics of Coal Ash Deposits, NERDDP Project No. 1181 Final Report Volume III, Secv R & D Dept, Australia.

Huffman, P.G., Huggins F.E., Dunmyre G.R. (1981), Investigation of the High Temperature Behaviour of Coal Ash in Reducing and Oxidising Atmospheres, Fuel, 60, 585-597.

Huggins, F.E., Kosmack, D.A. and Huffman, G.P. (1981), Correlation between Ash-Fusion Temperatures and Ternary Equilibrium Phase Diagrams, Fuel, 60, 577-584.

Juniper, L. (1995), Applicability of Ash Slagging "Indices" Revisited, Combustion News February, Australian Combustion Technology Centre. pp.1-4.

Khan, R.M. and Williford, C.A. (1989), Novel Technique for Direct Measurement of Ash Fusion and Sintering Behaviour at Elevated Temperature and Pressure - International Conference on Coal Science, Tokyo.

Gibson, J.R. and Livingston, W.R. (1991), The Sintering and Fusion of Bituminous Coal Ashes, Engineering Foundation Conference on Inorganic Transformations and Ash Deposition During Combustion, Palm Coast, Florida, p425-447.

Lee, G.K. et al, (1991), Assessment of Ash Sintering Potential by Conductance and Dilatometry, Pittsburgh Coal Conference.

Raask, E. (1979), Sintering Characteristics of Coal Ashes by Simultaneous Dilatometry - Electrical Conductance Measurements, J. Thermal Analysis 16, 91.

Sanyal, A. Mehta, A.K. (1994), Development of an Electrical Resistance Based. Ash Fusion Test. The Impact of Ash Deposition in Coal Fired Furnaces, p445-460, Taylor and Francis, Washington.

Sanyal, A., and Cumming I.W. (1981), An Electrical Resistivity Method for Deterring the

Onset of Fusion in Coal Ash, US Engineering Foundation Conference on Slagging and Fouling from Combustion Gases, Henniker, p329-341.

Wall, T.F. et al, (1995), Demonstration of the True Ash Fusion. Characteristics of Australian Thermal Coals. ACARP Project C3039 Final Report (in preparation).

SLAGGING PREDICTION WHEN USING THE CHEMICAL COMPOSITION OF FLY PARTICLES AND THE SLAGGING PROBABILITY MODEL

A.N. Alekhnovich, V.E. Gladkov, and V.V. Bogomolov

Ural Heat Engineering Institute (UralVTI), Chelyabinsk, Russia

ABSTRACT

The deposit growth model has been developed by UralVTI, wherein the fly ash particles are endowed by various properties depending on their part in the deposition. According to the model the contact probability of the particles with various properties which are on the deposit surface and in the fly ash flow is calculated. It was assumed that non-sticky particle can be bonded in the deposit if they come into contact with the matrix (captive) particle on the deposit surface.

The chemical composition of the individual particles of a fly ash has been measured with the help of the SEM/EMPA instrumentation for three coals with a different ash types. The relative deposition rate and the deposit composition have been calculated on the basis of the model. There is a definite accordance between the predicted data and the results, received in the industrial boilers.

1. INTRODUCTION

Dozens of empirical indexes have been formulated for estimating coal slagging properties. A readily available analysis of an average mineral matter and its fractions are used for their construction. The indexes provide a certain authenticity within the range of coals used for the index construction, but they don't represent real slagging processes and have low extrapolation possibilities.

It is known, that the composition of individual ash particles varies substantially and the particles contact little in a flue gas. Thus, the deposition is interaction between a forming non-homogeneous surface of deposits and various particles.

A slagging prediction has reached a new qualitative level during the recent years. It had gone from the coal classification and the usage of average data about the mineral matter to the prediction of deposit properties at the base of composition distribution of mineral inclusions. Mathematical models based on microscope analysis of the mineral matter have been developed, for example (Helble et al., 1992; Barta et al., 1993; Borio et al., 1993).

Applications of Advanced Technology to Ash-Related Problems in Boilers
Edited by L. Baxter and R. DeSollar, Plenum Press, New York, 1996

2. THE POINTS OF THE SLAGGING "PROBABILITY" MODEL

In the above - mentioned slagging models it is assumed, that slag deposits are formed from all particles which have reached the deposit surface and whose t viscosity is less, than a critical value. However, the deposits growth can be accompanied by the destruction of their surface; the deposition rate does not vary linearly against the sticky particles mass in a fly ash; solid, non-sticky particles can be bonded in the deposits as a result of the stickiness of the others. Such results have been obtained during of tests in rigs, presented in the report by Ural VTI, dedicated to the rigs at that conference and they are published in Russia (Alekhnovich et al., 1986 and 1995).

The model of the slag deposit growth which we chose to call the "probability" model has been worked out on the basis of those results. The first version of the model was developed ten years ago in order to explain the existence of the temperature of the start of slagging in spite of the presence of sticky particles below it and a various forms of a slagging rate dependence as a function of the temperature (Alekhnovich et al., 1986). During the last two years the model has been up-dated to the data concerning the chemistry of individual particles of a fly ash (Alekhnovich, 1995). The model is not full in the contrary of above - mentioned ones and deals only with formation of a slag deposits. It assumes that :

all particles reaching the deposit surface play this or that part in the formation of deposits;

the fly ash and the outer layer of the deposits contain particles with adhesive properties high enough which are sufficient to bond less sticky ones in certain conditions. We call them "matrix" or "captive";

if the stickiness of a particle and a deposit surface are not high enough, the particle adheres to the deposit temporary and is removed from the deposit as a result of an impact of an next particle irrespective of what type the letter is. A number of sticky particles do not adhere to the deposit and form a rebound;

a deposit growth takes place by a non-branchy chain.

3. THE PROCEDURES USED IN THE MODEL

The input data for the slagging "probability" model are the chemistry of individual particles of fly ash and their conventional size determined by scanning electron microscopy / electron microprobe analysis techniques (SEM/EMPA).

The procedural steps in the model include:

- grouping of particles into certain types according their role in the deposition;

- calculation of a contact probability of the various particle types on the deposit surface and in fly ash flow and the result of the contact;

- calculation of the deposition rate, the temperature of the start of slagging, deposit chemistry depending on the temperature.

The number of the particle types, the grouping procedure and boundary conditions in model versions can be changed. We used from four to eight particles types. If it is used four types, we call them "matrix" (symbol **m**), sticky (**s**), inert (**i**) and abrasive (**a**). The fifth type can be Ca - enriched. It is assumed, that matrix particles form strength bonds with any other particles except abrasive ones. Similar bonds are also formed by sticky particles contacting with each other, but, contacting with the inert ones they form temporary bonds. The endowed role of various particles types in deposition is shown in Table 1. An illustration of the deposit growth model is shown in Fig.1.

TABLE 1. Characteristic of various particle types according to their role in the deposition when grouping into four types.
    "+" formation of the strength bond, "o" formation of the temporary bond, "-" rebound.

matrix-m sticky-s inert-i abrasive-a		particles in flue gas flow			
		m	s	i	a
particles on a deposit surface	m	+	+	+	0
	s	+	+	0	–
	i	+	0	–	–
	a	0	–	–	–

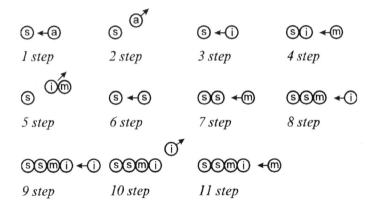

FIGURE 1. Illustration of the deposit growth .

    The following slagging mechanisms are taken in consideration in the model:
    adherence of low viscosity melts or supercooled liquids;
    adherence of the particles which contain a iron-enriched liquid phase at the temperature above the eutectic temperature;
    holding of the Ca - enriched particles owing their chemical activity at rather low temperatures.

In accordance with this, the particle size, CaO and Fe_2O_3 content, the critical temperature of sticking are used for particles grouping. The critical temperature of sticking of individual particles is calculated by means of the same empirical equations as those which had been inferred for the temperature of the start of slagging in dependence on the average ash chemistry (Alekhnovich et al., 1988). The usage of the critical viscosity value is planned for the future. In this model version a particle is counted as matrix if its temperature is more than 100 C above the critical temperature or if particle contains more than 65 % Fe_2O_3 at temperatures above 1200 C. The particle is counted as sticky one if its

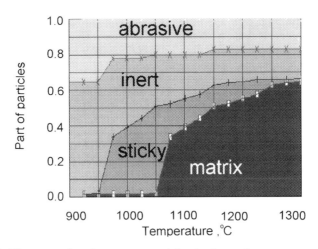

FIGURE 2. The part of various type particles in dependence on a temperature in the fly ash of Irsha-Borodinsky coal.

temperature is less, than 100 C above the critical temperature. If the temperature is below the critical temperature the fine particles are considered as "inert" and large-sized particles are considered as abrasive ones. The boundary conditions can be changed and used as latent empirical coefficients in order to get a quantitative agreement between calculated and experimental results.

An example of fly ash particles grouping in dependence on the temperature for Irsha-Borodinsky coal is shown in Fig. 2. The information about the chemistry of fly ash individual particles used for the grouping is presented in the next part of this article.

Two versions of a computer program of the deposition model have been worked out. On the basis of the first one the probability of contact between particles of various types in fly ash flow and on the deposit surface is calculated.

The examples of equations from the equation system are cited below.

$$k = M* (a_m + a_s + a_i - a_a * c) - S* b* (a_m + a_s - a_a) + I* d* (a_m - a_a);$$

$$r_s = [a_s* (M + L *b) - a_a* M *c] / k$$

$$M = [a_m*M + (a_m* b + a_a* e)* S + (a_m + a_a) *d *I + A] / (M + S + I + A).$$

Here,
k - the portion of particles bonded in a deposit;
a_i - the portion of particles of one of the types in fly ash;
r_i - the portion of particles in the deposit;
M, S, I, A - the portion of particles on a deposit surface;
b, c, d, e - relationships like $d = r_m / (r_m + r_s)$.

According to the other program the deposit growth and chemistry are calculated by "particle by particle " method. A sequence of particles is chosen by random numbers method and a surface feature are changed depending on the result of the previous particle contact.

4. THE CHEMISTRY OF INDIVIDUAL PARTICLES OF A FLY ASH

The chemistry of individual particles of a fly ash has been studied with the help of a "Comebax" unit. A fly ash of some coals with different ash composition has been studied. The number of analysed particles was not more, than 200 for each of the coal as far as one channel analyser was used and a treatment of initial data was not automatised. Such a number of particles is not enough for correct statistic calculations, but nevertheless it makes some useful information available.

The examples of A - B - Fe_2O_3 ternary diagrams for the fly ash from low slagging (Ekibastuz) and high slagging (Irsha-Borodinsky) coals are shown in Fig. 3.

The composition of individual particles varies over a wide range as it has been expected. Small particles have a more varied composition than large-scale ones. The CaO content increases in fine particles for the three studied coals including Irsha-Borodinsky coal with a high content of that component; SiO_2 and/or Al_2O_3 content increases in large-scale particles. Among three studied coals there was no coal with a high pyrite content. For these conditions the Fe_2O_3 content decreases in large-scale particles. The Na_2O content increases in fine particles of Irsha-Borodinsky coal and in large particles for Kuznetsk coal. Earlier it has been established, that Kuznetsk coal, predominantly, contained sodium in the form of feldspar and had a small amount of water - soluble sodium. It is assumed, that this explains the results.

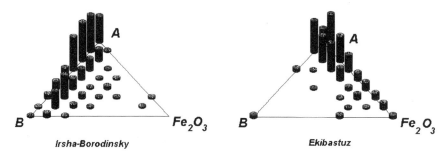

Irsha-Borodinsky   Ekibastuz

FIGURE 3. Ternary diagram showing   A - B - Fe₂O₃   fly ash particles. The
height of peaks represents the volume of ash particles of a given composition.
A = SiO₂ + Al₂O₃;   B = CaO + MgO + K₂O + Na₂O.

The deposits enriched by selective constituents (iron - enriched,
high calcium, alkali - matrix - type deposits) are formed by particles enriched
with the elements. It is possible to suppose that the index of those deposits is
depends, among other factors, on the amount of particles in fly ash enriched by
corresponding elements. The amount of particles enriched by Fe₂O₃, CaO to
this or that degree is shown in Figure 4. The conclusion from the data for
three coals can't be drawn, however, the results are not inconsistent with the
observed facts.

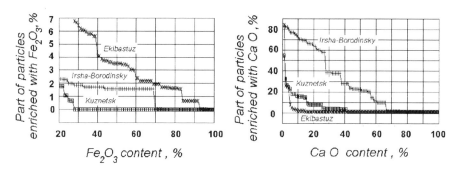

FIGURE 4. The amount of particles in fly ash enriched by elements depending
on the enrichment degree.

5. DEPOSITION PREDICTION

The models based on a composition and properties of individual
particles present the deposition process better, than empirical indexes and they
make it possible to predict the slagging properties, that are not predicted by
other methods. First of all it means the prediction of the rate and composition of
deposits.

The suitability of the "probability" model, in the first place, has been verified by the test results, received in the rigs using synthetic materials. Under certain additional conditions concerning the grouping of particles into types the correspondence with experiments in kind as well as in degree has been achieved a (Alekhnovich et al., 1986).

As for the real fly ash, a close correspondence with experiments data in kind for all parameters and a quantitative correspondence in prediction of the temperature of the start of slagging has been achieved. Examples of some predicted and experimental results are shown in Fig. 5 and Fig. 6.

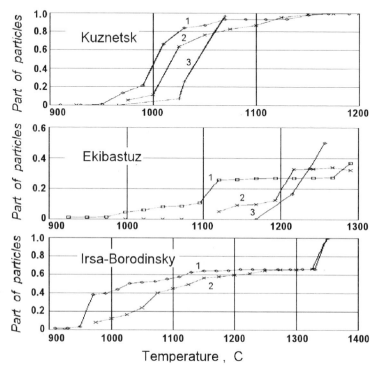

FIGURE 5. Part of fly ash particles depending on temperature:
1) matrix plus sticky types; 2) bonded in deposit as calculated; 3) bonded in deposit in tests on PK-39 and PK-40 boilers.

The experimental data plotted in Fig. 5, Fig. 6 have been obtained in full scale boilers with the help of deposition probes equipped with an uncooled test specimens. The information about the slagging rate and the deposit chemistry for Ekibastuz coal has been obtained in the PK-39 boiler of an 300 MW unit. The probe was located in a furnace on the level of a upper burners row (line 2 in Fig. 6) and on the level of the lower row (line 3 in Fig. 6). The information about slag deposits for Kuznetsk coal has been get in the upper part of a furnace of the PK-40 boiler of an 200 MW unit.

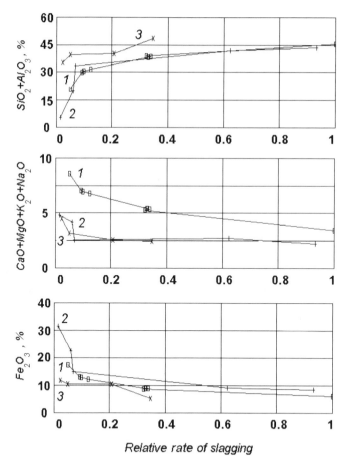

FIGURE 6. Chemistry of deposits of Ekibastuz coal depending on a slagging rate.
1) data calculated by the model; 2) , 3) experimental data.

6. CONCLUSION

All particles that reach the deposit surface participate in the deposition and play various part. In certain conditions the bonding of non - sticky particles and the rebound of sticky ones is possible. That is taken into consideration in the
slagging model worked out by Ural VTI.

The chemistry of the fly ash individual particles is used in the model as input data. Besides that data are useful for an explanation and prediction of slagging and fouling.

The suitability of the model has been verified by experiments with synthetic materials and a real fly ash.

7. REFERENCE

Alekhnovich, A. N., Bogomolov, V.V., Ivanova, N. I. and Verbovetsky, E. Kh. (1986). "The investigation of Slagging Problems by using synthetic mixtures": The effect of mineral matter of power - generating fuels on operational conditions of steam boilers. (Issledovanie voprosov shlakovanija pri primenenii sinteticheskih smesey : Vliyanie mineral'noi chasti energeticheskikh topliv na uslovija raboty parovikh kotlov). Tallinn, **I**, 77 - 82.

Alekhnovich, A. N. and Bogomolov, V. V. (1988). " Temperature Conditions of the Start of Slagging when Burning Coals with Ash of Acidic Composition". **Thermal Engineering, 35**(I), 20-24.

Alekhnovich, A. N. (1995). "The Probability Model of a Slagging Deposit Formation. (Verojatnostnaja model formirovanija shlakovih otlojeniy)". **Elektricheskie stantsii, 2**, 16-22.

Barta, L. E., Beer, I. M., Sarofim, A. F., Teare, V. D. and Toqan, M. A. (1993). "Coal Fouling Tendency Model". In I. Williamson and F. Wigley (Eds.). The Impact of Ash Deposition on Coal Fired Plants. Washington: Taylor & Francis, 177-188.

Borio, R. W., Patel, R. L., Morgan, M. E., Kang, S. G., Erickson, T. A. and Allan, S. E. (1993) " The Coal Quality Expert : A Focus on Slagging and Fouling". Pre-print of second Annual Clean Coal Technology Conference. Atlanta, USA. - Sept. 7-9.

Helble, I. I., Srinivasachar, S., Senior, C. L., Wilemski, G., Baker, I. E., Simons, G. A. and Johnson, S. A. (1992). Pre-print of the Second Australian Workshop on Ash Deposition. Brisbane. Australia. - July 13-14

SLAGGING: PROBLEM DEFINITION

John H. Pohl

Energy International
Laguna Hills, CA, USA

Lindsay A. Juniper

Ultra-Systems Technology Pty Ltd
Indooroopilly, Queensland, Australia

ABSTRACT

Ash related problems cost the US utility industry up to $8.7 billion per year. A part of this cost is caused by slagging. Empirical techniques to predict slagging have not been successful. More fundamentally based techniques are being developed. At this point, the practical application of the more fundamentally based techniques need to be demonstrated.

This paper reviews the causes of difficult deposits and recommends work that might help control slag in boilers. This paper starts with a deposit that can not be removed and analyzes the cause of this difficulty working backward. We start with soot blowing, proceed to deposit temperature, deposit strength, deposit building, and end with deposit initiation.

Finally, we recommend studies on soot blowing. This effort is aimed not only at determining the mechanism of soot blowing but optimizing practical soot blowing. We also recommend studies on the causes of difficult, plastic, and hard to remove deposits.

INTRODUCTION

Slagging in furnaces remains a major problem and increases the cost of power generation. Pohl (1995) estimated that, for nominal conditions in US power plants, five absolute percent reduction (an increase from 85 - 90 % availability) in outage would save $1.4 billion dollars per year, a absolute one percent reduction in operating and maintenance costs would save $2.8 billion dollars per year, a one percent gain in efficiency would save $2.8 billion dollars per year, and a reduction in coal cost by $0.10 per million BTU's would save $1.7 billion dollars per year. The total savings, to the US power industry, for reducing ash related problems could be $8.7 billion dollars per year. These amount are higher but on the same order as estimates by Penner (1988), $4. billion dollars per year and by Couch (1994), $0.4 billion dollars in maintenance costs.

Much work has been done to predict slagging, to design boilers, and specify operating conditions to control slagging. In the past, efforts have been empirical and relied on field and pilot scale data. These efforts have not provided a universal solution to slagging. Current efforts to control slagging have concentrated on a more fundamental approach. These efforts continue and it is to early to asses the results.

Applications of Advanced Technology to Ash-Related Problems in Boilers
Edited by L. Baxter and R. DeSollar, Plenum Press, New York, 1996

567

This paper reviews the data available on the properties of slagging deposits and recommends taking data that is needed to compliment the fundamental approach to slagging. This paper starts with a slagging deposit that can not be removed, investigates the reason the deposit is difficult to remove, reviews how the deposit was formed, and discusses how the deposit is initiated.

SOOT BLOWING

Deposits in a boiler are sprayed with steam, air, or water approximately every eight hours to remove deposits. Figure 1. shows a typical soot and a water blower. Any attempt to control slagging must prevent formation of a deposit that can not be removed within a normal soot blowing cycle.

Steam Blowing

Wall blowing is usually done with steam, but air is sometime used, Moore and Ehrler (1973). The blower penetrates the wall approximately 3-4 inches (76 - 102 mm). The blower will typically have one 1/4 inch (6 mm) hole and spray parallel to the wall. The blower rotates to impact a circle approximately 3-4 feet (914 - 1220 mm) in diameter. The blower uses steam or air at 100-300 psig (6.9-20.7 bar).

Water Blowing

Water blowers are becoming more popular, particularly for difficult to remove deposits. Operators previously were concerned with thermal shock to the wall. Juniper, et al. (1994) reports that the thermal shock varied from 70- 180 F (21 - 82 C). The water blower sprays water back at the wall at an angle of 10-20 ° from the horizontal. The blower advances into the boiler and rotates at a controlled rate to avoid prolonged spraying of the bare wall. The water blower operates at about 300 psig (20.7 bar).

Mechanical Force

The blowers exert a compression and shear force on the deposit. These forces will remove the deposit provided the deposit adhesion forces are less that the force of the soot blower.

A number of measurements of compression strength have been made using a penetrometer first introduced by Borio. By this method, deposits with compression strength of greater than 15 psig (1 bar) are considered difficult. The authors have encountered deposits with compression strengths of greater than 100 psig (6.9 bar) in pilot scale furnaces. One problem with this technique is that a plastic deposit in the furnace measures almost zero compression force, but is the most difficult deposit to remove.

Barnhart developed the Sinter Strength Test. In this test, fly ash or ASTM ash is compressed in to pellets, sintered in a furnace at different temperatures for different times, and the compression strength is measured. The strength must be measured on a number of pellets because of statistical variation in pellet defects. Pellets with compression strength greater than 5,000 psig (344.5 bar) are considered difficult.

The authors are aware of only limited data reported on the shear or tensile strength of deposits. The shear and tensile strength may be more important than the compression strength, but are more difficult to measure.

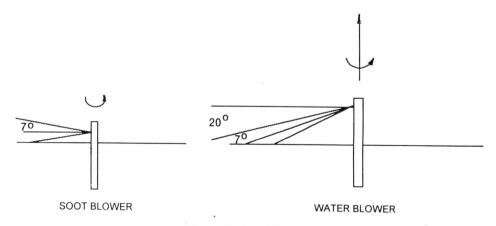

Figure 1. Soot Blowers

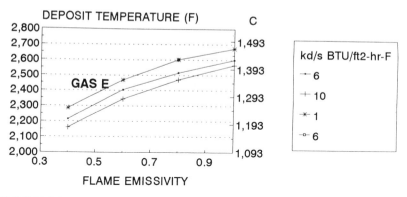

LITTLE DEPOSIT, Ew=Ed=0.7
DEPTEM

Figure 2. Deposit Temperature

Thermal Fracture

Thermal fracture contributes to removal of slag deposits. The deposit has a temperature gradient. This gradient is from a tube temperature at 800 F (427 C) up to a deposit surface of 2600 F (1427 C). The blowing media cools the surface of the deposit through convective cooling. This cooling reduces the deposit surface temperature below the internal temperature and produces a reverse temperature gradient. The sudden changes in the temperature gradient of the deposit induces thermal stress due to the thermal expansion of the deposit. The deposit will fracture when this stress exceeds the strength of the deposit.

A number of investigators have used simulated soot blowing as relative measures of the difficulty of removing the deposit. Some have used the difference in flow rate for a fixed position soot blower and some have measured the area cleaned as a function of soot blowing pressure in pilot scale rigs. Barnes, et al. (1994) reported that approximately 3 ft2 (91 mm2) could be cleaned with a unspecified soot blower at about 35 psig (2.4 bar).

The authors have determined the acceptable soot blowing cycle in a pilot scale furnace. The cycle is determined by measuring the air flow required to recover heat transfer with a fixed geometry air blower at different cycles. This procedure was only used on fouling deposits, but could be used on slagging deposits.

Wain, et al. (1994) has used the equation for crack propagation to estimate the difficulty of removing slag deposits. They have rearrange this equation into two expressions that estimate thermal shock. One expression contains the thermal conductivity; one does not. These shock parameters are the change in the surface temperature. The properties of the deposits are measured as functions of porosity.

Removal

The slag deposit will be removed when the soot blowing force exerted exceeds the strength of the material or adhesion. The soot blower can exert a mechanical shear or compression force or induce a temperature gradient which causes shear forces to develop in the deposit.

Raask (1986) has estimated that the van der Waals force, within 0.02 um, holding a 1 um particle is 100 time the gravity force. Raask measured the tensile force of a hot deposit on ferric and austentic steels. Raask measured the strength over 25 days. Extrapolation to an 8 hour soot blowing cycle indicates a tensile force of about 1 LB/in2 (0.07 bar). Raask has also concluded that deposits are easier to remove from austenitic than ferric steels because of the differences in thermal expansion between the tube metals and the deposit.

Van der Vort (1992) has also estimated the force of deposit adhesion to the wall. His estimates indicate that the capillary force is 2 LB/in2 (0.14 bar) for a 10 um capillary, the van der Waals force is 20 LB/in2 (1.4 bar) for a 1 um diameter particle, and the thermophoresis force is 2×10^{-6} LB/in2 (1.4×10^{-6} bar).

Barnes, et al. (1994) measured compressive forces of deposits. The compression force at zero porosity varied from 14,000 psig (965 bar) to greater than 400,000 psig 27,400 bar). The compression strength of the deposits was reduced to less than 10,000 psig (685 bar) at porosities greater than 25 percent. The elastic modules of glasses and ceramics 5-15 $X10^6$ psig (0.344×10^6 - 1.03×10^6 bar), modules of rupture between 3,000- 15,000 psig (206 - 1028 bar), and compression strength between 40,000-90,000 psig (2756 - 6165 bar).

570

Barnes, et al. (1994) also measured thermal expansion of 2-3 X10^{-6} 1/F (3.6 - 5.4X10^{-6} 1/C) for deposits. Thermal expansion of gases and ceramics vary between 0.5-3 X10^{-6} 1/F (0.9 - 5.4X10^{-6} 1/C).

Finally, Barnes ,et al. estimated the thermal shock parameters for deposits. The estimated thermal shock parameters varied form 1,000-10,000 F (538 - 5538 bar) at zero porosity. At 25 percent porosity, the estimated thermal shock parameters of 1000-2000 F (538 - 1093 C). Glasses and ceramics have thermal shocks between 300 and 500 F (149 - 260 C).

DEPOSIT TEMPERATURE

Fewster, et al.(1994) have measured gas temperatures near the wall of a slagging boiler and extrapolated these temperatures to the wall. The wall temperatures at the top burner ranged between 2430-2480 F (1330-1360 C) for non-slagging coal and 2250-2430 F (1230-1320 C) for a slagging coal. Fewster, et al. (1994) measured and extrapolated a temperature to the wall, 15 ft (457 mm) above the top burner, of 2570 F (1410 C).

A one-dimensional calculation procedure was used to estimate the flame temperatures, deposit temperature, and tube temperatures. Figure 2. shows the calculated wall temperatures above the burner, at a gas emissivity of one, varied form 2560-2580 F (1400-1470 C)

Flame Emissivity

The emissivity of gas components is well established. For the large boiler in this calculation, the gas emissivity is approximately 0.5. However, particulates and soot will raise the emissivity to close to one. Singer (1981) reports that the emissivity of large boiler flames approaches 1.0.

The difference in emissivity between the boiler and the pilot scale furnace can make simulation of slagging difficult. To simulate slagging, we use the same coal, same particle size, and same excess air level. The furnace velocity is reduced to provide the same residence time. The furnace is heavily insulated to provide the same flame temperature. A portion of the wall is equipped with a cooled plate or tubes at the same temperature as the water wall. We know we can not simulate flow patterns. The best that we can do is to use slag panels in different positions and orientations.

Everything, except the flow field is simulated at the start of the test. However, the flame emissivity in the pilot scale is lower that in the boiler. This causes the heat flux from the flame to be lower. As the deposit builds the surface temperature of the deposit drops below that in a boiler. The deposit no longer simulates the boiler deposit temperature. Figure 3. shows the difference between the furnace and boiler deposit. There are two possible solutions: 1. program an increase in the slag panel temperature or 2. increase the flame emissivity to 1.0. Increasing the panel temperature will alter the temperature and temperature gradient in the deposit. The flame emissivity of the flame can be increased by adding a couple percent of soot to the flame and adjusting the flame temperature to compensate. Alternatively an inert radiator can be added to the flame and the temperature adjusted.

Deposit Emissivity

Creek and Becker (1985) and Wain, et al. (1994) have measured the properties of deposits and Singer (1981), Wall et al.(1994), and have tabulated the results. The emissivites of deposits ranged from 0.5-0.95. The authors own measurements agree with those reported except we have measure an emissivity of 0.4 for a Western Coal in a pilot scale furnace.

Deposit Thermal Conductivity

Creek and Becker (1985), Singer (1981), Wall, et al. (1994), and Wain et al. (1994) have reported thermal conductivities between 1 and 10 BTU-in/ft2-hr-F (0.5-1.5 W/m-K). The values of thermal conductivity quoted in Singer (1981) and other Combustion Engineering work need to be divided by 12 to be consistence with the values and units reported here..

DEPOSIT STRENGTH

The deposit can stick to the wall and develop strength in place. The strength develops through sintering, viscous flow, and reaction.

Sintering

Sintering is the growth of necks between particles. This sintering is related to the ratio of surface tension and viscosity through the Frankel Equation. Nowok and Benson (1994) report that the surface tension of slags range from 400 to 550 dynes/cm, viscosity from 5-2000 poise, and the ratio of surface tension to viscosity from 1-10 cm at deposit temperatures.

The particles can sintered at high viscosity. Porosity is a partial measurement of the extent of sintering and strength. However, the strength depends on the uniformity of the deposit and the thickness of the necks.

Viscous Flow

Plastic deposits are difficult to remove. In a plastic deposits, the deposit yields under mechanical force and thermal shock before the deposit freezes and can develop stress. Once the deposit has partially relaxed, it will not develop high stress and fracture. It may be possible to freeze the deposit by slight cooling, and then subjecting the frozen deposit to thermal or mechanical fracture.

Plastic deposit form when part of the material has a low viscosity. This material flows into pores. The deposit becomes a two phase fluid consists of particulate fly ash in a viscous fluid. The plastic material yields when thermal or mechanical forces are applied. We do not know how much of the material must be fluid to develop a difficult deposit. Fewester, et al. (1994) suggests than 10-15 percent of the deposit may be sufficient. Figure 4. shows the fraction melting at different temperatures for samples from an Australian Boiler. Viscous flow brings materials to sites where they can react to form more low viscosity material.

Phase Diagrams

Phase diagram can be useful in working with slagging problems, but must be used with care. Seldom are deposits homogenous. However, equilibrium phase diagrams can be used with local compositions provided that reaction does not limit the composition.

Phase diagrams can be calculated from equilibrium. However, no good way exists to estimate the activity coefficient. Few activity coefficients have been measured for complex slags. Blander and Pelton (1986) has measured FeO activity coefficient in CaO-FeO-SiO2 solutions. The FeO activity coefficients are within 40 percent of 1. Boni, et al. (1990) have measured activity coefficient of SiO2 and Na2O in SiO2-Na2O. The activity coefficients of SiO2 can be 0.2 and the activity coefficients of Na2O can be 10^{-6} to 10^{-8}.

While, phase diagrams are helpful to understand slagging, no simple way has been found to predict

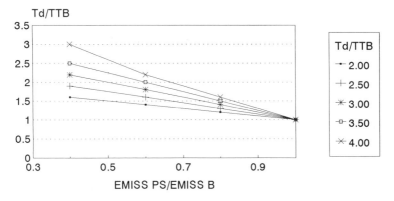

$$TTPS/TTB=Td/TTB(1-EPS/EB)+EPS/EB$$
TD
CORRECTION FOR DEPOSIT TEMPERATURE.

Figure 3. Scaling

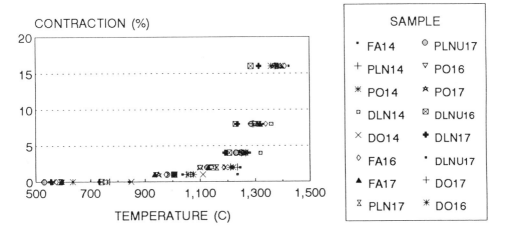

LOW NOx BURNER (LN), LN U (LN+3.2 M),AND ORIGINAL BURNER (O)
14, 16,AND 17 NONMINAL % Fe_2O_3 in ASH
TMAC

Figure 4. Fraction Melted - Contraction of Samples: Fly Ash (FA),
Particulates (P), and Deposits (D)

slagging based on phase diagrams. Figure 5. shows the primary crystallization phases in complex solutions measured by Kalmanovitch and Williamson (1986). The percent of the major phase is plotted against crystallization temperature. The temperature are accurate to about 100 C. In addition the activity coefficients may vary widely with nature of the solution.

DEPOSIT BUILDING

The deposits build by collecting sticky particles, dry particles in a sticky deposit, and reactions in the deposit.

Sticky Particles

Most of the work has been done on collection of sticky particles. Particles will become sticky when the rebound energy is less than the energy of adhesion. Pohl (1992) has outlined the causes of particles adhering to the wall. Particles can adhere to the wall or the deposit by condensation, van der Waals forces, thermophoresis forces, electrostatic forces, interfacial tension, and capillary forces. The order of magnitude of some of the forces was discussed above.

In addition, low viscosity particles will deform and absorb some of the rebound energy. Boni, et al. (1990) indicates that particle with viscosities less than 10^8 poise will stick to the wall.

Sticky Deposit

Dry particles can also add to sticky deposits. Again, deposition depends on the rebound force of the particle being reduced to less than adhesion force. Wagoner and Yan (1991) have experimentally investigated some simple systems where the particle becomes imbedded in the deposit

Reaction in Deposit

Finally reactions can take place on the particle and or in the deposit to form more low viscosity material.

Reaction in the deposit will likely be controlled by liquid or solid diffusion. If the reaction is solid phase, with a diffusion coefficient of 10^{-10} cm/sec, moving 10 um, about the size of a neck or bridge, will require about 3 hours, well below the time required in a soot blowing cycle.

Liquid phase reaction will be faster. If the liquid phase diffusion coefficient is 10^{-5} cm/sec, the time to move 10 um is 0.1 second.

Gases can also react with the particle or the deposit to form more low viscosity material. To be rapid the gas phase material must condense. Sinha, et al. (1994) predicts that Na2SO4 and CaSO4 will be stable below 930 C (1650 F). The tube temperatures are around 800 F (427 C) and CaSO4 and Na2SO4 can condense near the tube.

DEPOSIT INITIATION

The deposit can be initiated by direct impaction of large particles or by small particles driven to the wall. The authors favor the small particle hypothesis. However, both could take place depending on conditions.

Dry Particles

In the small particle hypothesis, small clay particles high in potassium, possibly illite, are driven to the wall by thermophoresis forces. These particle stick to the wall through a combination of van der Waals forces in interfacial tension. This thin layer of particles forms the base on which other particles land to build the deposit. Finally, the deposit grows into the boiler until its surface temperature is high enough so the outside of the deposit flows.

Molten Particles

The large sticky particle is another hypothesis of slag initiation. In this hypothesis, large, possible

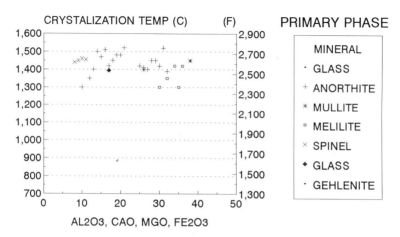

Figure 5. Crystallization Temperature

partially unreacted particles, are driven by inertia or eddies to the wall. These large particle have low viscosity, deform, and are held on the wall by interfacial tension. Other particles build on the particles that have already arrived.

RECOMMENDATIONS

The most important information needed to advance our knowledge of slagging is information on deposit removal and the relationship between ash chemistry and strength of the deposit.

The information needed on soot blowing should be determined on deposits extracted from the boiler. These deposits can be reheated in a furnace and the mechanical and thermal forces measured. The sequence and amount of force should be varied to determine the optimum soot blowing parameters. The size of the fragmented pieces of slag should also be determined because hopper blocking is an important problem.

The strength of the deposit should also be investigated. This investigation should be done on deposits withdrawn form slagging boilers. These deposits should be examined using the scanning electron microscope and the micro probe. The size and composition of bridges should be carefully examined and related to the porosity and strength of the deposit.

REFERENCES

Barnes, D.I., Lewit, M.W., and Smith, M. (1994), " The Slagging Behavior of Three UK Power Station Coals in an Ash Deposition Rig." in J. Williamson and F. Wigley (Eds.), *The Impact of Ash Deposition on Coal Fired Plants,* London: Taylor and Francis.

Blander, M. and Pelton, A.D. (1986), " Analyses of Thermodynamic Properties of Molten Slags." In K.S. Vorres (ed.), *Mineral Matter and Ash in Coal* Washington, DC, American Chemical society Symposium Series 301.

Boni, A.A, Beer, J.M., Bryers, R.W., Flagan, R.C., Helble, J.J., Huffman, G.P., Peterson, T.W., Sarofim, A.F., Srinivasachar, S. Wendt, J.O.L. (1990), " Transformations of Inorganic Coal Constituents in Combustion Systems." *DOE Contract No. DE-AC22-86PC90751.*

Creek,R.C. and Becker, H.B. (1985), " Thermal Properties of Boiler Ash Deposits." *NERDDP Project 593 Report ND/85/014* Melbourne, Australia, State Electricity Commission of Victoria.

Fewster, M., Juniper, L. Ottrey, A., Creelman, R. and Pohl, J. (1994), " Plant Investigation of Slagging at Callide Power Station." *Pacific Rim International Conference on Environmental Control of Combustion Processes* Maui, HI: American and Japanese Flame Research Committees, October.

Juniper, L.A., Pohl, J., Ottrey, A., Creelman, R., Wall, T., Kent, J., Griss, R. (1993), " Slagging Investigations at Callide B Power Station." *Queensland Electricity Commission, Contract No. CC/92/269* Brisbane, Australia.

Juniper, L.A. and Pohl, J.H. (1995), " Design of Pilot Scale Rigs to Simulate Combustion Processes." *The Australian Combustion Symposium including The Fourth Australian Flame Days* Adelaide, Australia.

Kalmanovich, D.P. and Williamson, J. (1986), " Crystallization of Coal Ash Melts." In K.S. Vorres (ed.), *Mineral Matter and Ash in Coal* Washington, DC, American Chemical society Symposium Series 301.

Moore, G.F. and Ehrler, R.F. (1973), " Western Coals-Laboratory Characterization and Field Evaluations of Cleaning Requirements." *ASME Paper 73-WA/FU-1.*

Nowok, J.W. and Benson, S.A. (1992), " Correlation of Interfacial Surface Tension/Viscosity Ratio with Base/ Acid Ratio." In S.A. Benson (Ed.), *Inorganic Transformations and Ash Deposition During Combustion* New York: Engineering Foundation.

Pohl, J.H. (1992), " The Causes, Prediction and Control of Slagging in Utility Boilers." *2nd Australian Workshop on Ash Deposition* Brisbane, Australia.

Pohl, J.H. (1995), " Electricity Cost Related to Ash." *Prepared for the American Society of Mechanical Engineers Research Committee on Corrosion and Deposits from Combustion Gases* Washington, DC.

Raask, E. (1986), " Deposition Constituent Phase Separation and Adhesion." In K.S. Vorres (ed.), *Mineral Matter and Ash in Coal* Washington, DC, American Chemical society Symposium Series 301.

Singer, J.G. (1981), *Combustion: Fossil Power Systems.* Windsor, CT: Combustion Engineering, Inc.

Sinha, S., Natesan, K. and Blander, M. (1990), " A New Calculation Method for Deducing the Complex Chemistry of Coal Ash Deposits." R.W. Bryers and K.S. Vorres (eds.), *Mineral Matter and Ash Deposition from Coal* New York: United Engineering Trustees, Inc.

Wagoner, C.L. and Yan, X.-X. (1992), " Deposit Initiation via Thermophoresis: Part 1, Insight on Deceleration and Retention of Inertially Transported Particles." In S.A. Benson (Ed.), *Inorganic Transformations and Ash Deposition During Combustion* New York: Engineering Foundation.

Wain, S.E., Sanyal, A., Livingston, W.R., and Williamson (1992)," Thermal and Mechanical Properties of Boiler Slags of Relevance to Soot Blowing." In S.A. Benson (Ed.), *Inorganic Transformations and Ash Deposition During Combustion* New York: Engineering Foundation.

Wall, T.F., Bhattacharya, S.P., Zhang, D.K., Gupta, R.P. (1994), " The Properties and Thermal Effects of Ash Deposits in Coal-Fired Furnaces: A Review." in J. Williamson and F. Wigley (Eds.), *The Impact of Ash Deposition on Coal Fired Plants,* London: Taylor and Francis

van der Vort, C.L. (1992), " Factors Influence Ash Particle Depositions in a Separator at High Velocities Temperatures." S.A. Benson (ed.), *Inorganic Transformations and Ash Deposition During Combustion* New York: Engineering Foundation.

MODELLING FLY ASH GENERATION FOR UK POWER STATION COALS

F. Wigley and J. Williamson

Department of Materials, Imperial College, London SW7 2BP, UK

ABSTRACT

An in-depth characterisation has been made of three UK bituminous coals and the combustion products from these coals when burned at a power station and on a range of experimental combustion facilities. The coals were chosen to represent the range of ash compositions and slagging propensities found at UK power stations.

CCSEM analysis of the pulverised coals has been performed to provide quantitative data on the size and chemical composition of individual mineral occurrences, and to determine the nature of the mineral-mineral and mineral-organic associations in the pulverised fuel. In a similar way the size and chemical composition of individual fly ash particle has been determined.

The mineral-mineral association information has been used predict the effects of mineral coalescence, the dominant mineral transformation process for UK power station coals. The CCSEM information correctly identifies the types of mineral-mineral association that are present, but strongly underestimates the degree of association and hence the predicted effects of coalescence. The limitations of the information are inherent in the analysis of a cross-section, but useful information for the modelling of ash generation may still be obtained.

INTRODUCTION

As part of a UK collaborative research programme entitled 'Minimising the effect of high temperature coal ash deposition' (Gibb, 1995), work was carried out to produce a simple model that could predict the properties of fly ash particles from the parent coal minerals. The objective of this work was to be able to provide the fly ash information required for the complex computer modelling of pf combustion and ash transport and deposition in a boiler, work that formed another part of the collaborative research programme (Lee et al, 1995). A particular aim of the work reported here was to make maximum use of the information on mineral-mineral associations available from computer-controlled scanning electron microscopy (CCSEM). The characterisation of mineral-mineral associations by CCSEM has been carried out by other workers (Yu et al, 1994), but the use of this information to date in the modelling of ash generation is

Applications of Advanced Technology to Ash-Related Problems in Boilers
Edited by L. Baxter and R. DeSollar, Plenum Press, New York, 1996

579

limited (Harb et al, 1994). The work was focussed on the slagging of UK bituminous coals, in line with the overall aims of the research programme.

The changes that occur to coal minerals during the combustion of pulverised coal can be considered from two viewpoints: the physical and chemical processes operating on the minerals, or the differences between coal minerals and ash particles in terms of their particle size and chemical distributions. Each of these viewpoints is discussed in the following sections. The CCSEM analytical procedure and the results obtained are then described, followed by a discussion of the limitations of this type of information.

ASH GENERATION PROCESSES

A relatively small number of physical and chemical processes operate on coal minerals during combustion.

Coalescence

When a pulverised coal particle contains two or more occurrences of mineral matter, this mineral matter may be expected to show some degree of coalescence during the burnout of the parent char particle. This coalescence will cause an increase in size during the transformation from mineral occurrence to ash particle. In addition, coalescence will modify the mineral chemical distribution by producing ash particles with intermediate chemical compositions. Because clay minerals dominate the mineralogy of UK power station coals, coalescence will increase the proportion of ash particles containing a significant aluminosilicate component.

Volatile Loss

Pyrite (FeS_2), calcite ($CaCO_3$), ankerite [$Ca(Fe,Mg,Mn)(CO_3)_2$] and gypsum ($CaSO_4.2H_2O$), all of which are present as major or minor minerals in UK power station coals, lose volatile components as they transform. This volatile loss changes the chemical composition and reduces both the mass and the size of the resulting ash particles. The effects of volatile loss on the size and chemical distributions of the resulting ash particles can readily be calculated for the relevant minerals. In particular, if chemical analyses are presented as oxides then there appears to be a significant mass loss during the transformation of pyrite to iron oxide ash particles.

Fusion

The main effect of fusion is in the change in shape from irregular coal minerals to mainly spherical ash particles. Because of the way in which CCSEM data on particle cross-sectional area is interpreted to give a measurement of size, this shape change need not be considered unless particle shape is required as an output from the model. Another effect of fusion is in the formation of small ash particles, as fused material separates from the main mass of mineral matter. This process has the opposite effect to mineral coalescence. In the absence of sufficient information to make independent

580

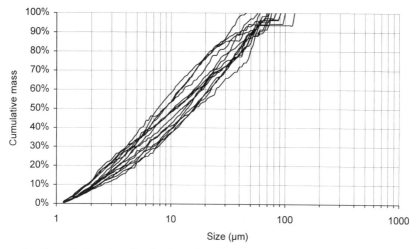

Figure 1. Cumulative size distributions for mineral occurrences in fourteen UK power station coals, determined by CCSEM.

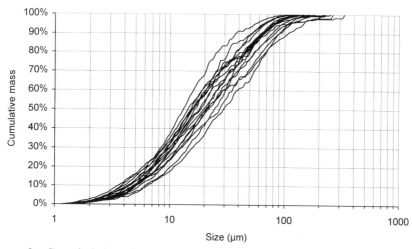

Figure 2. Cumulative particle size distributions for seventeen samples of UK power station ash, determined by CCSEM.

estimates of the magnitudes of these conflicting effects the size reduction due to fusion was considered within the effects of mineral coalescence.

Vapourisation and Condensation

There is evidence from the analyses of coal minerals and related ash samples from this collaborative research programme of only limited vapourisation and condensation of potassium, sodium and possibly calcium. The slightly higher K_2O and Na_2O concentrations in the ashes probably result from the capture by ash particles of alkalis volatilised from combusting macerals, and slightly lower CaO concentrations probably result from the loss of some carbonate mineral matter to the gas phase. Given the small and consistent level of these changes, the overall interest in slagging rather than fouling, and also the problems involved in modelling some of the other ash generation processes, a numerical simulation of vapourisation and condensation was not attempted.

COMPARISON OF ASH PARTICLES WITH COAL MINERALS

There are strong similarities between the properties of minerals in pulverised coal and the properties of related ash particles; the mineralogy of the parent coal is the dominant factor in determining the particles chemical distribution of the ash produced on combustion (Wigley and Williamson; 1994a, 1994b). However, there are consistent changes in size and chemical distribution that occur during ash generation.

Size Distribution

Both the mineral occurrences in pulverised coal and fly ash particles share a lower size limit at, or slightly below, about one micron diameter. The coal minerals have a median size of about 13µm and a top size that is usually below 100µm (Figure 1), whilst the ash particles have a median size of about 20µm and a top size that is usually in the range 150-200µm (Figure 2). The ash particles are consistently coarser than the coal minerals, and this is due to the effect of mineral coalescence. The variation in size distribution in the samples analysed is attributed to analytical imprecision and sampling inaccuracy; no consistent coal-related variation was observed.

Chemical Distribution

Compared to coal minerals, samples of ash consistently show both a higher mass proportion of aluminosilicate particles and a higher proportion of particles with intermediate chemical compositions that contain an aluminosilicate component (Figures 3 and 4). Part of the increase in abundance of clay-derived ash particles is caused by the reduction in mass of other minerals during volatile loss. However, much of the shift in chemical distribution towards aluminosilicate compositions is produced by mineral-mineral interactions that occur during coalescence, where clay minerals are usually involved in each interaction.

582

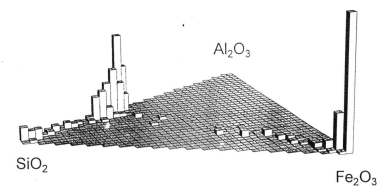

Figure 3. Chemical distribution of mineral occurrences in Silverdale coal, plotted on the Fe$_2$O$_3$-Al$_2$O$_3$-SiO$_2$ ternary diagram. The height of each column indicates the relative abundance of minerals with chemical compositions lying within the base of the column. Only minerals in which Fe$_2$O$_3$, Al$_2$O$_3$ and SiO$_2$ sum to over 80wt% of the chemical analysis have been included.

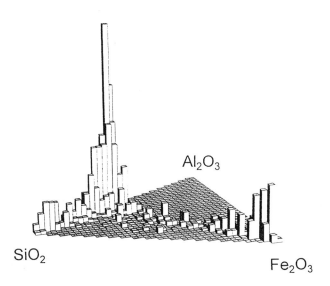

Figure 4. Chemical distribution of ash particles from Silverdale coal, plotted on the Fe$_2$O$_3$-Al$_2$O$_3$-SiO$_2$ ternary diagram. The height of each column indicates the relative abundance of particles with chemical compositions lying within the base of the column. Only particles in which Fe$_2$O$_3$, Al$_2$O$_3$ and SiO$_2$ sum to over 80wt% of the chemical analysis have been included.

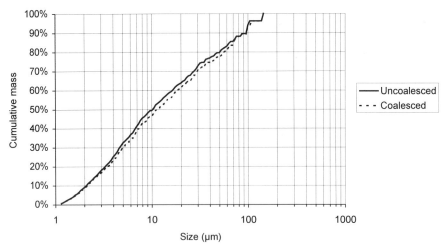

Figure 5. Cumulative size distributions for uncoalesced and coalesced minerals in Daw Mill coal.

Table 1. Types of mineral-mineral association in Silverdale coal, determined by CCSEM and listed in decreasing order of abundance. (Alsil ≈ clay, Si-rich ≈ quartz, Fe-rich ≈ pyrite, etc.)

Association	Abundance (%)
1 Alsil	37.1
2 Alsil + other	14.0
1 Fe-rich	12.8
2 Alsil	7.8
1 Si-rich	3.8
1 Fe-alsil	3.3
3 Alsil	3.0
1 non-CFAS	2.4
Alsil + Fe-rich	2.2
1 Ca-rich	2.1
2 Fe-rich	1.2
Alsil + non-CFAS	1.0
Alsil + Fe-alsil	1.0
Others	8.1

MINERAL-MINERAL ASSOCIATIONS

Because of the important role that mineral coalescence plays in the ash generation processes, a CCSEM analytical procedure was developed and used to characterise the mineral-mineral associations within pulverised coal particles for three UK power station coals. This analytical procedure, and the results obtained, are described in the following sections.

Analytical Procedure

Coal samples were mounted in an iodoform-doped epoxy resin, and prepared as polished cross-sections perpendicular to the settling direction. Each cross-section was analysed at multiple randomly-scattered fields of view and at two magnifications. For each field of view and at each magnification, the following automated procedure was performed:

1. Acquire a high-quality digital back-scattered electron (BSE) image.

2. Identify the mineral occurrences by locating clusters of pixels within the BSE image that have intensities corresponding to mineral matter.

3. Determine the size and shape of the mineral occurrence cross-sections from the BSE image.

4. Scan the electron beam over the mineral occurrence cross-sections to determine the chemical compositions.

5. Identify the coal particles by combining pixels with intensities corresponding to mineral matter and pixels with intensities corresponding to macerals.

6. Determine the size and shape of the coal particle cross-sections from the BSE image.

This analytical procedure produces two sets of data, one describing the size, shape and chemistry of individual mineral occurrences, and the other describing the size and shape of individual coal particle cross-sections. Since both sets of data were derived from the same BSE image, mineral occurrences can be re-united with the coal particle that contains them using their location on the image. This step has been automated using a relational database package. Although small errors do occur during this step they have a negligible effect on the final analysis. Once the links between mineral occurrence and coal particles have been established, the database package can also be used to analyse and interpret the mineral-mineral associations for coal particles that contain multiple mineral occurrences.

This procedure has the deficiency that mineral inclusions under 2μm in size are not identified in pf particles analysed at the lower magnification. However, observation has shown that almost all of the mineral occurrences under 2μm in size are excluded in the coals analysed.

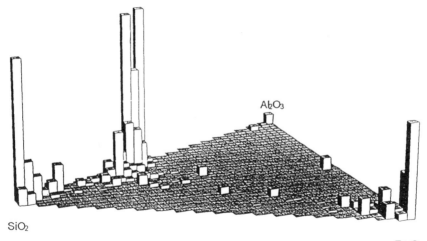

Figure 6. Chemical distribution of uncoalesced mineral occurrences in Daw Mill coal, plotted on the Fe_2O_3-Al_2O_3-SiO_2 ternary diagram. The height of each column indicates the relative abundance of minerals with chemical compositions lying within the base of the column. Only minerals in which Fe_2O_3, Al_2O_3 and SiO_2 sum to over 80wt% of the chemical analysis have been included.

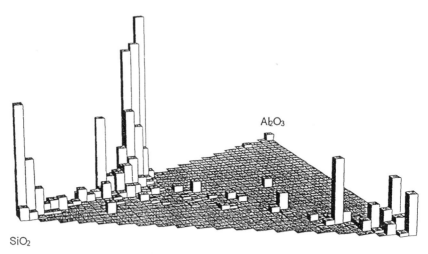

Figure 7. Chemical distribution of coalesced mineral occurrences in Daw Mill coal, plotted on the Fe_2O_3-Al_2O_3-SiO_2 ternary diagram. The height of each column indicates the relative abundance of minerals with chemical compositions lying within the base of the column. Only minerals in which Fe_2O_3, Al_2O_3 and SiO_2 sum to over 80wt% of the chemical analysis have been included.

586

Results

To illustrate the results obtained from the above procedure, the mineral associations found in a UK power station coal are listed in Table 1 in descending order of abundance. Single mineral occurrences within a coal particle are common, with clays, pyrite and quartz being most abundant. Clay minerals are ubiquitous amongst the mineral-mineral associations, with all significant multiple mineral associations containing at least one clay mineral occurrence. On a number basis about one third of the mineral matter is present as multiple mineral occurrences within individual coal particles, although this proportion is significantly lower when calculated on a mass basis.

DISCUSSION

A comparison between coal mineral and ash particle size and chemical distributions has indicated that mineral coalescence is the most important process during ash generation for UK power station coals. Information on mineral-mineral associations is available from the CCSEM analytical procedure described above. Since the mineral-mineral associations of individual mineral occurrences have been characterised, the effects of complete coalescence on the CCSEM data can be studied by combining mineral occurrences if they occur within a single coal particle.

Predicted Effects of Coalescence

The predicted effect of coalescence on size distribution is illustrated in Figure 5 for one of the coals analysed. Although there is an increase in median size, the coarsening in size during the transformation of coal minerals to ash particles has not been reproduced. The predicted change in chemical distribution with coalescence can be seen by comparing one representation of the uncoalesced chemical distribution (Figure 6) with the coalesced chemical distribution (Figure 7). No significant change is apparent in this graphical representation, but tabular data indicates that there has been a slight increase in the proportion of aluminosilicate-containing compositions.

Clearly, for both the size distribution and the chemical distribution, the predicted effects of coalescence have the correct sense but are much too small when they are based on the CCSEM characterisation of mineral-mineral associations. The reason for the underestimation of the degree of mineral-mineral association is comparatively simple. If a coal particle contains two mineral occurrences, then the probability that a cross-section through the particle contains both mineral occurrences is rather low. Cross-sections containing only one of the two mineral occurrences are much more likely.

Numerical Simulation

Two-dimensional numerical simulations to calculate the magnitude of this effect have been carried out. Two mineral occurrences (A and B) were represented by circles of equal diameter d, and a large number of lines with uniformly random position and orientation were used to calculate the probability of a cross-section intersecting one or both of the mineral occurrences. Simulations were calculated with the minerals touching, and separated by distances of d and $2d$. The calculated probabilities of intersecting mineral

Table 2. Probability of mineral associations occurring in a random cross-section, for different mineral separations.

Separation	A only	B only	A and B
0	0.39	0.39	0.22
d	0.46	0.46	0.09
$2d$	0.47	0.47	0.06

A, mineral B or both minerals A and B are shown in Table 2. It can be seen that there is only a small chance of both mineral occurrences being present in the cross-section, especially if the mineral occurrences are slightly separated. The mineral size and shape will affect the probability of a cross-section intersecting both mineral occurrences, and some geometries will be more favourable than others. For planar minerals, such as clays, the probability of intersection will be higher than for the circular minerals used in this two-dimensional calculation. However, the extra degree of freedom present in the real-life three-dimensional situation will cause a further significant reduction in the probability of a cross-section intersecting two mineral occurrences. The probability of a cross-section intersecting three or more mineral occurrences within a single coal particle is even less likely than for two minerals.

Limitations of CCSEM Data

The numerical simulation described above strongly indicates that, for coal particles containing two mineral occurrences, CCSEM analysis of a cross-section will only characterise the mineral-mineral association correctly for one particle in every ten or twenty. Although the type of mineral-mineral association will be determined correctly, the degree of mineral-mineral association will be strongly underestimated. Where a coal contains several different mineral-mineral associations, the relative proportions of these associations should be correctly determined unless there are strong differences in coal particle and mineral geometry, but the magnitude of all the associations will still be strongly underestimated. This explains why the predicted effects of mineral coalescence, based on CCSEM characterisation of mineral-mineral associations, have the correct sense but are too small in magnitude.

Further Developments

The model for ash generation, as it currently stands, is not successful in predicting the size and chemical distribution of ash particles. However, the CCSEM data does clearly identify which mineral-mineral associations are present in coal particles, and this information could be used to improve on existing models of ash generation by removing the assumption of randomness from, for example, 'mineral redistribution' procedures. Another possible way forward would be to use the CCSEM information to identify which mineral-mineral associations are present in a coal, but to assume a much higher degree of association when calculating the effects of mineral coalescence.

CONCLUSIONS

The coalescence of minerals during combustion is a dominant process during the transformation of coal minerals to ash particles. Mineral coalescence produces a coarsening of the particle size distribution and an increased proportion of aluminosilicate-containing particles in ash samples. A knowledge of the mineral-mineral associations in coal particles can be used to predict the effects of coalescence.

Although the types of mineral-mineral associations present in pulverised coal particles can be successfully characterised by computer-controlled scanning electron microscopy, the degree of mineral-mineral association is strongly underestimated in the current work. This limitation is inherent in the analysis of coal samples presented as a cross-section, and results from the possibility that a cross-section through a coal particle containing multiple mineral inclusions will not intersect all of the inclusions.

Although CCSEM information on mineral-mineral associations may be insufficient to correctly model the effects of mineral coalescence, there is scope for using this information to remove the assumption of randomness from ash generation models

ACKNOWLEDGMENTS

This work was funded by the UK Department of Trade and Industry, as part of a collaborative research programme entitled 'Minimising the effect of high temperature coal ash deposition'. Samples for analysis, and discussion of the results, were provided by the other members of the collaborative programme.

REFERENCES

Gibb, W.H. (1995). "The UK collaborative research programme on slagging in pulverised coal-fired boilers: Summary of findings", this conference.

Harb, J.N., Richards, G.H., and Baxter, L.L. (1994). "Investigation of mechanisms for the formation of fly ash and ash deposits from two low-sulfur Western coals", AIChE national meeting, San Francisco, California, November 1994.

Lee, F.C.C., Riley, G.S. and Lockwood, F.C. (1995). "Prediction of ash deposition in pulverised coal combustion systems", this conference.

Wigley, F., and Williamson, J. (1994a). "The characterisation of fly ash samples and their relationship to the coals and deposits from UK boiler trials", in *Proceedings of the 1993 Engineering Foundation conference on The Impact of Ash Deposition in Coal-Fired Plants*, J. Williamson and F. Wigley (Eds), Taylor and Francis, Washington, pp 385-398.

Wigley, F., and Williamson, J. (1994b). "The UK collaborative programme on slagging in pulverised coal furnaces: Sample characterisation", in *Proceedings of the 1994 International EPRI Conference on The Effects of Coal Quality on Power Plants*, Mehta, A.K. (Ed), Electric Power Research Institute, USA.

Yu, H., Marchek, J.E., Adair, N.L. and Harb, J.N. (1994). "Characterization of minerals and coal/mineral association in pulverised coal", in *Proceedings of the 1993 Engineering Foundation conference on The Impact of Ash Deposition in Coal-Fired Plants*, J. Williamson and F. Wigley (Eds), Taylor and Francis, Washington, pp 361-372.

MODELLING ASH DEPOSITION DURING THE COMBUSTION OF LOW GRADE FUELS

Jorma K. Jokiniemi[1], Jouni Pyykönen[1], Jussi Lyyränen[1],
Pirita Mikkanen[2] and Esko I. Kauppinen[2]

VTT Aerosol Technology Group
P.O.Box 1401, FIN-02044 VTT, Finland
[1]VTT Energy, [2]VTT Chemical Technology

ABSTRACT

Deposit formation on external surfaces, such as heat exchangers, is a serious problem in many combustion processes. Deposits form from residue ash particles and from volatilized species that deposit either as tiny condensed particles or as vapour. With modelling it is possible to make predictions about the rate and the chemical composition of deposition resulting from a specified fuel fired in different process conditions. This paper discusses the modelling of different deposit formation mechanisms that are used in the ABC (Aerosol Behaviour in Combustion) code (Jokiniemi et al., 1994). When modelling deposition it is essential to know the ash particle size distribution and chemical composition and well as the composition of the flue gases. The emphasis in the ABC code and in this paper is in the deposition of species that have volatilized during combustion. The chemical composition of these species is often such that in deposits they make surfaces sticky for residue ash particles or their presence may lead to the sintering of the deposited layer or their presence can cause high temperature corrosion. The ABC model couples the description of deposition mechanisms with the description of ash particle formation from volatilized species and the respective gas phase chemistry. The amount of residue ash particles is given in input for the ABC code. Detailed descriptions of the deposition models applicable to the process conditions are given in this paper. As example cases we present model results of deposition in a kraft recovery boiler and in a medium speed diesel engine.

INTRODUCTION

Slagging and fouling can be a serious problem when using low grade fuels. Often the cause is the species which have volatilized during combustion and later condensed and formed tiny aerosol particles that deposit on flow channel surfaces. Alternatively, the volatilized species may deposit as vapour either via direct condensation or due to

Applications of Advanced Technology to Ash-Related Problems in Boilers
Edited by L. Baxter and R. DeSollar, Plenum Press, New York, 1996

591

chemical reactions. Alkali species are specially harmful, as they lower the melting points of deposit layers and make surfaces sticky for deposition of larger residue ash particles. If the deposited layer gets sintered, it will be hard to remove. In some cases heavy deposition may lead to the blockage of the flow channels. Deposit formation on heat transfer surfaces also reduces the effectiveness of heat transfer. Alkali species are very important in biomass combustion, but they also have an effect on deposit formation in coal combustion or in the combustion of heavy fuel oil with a high ash content. Whether alkali species are likely to cause fouling problems in coal combustion, depends not only on their concentration in coal, but also on the way they are bound to the coal matrix. The presence of elements, such as chlorine, vanadium or nickel, in fuels can be important when considering the corrosive effect of the deposits.

Studying aerosol dynamics both experimentally and with a modelling approach, the focus of this paper, helps understand the processes which lead to deposit formation. As particles of different sizes can have different average chemical compositions and different tendencies to deposit, models of different deposition mechanisms should be used together with models of particle size spectrum formation. The coupling makes it possible to study the effects of fuel composition and combustion conditions on deposit formation. Vapour phase species release, gas phase chemistry and ash particle formation have to be taken into account when deposition is modelled. In addition, flow fields, temperature distributions and heat transfer phenomena must be considered. In our aerosol dynamics computer model ABC (Aerosol Behaviour in Combustion) (Jokiniemi et al., 1994) particle and vapour deposition models are coupled with models of particle formation from volatilized species and models of gas phase species chemistry. The amount of residue ash particles is given in input. A plug flow model is used to describe the main flow and a separate model is used for the description of the flow channel surface boundary layers.

The basic idea in particle and vapour deposition modelling in ABC is to compute the structure of the boundary layer on the basis of the flow rate and the dimensions of the flow channel in a way similar to that of Im and Chung (1983). The flow can be laminar or turbulent. In the case of turbulent flow the flow field in the boundary layer is divided into three regions: (I) the fully turbulent region, (II) the buffer layer and (III) the laminar sublayer (e.g. Schlichting, 1951) (Fig. 1).

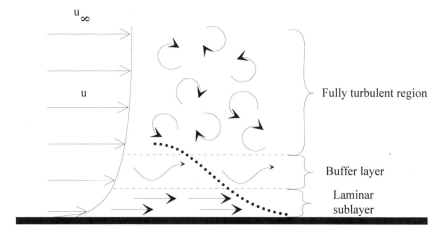

Figure 1. Schematic figure of the flow regions near the surface parallel to flow.

The particle concentration in the fully turbulent region is considered constant as a function of distance from the wall. In the buffer layer the particle flux towards the wall surface is determined by particle turbulent diffusion (Hinze, 1975). In the laminar sublayer turbulent diffusion is assumed to decay in a cubic fashion. Submicron particles do not usually have a large enough momentum to penetrate through this layer. The additional mechanisms for submicron particle deposition over the laminar sublayer are thermophoresis and Brownian diffusion. Supermicron particles, in contrast, can gain enough momentum to deposit by turbulent eddy impaction (flow is parallel with the wall surface) or inertial impaction (flow across an obstacle). Also deposition by gravitational settling is possible for these particles.

Vapour condensation on surfaces is calculated using mass transfer equations. Vapour deposition due to chemical reactions with surfaces is considered to be a combination of mass transfer from the gas to the surface, surface reaction kinetics and diffusion through the reacted layer.

Deposition in combustion processes has been modelled for example by Baxter (1993), Beér et al. (1992) Helble et al. (1992) and Lee et al. (1993). They, however, do not model in detail the deposition of a large range of particle sizes from a turbulent flow and they do not include a model for predicting partitioning of condensable species between fine particles and vapours. The novelty in this work is the application of detailed mechanistic deposition models presented in literature to combustion processes. Also a novelty is the link between the models of fly ash formation in volatilisation-condensation processes together with deposition models.

As example cases we have studied deposition in a kraft recovery boiler and in a medium speed diesel engine used in marine and land based power generation applications. The recovery boiler results presented here are a summary of deposition results presented in Jokiniemi et al. (1995). As compared with the existing modelling capability concerning slagging and fouling, the ABC code improves our ability to predict deposit formation, growth and elemental composition, especially as far as the deposition of volatilized species is concerned. It gives us the possibility to analyse the dependence of deposition formation on temperatures, elemental volatilisation and flow conditions.

AEROSOL PARTICLE FORMATION MECHANISMS

Aerosol particles are small liquid or solid particles suspended in air. Generally 100 μm is considered to be the maximum size of an aerosol particle. Particles smaller than 20 μm follow the fluid flow very closely. In combustion processes aerosol particles i.e. fly ash particles are formed either from volatilized species that condense to form new particles, from residue mineral inclusions or from unburned carbonaceous matter. We have concentrated in our aerosol particle formation modelling on new particle formation by the condensation of the volatilized species. So far we have not modelled the volatilisation process. Instead we have taken the extents of releases from measurements and started the simulation with the given amount of the volatilized species in the gas phase. For coal combustion there are models, such as the slagging advisor (Helble et al., 1992), that predict the size distribution and the chemical composition of ash particles resulting from mineral inclusions in coal. In our simulations we have just taken the amount of residue particles from measurements and included these particles in the model input and thus in the models of deposit layer formation.

To model the formation of the size distribution spectrum from volatilized species we have to consider the gas phase reactions of condensable species, homogeneous and heterogeneous condensation of these species to form and to increase the size of aerosols particles and agglomeration of the particles by different mechanisms. In the following we briefly describe these phenomena.

Homogeneous Condensation (nucleation)

In the combustion chamber volatilized vapours can become supersaturated due to flue gas cooling or due to chemical reactions. Saturation ratio of a certain vapour is defined as the ratio of its partial pressure to its equilibrium pressure. If the saturation ratio is larger than one, the vapour is supersaturated. For example, if gaseous NaOH reacts with SO_2, gaseous sodium sulphate is formed. Because of the very low equilibrium vapour pressure of $Na_2SO_4(g)$, the vapour becomes supersaturated immediately after formation and it starts to condense. Condensation decreases the partial vapour pressure toward its equilibrium value. If the flue gas does not contain enough particles to provide surfaces for the condensation vapour molecules stick together and form new aerosol particles. This process is called homogeneous nucleation and it requires a critical supersaturation, which is much larger than one. When this critical supersaturation level is reached, tiny aerosol particles are formed at a rate, which can be predicted with the use of thermodynamics and kinetic considerations (e.g. Friedlander, 1977).

Heterogeneous Condensation

If the gas contains particles, such as metal oxide seeds when supersaturation is reached, the vapour starts to condense on the surface of these particles before any new particles can be formed by homogeneous nucleation. This is due to the fact that condensation on surfaces starts at lower supersaturation ratios than homogeneous nucleation. The growth rate of the particles can be solved directly from the heat and mass transfer equations for single particles with the use of numerical methods. In coal combustion it has been observed that tiny metal oxide seed particles are produced even if only a minor fraction of the metals in coal is volatilized (e.g. Helble et al., 1986). In biomass combustion the importance of seeds is not known so well.

Agglomeration

Agglomeration is a process in which particles collide with each other and stick together. Collision rate is determined by Brownian and turbulent diffusion. When submicron particles collide, they practically always stick together. Particle size, chemical composition and process conditions determine the properties of these agglomerated particles. If the colliding particles are liquid they form a new spherical liquid droplet and in the other extreme the colliding solid particles stay together by Wan der Vaals attraction. In reality agglomerates formed in combustion processes tend to sinter together and form particles with a complex morphology.

Modelling aerosol dynamics in the ABC code

In the ABC code we simulate a combustion process using elemental volatilised fractions and a possible initial seed particle size distribution as input data. The change in the particle size and chemical composition spectrum is due to the mechanisms of chemical reactions, homogeneous nucleation, vapour condensation, agglomeration and deposition. The whole process is described by the General Dynamic Equation (GDE) (e.g. Friedlander, 1977). In combustion processes steady-state conditions can be assumed. Thus we solve the aerosol GDE in one-dimensional stationary form along the flow direction. Mass size distributions of different species are calculated by solving the condensed phase species GDEs, where the particle size spectrum is divided into a number of grid points:

$$\frac{d(m_{jk})}{dx} = \left[\frac{d(m_{jk})}{dx}\right]_{gtp} + \left[\frac{d(m_{jk})}{dx}\right]_{agg} - \frac{V_{dk} A_d}{u \Delta V} m_{jk} \tag{1}$$

Here m_{jk} is the mass concentration of species j at the kth grid point corresponding to the diameter $D_{p\,k}$. The first term on the right (gtp) corresponds to particle formation due to homogeneous nucleation and growth by condensation and chemical reactions on aerosol particle surfaces. The second term (agg) describes agglomeration and the third term the rate of particle removal due to deposition on boundary surfaces. V_{dk} is the particle deposition velocity, A_d is the deposition area and ΔV is the axial volume step. Particle velocity (u) is calculated by taking into account gas velocity, Stokes drag and gravitation.

The rate of change in the molar concentration of a gas phase species is given by the following gas phase species equation:

$$\frac{dc_j}{dx} = \left(\frac{dc_j}{dx}\right)_{form} - \left(\frac{dc_j}{dx}\right)_{gtp} - \frac{V_v A_d}{u_g \Delta V} c_j \tag{2}$$

where c_j is the gas phase molar concentration of species j. The first term (form) on the right represents the formation rate of these species, which can be calculated from local gas phase chemical equilibrium or from reaction kinetics. The second term (gtp) describes the vapour depletion rate by nucleation, condensation and chemical reactions on aerosol particle surfaces and the third term represents the depletion rate due to vapour deposition by condensation and chemical reactions on structures. V_v is the vapour deposition velocity and A_d is the deposition area. Time is related to the axial position through the gas velocity (u_g). A more detailed description of the ABC model can be found in Jokiniemi et al. (1994).

PARTICLE DEPOSITION MODEL

Deposition of particles on surfaces parallel to the flow is modelled on the basis of the boundary layer theory. Inertial deposition on heat transfer tubes perpendicular to the flow direction and deposition due to gravity are considered separately. For submicron particles deposition on heat transfer tubes perpendicular to the flow direction can be calculated in an approximate way with the parallel flow model.

All particles reaching the surface do not necessarily stick. The sticking efficiency is a function of particle size, density and viscosity. At present there are no general models for re-entrainment and therefore sticking efficiencies must be determined experimentally for specific applications. So far we have just assumed that no re-entrainment takes place, and neither have we modelled the effects of sootblowing.

Boundary Layer Theory for Particle Transport

In the boundary layer analysis the flow field in the channel is divided into three regions: (I) the fully turbulent region, (II) the buffer layer and (III) the laminar sublayer (Fig. 2).

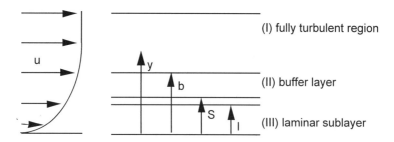

Figure 2. Schematic figure of the flow regions near the wall.

The particle concentration $n(x,y)$ in the fully turbulent region is considered to be constant as a function of distance from the wall. In the buffer layer the particle flux towards the wall surface is determined by particle turbulent diffusion:

$$J_p(y) = v_f(y)n(x,y) = D_t(y)(\partial n/\partial y) \qquad \text{when } b \geq y > l \qquad (3)$$

where D_t is the particle turbulent diffusivity and v_f is the effective particle turbulent diffusion velocity in the buffer layer.

In the laminar sublayer turbulent diffusion is assumed to decay in a cubic fashion. Here we treat sub- and supermicron particles separately, since supermicron particles can gain enough momentum from turbulent fluctuations to pass through the laminar sublayer (turbulent eddy impaction), whereas the submicron particles need additional mechanisms to complete deposition. These mechanisms are thermophoresis and Brownian diffusion. Actually, the terms sub- and supermicron are not quite accurate, but we use them here for brevity. Here a particle is called supermicron, if its stopping distance (S) is larger than the thickness of the laminar sublayer (l).

For submicron particles the total effective deposition velocity V_d is obtained from a series combination of the turbulent diffusion velocity v_f on the edge of the laminar sublayer (l) and the combined thermophoresis and Brownian diffusion velocity through the laminar sublayer v_{ThB}.

$$V_d = \left(\frac{1}{v_f(l)} + \frac{1}{v_{ThB}} \right)^{-1}$$

(4)

For supermicron particles we have the total effective deposition velocity as a series combination of the turbulent diffusion velocity v_f at the particle stopping distance and the velocity v_a supplied to the particle by turbulent fluctuations.

$$V_d = \left(\frac{1}{v_f(S)} + \frac{1}{v_a} \right)^{-1}$$

(5)

Mean velocities and lengths can be expressed as dimensionless parameters by defining a friction velocity u_*, so that (e.g. Davies, 1966):

$$u_* = \left(\frac{\tau_w}{\rho_g} \right)^{1/2}$$

(6)

where τ_w is the shear stress at the wall and ρ_g is the gas density. The following expression, where nondimensional parameters are used, can be found in Im and Chung (1983) for the effective particle turbulent diffusion velocity in the buffer layer:

$$v_f(y) = \left(\int_y^b \frac{dy}{D_t} \right)^{-1} = 0.4 \frac{u_*}{Sc_p} \left\{ 0.9 \ln\left(\frac{b^+ + k_s^+}{y^+ + k_s^+} \right) + \frac{13}{y^+ + k_s^+} + \frac{365}{\left(y^+ + k_s^+ \right)^2} \right.$$

$$\left. + \frac{13}{b^+ + k_s^+} + \frac{365}{\left(b^+ + k_s^+ \right)^2} \right\}^{-1} \quad when \quad l \le y < b$$

(7)

Sc_p is the particle Schmidt number, b^+ is the dimensionless distance from the wall, y^+ is the dimensionless distance from the wall and k_s^+ is the dimensionless surface roughness parameter. The superscript $^+$ denotes nondimensionalization with respect to the characteristic length of the ratio of kinematic viscosity to friction velocity v/u_* (i.e. $y^+ = y(u_*/v)$).

Thermophoresis and Brownian Diffusion

Thermophoresis, Brownian diffusion and rapidly decaying turbulence in the laminar sublayer have been considered simultaneously by Im and Chung (1983). They have integrated their effects through the sublayer and obtained the following expression for the effective deposition velocity through the laminar sublayer v_{ThB}:

$$v_{ThB} = \frac{v_{th}}{\exp(\frac{gv_{th}}{u_*})} - 1 \quad , \text{where} \tag{8}$$

$$g = 14.5 Sc_p^{2/3} \left\{ \frac{1}{6} \ln\left[\frac{(1+\psi)^3}{1+\psi^3} \right] + \frac{1}{\sqrt{3}} \tan^{-1}\left(\frac{2\psi-1}{\sqrt{3}} \right) + \frac{\pi}{6\sqrt{3}} \right\} \tag{9}$$

and

$$\psi = \frac{l^+}{14.5} Sc_p^{1/3} \tag{10}$$

v_{th} is the thermophoretic velocity. For the laminar sublayer the thermophoretic velocity in the free molecule region v_{th}^* is given by the following well-accepted formula:

$$v_{th}^* = -\frac{4}{45} CnD_p \frac{k_g / \mu}{(2\pi R_g T)^{1/2}} \frac{\partial T}{\partial y}\Big|_{wall} \tag{11}$$

where Cn is the Cunningham correction factor, D_p is the particle diameter, k_g is the thermal conductivity of gas, μ its viscosity, T its temperature and R_g is the universal gas constant.

Springer (1970) has proposed the following equation over the entire particle size spectrum:

$$v_{th} = \frac{v_{th}^*}{1 + f(Kn)} \tag{12}$$

where the interpolation formula f as a function of the particle Knudsen number Kn is given by:

$$f(Kn) = \left(\frac{2}{9Kn} \right) \left\{ \frac{1 + 2\alpha + Kn[2C_t + 3C_m(1 + 2\alpha + 2C_t Kn)]}{\alpha + C_t Kn + 3.2 C_m Kn(\alpha + C_t Kn - 1)} \right\} \tag{13}$$

where α is the ratio of gas to particle thermal conductivity, C_t and C_m are constants associated with the slip and temperature jump coefficients, respectively. The recommended values for them are $C_t =3.32$ and $C_m =1.0$.

An alternative formula for the thermophoretic velocity has been proposed by Talbot et al. (1980):

$$v_{th} = -\frac{Kv}{T}\frac{\partial T}{\partial y}\bigg|_{wall} \quad \text{,where} \tag{14}$$

$$K = 2C_s \frac{(\alpha + C_t Kn)Cn}{(1+3C_m Kn)(1+2\alpha +2C_t Kn)} \tag{15}$$

and

$C_s = 1.147, C_t = 2.20, C_m = 1.146$

Cn is the Cunningham slip correction factor. The Talbot formula is generally recommended. Both the Springer and the Talbot formulas are available in ABC.

Turbulent Eddy Impaction

Turbulent eddy impaction has received much interest because of its importance in many industrial processes. Experiments for pipes with smooth and rough surfaces have been carried out. Analytical expressions and numerical simulations for particle trajectories are available for modelling. In the following we describe a stopping distance model (originally proposed by Friedlander and Johnstone, 1957 and later improved by Im and Chung, 1983) and a correlation model derived from particle trajectory analysis in near-wall turbulent eddies (Fan and Ahmadi, 1993).

Stopping Distance Model

When a momentum $m_p v_a$ is imparted to a particle in a quiescent fluid, the stopping distance for the particle is (Friedlander and Johnstone, 1957):

$$S = \tau_p v_a, \tag{16}$$

where τ_p is the particle relaxation time.

The effective turbulent diffusion velocity of particles in the laminar sublayer is negligible. Thus only those particles that have a large enough velocity (v_a) to travel through the laminar boundary layer may reach the surface. This velocity has been determined by Hinze (1975) to be:

$$v_a = \left[v_r' \left(\frac{\tau_f}{\tau_f + \tau_p} \right)^{1/2} \right]_l \qquad (17)$$

where τ_f denotes the turbulence time scale.

Because the stopping distance (S) is usually close to the laminar sublayer thickness (l), τ_p and τ_f are evaluated at l for convenience. v_r' is approximated to be $0.8u_*$ (Hinze, 1975).

Correlations based on numerical simulations

The detailed analyses of particle trajectories (Fan and Ahmadi, 1993) have shown that the near-wall vortices of turbulence play a decisive role in the deposition processes. Fan and Ahmadi have analysed particle deposition from turbulent flows in vertical channels by looking at particle trajectories in the near wall region. Based on their numerical results Fan and Ahmadi (1993) have proposed the following correlation for the dimensionless deposition velocity V_d^+ derived with some approximations:

$$V_d^+ = \begin{cases} 0.084Sc_p^{-2/3} + 0.5 \left[\dfrac{\left[\left(0.64k^+ + \dfrac{D_p^+}{2} \right)^2 + \dfrac{\tau_p^{+2} g^+ L_1^+}{0.01085\left(1+\tau_p^{+2} L_1^+\right)} \right]^{1/\left(1+\tau_p^{+2} L_1^+\right)}}{3.42 + \dfrac{\tau_p^{+2} g^+ L_1^+}{0.01085\left(1+\tau_p^{+2} L_1^+\right)}} \\ \qquad \times \left[1 + 8\exp^{-\left(\tau_p^+ - 10\right)^2/32} \right] \dfrac{0.037}{1 - \tau_p^{+2} L_1^+ \left(1 + \dfrac{g^+}{0.037}\right)} \qquad \text{if} \quad V_d^+ < 0.14, \\ \\ 0.14 \qquad\qquad\qquad\qquad\qquad\qquad\qquad\qquad\qquad\qquad \text{othervise} \end{cases}$$

$$(18)$$

where the nondimensional Saffman lift force coefficient L_1^+ is computed as follows:

$$L_1^+ = 3.08\rho_g / (\rho_p D_p^+) \qquad (19)$$

Sc_p is the particle Schmidt number, k^+ is the dimensionless surface roughness, D_p^+ is the dimensionless particle diameter, τ_p^+ is the nondimensional particle relaxation time g^+ is the nondimensional acceleration of gravity, ρ_g and ρ_p are the densities of the gas and the particle, respectively and D_p^+ is the dimensionless particle diameter. Equation (18) is strictly valid only for relatively small surface roughnesses (about $k^+ < 1.5$).

Inertial Impaction

A detailed model of inertial deposition would require the simultaneous solution of the flow field, particle trajectories and the stickiness. In the ABC aerosol code this is not possible, because of the simplified flow field assumed. Thus we look for an approximate analytical solution for this problem. Langmuir's and Blodgett's (1948) analysis gave the following impaction efficiency η as a function of particle Stokes number Stk for spherical particles across a cylinder:

$$\eta = \left(\frac{Stk}{Stk + 0.06} \right)^2 \qquad Stk > 0.08 \qquad (20)$$

$$\eta = 0 \qquad Stk \leq 0.08$$

The effective deposition velocity V_d is now:

$$V_d = \frac{\eta u}{\pi} \qquad (21)$$

Alternatively, a more accurate model from Israel and Rosner (1983) could be used.

Gravitational Deposition

Usually gravitational deposition on horizontal surfaces is calculated from the terminal settling velocity V_{TS}:

$$V_{TS} = g\tau_p \qquad (22)$$

which is then simply added to the total deposition velocity. This, however, overestimates the deposition velocity of submicron particles, because they easily get caught in turbulent fluctuations. Thus, gravitational deposition should not be taken into account for particles smaller than 10 μm in turbulent flows. For supermicron particles an accurate gravitational deposition model should be included in the eddy impaction model, but such a model does not exist. So, in ABC we have simply added the gravitational deposition to the total deposition velocity of supermicron particles towards horizontal surfaces.

VAPOUR DEPOSITION ON HEAT EXCHANGE SURFACES

Vapours may deposit by direct condensation or due to chemical reactions with surfaces. We calculate the condensing vapour flux I_v with the following expression:

$$I_v = Sh(T_g) \frac{\left[D_v(T_g) D_v(T_w) \right]^{1/2}}{D_h R_g} \left(\frac{p_v(T_g)}{T_g} - \frac{p_{v,s}(T_w)}{T_w} \right) \tag{23}$$

where D_h is the hydraulic diameter of the flow channel, Sh is the Sherwood number, R_g is the gas constant, p_v is the actual vapour pressure and $p_{v,s}$ is the saturation vapour pressure. The vapour diffusivity D_v is expressed as the root mean square of wall and gas temperatures.

In the ABC code we also consider chemisorption on surfaces. The model incorporates together diffusion through the boundary gas film, diffusion through the reacted ash layer and chemical reactions. The change in vapour mass concentration due to chemical reactions with flat surfaces can be written as (e.g. Levenspiel, 1972):

$$\frac{dC_v}{dt} = \frac{C_v}{\dfrac{1}{k_g} + \dfrac{1}{k_r} + \dfrac{1}{k_s}} \tag{24}$$

where C_v is the vapour mass concentration in the gas phase, k_g is the mass transfer coefficient for the gas boundary layer, k_r is the first order rate constant for surface reaction and k_s is the mass transfer coefficient for diffusion through the reacted layer.

RECOVERY BOILER CALCULATIONS

Aerosol formation in recovery boilers

Black liquor is a waste sludge from the pulping process, which is burned in a kraft recovery boiler for the recovery of the pulping chemicals and for energy production. Black liquor does not contain mineral inclusions, but instead a large amount of sodium, potassium, sulphur and chlorine, which partially volatilise and condense to form small aerosol particles commonly referred to as fume particles. At high temperatures (1400-1600 °C) all volatilized Na and K convert rapidly to hydroxides and chlorides in the gas phase. As the temperature decreases, these convert to sulphates by reactions with SO_2 and condense. When there is not enough SO_2 to bind all Na and K, these species convert to carbonates by surface reactions. However, the surface reactions may also be kinetically limited. Some of the chlorides condense directly as the temperature decreases further. The fume formation mechanisms in the recovery boiler as well as ABC calculation results are discussed in more detail in Jokiniemi et al. (1995). Here we present a short summary of those results with the emphasis on results concerning deposition.

Input data for the calculations

The boiler geometry, the flue gas and the heat transfer surface temperatures at different locations, the amounts of elements volatilized from black liquor and the amounts of air injection were given as input to the model. The detailed fume particle characterisation studies that we have carried out at the selected boilers (Mikkanen et al., 1994 a,b,c and Mikkanen et al., 1995) offer a good set of validation data for the simulated results. The measurements were made before the electrostatic precipitator, when the sootblowing was turned off. We have modelled the situation prevailing in the boiler during these measurements.

We postulate that seed particles for the condensation of gaseous sodium and potassium sulphates are provided by nucleated metal oxides. Their input size distribution was given as a log-normal distribution with a number median diameter at 60 nm. We also add a mode of inert particles with an initial log-normal size distribution with a mass median diameter at 5 μm, and call them mechanically produced particles. The idea is to study their deposition behaviour as they are transported through the boiler. Currently we believe that these particles originate from re-entrainment of heat exchanger surface deposits.

Two cases with different black liquor compositions were calculated for the boiler under consideration (Table 1). In the base case we used the fuel elemental analysis of the hardwood black liquor (67% dry solids) fired during the measurements. In the low sulphur case we simulated aerosol dynamics in the same boiler with a liquor, which has a high chlorine content and a low sulphur content. Also the volatilized fractions were changed so that the Cl/S ratio in the gas phase becomes considerably high.

Table 1. Fuel composition and volatilized fractions (Jokiniemi et al., 1995).

Element	Base case		Low sulphur case	
	Dry solids content [wt-%]	Volatilized fraction [%]	Dry solids content [wt-%]	Volatilized fraction [%]
Na	19.9	10	20.6	10
K	2.6	10	2.7	10
Cl	0.18	5	1.3	50
S	6.4	35	2.6	15
C	30.0	97.5	30.8	97.5
N	0.15	100	0.15	100
O	37.0	100	39.2	100
H	3.0	100	3.3	100
Mech. prod.	0.78	-	0.80	-
Seeds	0.016	-	0.016	-

Calculation results

Particle mass size distributions and chemical compositions before the electrostatic precipitator are shown as differential mass size distributions in Fig. 3 for both of the considered cases. The total area in these graphs is proportional to the total normalised mass concentration at gas temperature 0 °C and each coloured area is proportional to the normalised mass concentration of the respective species. The simulated mass size

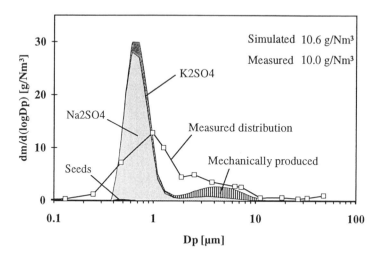

Figure 3a. Particle mass size distribution before the electrostatic precipitator in the base case calculation for a recovery boiler. The measured mass size distribution corresponding to the base case is also shown (Jokiniemi et al., 1995).

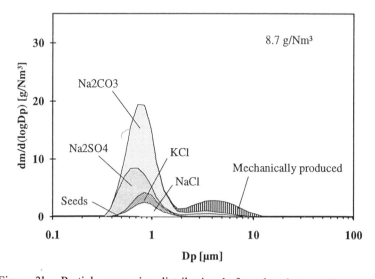

Figure 3b. Particle mass size distribution before the electrostatic precipitator in the low sulphur case (Jokiniemi et al., 1995).

distribution is in qualitative accordance with the measured distribution assuming spherical particles with a density of 2500 kg/m^3 i.e. the bulk density of Na_2SO_4. In the low sulphur case the mass size distribution peak is broader. Sodium sulphate, which is the first to condense, is enriched in smaller particles, while the chlorides, which are the last to condense, are enriched in larger fume particles.

Deposition layer growth rates are calculated assuming that all particles that deposit on external surfaces stick. For comparison with the measured results this is quite a good approximation, since the sootblowing was turned off during the measurements. In the bends, at least, re-entrainment can be significant even in these conditions. According to the calculated results most of the deposition can be attributed to thermophoresis. Turbulent impaction and inertial impaction are significant in the bends of the economiser section. Turbulent impaction has some significance in the economisers, too. We still need to do more work to validate the models used for bend impaction. The normalised particle concentration is decreased by more than half as the particles are transported through the boiler to the electrostatic precipitator. This is mainly due to deposition on the economiser surfaces. The effect of the bends is not very large in this respect because of the short flue gas residence times in them. Kraft recovery boilers usually have the biggest problems with deposition in the hot superheater section, where sticky deposits are hard to remove by sootblowing, and in the bends because large amounts of deposits in them can cause plugging of the flow channel. The geometries of the heat transfer surfaces and their effect on deposition were not considered in detail in these calculations.

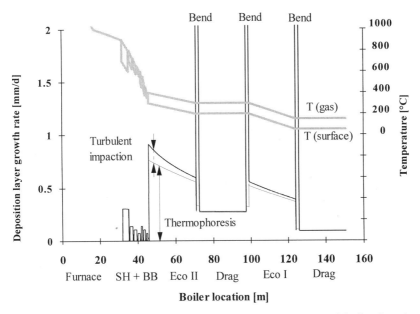

Figure 4. Deposition layer (fume) growth rate as a function of boiler location and different deposition mechanisms in the base case. Also gas and heat exchanger surface temperatures are shown (Jokiniemi et al., 1995).

The only qualitative difference between the base case and the low sulphur case is the importance of sodium and potassium chloride vapour deposition on superheater surfaces in the low sulphur case. Vapour deposition is more important than particle deposition in the first superheater (Fig. 5). Deposited chloride vapours also serve as a kind of a glue by decreasing the melting point of the deposited layer and making heat exchanger surfaces sticky for particle deposition even in the upper superheater section. Deposition velocities at different locations in the boiler have been computed on the basis of the flow conditions (Fig. 6). Deposition velocities in the bends seem to be high regardless of particle size. For deposition in economiser II there is a minimum for 3 μm particles. Particles smaller than 3 μm are mainly deposited by thermophoresis and particles larger than 3 μm by turbulent impaction. The steep rise in the deposition velocity above 10 μm explains why the mechanically produced particles with diameters larger than 10 μm seen in the mass size distribution before the injection of tertiary air are no longer there before the electrostatic precipitator. For the superheater II the model predicts a minimum deposition velocity for 10 μm particles.

The average elemental composition of the deposited fume particles is much the same as that of the gas phase fume particles. In addition to fume particle deposition, the composition of the deposited layer is affected by carryover deposition and vapour deposition, of which only the latter is currently considered in the model. Note the high fraction of chlorine and potassium in superheater II deposits due to vapour deposition in the low sulphur case (Table 2). The differences in composition, when only particle deposition occurs, arise from different deposition velocities for different particle sizes. For instance, the largest mechanically produced particles have the highest deposition rates and therefore they enrich in the deposition layers. In the low sulphur case the fraction of chlorine in the deposited layer in economiser II is 11% smaller than the chlorine content of the gas phase fume particles, while the fractions of sulphur are nearly the same. This is due to the fact that, compared with sulphur, chlorine is concentrated in larger fume particles or in other words in particle sizes nearer to the deposition rate minimum (Fig. 3b).

Table 2. Elemental composition (weight %) of fume deposits in superheater II (SH II) and economiser II (ECO II) and aerosol particle compositions at the same locations (Jokiniemi et al., 1995).

Ele-ment	Base case		Low sulphur case			
	Depo-sition SH II	Fume SH II	Depo-sition SH II	Fume SH II	Depo-sition ECO II	Fume ECO II
Na	23	26	21	34	30	32
K	4	3	20	0	3	3
Cl	0	0	30	4	8	9
S	17	19	3	7	6	6
C	0	0	2	6	5	5
O	35	39	15	36	31	32
Mech. Prod.	21	13	8	14	18	12

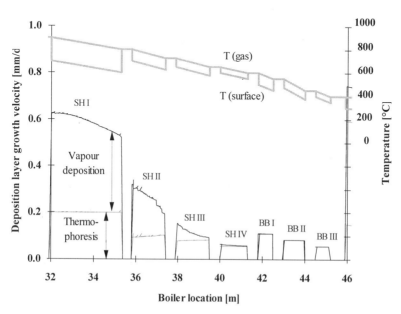

Figure 5. Fume deposition rate in the superheater (SH) and boiling bank (BB) section in the low sulphur case (Jokiniemi et al., 1995).

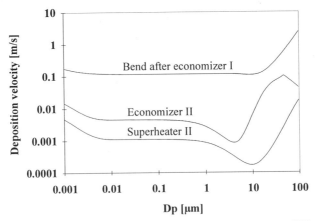

Figure 6. Deposition velocities as a function of particle size at different locations (Jokiniemi et al., 1995).

MEDIUM SPEED DIESEL ENGINE CALCULATIONS

Aerosol formation in medium speed diesel engines

Medium speed diesel engines are today frequently used in marine and land based power generation applications. Unlike gas turbines, diesel engines can successfully burn heavy fuel oil with a high ash content. In some extreme cases, especially in Central and South America, where the V and Ni content in the crude oil is very high, some engines, object to exceptional operating conditions, have experienced exhaust valve failures due to corrosive deposits on exhaust valves. At the moment the mechanisms that lead to deposit formation and subsequent burning of the valves are poorly understood. The understanding of how the V, Na and Ni contents in the fuel are related to the valve burning probability is not yet quite clear. To clarify this link, we have modelled with the ABC code how ash particles are formed in the diesel engine cylinder and how they deposit on the valve surfaces. Contrary to the typical vehicle diesel engines, very little stable soot is produced in power generation diesel engines because of a high air ratio during combustion and because of a higher compression ratio.

On the basis of chemical equilibrium calculations we have identified the reactions by which the ash particles are formed. We found that nickel oxide vapours are the first to condense. We assumed that the first ash particles are produced by homogeneous nucleation. In reality, volatilized aluminium and silicon and other such species that heavy fuel oil contains in small amounts can form particles, before nickel starts to condense. Other reactions are the formation of sodium and potassium vanadates on ash particle surfaces and direct condensation of vanadium oxide and sodium sulphate. As there is no information available on reaction kinetics for the above mentioned chemical reactions, we have assumed in our calculations that they are not kinetically controlled. As the modelled diesel engine cycle takes only about 80 ms, the dynamic aspects in aerosol processes, such as nucleation dynamics and supersaturated vapours, become important.

Input data for the calculations

As ABC is a one-dimensional model, we have modelled the average ash particle formation and growth situation in a Lagrangian way, and studied the sensitivity of the model to different parameters. In the base case we used the following typical problematic heavy fuel oil composition of the ash forming species: vanadium 300 ppm, sodium 50 ppm, nickel 50 ppm and sulphur 3.5 %. Some particles in the size range 1 - 10 μm were added in the input to represent unburned particles. The temperature and the pressure profiles in the cylinder and in the flow channel between the seat and the exhaust valve were obtained from the simulations of the engine manufacturing company. A flow speed of 150 m/s was used for the flow past the exhaust valve. The following sensitivity studies concerning the deposition behaviour were made:

- Half open exhaust valve. The flow velocities past the valve are the highest when the exhaust valve is in the process of opening. A velocity of 575 m/s was used.
- Bend in a flow channel. A channel with a 90° bend was studied to assess the importance of inertial impaction.

The assessment of the importance of different deposition mechanisms had to be based on the plug flow assumption, since our present model is unable to deal with the effects of complex geometries. This approach should, however, give us a rough picture of the essential phenomena. The geometry of the exhaust valve is such that we do not expect to get much inertial impaction on the surface under consideration. To get some kind of a preliminary idea of its importance we made a calculation with the bend model. At this stage, we have not considered the effects of the compressibility of the flow on deposition rates.

Calculated results

In the base case the geometric mass mean diameter of the ash particles formed by condensation is about 0.1 μm. VO_2 was found to be the most important species in these particles. If the V/Na ratio is lower, sodium sulphate becomes the dominant component. The temperature gradient during the nucleation phase was the most important parameter affecting the particle mass size distribution. At high temperature gradients saturation ratios become high and a large number of particles are formed and because of this, particles cannot grow as large as they do with a lower temperature gradient. At lower pressures the nucleation reactions speed up further.

According to the results, deposition takes place as particle deposition. The most important deposition mechanisms are turbulent impaction and thermophoresis. Turbulent impaction is more important for the larger particles of the unburned mode, while thermophoresis is more important for the smaller particles produced by the condensation of the volatilized species. The growth rates of the deposition layer that we calculated are presented in Table 3. The rates are presented for steady state conditions i.e. the time fractions of different phases in the engine cycle are not taken into account. In the base case two thirds of the deposition is due to turbulent impaction and one third is due to thermophoresis. In the low V/Na ratio case the proportion of the condensed particles is lower and therefore the proportion of thermophoresis decreases. Due to the very high flow velocity in the half open valve case, calculated deposition layer growth rates are quite large during this phase. The bend geometry used in the study of inertial impaction turned out to be rather extreme, since the deposit layer growth rate became very high in that case. This result only serves to show that it is worthwhile to consider the importance of inertial impaction in more detail for spots where it can have some importance. On the valve surface there should be no inertial impaction due to the geometry of the valve.

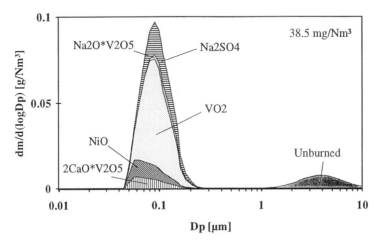

Figure 7a. Particle mass size distribution at the exhaust valve in the base case.

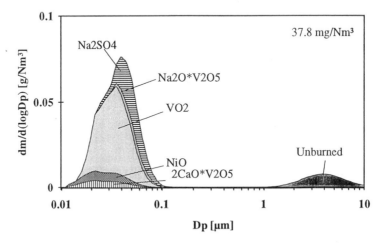

Figure 7b. Particle mass size distribution at the exhaust valve in a case with late combustion at a high temperature gradient.

The deposit layer rates growth rates modelled in this study were calculated with the assumption that that no re-entrainment takes place. In reality, however, re-entrainment to a considerable degree determines the deposit layer growth rate at least on a smooth valve surface not covered by a deposits. Because of the very high flow velocities past the valve surface the energy of the turbulent fluctuations reaching the surface is large enough to overcome the energy binding the deposited particles to the valve surface. As this is not modelled in our model, the calculated deposition layer growth rates describe, rather than the actual growth rate, the mass flux to the valve surface.

It can be seen from Fig. 8 that deposition velocities are much larger for submicron particles in the case of the half open exhaust valve compared to the base case. With high flow velocities, the boundary layers are thinner, and it is easier for particles to get carried into the laminar sublayer. Deposition velocity due to thermophoresis is higher and the effect of turbulent impaction extends further into the submicron range. All these things together make the period of valve opening important for the formation of corrosive deposits. If we assume that the particles of the unburned mode consist of

Table 3. Proportions of different deposition mechanisms and deposition layer growth rates on exhaust valve seat ring assuming stationary fully developed turbulent channel flow.

Case	Thermo-phoresis (%)	Turbulent Impaction (%)	Inertial impaction (%)	Deposition layer growth rate [mm/d]
Base case	34.0	66.0	0.0	0.115
Low V/Na ratio case	24.5	75.4	0.0	0.104
Half open valve case	48.9	51.0	0.0	0.401
Bend case	0.0	0.0	100.0	>> 1.0

carbon, which burns away on the surface of the deposit layers, the composition of the deposit layers is much the same as the average composition of the particles formed by condensation. No vapour deposition occurs according to our calculations. As our simulations were performed in a Lagrangian way, and the inhomogeneous conditions were taken into account by sensitivity studies, particle size distributions in individual simulations became rather narrow. Therefore, the differences in the deposition tendencies of different species do not become very significant. Considering the inhomogeneous conditions and the effect of reaction kinetics these differences could in reality be more significant with sodium sulphate enrichment in the deposit layers. In addition, re-entrainment may cause enrichment of sticky species in the deposit layers.

CONCLUSIONS

A fully mechanistic model for fine particle formation and deposition in combustion processes has been implemented in the ABC code. Here the model has been applied to recovery boilers and medium speed diesel engines. It does not include correlations from laboratory or field measurements. So far field data has been used only to verify the assumptions, data and theories embedded in the ABC code. The individual deposition mechanisms have been validated with laboratory scale experiments, but there is not accurate field data to validate the model in real process conditions. The future will show, whether experimental data indicate that adjustments to the model are required.

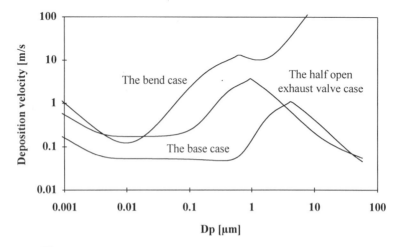

Figure 8. Deposition velocities as a function of particle size.

When looking at the deposit layer growth results, one should keep in mind that they have been calculated assuming that no re-entrainment takes place and that no sootblowing is used. If the focus of attention is on the mechanisms of deposit formation on surfaces after the combustion chamber (e.g. furnace or diesel engine cylinder) and their dependencies on fuel composition and combustion conditions, this is not a serious omission. Naturally, sootblowing has a dominant effect on the growth rate of the deposited layer. However, it is possible to gain an insight into the mechanisms that lead to deposit formation just by modelling the situation when the sootblowing is off. So far we have not studied the importance of surface roughness in our calculations i.e. we have assumed a smooth surface. Therefore, one interpretation of the calculated results is that the increase in the deposition rate obtained by taking surface roughness into account is compensated by the decrease in the net deposition rate by taking re-entrainment into account. In the future we are going to study the importance of these phenomena.

Modelling deposition in a kraft recovery boiler and in a medium speed diesel engine has served to identify the processes that lead to deposit formation. With the ABC code it was possible to study the link between the deposit formation and the fuel composition and the combustion conditions. This helps identify the key issues in deposition-related problems. In recovery boilers vapour deposition of supersaturated chlorides on superheater surfaces was identified to be important, if the Cl/Na ratio is high and the S/Na ratio is low in the volatilized material. The presence of chlorides in the deposited layer is known to be specially harmful. We also observed, that it is hard for particles larger than 10 μm to penetrate through the economisers. From the results of diesel engine calculations, we can see that the composition of the deposit layers seems to be much the same as that of the volatilized species. Deposition during the period of exhaust valve opening can be of great importance for the deposition of corrosive species, since in those circumstances submicron ash particles have a high mass flux towards the valve surface. To properly predict the deposit layer growth rate and its composition, we should take into account the role of re-entrainment. As the general modelling theory for re-entrainment is at the moment only in the early stages of development, the rates of re-entrainment should be determined with the help of suitable measurements both for a smooth valve surface and for a surface with a deposited layer. Further studies are also needed to assess the role of the compressibility of the flow.

The ABC modelling studies on deposition need to be complemented with suitable aerosol measurements. In this way the modelling results can help interpret the measured results and they can be of assistance when planning measurements. In kraft recovery boilers we have already performed a number of aerosol measurements. The comparison of the simulated fume particle mass size distribution with the measured one before the electrostatic precipitator shows that the results are in a relatively good agreement. It seems probable that the discrepancies between the simulated and the measured size distributions result mainly from the particle transport through the flue gas channel to the ESP. In this case the interpretation that the omissions of re-entrainment and surface roughness cancel out seemed to work surprisingly well, considering the good match in the total fume mass concentrations. In the near future we hope to be able to compare our modelling results with measurements made at the furnace exit. In that way we can avoid the errors caused by the flue gas transport and concentrate on deposition on the superheater surfaces, where the deposition problems are the most severe.

The deposition models need to be validated for each specific application before they can be used for reliable quantitative predictions of deposition. However, even now the results give a qualitative insight into the parameters affecting deposition. The ABC model, as a whole, can be valuable tool, when solving deposition problems and assessing the importance of different parameters when developing new concepts.

ACKNOWLEDGMENT

This study was funded by the Finnish combustion research programme LIEKKI and the companies Ahlstrom Corporation, Tampella Power, and Wartsila Diesel

International, whose support is gratefully acknowledged. In addition, the authors want to thank Veitsiluoto Oulu mills, Metsä-Botnia Kemi mills, Dr. Kauko Janka, Dr. Esa Vakkilainen, Mr. Pekka Siiskonen, Mr. Peter Svahn and Mr. Arto Järvi for their technical support.

REFERENCES

Ahluwalia, R. K., Im, K. H, Chuang, C. F. and Hajduk, J-C. (1986). "Particle and Vapour Deposition in Coal-Fired Gas Turbines", *31st Int'l Gas Turbine Conference*. Düsseldorf, West Germany, 86-GT-239.

Baxter, B. L. (1993). "Ash deposition during biomass and coal combustion: A mechanistic approach", *Biomass and Bioenergy* Vol. 4 No. 2, pp. 85-102.

Beér, J. M., Sarofim, A. F. and Barta, L. (1992). "From Properties of Coal Mineral Matter to Deposition Tendencies of Fly Ash - A Modelling route", *Journal of the Institute of Energy*, March 1992, 65, pp. 55-62.

Davies, C. N. (1966). "Aerosol Science", Academic Press Inc., London.

Fan, F. and Ahmadi G. (1993). "A Sublayer Model for Turbulent Deposition of Particles in Vertical Ducts with Smooth and Rough Surfaces", *J. Aerosol Sci.*, 24(45-64).

Friedlander, S. K. and Johnstone, H. F. (1957). "Deposition of Suspended Particles from Turbulent Gas Streams", *Ind. Eng. Chem.*, 49, 1151.

Friedlander, S. K. (1977). "Smoke Dust and Haze", John Wiley & Sons, New York.

Helble, J., Neville, M., and Sarofim, A. F. (1986). "Aggregate Formation from Vaporized Ash during Pulverized Coal Combustion", 21th Symp. (Int'l) on Combustion, pp. 411-417.

Helble, J. J., Srinivasachar, S., Wilemski, G., Boni, A. A., Kang, S. -G., Sarofim, A. F., Graham, K. A., Beér, J. M., Peterson, T. W., Wendt, J. O. L., Gallagher, N. B., Bool, L. Huggins, F. E., Huffman, G. P., Shah, N. and Shah A. (1992). "Transformations of Inorganic Coal Constituents in Combustion Systems", Final Report submitted to the U.S. DOE/PETC, November, Available from NTIS, Document DE/PC/90751-T15 (BE93013076). 3 volumes.

Hinze, J. O. (1975). "Turbulence", Chapter 5, McGraw Hill Book Co., New York.

Im, K. H. and Chung, P. M. (1983). "Particulate Deposition from Turbulent Parallel Streams", *AIChE J.* 29(3), 498.

Israel, R. and Rosner, D. E. (1983). "Use of a Generalized Stokes Number to Determine the Aerodynamic Capture Efficiency of Non-Stokesian Particles from a Compressible Gas Flow", *Aerosol Sci. and Tech.* 2, pp. 45-51.

Jokiniemi, J., Lazaridis, M., Lehtinen, K. and Kauppinen, E. (1994). "Numerical simulation of vapour-aerosol dynamics in combustion processes". *J. Aerosol Sci.*, Vol. 25(3). pp. 429 - 446.

Jokiniemi, J., Pyykönen, J., Mikkanen, P. and Kauppinen, E. (1995). "Modelling Alkali Salt Deposition on Kraft Recovery Boiler Heat Exchangers", Proceedings of the 1995 Int. Chem. Recovery Conference, Toronto, April 24-27, 1995. pp. 77-87.

Langmuir, I. and Blodgett (1946). U.S. Army TR 5418.

Lee, F. C. C., Ghobadian, A. and Rilet G. S. (1993). "Prediction of Ash Deposition in a Pulverised Coal-Fired Axisymmetric Furnace", In: Williamson, J. and Fraser W. (eds.) The Impact of Ash Deposition on Coal Fired Plants. The Engineering Foundation Conference Proceedings, Solihull , UK, June 20-25, 1993. pp. 247-258.

Levenspiel, O. (1972). *Chemical Reaction Engineering.* 2nd ed. John Wiley & Sons, New York.

Mikkanen, P, Kauppinen, E. I. and Jokiniemi, J. K. (1994 a) "The Effect of Black Liquor Characteristics on Alkali Salt Aerosols in Industrial Recovery Boilers", 30 Years Recovery Boiler Co-operation in Finland. Int. Conf., Baltic Sea, 24-26 May 1994.

Mikkanen, P., Kauppinen, E. I., Jokiniemi, J. K., Sinquefield, S. A., Frederick, W. J. (1994 b) "Bimodal Fume Particle Size Distributions from Recovery Boiler and Laboratory Scale Black Liquor Combustion", *Tappi J.* 77(12).

Mikkanen, P., Kauppinen, E. I., Jokiniemi, J. K., Sinquefield, S. A., Frederick, W. J. and Mäkinen, M. (1994 c) "The Particle Size and Chemical Species Distributions of Aerosols Generated in Kraft Black Liquor Pyrolysis and Combustion", The 1993 Forest Products Symposium. AIChE Symposium Series 90, pp. 46-54.

Mikkanen, P., Kauppinen, E. I., Jokiniemi, J. and Mäkinen, M. (1995), "Characteristics of Aerosols from Industrial Scale Black Liquor Combustion", To be submitted for publication in *J. Aer. Sci.*

Schlichting, H., "Boundary Layer Theory", Pergamon Press, London (1951).

Springer, G. S. (1970). "Thermal Force on Particles in the Transition Regime", J. of Colloidal Sci. 34(215-220).

Talbot, L., Cheng, R. K., Schefer, R. W. and Willis, D. R. (1980). "Thermophoresis of Particles in a Heated Boundary Layer", *J. Fluid Mech.* 101(4), 737.

DEVELOPMENT OF FIRESIDE PERFORMANCE INDICES
FOR COAL-FIRED UTILITY BOILERS

Christopher J. Zygarlicke, Kevin C. Galbreath,
Donald P. McCollor, and Donald L. Toman

Energy & Environmental Research Center
University of North Dakota
PO Box 9018
Grand Forks, ND 58202-9018

ABSTRACT

A series of eight fireside performance indices have been developed for pulverized coal-fired utility boiler systems. The indices are calculated to predict slagging, high-temperature fouling, low-temperature fouling, slag tapping ability, opacity, tube erosion, coal grindability, and sootblower necessity. The indices are most useful for screening new fuels or fuel blends for pulverized coal-fired systems. Coal data input required to calculate the indices includes proximate/ultimate analysis results, elemental oxide coal ash chemistry, coal mineral quantities, and juxtaposition as derived using computer-controlled scanning electron microscopy and quantities, of organically bound or submicron mineral components as determined by chemical fractionation analysis. Limited boiler inputs are also required, such as boiler type, power-generating capacity, percentage of load, gas velocities, type of particulate control, and basic furnace dimensions. The indices were formulated primarily by combining inorganic transformation and ash deposition theory with empirical correlations derived from bench-, pilot-, and full-scale combustion tests. The ash deposition and opacity indices have been validated at several full-scale utility boilers.

INTRODUCTION

Competition within the U.S. coal-fired power industry demands that individual utilities be poised to respond quickly and successfully to changing fuel markets. Long-term coal contracts are nearly obsolete, and spot-market purchases of coal are becoming the norm. The economics associated with sulfur emissions conrol have caused many utilities to switch fuels to the Powder River Basin (PRB) subbituminous coals. A problem with PRB coals, however, is that their overall or bulk compositional properties are not good indicators of fireside performance. PRB coals with similar bulk elemental oxide chemistries can often behave quite differently in a given boiler [Weisbecker et al., 1992]. Also, smaller boilers that are designed for bituminous coal may develop ash deposition problems when burning higher-alkali, lower-rank coals, either as a total primary feedstock or as a blend with bituminous coal [Zygarlicke et al., 1994]. Traditional indices such as the base-to-acid ratio, the slagging factor R_s, and fouling factor R_f, as described by Winegartner [1974], are more appropriately designed for eastern bituminous coals and not lower-rank coals such as the subbituminous PRB coals. The indices are inappropriate for lower-rank coals, because the type of input data is inadequate. Newer and more advanced techniques such as computer-controlled scanning electron microscopy (CCSEM) provide more detailed mineralogical data that goes beyond the conventional bulk elemental oxide, moisture, and carbon measurements. Such advanced analysis has allowed for better characterization of the components within coal that can cause adverse operational effects, such as tube erosion, furnace wall slagging, and tube fouling [Karner et al., 1994]. In this paper, a series of indices are described that use advanced and conventional coal analysis data to predict fireside performance in coal-fire utility boilers more reliably. The indices are part of a package designated as the Predictive Coal Quality Effects Screening Tool or PCQUEST (Erickson et al., 1994; Benson et al., 1993). Eight indices are calculated by PCQUEST to predict slagging, high-temperature fouling, low-temperature fouling, slag tapping ability, opacity, tube erosion, coal grindability, and sootblower necessity (Fig. 1). The indices were developed using knowledge of inorganic transformations, entrained ash formation, and ash deposition [Benson et al., 1993]. Formulation of the indices was accomplished using bench-, pilot-, and full-scale data to identify the primary coal inorganic properties that relate to the occurrence of main furnace slagging and convective pass fouling in utility boilers. The indices provide much more accurate diagnostic information for predicting the ash deposition behavior of coal than

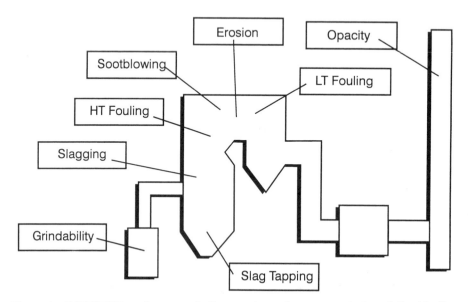

Figure 1. PCQUEST performance indices as they relate to a typical coal-fired boiler.

conventional indices, which are based on simplistic American Society for Testing and Materials (ASTM) coal analysis data. The accuracy of these indices has been demonstrated through their repeated use by several utilities in the midwestern United States for screening different coals, largely PRB, for use in their boilers.

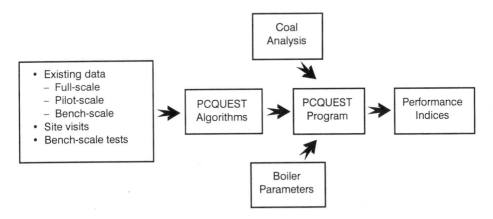

Figure 2. Development of PCQUEST algorithms.

RESULTS

Basis for Indices Formulation

The indices that are calculated by PCQUEST are based on coal inorganic content, limited combustion conditions, and theoretically and empirically derived relationships. The program that calculates the indices is not a comprehensive mechanistic engineering model. PCQUEST was developed using a plethora of existing data and mechanistic information [Benson et al., 1993] on coal ash formation (Fig. 2). The software package is designed to be relatively simple to use with a minimum amount of computer processing time. The key to formulating the indices is having quantifiable coal inorganic parameters or criteria, which have been found to govern ash formation and utility boiler ash deposition and have been assembled into algorithms that output index values that express the degree of fouling, slagging, and opacity. PCQUEST is most useful for comparing the relative fireside performance of two or more coals, and for determining optimum coal-blending ratios.

A numerical system is used to rank the indices from 0 to 100. A range of 1–33 is considered a low magnitude, 34–66 an intermediate or medium magnitude, and 67–100 a high or severe magnitude. This propensity-index classification scheme applies to all the indices except the coal grindability index, which is directly related to the ASTM Hardgrove grindability index. A value of 0 for any of the indices indicates that the proper input data required to calculate the index were not provided.

The coal input analysis is the key to the operation of the indices. Predictive models require quantifiable input data on inorganic constituents, including discrete minerals and dispersed cations. CCSEM and chemical fractionation analysis give a complete inorganic analysis of the coal, which allows for more mechanistic ash interactions to be predicted. These methods are discussed in great detail elsewhere [Karner et al., 1994; Benson et al., 1993; Zygarlicke and Steadman, 1990; Benson and Holm, 1985]. Mineral sizes and quantities (Fig. 3) are provided by CCSEM, and the quantities of organically bound inorganics such as Na, Ca, Mg, and K are determined by chemical fractionation. Chemical fractionation analysis is normally only necessary for lower-rank coals which can contain significant quantities of organically bound constituents. Input data formats for PCQUEST are listed in Table 1. The role of the boiler input is not to determine how to operate the boiler, but only to qualify the combustion behavior impacts of the same coal in different boilers. The same PRB coal can show different superheater fouling characteristics for two different-sized boilers, however, it is not the function of the indices to correspond to specific derates or maximum continuous ratings. Boiler input parameters have the primary role of relating boiler heat release rates and ash loadings to that of the design coal.

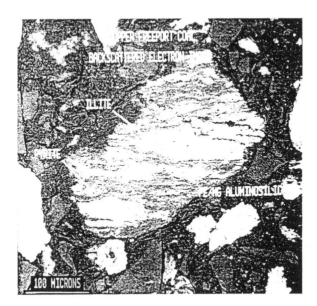

Figure 3.   SEM photograph of pulverized coal (dull light gray), minerals (bright white), and binding matrix (dark gray and swirling texture).

TABLE 1

PCQUEST Coal Analysis Input Parameters

Method	Parameter	Unit
CCSEM	Total mineral content	wt% coal basis
CCSEM	Quantitative mineralogy as a function of particle diameter	wt% mineral basis
CCSEM	Excluded mineralogy as a function of particle diameter	frequency %
Chemical Fractionation	Leachable Si, Al, Fe, Ti, P, Ca, Mg, Na, and K by H_2O, NH_4OAc, and HCl	% removed and % remaining
Proximate	Moisture, volatile matter, fixed carbon, ash	wt% as-received
Ultimate	H, total C, N, S, O	wt% as-received
Heating Value	Calorific	Btu/lb
Ash Chemistry	SiO_2, Al_2O_3, Fe_2O_3, TiO_2, P_2O_5, CaO, MgO, Na_2O, K_2O, SO_3	wt%

An example of the consistency of results from PCQUEST is given in Table 2 for Rochelle coal samples that were acquired from the mine every year for the past 4 years. Rochelle coal is mined from Campbell County in Wyoming, situated in the southern part of the Powder River Basin. The Rochelle coal shows very little variability in elemental and mineral composition over the past 4 years. The index values for fouling and slagging follow the trend of the coal compositional data, giving similar predictive results for each coal. As the table indicates, the Rochelle coal is predicted to be low-medium in its tendency to form slag and high-temperature deposits and low in its tendency to form low-temperature deposits. Communication with users of the Rochelle coal have confirmed low-to-intermediate fouling and slagging combustion performance of the Rochelle coal [Stuckmeyer, 1993; Albertson, 1995].

Fouling, Slagging, and Slag Tapping Indices

Three indices presently output by PCQUEST pertain to ash deposition and rank main furnace slagging, high-temperature fouling, and low-temperature fouling. Assumptions used in weighing the relationship between the key coal inorganic constituents are based on years of studying coal inorganic content and ash deposition mechanisms in a variety of coal combustion systems [Benson et al., 1993; Honea et al., 1981a, b; Hurley et al., 1991; Zygarlicke et al., 1992; McCollor et al., 1993; Hurley et al., 1994]. The slagging index is an indicator of main furnace slagging such as occurs on waterwalls. Slag deposits are

TABLE 2

PCQUEST Input Data and Indices for Rochelle Coal

	Rochelle 1990	Rochelle 1991	Rochelle 1992	Rochelle 1993
Slagging Index	51	52	44	34
High-Temp. Fouling Index	45	65	50	40
Low-Temp. Fouling Index	10	12	38	5
CCSEM Minerals, wt% minerals				
Quartz	33.5	44.5	62.5	48.7
Kaolinite	26.0	24.1	22.2	25.4
Montmorillonite	2.5	4.2	0.2	1.8
Pyrite	0.7	3.5	0.6	1.7
Ca Al-P	7.8	7.4	7.7	9.4
Organically Bound Inorg., wt% ash				
Al	5.1	3.7	7.3	3.6
Ca	17.4	18.3	23.3	20.3
Mg	5.0	7.3	9.6	5.0
Na	1.8	0.7	0.8	1.0
Coal Ash Chemistry, wt% ash				
SiO_2	33.0	27.2	30.9	37.5
Al_2O_3	16.7	18.4	17.1	19.8
Fe_2O_3	4.5	5.6	5.3	5.4
TiO_2	1.6	1.3	1.1	1.1
P_2O_5	1.7	1.4	1.4	1.0
CaO	18.9	21.1	24.1	21.6
MgO	5.6	8.0	9.8	5.3
Na_2O	1.9	0.9	1.0	1.1
K_2O	0.1	0.3	0.3	0.1
SO_3	15.8	15.8	9.1	7.1
Coal Composition				
Proximate				
Moisture	26.0	25.0	29.3	26.7
Volatile Matter	32.0	34.4	31.0	33.1
Carbon	37.0	36.2	35.8	36.1
Ash	4.5	4.4	3.9	4.2
Ultimate				
Hydrogen	6.4	6.4	6.6	6.2
Carbon	52.5	52.5	49.4	51.5
Nitrogen	0.6	0.7	0.6	0.8
Sulfur	0.3	0.3	0.2	0.3
Oxygen	35.7	35.8	39.4	37.1
Ash	4.5	4.4	3.9	4.2
Calculated Btu	8800	9085	8524	8910

usually the result of large, partially molten mineral particles or larger coalesced ash particles having the appropriate inertial energies and stickiness to break free from flue gas entrainment forces impacting and sticking to waterwalls. For bituminous coal-designed boilers, the primary slagging coals contain pyrite and clays which can form high inertial energy and sticky entrained ash particles that readily accumulate on the boiler waterwalls. If calcium is present in the coal and is in suitable juxtaposition to react with iron-rich particles, the slagging severity can be greatly magnified. The high-temperature fouling index pertains to fouling in higher-temperature zones, such as in the secondary superheater or reheater region of a conventional pulverized coal-fired boiler. Deposits in this region are generally silicate-rich. Organically bound alkali such as Na and K quickly enter the vapor phase as oxides, hydroxides, or sulfates and interact with silica or aluminosilicates in the entrained ash or deposits to potentially cause accelerated deposit growth and strength development. Organically bound Ca may also rapidly react with aluminosilicates and quartz within coal particles during combustion to form lower melting point phases that may enhance deposit formation. Aluminosilicate clay minerals such as kaolinite and illite can play a major role in the degree of ash deposition by facilitating production of lower melting point liquid phases if combined with organically bound cations such as K, Na, and Ca early during combustion. Production of lower melting point phases will influence both slagging and high-temperature fouling deposit formation. However, if clay minerals remain fairly pure, they can actually act as refractory compounds to diminish fouling and slagging by increasing melting points and viscosities.

The low-temperature index pertains to fouling occurring in lower-temperature regions of a boiler, such as in the primary superheater or economizer. Depending upon its association with other elements or compounds, calcium may react with sulfur in lower-temperature zones to form calcium sulfate-based deposits. The form of the calcium is important for how readily it will react with sulfur to form strong calcium sulfate deposits [Hurley et al., 1994]. Calcium oxides and pure silicates react more readily than calcium that is bound as an aluminosilicate. PRB coals with an abundance of fine included quartz and organically bound calcium will usually result in abundant calcium silicates [Zygarlicke et al., 1992] that can deposit in lower-temperature regions and readily begin sulfating.

Opacity Index

The opacity index predicts the propensity for a coal to produce large quantities of very fine particulate. The production of fine ash particles during coal combustion is largely dependent on the coal's inorganic properties, including the amount of volatile inorganic species that can vaporize and subsequently condense and the mineral grain size distribution. The direct relationship between the abundance of fine mineral grains in pulverized coal and the production of fine ash during combustion was documented by Holve [1986]. The results from the laminar flow drop-tube furnace (DTF) tests are generally consistent with this relationship, as shown in Fig. 4. A best-fit line to the particle-size trend was calculated using linear regression analysis. The data points for the test coals, except for an outlying coal, produce a well-defined linear trend, with a correlation coefficient of 0.94. A well-defined relationship also exists between the production of submicron fly ash and the concentrations of submicron minerals and organically bound inorganics in the coals (Fig. 5). The following ratio, designated as the

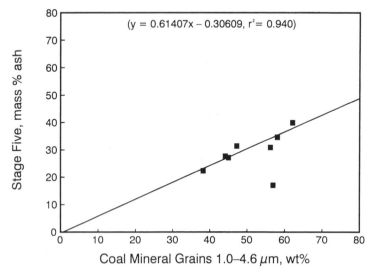

Figure 4. Correlation between the amount of very small-sized minerals and the amount of fly ash collected in the fifth stage (d_{50}=1.9μm) of a multicyclone during DTF fly ash production.

inorganic distribution coefficient, was devised to express the proportion of inorganic constitutents associated with discrete supermicron minerals relative to those associated with submicron minerals and macerals:

$$\text{inorganic distribution coefficient} = \frac{\text{supermicron minerals}}{\text{submicron minerals} + \text{organically associated inorganics}}$$

Figure 5 indicates that greater amounts of small minerals and organically bound inorganics correspond to more fine particulate ash generated during combustion of pulverized coal. Although a link has been established between coal inorganics and entrained ash that can lead to opacity problems in a boiler, the opacity index is not entirely fuel-specific. The index predicts the potential for fine particulate production before collection devices, and most control devices remove substantial amounts of particulate. Modification of the predicted index value can be applied to boilers that have electrostatic precipitators (ESPs) by accounting for fly ash resistivity and gas temperatures at the ESP inlet. Resistivity is determined using the Bickelhaupt [1985] equations and can be calculated for cold- or hot-side ESPs.

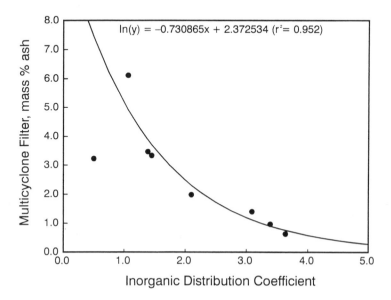

Figure 5. Submicron fly ash from a DTF multicyclone filter (d_{50} nominally 0.4 μm) versus an inorganic distribution coefficient (supermicron mineral concentration/[submicron minerals + organically associated inorganic concentration]).

Erosion Index

Quantity and size of excluded quartz and pyrite grains in the pulverized fuel are the primary criteria for deriving the erosion index. These mineral components transform to fly ash components that are usually the hardest and most erosive fly ash components. Quartz often retains its original particle angularity even after coal combustion. Boiler gas velocity is an input parameter as demonstrated in Fig. 6 for several PRB coals. Erosion is a very difficult parameter to define or verify. No experimental or full-scale utility boiler data was available to verify the PCQUEST erosion index.

Grindability Index

Several different approaches were investigated to formulate an improved grindability or hardness index. The PCQUEST grindability index is based only on total ash content and sulfur content. Very weak correlations were observed for macerals and minerals such as quartz and pyrite. This index is the only one of the eight that has a scale that follows a traditional index, in this case the Hardgrove grindability index (HGI). Higher values for the PCQUEST grindability index follow the numbering format for HGI, with softer coals or easier grindability giving higher values. Figure 7 shows grindability values for commonly mined subbituminous and lignite coals.

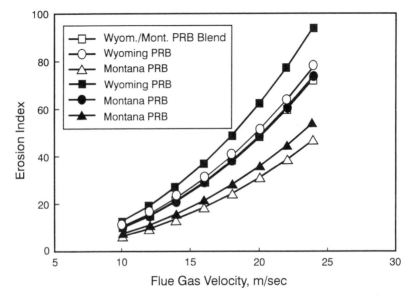

Figure 6. Erosion index values for several PRB coals as a function of boiler flue gas velocity.

Sootblowing Index

The sootblowing index estimates the removability of a deposit and effectiveness of sootblowers. Adhesion and crushing strength measurements of deposits generated in the DTF were the primary factor used to derive the index. Growth rate and quantity of deposit accumulation are irrelevant for the index calculation. Correlations were made between the inorganic components that can cause accelerated sintering and solidifying of deposits and deposit strength as related to removability. A higher index value indicates more difficulty for removing the deposit.

Demonstration of PCQUEST Indices

The fireside performance indices that make up PCQUEST were subjected to intensive verification and testing using bench-, pilot-, and full-scale testing. Some of the experimental work that was performed is discussed here in order to demonstrate the utility of the slagging, fouling, and opacity indices. The original goal of the PCQUEST indices was to provide a more reliable way of assessing ash deposition and opacity; therefore, much more time and effort has gone into developing these specific indices.

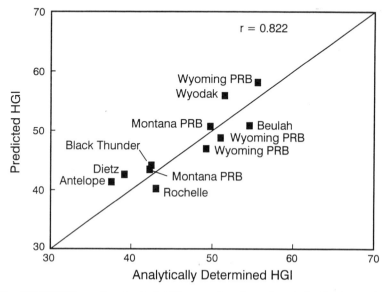

Figure 7. PCQUEST predicted grindability as a function of analytically determined HGI.

An optical access laminar flow DFT was used to produce fly ash and deposits for verification purposes. DTFs can be configured using key operational parameters to produce results similar to those observed in a full-scale boiler. Key parameters include gas temperature, gas velocity, particle residence time, pressure, excess air, gas cooling rates, and deposit surface substrate temperatures. The optical section allows visual and video monitoring of deposit formation and in situ deposit adhesion strength measurements. The optical access section allows in situ temperature measurements by optical pyrometry as well as the capacity to add additional nonintrusive optical diagnostics. The optical access DTF is especially designed to assess the fouling and slagging tendencies of coal. A cylindrical deposition probe with a controlled skin temperature can be inserted into the heated deposition zone and the initial slagging temperature, deposit adhesion strength, and deposit growth rates determined. The heated zones in the furnace reactor tube can be customized to reflect temperature regimes in a full-scale utility boiler. The maximum temperature that can be achieved is 1650°C.

PCQUEST results were derived for seven coals having boiler experience or known ash deposition behavior patterns. Compositions of the coals are given in Table 3. The Beulah lignite, Black Thunder subbituminous, and Illinois No. 6 bituminous coals have been studied through various programs and have well-known documented ash behavior characteristics. Beulah lignite can produce fairly severe fouling deposits because of the fluxing action of sodium vapor or fine particulate species on calcium–magnesium-rich aluminosilicates. Slagging may also be medium to severe on the bottom of secondary superheater platens. Black Thunder coal is a low-sulfur PRB coal that typically shows low-to-medium ash-fouling tendencies with some slagging possible, again more located at the bottom of secondary superheater platens. Fouling is generally driven by extensive interaction of organically bound calcium, magnesium, and some aluminum with aluminosilicate clays. Illinois No. 6 coal is generally not a problem for fouling, but may cause slagging problems in boilers designed for bituminous coals. Problematic slagging is usually the result of the abundant larger-sized pyrite minerals interacting with aluminosilicate clays such as kaolinite or illite to form lower melting point and sticky iron aluminosilicate material which has a tendency to stick and accumulate on main furnace waterwalls. The particular Illinois No. 6 coal sample that was analyzed (Table 3) contained an abundance of pyrite and clays for slag formation. PCQUEST results for four PRB coals labeled coals A–D are also presented in Table 3. Each of these coals was fired in a full-scale boiler as part of a field test program that aided the formulation of the PCQUEST indices. The names of the coals and boilers are not revealed at the utility sponsors' requests. Coals A and C are from Wyoming; Coal B is a blend of two Montana coals; and D is a blend of a Wyoming coal and a Montana coal. These coals have the typical PRB characteristics of high moisture, calcium, kaolinite, and quartz and low ash, sulfur, and pyrite.

High-temperature fouling predictions for the seven coals are shown in Fig. 8. The predictions for the Beulah, Black Thunder, Illinois No. 6, and Coal C coals were all based on firing in a typical 400-MW tangentially fired boiler designed for bituminous coal. Index predictions for Coals A, B, and D were based on different boilers that were using those specific coals at the time field observations of fouling and slagging were conducted. Coal C had experience in two different boilers. Comparison of high-temperature fouling indices for the four coals run under the same boiler configuration show the Beulah lignite coal as having the greatest potential for fouling, followed by Black Thunder and Coal C PRB coals. The least degree of fouling was exhibited by the Illinois No. 6 coal. This same trend is also seen in the coal power industry where greater potential exists for fouling problems to occur when a boiler that is designed for bituminous coal switches to a lower-rank coal, such as a PRB fuel. Actual boiler testing during the firing of Coal C in this boiler showed significant slagging and fouling located near the furnace exit on secondary superheater platens and on reheat surfaces of the convective pass (Fig. 9). Deposits collected in this section of the boiler were typical of Ca–Mg-rich aluminosilicates (Fig. 10) which form readily in the higher temperature 2000°–2300°F zones when

TABLE 3

Composition of Test Coals

	Beulah	Illinois No. 6	Black Thunder	Coal A	Coal B	Coal C	Coal D
Proximate/Ultimate, wt%							
Moisture	18.6	7.3	24.3	26.7	23.3	25.6	24.0
Vol. Matter	37.2	34.4	35.9	33.1	32.4	34.3	33.9
Fixed Carbon	33.8	43.7	35.3	36.1	38.6	34.9	35.5
Ash	10.4	14.6	4.5	4.2	5.7	5.2	6.6
Hydrogen	5.6	5.1	7.0	6.2	6.3	6.8	6.5
Carbon	48.5	61.5	52.8	51.5	54.1	51.5	51.8
Nitrogen	0.7	1.0	0.7	0.8	0.7	0.7	0.7
Sulfur	1.7	4.3	0.4	0.3	0.5	0.4	0.5
Oxygen	33.2	13.5	34.5	37.1	32.7	35.5	33.9
Ash	10.4	14.6	4.5	4.2	5.7	5.2	6.6
Btu	8433	11193	9619	8910	9438	9234	9201
Elemental Oxide, wt%							
SiO_2	16.5	49.8	32.6	37.5	39.1	32.2	37.5
Al_2O_3	13.3	19.3	16.8	19.8	20.0	16.4	19.8
Fe_2O_3	16.6	19.4	5.7	5.4	2.4	4.0	4.6
TiO_2	0.8	0.7	1.1	1.1	1.1	1.1	0.8
P_2O_5	0.0	0.0	1.2	1.0	0.7	1.1	0.8
CaO	19.5	5.3	22.1	21.6	13.3	21.9	18.6
MgO	7.4	0.8	4.8	5.3	5.8	6.5	6.1
Na_2O	5.2	1.2	0.9	1.1	1.8	1.0	1.0
K_2O	0.2	1.4	0.2	0.1	0.3	0.5	0.6
SO_3	19.8	2.1	14.7	7.1	15.6	15.3	10.2
% Element Organically Bound							
Al	12.0	0.3	20.0	18.0	0.9	22.0	8.0
Fe	7.8	1.5	31.1	86.4	23.5	73.8	31.7
Ti	12.0	10.4	20.1	22.1	15.9	17.4	18.0
Ca	96.4	27.5	89.9	94.2	87.7	89.9	91.9
Mg	92.0	1.2	93.7	93.8	91.7	96.0	92.5
Na	100.0	67.4	96.0	90.9	99.8	100.0	96.0
K	67.0	6.0	15.0	0.0	3.3	47.0	29.0
Minerals, wt% mineral basis							
Quartz	6.7	18.1	24.1	40.3	23.8	22.1	23.9
Iron Oxide	0.5	0.5	4.6	0.0	1.0	0.0	0.0
Calcite	0.0	6.5	0.9	0.5	2.8	0.0	2.2
Kaolinite	31.3	13.9	29.1	26.6	50.4	21.6	33.3
Montmorillonite	3.1	5.6	7.1	3.1	1.4	7.5	9.7
Illite (K Al-Silicate)	0.2	9.4	0.7	2.2	4.0	2.7	3.1
Pyrite	35.5	35.9	4.7	2.6	5.7	5.9	8.3
Barite	1.6	0.0	0.6	2.0	1.4	0.8	0.7
Ca–Al–P Mineral	3.0	0.0	6.1	8.0	0.1	9.8	4.1
% Ash as Minerals	6.9	13.9	2.6	1.8	4.3	3.3	3.2
% Ash Organically Bound	3.5	0.7	1.9	2.4	1.3	1.9	3.4

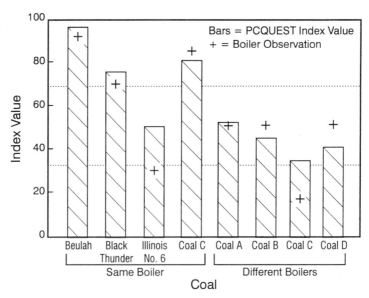

Figure 8. High-temperature fouling predictions for seven coals. The values for the Beulah, Black Thunder, Illinois No. 6, and Coal C coals were all determined based on firing in the same 400-MW tangentially fired boiler designed for bituminous coal. Index predictions for Coals A, B, and D were based on different boilers that were using those specific coals at the time field observations of fouling were conducted.

organically bound Ca and Mg react with abundant kaolinite within burning coal particles and in the deposit. Coals A, B, and D showed low-to-moderate high-temperature fouling deposition potential. Accompanying the high-temperature fouling indices in Fig. 9 are markings of full-scale boiler fouling assessments based upon known or measured observations, which generally agree with the predicted index values. Boiler observations were based on observed deposition rates and removability of deposits and were given values corresponding to the lower, upper, or middle boundaries of the three major PCQUEST ranking criteria of low (1–33), medium (34–66), or high (67–100) fouling.

For low-temperature deposition, the Beulah lignite again shows significant potential for forming the calcium sulfate-based type of deposits that occur in the cooler backpass regions of the convective pass (Fig. 11). No boiler observations could be made for this type of deposition primarily because these types of deposits require long periods of time to form, usually on the order of weeks. Beulah has been known to form these types of deposits in smaller, tighter boilers. Historically, Black Thunder and the PRB Coals A–D have not shown problems with low-temperature deposit buildup, even in cases where boilers designed for bituminous coals have switched to PRB coals.

Slagging indices were calculated for these same series of coals. Figure 12 shows that for the same bituminous-designed boiler, the Beulah, Black Thunder, and Illinois No. 6 coals could all produce potentially high levels of slagging deposition in the form of waterwall deposits or large molten deposits that could form on the bottom of superheater pendants, as earlier shown in Fig. 10. The other PRB coals show medium or high slag deposition

Figure 9. High-temperature fouling deposits at the inlet to the secondary superheater. Deposits in the center of the picture are approximately 12 inches in thickness. Deposits below this region were closer to the nose of the boiler and were more slaglike, consisting of molten and dripping silicate-rich phases.

potential. Boiler observations confirmed the slag indices for all but the Illinois No. 6 coal, which would be expected to produce more moderate slag deposition, and PRB Coal B, which utility experience shows has more moderate slagging and not heavy or severe.

Part of the demonstration and verification of the opacity index was accomplished by comparing measured opacity levels at the TransAlta Utilities Corporation Keephills Generating Station and another unnamed utility generating station with PCQUEST predicted values. The Keephills station was burning several different seams of

Figure 10. SEM micrograph of deposit material shown in Fig. 9, consisting of a variety of Ca–Mg aluminosilicate phases including plagioclase, clinopyroxene, and amorphous glass.

subbituminous coal from the Highvale mine, which is located in the central plains of Alberta, Canada. The other utility from which opacity data were obtained was burning Coal D (Table 2) and another PRB coal. Opacity levels were measured during several controlled test burns at both power stations. Figure 13 shows that a good correlation exists between the predicted and measured opacity levels, with the PCQUEST values being corrected to correspond more closely to the scale used to measure stack opacity. Using the quantity of fine clay and organically bound inorganics in the calculation of the opacity index is what make the index a reliable tool.

CONCLUSIONS

A series of fireside performance indices have been developed for coal-fired utility boiler systems based on both theoretical knowledge of inorganic transformations and ash deposition and empirical correlations derived from experimental data. PCQUEST is the name of the program which calculates eight indices, including slagging, high-temperature fouling, low-temperature fouling, slag tapping ability, opacity, tube erosion, coal

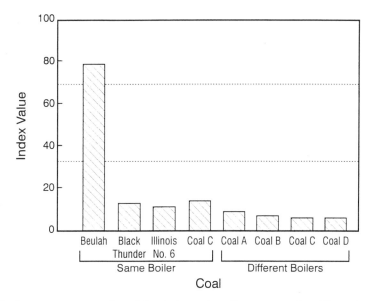

Figure 11. Low-temperature deposition index values for seven coals. The values for the Beulah, Black Thunder, Illinois No. 6, and Coal C coals were all determined based on firing in the same boiler, a typical 400-MW tangentially fired boiler designed for bituminous coal. Index predictions for Coals A, B, and D were based on different boilers that were using those specific coals at the time field observations of fouling were conducted.

grindability, and sootblower effectiveness. The indices are useful for screening fuels or fuel blends for performance. Input data consist of advanced CCSEM mineral data, chemical fractionation data, and standard coal characterization data from ASTM methods. Limited boiler inputs are also required, such as boiler type, power-generating capacity, percentage of load, gas velocities, type of particulate control, and basic furnace dimensions.

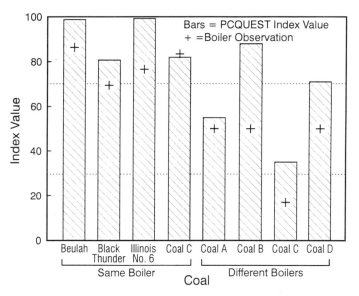

Figure 12. Slagging indices for the seven test coals showing that for the same bituminous-designed boiler, the Beulah, Black Thunder, and Illinois No. 6 coals could all produce potentially high levels of slagging.

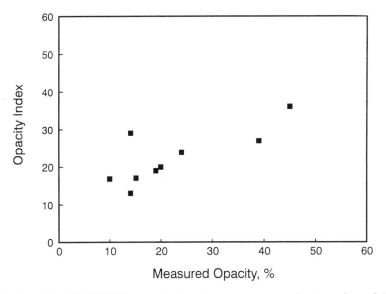

Figure 13. Predicted PCQUEST opacity levels versus measured values from full-scale utility boilers.

for a bituminous coal-designed boiler revealed that lignite and PRB coals showed greater potentials for high- and low-temperature regime fouling, while the bituminous coal showed potential for slagging problems. Several PRB coals with similar ASTM coal ash compositional similarities had varying fouling and slagging indices. The opacity index showed strong correlation with measured opacities at two utilities, one Canadian and the other American, both of which were burning various subbituminous fuels.

ACKNOWLEDGMENTS

The work presented here was supported by the United States Department of Energy, Cooperative Agreement Nos. DE-FC21-86MC10637 and DE-FC21-93MC30098, Electric Power Research Institute Agreement No. RP3579-02, Kansas City Power and Light, Minnesota Power Company, Northern States Power Company, and Union Electric. In addition, we also acknowledge the guidance provided by the following individuals: Philip Goldberg, Todd Albertson, Robert Brown, Gerald Goblirsch, Arun Mehta, George Nehls, and Kenneth Stuckmeyer.

REFERENCES

Albertson, T. (1995). Northern States Power Company, Personal Communication.

Benson, S.A. and Holm, P.L. (1985). "Comparison of Inorganic Constituents in Three Low-Rank Coals." *Ind. Eng. Chem. Prod. Res. Dev.*, *24*, 145.

Benson, S.A., Hurley, J.P., Zygarlicke, C.J., Steadman, E.N., and Erickson, T.A. (1993). "Predicting Ash Behavior in Utility Boilers." *Energy & Fuels*, *7*, 746–754.

Benson, S.A., Jones, M.L., and Harb, J.N. (1993). "Ash Formation and Deposition." In L.D. Smoot (Ed.), *Fundamentals of Coal Combustion for Clean and Efficient Use*. New York: Elsevier, pp 299–374.

Bickelhaupt, R.E. (1985). "A Study to Improve a Technique for Predicting Fly Ash Resistivity with Emphasis on the Effect of Sulfur Trioxide," U.S. Environmental Protection Agency, Report SoRI-EAS-85-841.

Erickson, T.A., O'Leary, E.M., Folkedahl, B.C., Ramanathan, M., Zygarlicke, C.J., Steadman, E.N., Hurley, J.P., and Benson, S.A. (1994). "Coal Ash Behavior and Management Tools." In *The Impact of Ash Deposition on Coal-Fired Plants*. Taylor and Francis Publishing Company, pp 271–284.

Holve, D.J. (1986). "In Situ Measurements of Fly Ash Formation from Pulverized Coal." *Combustion Science and Technology*, *44*, 269–288.

Honea, F.I. (1981a). "Studies of Ash-Fouling Potential and Deposit Strength in the GFTEC Pilot Plant Test Furnace." In R.W. Bryers and S.S. Cole (Eds.), *"Fouling and Slagging Resulting from Impurities in Combustion Gases."* Proceedings of the Engineering Foundation Conference, Henniker, NH, July 12–17, 1981, pp 117–142.

Honea, F.I. (1981b). "Survey of Ash-Related Losses at Low-Rank Coal-Fired Utility Boilers." In R.W. Bryers and S.S. Cole (Eds.), *Fouling and Slagging Resulting from Impurities in Combustion Gases."* Proceedings of Engineering Foundation Conference, Henniker, NH, July 12–17, 1981, pp 527–539.

Hurley, J.P., Benson, S.A, and Mehta, A. (1994). "Ash Deposition at Low Temperatures in Boilers Burning High-Calcium Coals," In *The Impact of Ash Deposition on Coal-Fired Plants.* Taylor and Francis Publishing Company, pp 19–30.

Hurley, J.P., Erickson, T.A., Benson, S.A., and Brobjorg, J.N. (1991). "Ash Deposition at Low Temperatures in Boilers Firing Western U.S. Coals." Presented at the International Joint Power Generation Conference, San Diego, CA, 8 p.

Karner, F.R., Zygarlicke, C.J., Brekke, D.W., Steadman, E.N., and Benson, S.A. (1994). "New Analysis Techniques Help Control Boiler Fouling." *Power Engineering, 98,* 35–38.

McCollor, D.P., Zygarlicke, C.J., Allan, S.E., and Benson, S.A. (1993). "Ash Deposit Initiation in a Simulated Fouling Regime." *Energy & Fuels, 7,* 761–767.

Stuckmeyer, K. (1993). Union Electric, Personal Communication.

Weisbecker, T., Zygarlicke, C.J., and Jones, M.L. (1992). "Correlation of Inorganics in Powder River Basin Coals in Full-Scale Combustion." In Benson, S.A., Ed., *Inorganic Transformations and Ash Deposition During Combustion.* New York: ASME for the Engineering Foundation, pp 699–711.

Winegartner, E.C. (1974). "Coal Fouling and Slagging Parameters." Prepared by the ASME Research Committee on Corrosion and Deposits from Combustion Gases, ASME, 34 p.

Zygarlicke, C.J., Benson, S.A., and Borio, R.W. (1994). "Pilot- and Bench-Scale Combustion Testing of a Wyoming Subbituminous/Oklahoma Bituminous Coal Blend." In R.W. Bryers and N.S. Harding (Eds.), *Coal-Blending and Switching of Low-Sulfur Western Fuels.* New York: American Society of Mechanical Engineers, pp 281–300.

Zygarlicke, C.J., Benson, S.A., Borio, R.W., and Mehta, A.K. (1992). "Examination of Ash Deposition in Full-, Pilot-, and Bench-Scale Testing." In Proceedings of the 3rd International Conference on the Effects of Coal Quality on Power Plants, La Jolla, CA, Aug. 25–27, 1992.

Zygarlicke, C.J., McCollor, D.P., Benson, S.A., and Holm, P.L.. (1992). "Ash Particle Size and Composition Evolution During Combustion of Synthetic Coal and Inorganic Mixtures." In 24th Symposium (International) on Combustion/The Combustion Institute, pp 1171–1177.

Zygarlicke, C.J. and Steadman, E.N. (1990). "Advanced SEM Techniques to Characterize Coal Minerals." *Scanning Microscopy, 4* (3), 579–590.

PREDICTION OF ASH DEPOSITION IN PULVERISED COAL COMBUSTION SYSTEMS

Francis C. C. Lee, Gerry S. Riley and Fred C. Lockwood*

National Power PLC, Windmill Hill Business Park, Whitehall Way, Swindon, SN5 6PB, UK.
*Imperial College of Science, Technology and Medicine, Department of Mechanical Eng., Exhibition Road., London, SW7 2AZ, UK

ABSTRACT

A predictive scheme based on CCSEM flyash data and Computational Fluid Dynamics (CFD) has been developed to study the slagging propensity of coals. The model has been applied to predict the deposition potential of three UK coals; Bentinck, Daw Mill and Silverdale, in a pilot scale single burner ash deposition test facility and an utility size multi-burner front wall-fired furnace. The project is part of a collaborative research programme sponsored by the UK Department of Trade and Industry and involved various industrial organisations and universities. The objective of the project is to understand the fundamental aspects of slagging in pulverised coal-fired combustion systems. This paper is a sequel to the poster paper entitled: The Prediction of Ash Deposition in a Coal Fired Axi-symmetric Furnace, presented in the last Engineering Foundation Conference [Lee et al, 1993]. The present model predicts the relative slagging propensity of the three coals correctly. The predicted deposition patterns are also consistent with the observations. The results from the model indicate a preferential deposition of iron during the initial stage of ash deposition. The average compositions of the deposits become closer to that of the bulk ash when the accumulation of ash deposits is taken into account.

1. INTRODUCTION

The accumulation of furnace deposits is one of the many difficulties faced by utility operators all over the world. Slagging not only reduces the thermal efficiency of a furnace, but also affects its integrity due to corrosion, erosion or impact on the bottom tubes. Slagging when severe can lead to substantial financial losses to an operator. This may be through loss of availability, increased maintenance and operating cost. At its worse, deposits can lead to ash bridges or slag flags requiring generator units to be dropped out of service for cleaning or repair. Although methods (e.g. high pressure jets of air, steam or water) have been developed to remove ash deposits from furnace walls, slagging can still occur at locations which the cleaning devices cannot reach, or when the bonding strength between the walls and the deposits is too strong for the cleaning equipment to be effective.

The slagging propensity of coal is usually measured by indices derived from chemical properties of coal and/or its ashes. Since external factors, such as furnace design and operating conditions, also play an important role in the deposition process, a 'chemistry only' approach is generally not sufficient to assess such a complex phenomenon. Recent advances in Computer Control

Applications of Advanced Technology to Ash-Related Problems in Boilers
Edited by L. Baxter and R. DeSollar, Plenum Press, New York, 1996

637

Scanning Electron Microscopy (CCSEM) and other analytical techniques have provided invaluable information, allowing more sophisticated models to be developed. During the last few years, prediction schemes have been constructed [Srinivasachar et al, 1992; Baxter, 1993; Walsh et al, 1990] which make use of CCSEM flyash data together with some simple flow models to determine the deposition rate of flyash particles. However, due to the lack of generality of the flow models employed by these schemes, the external factors that control slagging, such as burner arrangement or particle-turbulence interaction, are usually over-looked. Prediction schemes which emphasise flyash chemistry or particles' dynamics are available [Kalmonovitch, 1992; Wall, 1992], but they fail to integrate all the relevant factors to provide a complete description of coal ash deposition.

This paper presents an alternative approach to model ash deposition in coal fired systems. It allows for the effects of firing rate, burner arrangement, furnace geometry, wall conditions as well as coal ash properties. The model was validated against a single burner pilot scale ash deposition rig and the results are discussed in Section 3. Some preliminary predictions for a full scale multi-burner furnace application are presented in Section 4.

2.   METHODOLOGY

Since the pulverised fuel ash (pfa) phase is relatively dilute, it is reasonable to ignore the contribution of momentum and energy sources from the pfa particles to the carrier phase. Although this simplification will impair the accuracy of the gaseous phase solutions, it will drastically reduce the run time of the deposition calculations. However, it must be pointed out that this decoupling is only limited to the fly ash trajectory simulation. The pulverised coal (pc) particles are fully coupled to the flow field and temperature calculations, and the resulting conservation equations are solved by a Computational Fluid Dynamics (CFD) method. The dispersion of pfa particles and their subsequent deposition propensity are calculated by post-processing. Essential data for the ash deposition post-processor are, 1) flow field and temperature distributions as obtained from the initial CFD calculations and, 2) pfa properties as determined from Computer Control Scanning Electron Microscopy (CCSEM) analysis.

2.1   Flow Dynamics and Combustion Modelling

The CFD combustion code, CINAR, developed by CINAR Ltd under the supervision of Professor FC Lockwood is used to predict the flow patterns, temperature profile and other field variables. The code models the gaseous phase by solving the corresponding Eulerian conservation equations of mass, momentum and energy on a structured 3 dimensional staggered finite volume cartesian grid. Turbulence is modelled by using a Boussinesq effective viscosity approach in partnership with the conventional two-equation $k - \varepsilon$ turbulence model [Launder & Spalding, 1974] for closure. Radiation is modelled by the Discrete Transfer method developed by Lockwood and Shah [1981].

The size distribution of pulverised coal (pc) particles is divided into a finite number of size groups, each one representing the appropriate number of mono-sized particles having the same mass, momentum and temperature. Dispersion of the pf particles by the action of fluid turbulence is not currently accounted for, but will be incorporated in the next development stage using existing theory. The mean pf trajectories and temperature-time histories are computed by solving the corresponding equations of motion and energy balance equation for the representative particle of each size group. Coal pyrolysis is modelled by a single reaction rate model [Badzioch & Hawksley, 1970]. Heterogeneous char combustion is approximated by a first-order differential equation, with a combined rate coefficient that accounts for both the global diffusion of oxygen to the char surface and the chemical reaction rate of char oxidation [Field et al, 1967]. Char particles are considered to burn at constant size, producing carbon monoxide as the main initial product of surface oxidation. The CO thus formed is further assumed to be completely oxidised to carbon dioxide inside the particle boundary layer. The heat liberated in the formation of CO is transferred to the particle to raise its temperature, and the heat generated by the subsequent oxidation to CO_2 is supposed to enter into the bulk gas [Lockwood et al, 1984].

Two combustion schemes have been developed to model the global combustion rate of volatiles and char in the CINAR code. The first option involves the solution of a single mixing parameter which defines the concentration of volatiles and a fictitious carbon source derived from the heterogeneous oxidation of char [Lazoupolous, 1995]. The oxygen and fuels concentration are re-calculated according to a fast chemistry reaction scheme. The results are then fed into the next iterative loop until convergence is achieved.

The second option involves the solution of two mixing parameters representing respectively the volatiles from coal pyrolysis and the carbon atom in the carbon dioxide stream from char oxidation [Kandamby, 1995]. As in the first scheme, the oxygen and fuels concentration are re-calculated according to a fast chemistry reaction model. The results are then fed into the next iterative loop until convergence is achieved.

The presence of pf particles is made known to the gaseous phase by the Particle Source In Cell (PSIC) method [Migdal & Agosta, 1967]. This method couples the two phases by adding source terms, i.e. mass, momentum and energy, from the particulate phase to the corresponding Eulerian gas phase equations [Lockwood et al, 1978 and 1984]. An iterative procedure is used to obtained the final converged solution of the two phases.

2.2   Particle Trajectories and Flyash Properties

To calculate the trajectories of the flyash (pfa) particles, the equation of motion for a single particle moving in a viscous environment is solved.

$$\rho V \frac{d\vec{U}_p}{dt} = -\frac{3}{4}\frac{V}{d}\rho_f C_D (\vec{U}_p - \vec{U}_f)\left|\vec{U}_p - \vec{U}_f\right| + V(\rho - \rho_f)g$$

where,

ρ	= particle density
ρ_f	= density of the carrier phase
V	= particle volume
U_p	= instantaneous velocity vector of a particle
U_f	= instantaneous velocity vector of the carrier phase
d	= particle diameter
g	= gravitational acceleration
C_D	= particle drag coefficient

(1)

The Basset term, the added mass term and the temporal derivatives of the fluid fluctuating velocities which appear in the full equation [Berlemont et al, 1990] are omitted. The removal of these terms has been justified previously by other researchers [Faeth, 1983], and the results from a similar study carried out by one of the authors [Lee, 1995]. The solution of Equation (1) requires the instantaneous velocity of the carrier phase. This may be decomposed into a mean, U_m, and a fluctuating component, u. The mean velocity field is obtained from CFD calculations, whereas the fluctuating velocity is derived from a modified stochastic scheme as described below.

The turbulent flow field is represented by a series of characteristic eddies as shown in Fig.1. As the particle traverse the flow field, it interacts with the eddies successively. Two factors have been identified to characterise these interactions: 1) the instantaneous velocity of an eddy and, 2) the interaction time.

According to the definition of the Lagrangian auto-correlation function, $R_L(\varsigma)$, of a fluid particle,

$$R_L(\varsigma) = \frac{\overline{u(t)u(t+\varsigma)}}{\overline{u(t)^2}}$$

(2)

the correlation becomes zero only when the time lag, ς, tends to infinity. As a result, in order to determine the instantaneous velocities of the fluid phase, the temporal correlation of the fluctuating velocities must also be taken into account.

In the present model, the integration time interval, ς, is calculated dynamically at each iteration interval as the lesser of the courant timescale, t_c, the eddy lifetime, t_e, and the particle transit time, t_r. $R_L(\varsigma)$ is considered to be negligible when ς is greater than the minimum of t_e and t_r. In this case, a different correlation domain is computed by using a locally homogeneous and isotropic turbulence assumption as described by Gosman and Ioannides, 1981. On the other hand, when ς is smaller than the minimum of t_e and t_r, the contribution from the temporal correlation term u(t)u(t+ς) is not negligible. In the case, the correlation is incorporated as follow,

$$u_i(t+\varsigma) = \varphi_i(\varsigma)u_i(t) + \alpha_i$$

(3)

where $\varphi_i(\varsigma)$ is a coefficient to be determined, and α_i is a function characterising the randomness at the corresponding fluid location. The derivations of φ_i and α_i and the final forms of Equation (3) are described elsewhere [Lee, 1995].

The continuous flyash size distribution is divided into a number of groups according to their relaxation times. The aerodynamic properties of each group are represented by the group's mass averaged diameter and density. A number of flight paths for each representative particle from each group are then simulated by solving equation (1) to model their dispersion by fluid turbulence. For example, if the pfa size distribution is divided into 8 groups and the representative particle from each group is tracked for a thousand times, the total number of particle tracks from each injection point will then be equal to 8,000. As a result, a total of 32,000 particle tracks from each burner will be simulated if there are 4 injection points per burner.

2.3 Arrival Rate of Flyash Particles

The net rate of deposition depends on both the rate of transport of the particles to the wall and their propensity to stick once the wall is reached. This section describes the rate of transport of flyash particles to the substrate surfaces.

The mechanism governing the transport of particles onto a surface depends on their sizes. For very fine particles (<1 μm), Molecular diffusion and Thermophoresis are dominant. For larger particles, the inertia of the particles normal to a surface becomes more significant, and controls their arrival rate on the substrate surface. In regimes where the mean flow is parallel to a surface, no deposition will occur if the particles were to follow the mean flow streamlines perfectly. In reality, deposition does however occur. Previous studies suggested that this is caused by the turbulence in the free stream, which causes the particles to deviate from the mean flow streamlines.

In general, particles obtain energies from the turbulent eddies when they are caught up by them. If the energies given to the particles are not fully dissipated in the boundary layer, they will traverse the layer and reach the substrate surface. This arrival rate mechanism was first observed by Friedlander & Johnstone [1957] and is known as the Eddy Diffusion-Impaction regime. Since the particles cannot traverse the boundary layer faster than the turbulent eddies, the maximum deposition velocity of particles via this mechanism is therefore roughly equal to the magnitude of the eddy velocity. When the particles size is bigger than the characteristic size of the eddies, the influence of the eddies on the particle's motion diminishes. When this happens, the particle will follow the mean flow streamlines again, causing a decline in arrival rate [Liu & Agrawal, 1974].

In impinging flows, the particles are propelled by the mean flow towards the substrate surface. The particles' momentum is therefore the driving force for them to deviate from the mean flow streamlines. Particles will reach the substrate surface when their inertia are high enough for them to penetrate the wall boundary layer completely. As a result, the arrival rate increases with particle size under this type of flow condition.

Since the instantaneous velocity of a particle is the true measure of it's motion in all the flow regimes mentioned above, the particle trajectory model discussed in the previous section can therefore be extended to unify all the transport mechanisms up to the edge of a boundary layer. It remains then to model the penetration efficiency of particles through this layer for the determination of the particles arrival rate.

The arrival rate of flyash particles are characterised by a 'critical velocity' and a predefined displacement thickness, y+. The 'critical velocity' is defined as the minimum initial velocity required by a particle to traverse a distance y+ in a stagnant and viscous environment within a time period corresponding to the particle's relaxation time. In the present study, the representative flyash particles are tracked until they reach the edge of the viscous sublayer extending to a value of y+= 5 [Reeks & Skyrme, 1976; Lee, 1995]. The particles are then

MODEL: CRESIL

Flyash Particle   Fluid Eddies

Figure 1. Particle-Eddy Interaction

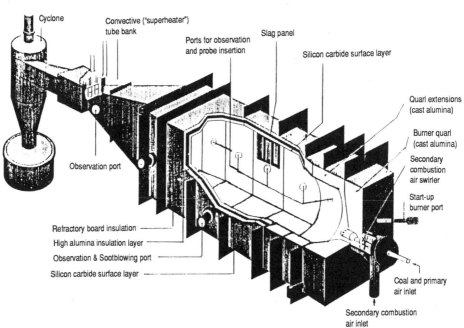

MODEL: CREADR

Cyclone

Convective ("superheater")
tube bank

Ports for observation
and probe insertion

Slag panel

Silicon carbide surface layer

Quarl extensions
(cast alumina)

Burner quarl
(cast alumina)

Secondary
combustion
air swirler

Start-up
burner port

Observation port

Refractory board insulation

High alumina insulation layer

Observation & Sootblowing port

Silicon carbide surface layer

Coal and primary
air inlet

Secondary combustion
air inlet

Figure 2.1

642

MODEL: CREADR

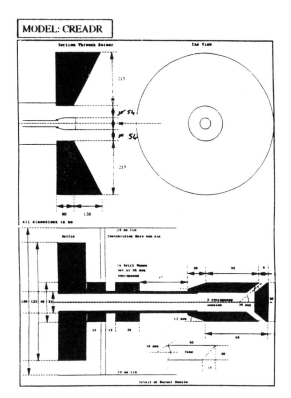

Figure 2.2

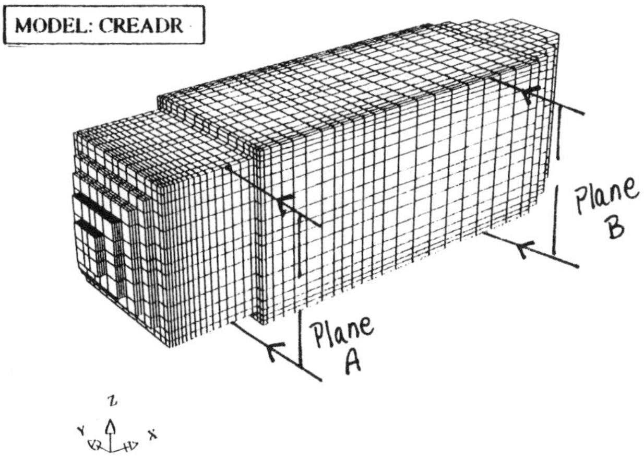

MODEL: CREADR

Figure 2.3

Plane B

Plane A

MODEL: CREADR

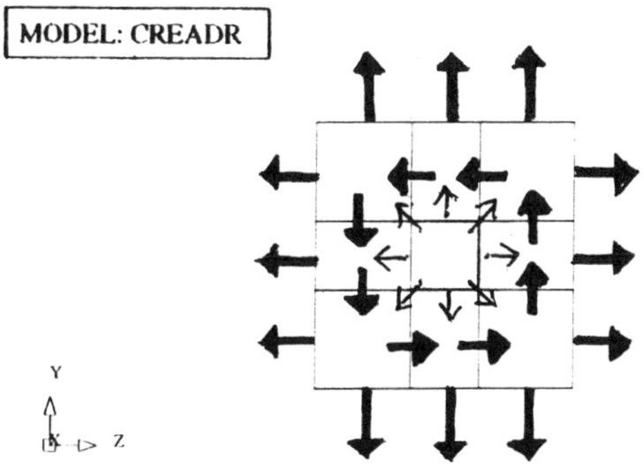

Figure 2.4

643

divided into 8 chemical classes according to their composition as described in Table 1. The mass average diameter and density from each chemical class are then used to determine the penetration depth of the particles through the sublayer. Particles that posses the required 'critical velocity' at the edge of the sublayer are assumed to be able to traverse the sublayer and arrive on the surface. Those that do not have the required minimum velocity are considered to be trapped and carried by the boundary layer to the outlet.

2.4   Sticking Efficiency of Flyash Particles

As mentioned in the previous section, the net rate of deposition depends on both the arrival rate of particles and their ability to stick to the walls. Amongst all the possible qualities, particle viscosity seems to be the most appropriate and quantifiable ones that can be use to characterise the sticking probability of flyash particles [Walsh et al, 1990; Srinivasacher et al, 1990]. Since the objective of the present study is to design a scheme which can be apply to a wide range of coals, the more general Urbain viscosity model [Frank & Kalmanovitch, 1988] is therefore chosen to determine the viscosity of fly ash particles. However, detailed viscosity models for non-silicate systems (e.g. iron-rich pyrite particles) can be incorporated easily at a later stage if necessary.

The inputs to the Urbain viscosity model are the temperature and chemical properties of the fly ash particles. The fly ash chemistry is obtained from CCSEM analysis. The energy balance equation (4) of a fly ash particle is solved along its trajectory to obtain the corresponding temperature at the point of impact.

$$\frac{d(m_p C_p T_p)}{dt} = m_p(Q_c + Q_r)$$

where,

m_p = mass of a fly ash particle
C_p = particle specific heat
T_p = particle temperature
Q_c = rate of heat gain or loss by the particle per unit mass by convection
Q_r = rate of heat gain or loss by the particle per unit mass by radiation

(4)

To assess the sticking probability of flyash particles, a 'critical viscosity' of 10^5 Pa.s is chosen for complete sticking to occur [Srinivasachar et al, 1992]. The sticking probability of particles having viscosities less than this critical value is therefore straightforwardly assigned the sticking probability of unity. The sticking probability of particles with viscosities greater than the critical values is determined by the ratio between the critical value and their actual viscosities. The overall scheme may be expressed as follow,

| Viscosity < Critical Value; | Sticking Probability | = | 1 |
| Viscosity > Critical Value; | Sticking Probability | = | μ_{crit}/μ_{actual} |

2.5   Deposits Accumulation Model

Since deposits accumulation rate is affected by many factors (e.g. erosion by impacting fly ash particles, viscous flow, external removing mechanisms or load cycling), development of a detailed ash accumulation model is generally not feasible. In order to estimate the effect of wall deposits on subsequent deposition rates, a simple but pragmatic scheme has been proposed [Lee, 1995] to account for the changes in thickness and surface temperatures of the wall deposits.

The amount of fly ash deposits over a specific time interval is calculated by multiplying the initial deposition rate (based on the initial assumption of a clean wall) by the corresponding time interval. According to the estimated wall deposits properties and the corresponding wall and adjacent gas temperatures, the thickness of the deposit layer and the surface temperatures of the layer are calculated. In the subsequent time interval, if the calculated surface temperatures of the wall deposits are higher than the fusion temperature of the bulk ash, the local average composition of the deposits and the new surface temperatures are then used to calculate the surface viscosity of the wall deposits. The sticking probability of flyash particles is then re-examined by simultaneously comparing the surface viscosity of the deposits and the particles according to the criteria mentioned above.

3.   VALIDATION CASE

The model has been applied to predict the slagging propensities of three UK coals: Bentinck, Daw Mill and Silverdale, in a single-burner pilot scale ash deposition rig [Lewitt et al, 1993a-c]. A 3D cutaway view of the rig is shown in Fig.2.1. The rig is designed to study the ash deposition characteristics of power station grade coals and to simulate the temperature-time history of pf particles in a front wall fired furnace. The coal/primary air stream is conveyed to the burner by a flexible hose. The burner is similar to a gas burner in design with a closed end and 8 nozzles spaced equally on the circumference, Fig.2.2. Secondary air, preheated to temperatures of up to 450°C is introduced at a swirl angle of 45° to the coal/primary air stream. The burner fires into a refractory lined chamber where an air cooled mild steel plate serves as the 'slag panel', simulating a section of a boiler wall. The mass flow rate of pf and the corresponding inlet combustion air are presented in Table 2. The ultimate and proximate analyses of the three coals are given in Table 3.

The rig is represented by a computational grid consisting of 58 x 17 x 17 cells (Fig.2.3). In order to simplify the mesh structure, all the circular sections of the rig (i.e. burner, burner quarl and quarl extension) are represented by square equivalents of the same cross-sectional area. The burner is represented by 9 rectangular cells as shown in Fig.2.4. The secondary air is entered into the computational domain through the cell faces representing the secondary inlet at constant velocity and at the corresponding swirl angle. The primary jets are modelled by the introduction of sources (i.e. momentum, mass, enthalpy, turbulence kinetic energy and its dissipation rate) into the 8 scalar inlet cells as indicated by the thick arrows in Fig.2.4. The pf sample are divided equally into 4 groups with mass averaged representative diameters of 16, 32, 51 and 64μm respectively. These representative particles are introduced into the computational domain at the same velocities as the inlet primary air. The thin arrows shown in Fig.2.4 represent their injection points. Referring to the figure, a total of 32 pulverised coal particles tracks are thus simulated. This, however, should not be confused with the number of tracjectories used for fly ash deposition calculations. In this study, a total of 64,000 pfa

Table 1. Chemical Classification Scheme

Chemical Groups	Definition
1)Alumino-Silicates	$Al_2O_3 + SiO_2 > 80$ wt% and $SiO_2 < 80$ wt%
2)Iron-Calcium	$CaO > 20$ wt% and $Fe_2O_3 > 20$ wt%
3)Calcium-Alumino-Silicates	$Al_2O_3 + SiO_2 < 80$ wt% and $CaO < 80$ wt% and $Fe_2O_3 < 20$ wt% and $CaO > Fe_2O_3$
4)Calcium Rich	$CaO > 80$ wt%
5)Iron-Alumino-Silicates	$Al_2O_3 + SiO_2 < 80$ wt% and $CaO < 20$ wt% and $Fe_2O_3 < 80$ wt% and $Fe_2O_3 > CaO$
6)Iron Rich	$Fe_2O_3 > 80$ wt%
7)Non CFAS	$(CaO + Fe_2O_3 + Al_2O_3 + SiO_2) < 80$ wt%
8)Silica Rich	$SiO_2 > 80$ wt%

Table 2. Operating Conditions of the Test Rig as Specified in the Present Simulation

	Operating Conditions of the ADR as specified in the furnace simulation
Mass flow rate of pf	0.0059 kg/s
Mass flow rate of primary air @ 30°C	0.0161 kg/s
Mass flow rate of secondary air @ 404°C	0.0463 kg/s

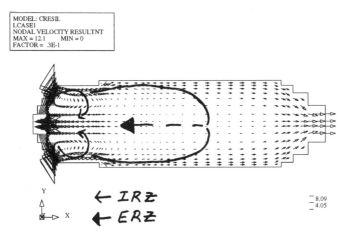

Figure 3. Velocity Vectors Showing IRZ and ERZ

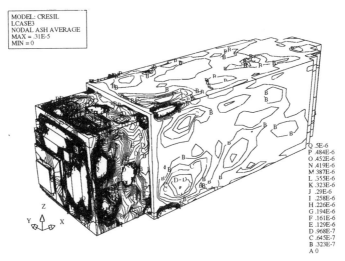

Figure 4. Predicted Deposition Pattern (Silverdale Coal)

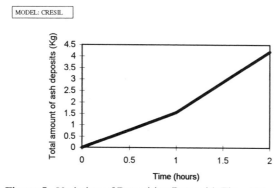

Figure 5. Variation of Deposition Rate with Time (Silverdale Coal)

647

trajectories are used (i.e. 8 nozzles x 8 size groups per nozzle x 1,000 trajectories per size group). The temperature profile of the slag panel is set to the measured values. A constant wall temperature profile of 1250°C is specified for the refractory lined chamber walls.

3.1   Results and Discussions

The predicted flow pattern on a horizontal plane across the centre of the Ash Deposition Rig (ADR) is shown in Fig.3. Also shown in the figure are the outlines of the internal (IRZ) and external (ERZ) recirculation zones formed in the rig enclosure. The formation of an IRZ in the near burner region (NBR) is the direct result of the negative pressure gradient developed by the swirling secondary air and the high velocity primary air jets. The IRZ is enclosed by the larger ERZ which extends downstream to about 7 times the secondary air inlet diameter. This flow pattern results in a stable flame and rapid heat release. Coal devolatilisation takes place at the oxygen-rich IRZ boundaries and burnout is achieved by most of the pf particles before they leave the quarl extension (i.e. about 3 burner diameter downstream from the inlet).

The predicted heat flux to the slag panel resembles the measured values reasonably well (within 8%). The flow pattern and temperature distributions are also consistent with those measured previously [Lewitt, 1994]. The predicted results were also assessed by the rig operators [Lewitt, 1994] qualitatively. It has been concluded that the accuracy of the predictions are adequate for the subsequent ash deposition calculations.

In general, the predicted deposition patterns for the three coals, i.e. Bentinck, Daw Mill and Silverdale, are in reasonable agreement with observations. The predictions show that for the Bentinck coal, the two side walls and the roof are entirely covered by a thin layer of ash deposits with the floor exhibiting a thicker deposit layer than for the other three walls. The predicted deposition rate on the roof at the converging section near the outlet is also consistent with the observations, which suggest a relatively low deposition rate in this region. Both the predictions and the observations suggest localised deposition spots in the quarl extension. This is believed to be caused by the direct impaction of flyash particles on the quarl as they were blown from the burner. The deposition patterns predicted for the Daw Mill run are similar to those for the Bentinck fuel. The two side walls are covered by a thin layer of ash with little deposit being formed downstream from the slag panel. Localised deposition spots are predicted correctly to form on the quarl extension. A relatively high deposition rate on the burner quarl is predicted. For the Silverdale coal, the model correctly predicts it's deposition rate relative to that from the Bentinck and Daw Mill coal. The overall deposition pattern is similar to the other two runs. Nevertheless, a thin layer of ash deposits is predicted to cover the entire roof of the rig for the Silverdale coal. This prediction agrees very well with the observations. Figure 4 shows a typical contour plot of the ash deposition pattern for the one of the three coals under investigation. Detailed results and photographic records from the experimental study are presented elsewhere [Lewitt et al, 1993a,b,c].

Table 4 summarises the predicted results for the three coals during the initial deposition stage (i.e. start from a clean wall condition). The amount of non-sticky flyash shown in the Table corresponds to the percentage of ash particles from the inlet that are able to penetrate the boundary layer completely, but are unable stick to the walls due to their high viscosities. Also shown in the Table is the percentage of flyash particles that are trapped in a boundary layer. This refers to the particles that do not have sufficient momentum to traverse the layer. As a result, they are considered to be trapped in the layer, and are carried by the layer to the furnace outlet. Finally, the percentage amount of flyash particles that go straight to the outlet corresponds to the amount of particles that have not collided with any interior surfaces of the rig before leaving.

In general, the overall predictions are consistent with the observations. Since the Bentinck flyash sample had a higher proportion of large particles (> 50μm), and the density variations are relatively small amongst the three samples, the average inertia as possessed by the Bentinck flyash particles are therefore higher than the other two coals. Consequently, the Bentinck flyash particles are more able to penetrate the boundary layer, resulting in the least quantity of trapped particles. Because of their high inertia, the Bentinck flyash particles are also less affected by the flow field. This allows more of them to traverse the IRZ, and reach the outlet before hitting the wall boundaries.

Referring to the results shown in Table 4, approximately 42% of the input flyash particles from Silverdale are deposited during the initial stage of ash deposition. When the effect of the wall deposits was taken into account by the accumulation model mentioned in Section 2.5, this increased to nearly 55% after 1 hour of prolonged firing. Examinations of the results suggested that this increase is principally caused by the rise in particles arrival rate to the substrate surfaces and the reduction of non-sticky flyash particles. The increase in particles arrival rate is thought to be the direct result of the increasing thickness of the wall deposits as computed by the accumulation model. Since the viscosity of flyash particles should remain constant before and after the accumulation model is used, the reduction in non-sticky flyash particles is therefore mainly caused by the formation of a sticky layer on the surface of the wall deposits which retains the otherwise non-sticky particles.

Figure 5 shows the deposition rate of Silverdale coal with respect to time. The graph essentially consists of two straight lines corresponding to the initial and the subsequent deposition rate after 1 hour of prolonged firing. By using the procedures mentioned in Section 2, a series of lines can be calculated to shown the variations of deposition rate with time. Since the present model has not incorporated a deposits removal mechanism, the predicted deposition rate will ultimately increase to its maximum value, i.e. the flyash generation rate. In reality, this is unlikely to occur as the ash deposits will be removed by natural or mechanical mechanisms during the course of operation.

Coal ash deposition is a very complex process, and it is not sufficient to judge the slagging propensity of coal from their deposition rate alone. In order to better understand the slagging performance of a coal, the properties of its deposits must also be taken into account. For example, coals which produce a large amount of runny slag may not be as problematic as those which generate a moderate amount of highly fused and viscous slag that is very difficult to remove. As a result, a further examination of the chemical properties of the slagging deposits was performed.

The predicted results in terms of the chemical compositions of the initial deposits formed on the slag panel are shown in Table 5. The 'furnace ash' refers to in the Table represents the flyash sample collected in the furnace. 'Back end' ash, on the other hand, refers to the sample collected from the cyclone at the back end of the rig. In order to be more consistent with the present predictions, the compositions of the 'furnace ash' and the 'back end ash' were both obtained from CCSEM analysis.

By comparing the model predictions with the 'furnace ash' data, it has be shown that all three deposits have an initial enrichment in Fe and a corresponding reduction in Al_2O_3 and SiO_2. The enrichment in Fe relative to the flyash is expected and confirms the known role of low-melting point pyrite derived particles in the initial stage of deposition. Examination of the predicted viscosity of the slag panel deposits shown in Table 6 reveals that the ash deposits on the slag panel from the Silverdale coal are generally more fluid than those of the other two coals. This can be explained by

Table 3. Proximate and Ultimate of the Three Coals Under Investigation

Properties	Bentink	Daw Mill	Silverdale
Ash (% ad)	17.1	17.4	17.6
Moisture (% ad)	3.6	5.2	2.2
Volatile Matter (% ad)	29.2	31.4	32.1
Volatile Matter (% daf)	36.8	40.5	40.0
Calorific Value (kJ/kg daf)	33660	32800	33860
Carbon	67.4	65.18	67.8
Hydrogen	4.6	3.6	4.7
Nitrogen	1.4	1.13	1.6
Oxygen	N/D	7.52	5.8
Sulphur			
Pyritic	N/D	0.33	2.54
Organic	N/D	1.18	0.83
Sulphatic	N/D	<0.1	<0.1
Total	1.7	1.51	3.37
Chlorine	0.3	0.16	0.23

Table 4. Summary of the Deposition Results for the Test Rig

Percentage of the incomming ash by weight	Bentinck	Daw Mill	Silverdale
Total amount of Deposits	36.8	38.3	41.8
Amount of pfa that go straight to Outlet	30.0	28.4	25.2
Amount of pfa that are trapped in a boundary layer	25.8	28.2	29.1
Amount of non-sticky flyash particles	7.4	5.1	3.9

Table 5. Predicted Chemical Properties of the Deposits on the Slag Panel in ADR

	BENTINK				DAW MILL				SILVERDALE			
	P	**B**	**S**	**M**	**P**	**B**	**S**	**M**	**P**	**B**	**S**	**M**
SiO_2	55.6	57.0	47.6	53.1	53.7	55.1	46.8	53.0	50.0	48.4	48.4	45.7
Al_2O_3	26.8	25.6	21.7	25.7	25.2	22.8	10.9	25.0	26.7	25.0	25.0	25.0
Fe_2O_3	9.1	8.3	18.3	12.0	8.7	7.9	30.3	9.9	17.4	18.9	18.9	25.0
CaO	1.1	1.7	3.3	1.9	4.8	7.6	7.5	5.4	1.5	3.3	3.3	1.1
MgO	0.9	1.0	1.3	1.1	1.8	1.4	1.0	1.9	0.6	0.6	0.6	0.6
Na_2O	1.2	1.2	0.9	1.3	0.7	0.5	0.4	0.6	0.4	0.4	0.4	0.2
K_2O	3.4	3.4	3.3	3.7	3.2	2.9	1.4	3.2	2.1	2.0	2.0	1.8
TiO_2	1.2	1.1	1.4	1.2	1.1	1.1	0.5	1.1	0.8	0.9	0.9	0.6
Mn_3O_4	0.2	0.3	0.8	/	0.4	0.4	0.7	/	0.2	0.3	0.3	/
P_2O_5	0.4	0.6	0.7	/	0.5	0.3	0.4	/	0.3	0.2	0.2	/

P = Probe ash, B = Backend ash, S = Slag Panel Deposits, M = Predicted compositions of the slag panel deposits

Table 6. Predicted Deposition Rate on the Slag Panel in the ADR

	Bentink	Daw Mill	Silverdale
Predicted Deposition Rate (g/hr)	1.28	1.93	2.90
Viscosity of the slag panel deposits (Pa.s)	9.00E+5	6.11E+5	1.26E+4

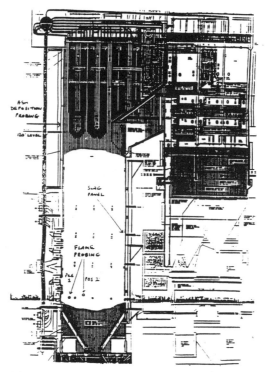

Figure 6.1. RATCLIFFE U2: Boiler Diagram Showing Probing Positions

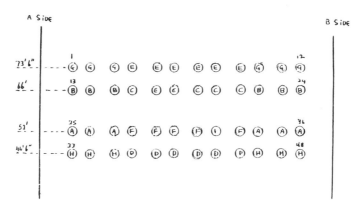

Figure 6.2. RATCLIFFE P5: Burner Configuration

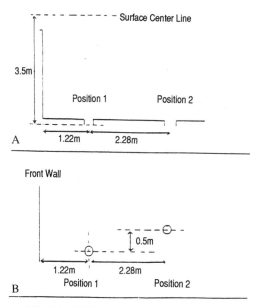

Figure 6.3. RATCLIFFE U2: Burner Probing Sample Points.
A) Plan View B) Elevation View

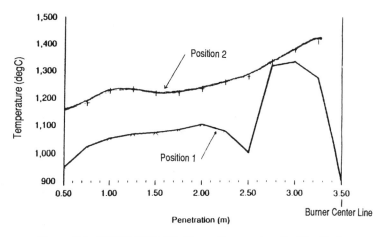

Figure 6.4. RATCLIFFE U2: Flame Temperature Profiles During
Silverdale Slagging Trial (H 48 Burner)

the relatively large proportion of iron found in the Silverdale deposits, which substantially lowers their viscosities. Since the present sticking model is solely dependent on viscosity, the lower the viscosity the higher will be the predicted deposition rate. As a result, the predicted deposition rate on the slag panel is higher for the Silverdale coal than the other two coals.

On the whole, the predicted compositions of the deposits on the slag panel are in reasonably good agreement with the measurements. However, the dramatic increase in the measured CaO level in the three samples is not immediately anticipated. Further examinations of the furnace ash and the back end flyash (Table 5) indicate that all the flyash samples collected at the back end are enriched in CaO. The reasons for this enrichment are not fully understood at present. However, the depletion of CaO in the samples taken from the main furnace may be due to the vaporisation of the calcium-rich phase present in the parent coals. Furthermore, because the combustion gases are sucked sideways into the end of the collection probe in the main furnace, the composition of the collected ash samples is undoubtedly affected by the aerodynamic properties of the flyash particles. Apart from the CaO level, the measured Fe_2O_3 levels on the slag panel as obtained from CCSEM analysis for the Daw Mill trial are also exceptionally high when compared with the parent flyash sample (almost 250% increase). Examination of the slag panel deposits data as obtained from atomic absorption induced couple plasma technique shows a different trend and suggests only a 30% increase in Fe content, which is more consistent with the current predictions. Finally, the deposit composition is also found to be closer to the bulk ash when the effects of wall deposits have been taken into account. This trend is generally consistent with the measurements from the full scale plant trial as described in the next section.

4   APPLICATION - FULL SCALED FRONT WALL-FIRED FURNACE

The side view of the furnace is shown in Fig.6.1. Forty-eight low-NO_x burners are arranged in the lower front wall of the furnace in four tiers with 12 burners per tier. Thirty-six burners are required to generate the design output of 500MW. Burner groups A, B, C, F, G and H are in operations for this study and they are arranged in a way that encourages a symmetrical distribution of heat over the furnace (Fig.6.2). The mass flow rate of pf and other operating variables are presented in Table 7.

The furnace is represented by a computational grid consisting of 13 x 124 x 65 cells (Fig.7.1). Each of the low-NO_x burner is represented by a 8 x 8 grid as shown in Fig.7.2. In the present study, 4 representative pf size groups consisting of 16, 32, 51 and 64μm particles were introduced into the computational domain with the same initial velocity as the primary air. Since 4 injection points are defined for each burner, this gives a total 576 pc trajectories (i.e. 4 injection points x 4 representative particles per injection point x 36 burners). The fly ash trajectory calculations, on the other hand, simulate approximately 1.3 million trajectories (i.e. 4 injection points x 8 or 9 representative pfa particles per injection point x 1,000 trajectories per representative particle x 36 burners) to cover the computational domain. A constant wall temperature of 300°C is specified for this simulation.

4.1   Results and Discussions

Predictions for the Silverdale coal are presented in this section. The predicted gas flow patterns in the vertical planes indicated in Fig.7.2 are displayed in Fig.7.3 and Fig.7.4. A recirculation zone is formed in the hopper region. The size of this recirculation zone varies slightly across the

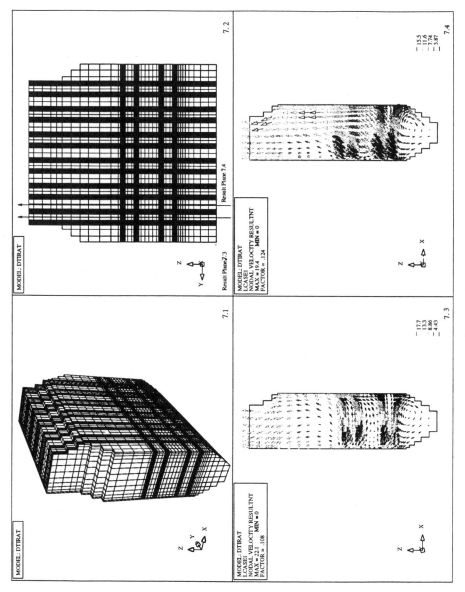

Figures 7.1 to 7.4

655

width of the furnace, but never extends above the bottom burner row level. The centre of this zone is found to be slightly skewed towards the rear wall. Significant impingement of the burner jets on the rear wall is apparent. The impingement is more noticeable on the inter-burner plane (Fig.7.3) than on the plane between two burner groups (Fig.7.4).

The temperature distribution on a horizontal plane through the centre of the burner belt between the second and the third burner row is presented in Fig.8.1. The temperature pattern is generally symmetrical about the furnace centre-line. The peak temperature on this plane is about 1,450°C, and the regions near the rear wall are generally hotter than those near the front wall.

Temperature measurements have been made at the positions shown in Fig.6.3 and are presented in Fig.6.4. The predicted results on the corresponding horizontal planes are displayed in Fig.8.2 and 8.3 respectively. As shown in Fig.8.2, the predicted temperature at Position 1 varies from 300°C (i.e. wall temperatures) to approximately 1,400°C three metres away from the side wall. The temperatures then drop to about 1,000°C near the burner centreline. This profile generally agrees with the measured one, considering that the measurements normally under-estimate the true temperatures by an average of 100°C due to the use of an unshielded temperature probe. The temperature drop at 2.5 metres from the side wall is not predicted by the present calculations. However, further refinement of the grid in this region will allow these small variations to be picked up correctly. The predictions shown in Fig.8.3 shows that the temperature profile at Position 2 varies gradually from the wall temperatures to a maximum of 1420°C at the burner centreline. Again, this agrees reasonably well with the measurements, especially considering the standard deviation of the measured data.

Figure 8.4 shows the temperature distribution on a vertical plane across the burners adjacent to the side wall. The predicted temperature pattern is generally consistent with the given furnace design and firing conditions. The pf particles are seen to devolatilise very rapidly in the near burner regions. The combustion of volatiles raises the temperatures in the near burner regions to approximately 900°C. This promotes coal pyrolysis and helps to stabilises the flames. Char combustion takes place further downstream from the burner inlets, forming a very hot zone, with a peak temperature of approximately 1500°C, in the burner belt at the centre of the furnace.

The predicted ash deposition patterns on the side walls, the front wall and the rear wall of the furnace are presented in Fig.9.1 - 9.4. The front wall is generally free from ash deposits apart from a few light patches at the shoulder level. The rear wall is covered by ash deposits in the regions corresponding to the inter-burner spacing on the front wall. Long thin deposits strips can also be found above the burner belt on the rear wall. Contrary to expectations, these predictions indicate that direct impaction of flyash particles is not a dominant deposition mechanism on the rear wall in this set-up.

Deposition rate is found to be quite high in the furnace nose. The hopper is found to be heavily slagged on the rear wall slope compare to that on the front wall slope. Deposition hot spots are formed on the rear wall slope at approximately 5 meters from the side walls. This may be cause by the recirculation zones formed at the corners in the hopper region due to the swirling flow created by the burners on the bottom row.

Two deposition hot spots are predicted on the side walls. One just above the burner belt and one on the convergent section. These deposition hot spots were, however, not observed during the

MODEL: DTIRAT
LCASE2
NODAL TEMP AVERAGE
MAX = .146E4
MIN = 442

J .137E4
I .128E4
H .118E4
G .109E4
F .998
E .905
D .813
C .720
B .627
A .534

8.1

MODEL: DTIRAT
LCASE2
NODAL TEMP AVERAGE
MAX = .171E4
MIN = 334

P .166E4
O .163E4
N .153E4
M .144E4
L .135E4
K .126E4
J .119E4
I .109E4
H .999
G .910
F .821
E .733
D .644
C .555
B .467
A .378

8.2

MODEL: DTIRAT
LCASE2
NODAL TEMP AVERAGE
MAX = .174E4
MIN = 353

P .171E4
O .166E4
N .157E4
M .148E4
L .139E4
K .130E4
J .121E4
I .112E4
H .103E4
G .937
F .847
E .757
D .668
C .578
B .488
A .398

8.3

MODEL: DTIRAT
LCASE2
NODAL TEMP AVERAGE
MAX = .174E4
MIN = 346

P .169E4
O .165E4
N .156E4
M .147E4
L .138E4
K .129E4
J .120E4
I .111E4
H .102E4
G .930
F .840
E .751
D .661
C .571
B .481
A .391

8.4

Figures 8.1 to 8.4

MODEL: DTIRAT
LCASE1
NODAL ASH AVERAGE
MAX = 124E-1
MIN = 0

P .117E-1
N .10E-1
M .923E-2
L .87E-2
K .805E-2
J .732E-2
I .659E-2
H .586E-2
G .513E-2
F .439E-2
E .366E-2
D .293E-2
C .22E-2
B .146E-2
A .732E-3

9.1

MODEL: DTIRAT
LCASE3
NODAL ASH AVERAGE
MAX = 124E-1
MIN = 0

P .117E-1
N .10E-1
M .923E-2
L .87E-2
K .805E-2
J .732E-2
I .659E-2
H .60E-2
G .513E-2
F .439E-2
E .366E-2
D .293E-2
C .22E-2
B .146E-2
A .732E-3

9.2

MODEL: DTIRAT
LCASE3
NODAL ASH AVERAGE
MAX = 124E-1
MIN = 0

P .12E-1
O .110E-1
N .108E-1
M .1E-1
L .923E-2
K .84E-2
J .76E-2
I .68E-2
H .602E-2
G .522E-2
F .442E-2
E .361E-2
D .28E-2
C .20E-2
B .12E-2
A .402E-3

9.3

MODEL: DTIRAT
LCASE3
NODAL ASH AVERAGE
MAX = 124E-1
MIN = 0

P .12E-1
O .110E-1
N .108E-1
M .1E-1
L .923E-2
K .84E-2
J .76E-2
I .68E-2
H .602E-2
G .522E-2
F .442E-2
E .361E-2
D .28E-2
C .201E-2
B .12E-2
A .402E-3

9.4

Figures 9.1 to 9.4

furnace trial. Further examinations of the flow field near the side walls suggest that these hypothetical deposition hot spots could have been caused by the use of a relatively coarse grid between the side wall and the adjacent burners, which makes the particles unable to follow the streamlines and results in an over-prediction of the particles arrival rate. Furthermore, swirling components from the burners adjacent to the side walls may have been over-specified in the present simulation, causing too much flow impingement on the side walls. Similarly, the deposition hot spots formed in the converging section may be caused by the inability of the flyash particles to follow the flow streamlines due to use of insufficient cells to resolve the flow field in this region.

A summary of the current predictions is presented in Table 8. As shown in the table, approximately 40% of the input flyash particles became ash deposits during the initial stage of deposition. Of all the flyash particles found at the outlet (60% of the input amount), most of them represent the particles that travel to the outlet without striking the boundary surfaces. Only about 10% of the particles at the outlet are found to originate from trapped or non-sticky particles. The fate of the input flyash particles is further broken down according to their sizes and chemical compositions in Tables 9 and 10 respectively. Some interesting trends may be observed :

- The percentage of flyash particles escaping from the furnace decreases with particle size,
- The arrival rate of particles increases with particle size,
- The percentage of trapped particles varies inversely with particle size,
- The percentage of non-sticky particles is generally independent of their size,
- Most of the Si rich species from the depositing flyash go to the furnace outlet,
- Since the Silverdale flyash sample has a high proportion of Al-Si and Fe-Al-Si species, the composition of the deposits is therefore mainly comprised of these elements,
- Both the amount of trapped particles and the particles arrival rate are relatively insensitive to the variations in chemical groupings.

5.   CONCLUDING REMARKS

A predictive scheme based on CCSEM flyash data and a CFD technique has been developed to model ash deposition in a pf-coal fired furnace. The model has been successfully applied to a single-burner pilot scale ash deposition rig and to a multi-burner utility size furnace to predict the deposition behaviour of three UK coals (Bentinck, Daw Mill and Silverdale) exhibiting different degrees of slagging tendencies.

The predicted slagging potential and deposition patterns of the three coals are in reasonable agreement with observations. The model also correctly predicts the chemical partitioning in the initial deposits layer of the three coals, that is the enrichment in iron and a depletion in silicates and aluminium. The deposit compositions are found to resemble the bulk ash chemistry when the effects of wall deposits are taken into account. This finding is consistent with general observations, and agrees with the measurements from the full scale plant trails. Furthermore, the results from this study indicate that large flyash particles are more likely to deposit on the furnace walls than smaller ones. The sticking efficiency of particles are found to be mainly controlled by their thermal (e.g. temperature) and chemical properties. The quantity of flyash particles trapped in the boundary layer and the particles arrival rate is very sensitive to their size and density, but it is more sensitive to the particle chemistry. In general, the model predictions

Table 7. Operating Conditions of the Furnace as Specified in the Present Simulation

	Operating Conditions
Total mass flow rate of pf	51.3 kg/s
Mass flow rate of primary air per burner @ 85°C	2.574 kg/s
Mass flow rate of secondary air per burner @ 275°C	2.405 kg/s
Mass flow rate of tertiary air per burner @ 275°C	9.621 kg/s

Table 8. Summary of the Deposition Predictions for the Ratcliffe Furnace

% Deposits	% Straight to Outlet	% Trapped	% Non-Sticky
40.3	52.0	4.4	1.9

Table 9. The Fate of Flyash Particles According to their Sizes

Representative Size (μm)	Inlet Flow (Kg/hr)	% Total Outlet	% Deposit	% Arrival	% Trapped	% Non-Stick
1.14	5.65	97.5	2.5	39.0	36.5	0.0
1.82	349.40	97.6	2.4	38.9	36.4	0.1
3.48	1411.08	96.0	4.0	38.9	34.7	0.2
6.93	2721.08	77.4	22.6	38.9	15.0	1.3
13.56	5048.18	66.4	33.6	39.3	2.7	2.9
26.13	6616.86	62.0	38.0	40.7	0.7	1.9
54.10	6856.36	52.9	47.1	50.0	0.3	2.5
96.06	5084.47	37.4	62.6	63.4	0.1	0.7
154.54	78.80	22.1	77.9	78.0	0.0	0.1

Table 10. The Fate of Flyash Particles According to their Chemistry

Chemical Groups	Inlet Flow (Kg/hr)	% Outlet	% Deposit	% Arrival	% Trapped	% Non-Stick
Al-Si	17396.6	61.4	38.6	45.6	5.3	1.8
Fe-Ca	22.36	59.5	40.5	43.2	2.7	0.0
Ca-Al-Si	258.99	54.2	45.8	49.5	3.7	0.0
Ca Rich	50.77	53.7	46.3	56.6	10.2	0.0
Fe-Al-Si	8625.65	53.1	46.9	49.3	2.4	0.0
Fe Rich	963.82	58.2	41.8	43.4	1.6	0.0
Non CFAS	160.58	75.5	24.5	41.0	16.5	0.0
Si Rich	692.96	98.4	1.6	42.1	8.5	31.9

are found to be quite sensitive with respect to the physical and chemical properties of the input flyash data. As a result, the accuracy of the predictions are dependent on a proper knowledge of the flyash properties.

As for future work, there are some features of the model that could benefit from further development. In this respect, an improved sticking mechanism based on the more sophisticated viscosity and liquidus temperature of the depositing ashes is currently being investigated. The incorporation of a flyash generation model will also reduce the uncertainties inherent in the measured flyash data. The inclusion of deposits removal mechanisms, such as soot blowing and viscous flow, are desirable as well. Finally, the present study has revealed a serious lack of good quantitative data essential to code development and validation for the furnace aerodynamics and ash deposition.

6. ACKNOWLEDGMENTS

Most of this work is conducted at National Power Plc Research & Engineering Division and is published with the permission of National Power Plc.

The authors would like to thank Mr. F. Wigley (Imperial College), Dr. W. Gibb (PowerGen Ltd) and Mr. M. Lewitt (CRE) for the provision of flyash data and other related information for the two test cases.

One of the author, Francis Lee, would also like to thank Dr. G. Lazoupolous and Mr. N. Kandamby for their valuable advice and assistance at various stages of this research.

7. REFERENCES

Badzioch, S. and Hawkesley, P.G.W., Kinetics of thermal decomposition of pulverised coal particle" Ind. Eng. Chem. Process Des. Develop., Vol. 9, 1970

Faeth, G.M., Evaporation and combustion of sprays, Prog. Energy Combustion Science., Vol. 9, 1983

Field, M.A., Gill, D.W., Morgan, B.B. and Hawksley, P.G.W., Combustion of Pulverised Coal, The British Coal Utilisation Research Association, 1967

Frank, M. and Kalmonovitch, D.P., An effective model of viscosity for Ash Deposition Phenomena, Proceedings of the 1987 Engineering Foundation Conference, 1988

Friedlander, S.K., and Johnstone, H.F., Deposition of suspended particles from turbulent gas streams, Ind. Eng. Chem., Vol 49, 1957

Gibb, W. H., PowerGen Technical Memo, PT/93/220032/M, 1993

Gosman, A.D. and Ioannides, E., Aspects of computer simulation of a liquid-fuelled combustors, AIAA 19th Aerospace Science Meeting, 1980

Kalmonovitch, D.P., Predicting Ash Deposition From Flyash Charateristics, Proceedings of the 1991 Engineering Foundation Conference, 1992.

Kandamby, N., Private Communications, 1995

Launder, B.E. and Spalding, D.B., Mathematical Models of Turbulence, Academic Press, 1972.

Launder, B.E. and Spalding, D.B., The Numerical Computation of Turbulent Flows, Computer Methods in Applied Mechanics and Engineering, Vol 3, 1974

Lazouplous, G., PhD Thesis, Imperial College of Science Technology & Medicine, Dept. of Mech. Eng., 1995.

Lee, C.C.F., MPhil to PhD Transfer Report, Imperial College of Science Technology & Medicine, Dept. of Mech. Eng., 1995

Lee. C.C.F., Ghobadian, A. and Riley, G.S., Prediction of ash deposition in a coal-fired axi-symmetric furnace, Proceedings of the 1993 Engineering Foundation Conference, 1994

Lewitt, M.W., Keeling, J and Hamlin, C.J., Collaborative Slagging Research Programme - Tests on Bentinck Coal in the CRE Ash Deposition Rig, CRE report No. BC 21, 1993a

Lewitt, M.W., Keeling, J and Hamlin, C.J., Collaborative Slagging Research Programme - Tests on Daw Mill Coal in the CRE Ash Deposition Rig, CRE report No. BC 20, 1993b

Lewitt, M.W., Keeling, J and Hamlin, C.J., Collaborative Slagging Research Programme - Tests on Silverdale Coal in the CRE Ash Deposition Rig, CRE report No. BC 22, 1993c

Lewitt, M.W., Private Communications, 1994

Liu, B.Y. and Agarwal, J.K., Experimental observation of aerosol deposition in turbulent flow, J. Aerosol Sci., Vol 5, 1974

Lockwood, F.C. and Shah, N.G., A New Radiation Solution Method for Incorporation in General Combustion Prediction Procedures, 18th Symposium on Combustion, 1981.

Lockwood, F.C., Rizvi, S.M.A., Lee, G.K. and Whaley, H., Coal combustion model validation using cylindrical furnace data, 20th Symposium on Combustion, 1984

Lockwood, F.C., Salooja, A.P. and Syed, S.A., Some Exploratory Calculations of the flow and combustion in a cement kiln, Report to the associated Portland Cement manufacturers Ltd., 1978.

Migdal, D. and Agosta, V.D., A Source flow model for continuous gas-particle flow, J. Applied Mechanics, Vol 34, 1967

Reeks, M.K. and Skyrme, G., The dependenc of particle deposition velocity on particle inertia in turbulent pipe flow, J. Aerosol Sci., Vol 7, 1976

Srinivasachar, S., Helble, J.J. and Boni, A.A., An experimental study of the inertial deposition of ash under coal combustion conditions, 23rd Symposium on Combustion, 1990.

Srinivasachar, S., Senior, C.L., Helble, J.J. and Moore, J.W., A Fundamental Approach to the prediction of coal ash deposit formation in combustion systems, 24th Symposium on Combustion, 1992.

Wall, T.F., Bhattacharya, S.P., Zhang, D.K., Gupta, R.P. and He, X., The properties and thermal effects of ash depositions in coal fired furnaces: A Review, Proceedings of the 1993 Engineering Foundation Conference, 1994.

Wall, T.F., Mineral Matter transformation and ash deposition in pulverised coal combustion, 24th Symposium on Combustion, 1992

Walsh, P.M., Sayre, A.N., Loehden, D.O., Monroe, L.S., Beer, J.M. and Sarofim, A.F., Progr. Energy Combust. Sci, Vol 16, 1990.

INDEX